DISCARDED

Global Events and Event Stratigraphy in the Phanerozoic

Results of the
International Interdisciplinary
Cooperation in the IGCP-Project 216
"Global Biological Events
in Earth History"

Springer
Berlin
Heidelberg
New York
Barcelona
Budapest
Hong Kong
London
Milan
Paris
Santa Clara
Singapore
Tokyo

Otto H. Walliser (Ed.)

Global Events and Event Stratigraphy
in the Phanerozoic

With 86 Figures

Springer

Prof. Dr. Otto H. Walliser
Institut und Museum für Geologie und Paläontologie
Goldschmidtstrasse 3
D-37077 Göttingen
Germany

```
            Library of Congress Cataloging-in-Publication Data

Global events and event stratigraphy in the Phanerozoic : results of
  international interdisciplinary cooperation in the IGCP Project 216
  "Global Biological Events in Earth History" / edited by Otto H.
  Walliser.
       p.   cm.
    Includes bibliographical references.
    ISBN 3-540-59056-0 (hardcover : alk. paper)
    1. Extinction (Biology)--Congresses.  2. Paleontology,
  Stratigraphic--Congresses.  3. Evolution--Congresses.  I. Walliser,
  Otto H.  II. IGCP Project 216--"Global Biological Events in Earth
  History"
  QE721.2.E97G57   1995
  560'.172--dc20
```

ISBN 3-540-59056-0 Springer-Verlag Berlin Heidelberg New York

This work is subject to copyright. All rights are reserved, whether the whole or part of the material is concerned, specifically the rights of translation, reprinting, reuse of illustrations, recitation, broadcasting, reproduction on microfilm or in any other way, and storage in data banks. Duplication of this publication or parts thereof is permitted only under the provisions of the German Copyright Law of September 9, 1965, in its current version, and permission for use must always be obtained from Springer-Verlag. Violations are liable for prosecution under the German Copyright Law.

© Springer-Verlag Berlin Heidelberg 1996
Printed in Germany

The use of general descriptive names, registered names, trademarks, etc. in this publication does not imply, even in the absence of a specific statement, that such names are exempt from the relevant protective laws and regulations and therefore free for general use.

Typesetting: Camera ready by editor
SPIN 10495778   31/3136 – 5 4 3 2 1 0 – Printed on acid-free paper

Contents

Introduction

The Idea of Global Events: A Prologue
Otto H. Walliser   1

Part I   General Themes

Patterns and Causes of Global Events
Otto H. Walliser   7

Evaluating Paleontologic Data Relating to Bio-Events
J. John Sepkoski, Jr. and Carl F. Koch   21

Patterns of Phanerozoic Extinction: A Perspective from Global Data Bases
J. John Sepkoski, Jr.   35

Phanerozoic Development of Selected Global Environmental Features
Jared R. Morrow, Eberhard Schindler and Otto H. Walliser   53

Global Isotopic Events
William T. Holser, Mordeckai Magaritz † and Robert L. Ripperdan   63

Part II   Phanerozoic Global Bio-Events

Geobiological Trends and Events in the Precambrian Biosphere
Mikhail A. Fedonkin   89

The Basal Cambrian Transition and Cambrian Bio-Events (From Terminal Proterozoic Extinctions to Cambrian Biomeres)
Martin D. Brasier   113

The Pattern of Global Bio-Events during the Ordovician Period
Christopher R. Barnes, Richard A. Fortey and S. Henry Williams   139

Silurian Bio-Events
Dimitri Kaljo, Arthur J. Boucot, Richard M. Corfield, Alain Le Herisse, Tatyana N. Koren, Jiri Kriz, Peep Männik, Tiiu Märss, Viiu Nestor, Robert H., Shaver, Derek J. Siveter and Viive Viira   173

Global Events in the Devonian and Carboniferous
Otto H. Walliser   225

Permian Global Bio-Events
Douglas H. Erwin   251

Major Bio-Events in the Triassic and Jurassic
   Anthony Hallam   265

Cretaceous Bio-Events
   Erle G. Kauffman and Malcolm B. Hart   285

The Man-Made Global Disaster: An Epilogue to the Subject of Global Bio-Events
   Otto H. Walliser   313

Part III   Phanerozoic Global Event-Stratigraphy

Christopher R. Barnes et al. (compiled by all authors of this volume)   319

List of Contributors

Barnes, Christopher R.: Centre for Earth and Ocean Research, University of Victoria, P.O. Box 1700, Victoria, B.C. V8W 2Y2, Canada

Boucot, Arthur J.: Department of Zoology, Oregon State University, Corvallis, OR 97331-2914, U.S.A.

Brasier, Martin D.: Department of Earth Sciences, University of Oxford, Parks Road, Oxford OX1 3PR, U.K.

Corfield, Richard M.: Department of Earth Sciences, University of Oxford, Parks Road, Oxford OX1 3PR, U.K.

Erwin, Douglas H.: Department of Paleobiology, National Museum of Natural History, NHB-121, Smithsonian Institution, Washington D.C. 20560, U.S.A.

Fedonkin, Mikhail A.: Paleontological Institute, Russian Academy of Sciences, Profsoyuznaya ul., 123, Moscow 117647, Russia

Fortey, Richard A.: Department of Palaeontology, The Natural History Museum, Cromwell Road, London SW7 5BD, U.K.

Hallam, Anthony: School of Earth Sciences, University of Birmingham, Edgbaston, Birmingham B15 2TT, U.K.

Hart, Malcolm B.: Department of Geological Sciences, University of Plymouth, Drake Circus, Plymouth PLA 8AA, U.K.

Holser, William T.: Departments of Geological Sciences, Cornell University, Ithaca, NY 14853, U.S.A., and University of Oregon, Eugene, OR 97403, U.S.A.

Kaljo, Dmitri: Institute of Geology, Estonian Academy of Sciences, 7 Estonia Ave., EE 200 105 Tallinn, Estonia

Kauffman, Erle G.: Department of Geological Sciences, University of Colorado, Boulder, CO 80309, U.S.A.

Koch, Carl F.: Department of Geological Sciences, Old Dominion University, Norfolk, Virginia 23529, U.S.A.

Koren, Tatyana N.: Geological Institute, Sredni prosp. 74, St. Petersburg, 199026, Russia

Kriz, Jiri: Czech Geological Survey, P.O. Box 85, Praha 011, 11821 Czech Republic

Le Herisse, Alain: Laboratoire de Paléontologie et de Stratigraphie du Paléozoique, Université de Bretagne Occidentale, 6 avenue Le Gorgeu, 29287 Brest, France

Männik, Peep: Institute of Geology, Estonian Academy of Sciences, 7 Estonia Ave., EE 200 105 Tallinn, Estonia

Märss, Tiiu: Institute of Geology, Estonian Academy of Sciences, 7 Estonia Ave., EE 200 105 Tallinn, Estonia

†Magaritz, Mordeckai: formerly Department of Environmental Sciences and Energy Research, The Weizmann Institute, 76100 Rehovot, Israel

Morrow, Jared R.: Department of Geological Sciences, University of Colorado, Boulder, Colorado 80309, U.S.A.

Nestor, Viiu: Institute of Geology, Estonian Academy of Sciences, 7 Estonia Ave., EE0001 Tallinn, Estonia

Ripperdan, Robert L.: Department of Environmental Sciences and Energy Research, The Weizmann Institute, 76100 Rehovot, Israel. Present address: Department of Geological Sciences, University of California, Santa Barbara, CA 93106, U.S.A.

Schindler, Eberhard: Forschungsinstitut Senckenberg, Senckenberganlage 25, 60325 Frankfurt am Main, Germany

Sepkoski, J. John, Jr.: Department of the Geophysical Sciences, University of Chicago, 5734 South Ellis Ave., Chicago, Illinois 60637, U.S.A.

Shaver, Robert H.: Indiana Geological Survey, 611 North Walnut Grove, Bloomington, Indiana 47405, U.S.A., and Indiana University Department of Geological Sciences, 1005 East Tenth Street, Bloomington, Indiana 47405, U.S.A.

Siveter, Derek J.: Geological Collections, University Museum, University of Oxford, Parks Road, Oxford OX1 3PW, U.K.

Viira, Viive: Institute of Geology, Estonian Academy of Sciences, 7 Estonia Ave., EE0001 Tallinn, Estonia

Walliser, Otto H.: Institut und Museum für Geologie und Paläontologie, Goldschmidt-Strasse 3, 37077 Göttingen, Germany

Williams, S. Henry: Department of Earth Sciences, Memorial University of Newfoundland, St. John's, Newfoundland, A1B 3X5, Canada

The Idea of Global Events
A Prologue

Otto H. WALLISER

Abstract. A few spotlights are directed to the development of knowledge on global events. In addition, the activity and results of the IGCP Project 216, Global Biological Events in Earth History, are summarized.

Introduction

The idea of global biological events has been under discussion since the very beginning of the science of Earth History, that is far about 180 years. Georges de Cuvier (1769-1832) already recognized that short intervals occurred in Earth history from time to time, each of them with a strong faunal overturn, later called catastrophe, extinction event, Faunen-Schnitt (faunal cut), faunal change, bio-event, etc.

Bio-events, often in connection with a significant change in lithology, were used by Cuvier and by most of the scientists concerned in the first three quarters of the 19th century in order to subdivide the Phanerozoic. According to the significance of the overturn, they distinguished between larger or smaller units, nowadays called Systems, Series, and Stages. It must be emphasized that most of these still valid units were originally defined by such "natural" boundaries (Walliser, 1985), i.e. more or less sharp changes which separate an "older" fossil community from a subsequent and more "modern" one.

With respect to bio-events as well as to the larger stratigraphic units, the principle recognitions were made in those early days of the science. Thus it is legitimate to ask why it was necessary to enforce an international research programme on global biological events. The answer to this question is evident if we recall that only two decades before, even the fact of bio-events was vehemently disputed; but even if these have now been accepted, the global synchroneity of the events has often been doubted, and the question as to causation has been discussed very controversially.

Cuvier was biased and influenced by the doctrine of the Christian Church at that time, i.e. by the book of Genesis, and therefore explained the faunal overturns as catastrophes, each of them caused by some kind of Deluge. He thought that the organisms thereby extinct became replaced by a new Creation, which resulted in more "modern" types.

Under the directorship of Cuvier at the Jardin des Plantes in Paris, Jean-Baptiste de Lamarck (1744-1829) developed the first scientific evolutionary theory. Obviously, the conflict of personality of these two great scientists forced Cuvier to hold fast stubbornly to his archaic interpretation. If we nevertheless still — or again — use the term catastrophism or cataclysm, we are not agreeing with Cuvier's wrong hypothesis, but accepting his valid observation of the existence of bio-events.

Alcide d'Orbigny (1802-1857) recognized a total of 27 revolutions, each of them separating two different worlds of organisms. In his opinion, only in a few cases did one or the other species survive the revolution.

The time of Cuvier and d'Orbigny was the period in which numerous of the first monographs on fossil vertebrates, invertebrates and plants were published.

Some of these works were restricted to the description of taxa, without comments in respect to major changes or the idea of evolution, but often demonstrating sequences of species which are related and following in time one after the other. Certainly, some of the authors already recognized the variability of species as well as the fact of evolution, thus acting as forerunners and pioneers for the evolutionary theory. This theory finally became generally accepted after Charles Darwin's (1809-1882) On the Origin of Species, published in 1859.

Darwin's theory comprises gradual steps in the evolution of organisms. Although he already recognized that the speed of evolution can change within one group of organisms, not only the existence of new creations was rejected but also the occurrence of major changes and catastrophes. The observation of the latter was explained by subsequent authors either by gaps in the fossil record or by local or regional facies changes.

Based on the pioneer works of the 19th and early 20th century, a period of biostratigraphic refining started, thereby the existence of major changes in the biota became even more evident. Thus the discussion started again, but now from a modern viewpoint, including the increased knowledge of evolutionary mechanisms and geological processes.

The controversial discussion included the old question, whether the major faunal changes could be pretended by gaps. Also the old struggle for global synchroneity of events was revived. With regard to causation, a great variety of explanations was offered, including a combination of "normal" geological processes, or simply regressions or transgressions, in addition extratelluric influences, such as the impact of meteorites, comets or planetoids, as well as an extraordinary increase in hard cosmic radiation. In addition, reasons which are inherent in the evolution itself have been claimed. The basis for the latter was the recognition of a general pattern of evolution, of which the main elements are radiation and stasis. Although this pattern has long been recognized, the discussion culminated with Otto H. Schindewolf's (1896-1971) typostrophism and the gradualistic interpretation of his contrahent George Gaylord Simpson (1902-1984).

The IGCP Project 216

In the course of all the above-mentioned disputes it became quite clear that better research strategies are needed in order to achieve an improvement in basic data and thus to reach a better approach to the recognition of the real patterns of short-term biotic overturns. The required new research strategy had to be based on three essentials: (1) international cooperation to investigate the critical time intervals all over the world; (2) interdisciplinary investigations, i.e. the close cooperation of palaeontology, sedimentology, geochemistry, oceanography, structural geology, etc.; (3) a holistic approach to data collecting, analysis and interpretation.

Being Secretary-General of the International Palaeontological Association (IPA) in the late 1970s, I therefore proposed a research programme on G l o - b a l   B i o l o g i c a l   E v e n t s   i n   E a r t h   H i s t o r y. This proposal was strongly supported by the then President of IPA, Kurt Teichert, and actively accepted by many colleagues.

This idea and the research programme received a strong impetus through Alvarez et al.'s 1980 proposal of a giant bolide impact at the K-T boundary as the cause of catastrophic mass extinctions. In combination with Raup and Sepkoski's (1984) proposal of a 26-Ma cyclicity of mass extinction events, the Nemesis hypothesis was created (Davis et al., 1984). This spectacular and fascinating hypothesis triggered in the following few years — and partially even to the present — a huge number of publications on bolide-caused catastrophic bio-events throughout the Phanerozoic.

Many of these papers overshot their target, so that it was even more necessary to continue and even to improve the current investigations of our project. The popularity which the event idea received in the meantime enabled the IPA to apply for a project within the International Geological Correlation Programme (IGCP). The then President of the IPA, Boris S. Sokolov, successfully promoted the project and effected its wide acceptance during the XXVII. International Geological Congress in Moscow in 1984.

Objectives

The objectives of the former IPA Research Programme have been maintained in the new Project 216. These objectives are:
1. G e o l o g i c a l   A s p e c t. Integrated study of the geological aspects of major bio-event intervals - i.e. the abiotic environments of biological change, including tectonic/volcanic phenomena, sea level and oceanographic history (especially including careful geochemical analyses), climatic history (e.g. effects of cooling events, glaciation, greenhouse conditions, Milankovitch climate cycles, etc.), and sedimentary dynamics.

2. Ecological Aspect. Reconstruction of the overall effect of global events on the biosphere, by ecological analyses of change in population structure, species-level dispersion, community and ecosystem structures and survival strategies across bio-event intervals, as well as the relation of ecological patterns of change to the pattern and intensity of physico-chemical changes ("selection pressures, and stress levels") in these intervals, or to new opportunities (e.g. new or vacated ecospace) inherent in these bio-events.

3. Evolutionary Aspect. Evolutionary factors related to bio-events as compared to background patterns of biological change. Of special interest were the relative importance of macro- vs. micro-evolutionary processes in response to changing stress levels during bio-events, changing rates and patterns of speciation and extinction and their regulating mechanisms or taxonomic expression, and the rates and patterns of recovery/radiation during and following mass extinctions.

4. Chronological Aspect. Development of a new international scale of chronology for bio-event intervals, integrating biostratigraphic, magnetostratigraphic, event-stratigraphic, and geochronologic data into a single holostratigraphic high-resolution (< 100-500 ka) matrix of dating and correlation. Application of this chronology to testing of the duration, rates, relative synchroneity, and regional variation in bio-event intervals.

The programme of the bio-event project received wide acceptance and support. Scientists of 49 countries became members. In several countries, national working groups have been established in order to coordinate and forward the investigation on global bio-events.

Numerous project members also participated actively in subcommissions and boundary working groups of the International Commission on Stratigraphy, thus contributing to a cooperation which was very fruitful in reaching the necessary high resolution stratigraphy as well as a very detailed picture of the pattern of bio-events.

Within the Phanerozoic, the research activity of Project 216 was mainly dedicated to the pre-Cenozoic time, thereby especially concentrating on the early and middle Palaeozoic as well as on the Cretaceous. Therefore the Cenozoic events are treated only in Part III of this volume.

A special part of the research strategy was the effort to analyze in addition global events of minor order in the Devonian as well as in the Cretaceous. This was based on the assumption that (a) in cases of smaller changes the causation and involved processes would be more clearly recognizable, and (b) the recurring occurrence of parameters in a larger number of smaller events will elucidate their importance better than a catastrophic, perhaps singular event.

Conferences and Symposia

The results of the Bio-Event Project have been discussed and presented to the international scientific community in numerous conferences and symposia as listed in the following.

Main Conferences

1984: 26th International Geological Congress in Moscow, August 4-18, Symposium on Global Bio-Events.

1986: 1. International Conference on Global Bio-Events; Göttingen, May 21-24. Organizer: O.H. Walliser.

1987: 2. International Conference on Global Bio-Events: Palaeontology and Evolution: Extinction Events; Bilbao, October 20-23; Organizer: M.A. Lamolda.

1988: 3. International Conference on Global Bio-Events: Abrupt Changes in the Global Biota; Boulder, May 16-22; Organizer: E.G. Kauffman.

1989: 28. International Geological Congress in Washington, D.C., July 9-19, with the following symposia: L7: Global Biological Events in Earth History. B10: Global changes indicated by stratigraphic boundaries. In addition, members of the project contributed to the following project-related symposia: A2: Major dispersal events. A6: Influence of extraterrestrial impact phenomena on the course of geologic history. B4: The record of sea-level fluctuations. B5: Event stratigraphy: catastrophism, gradualism, and evolution in the geosphere-biosphere. B8: Black shales and oceanic anoxia in the Paleozoic and Mesozoic. L1: Adaptive radiations.

1990: 4. International Conference on Global Bio-Events: Innovations and Revolutions in the Biosphere; Joint meeting of Project 216 and the "daughter" Project 303; Oxford, September 25-27; Organizer: M. Brasier.

1991: Geochemical Event Markers; Joint Meeting of Projects 216, 293 (Geochemical Event Markers in the Phanerozoic) and 303 (Precambrian/Cambrian Event Stratigraphy); Calgary, August 28-30; Organizer: H. Geldsetzer.

1992: 5. International Conference on Global Bio-Events: Phanerozoic Global Bio-Events and Event Stratigraphy; Göttingen, February 16-19; Organizer: O.H. Walliser.

Additional Symposia

In the scope of diverse international meetings, IGCP 216 held special Symposia in order to trigger further investigations.

1986: January 27-31, Tallinn/Estonian SSR: The Annual Meeting of the Palaeontological Society of the USSR was devoted to Global Biological Events in Earth History.

1987: April 13-16, Strasbourg (France): Symposium on Climatic History, Evolution of Plankton, and Events; during the Conference of the European Union of Geosciences. May 20-30, Rohanov (CSSR): Symposium on Evolution, Phylogeny and Biostratigraphy of Arvicolids. July 27-August 1, Berlin (FRG): Workshop on Palaeofloristic and Palaeoclimatic Changes in the Cretaceous and Tertiary; during the XIV. International Botanical Congress. August 17-20: Calgary (Canada): Symposia on Transgressive-Regressive Cycles and Event Stratigraphy and on Global Biological Events (IGCP 216); during the International Symposium on the Devonian System. August 31-September 4, Beijing (P.R. China): Symposium on Global Events and Evolution during Permo-Carboniferous Time; during the XI. International Congress of Carboniferous Stratigraphy and Geology.

1988: July 21-August 1, Frankfurt (FRG): Symposium on Conodonts and Event-Stratigraphy; during the V. International Conodont Symposium held in Europe (ECOS V). July 25-30, Aberystwyth (Wales, U.K.): X. International Symposium on Ostracoda: Ostracoda and Global Events. August 9-12, St. John's (Newfoundland, Canada): Symposia on Eustasy and Event Stratigraphy, Biostratigraphy and Bio-Events, and a workshop of Project 216; during the 5. International Symposium on the Ordovician System. August 18-24, Blagoveshensk (USSR): Symposium on Phanerozoic Events within the Circumpacific Province; during the Symposium on Deep Structure of the Pacific Ocean and its Continental Surroundings.

1989: March 20-23, Strasbourg (France): Session on Evolution and Global Events; during the 1. International Conference on Charophytes. August 29-September 2, Prague (CSSR): Symposium on Palaeofloristic and Palaeoclimatic Changes in the Cretaceous and Tertiary.

1990: August 24-30, Tallinn (Estonia): Workshop: Wenlock/Ludlow Event; during the meeting of the International Subcommissions on Ordovician and Silurian Stratigraphy.

Publications

Far more than 1000 articles have been published by project members, either in various scientific journals or in one of the following special volumes.

Walliser, O.H., 1986 (ed.). Global Bio-Events - A Critical Approach. Lecture Notes in Earth Sciences 8, VII+442 p., Springer, Berlin Heidelberg New York.

Lamolda, M.A., Kauffman, E.G. and Walliser, O.H., 1988 (eds.). Palaeontology and Evolution: Extinction Events. III Jornadas de Paleontologia. 2nd International Conference on Global Bioevents, Leioa, 20-23 October 1987. 155 pp., Bilbao.

Kauffman, E.G. and Walliser, O.H., 1990 (eds.). Extinction Events in Earth History. Lecture Notes in Earth Sciences 30, VI + 432 pp., Springer, Berlin Heidelberg New York.

Knobloch, E. and Kvacek, Z., 1990 (eds.). Proceedings of the Symposium Paleofloristic and Paleoclimatic Change in the Cretaceous and Tertiary. Prague, August 28 - September 1 1989, 323 pp., Prague.

Brasier, M., 1991 (ed.). Innovations and Revolutions in the Biosphere. Historical Biology 5 (2-4), 367 pp.

Walliser, O.H., 1995 (ed.). Global Events and Event Stratigraphy in the Phanerozoic. 324 pp. Springer, Berlin Heidelberg New York.

Results

In this volume the results of the bio-event project are discussed in detail. Therefore generally only the main results are mentioned here:

1. Global bio-events do exist. They even occur in great numbers within the Phanerozoic.

2. With numerous global events it has been proven that they occur synchroneously.

3. Most of the global bio-events are connected with a litho-event, i.e. a strong facies change.

4. Each of the numerous global events has its individual pattern.

5. In general, global bio-events are caused by "normal" telluric factors or processes, respectively. The impact of an extratelluric body as causation for a global bio-event is, if it happened at all, an exception.

Acknowledgement

The Bio-Event Project would not have achieved all the results presented in this volume without the encouraging cooperation and the contributions of its members. I am especially grateful to those colleagues who established and directed national working groups or organized our conferences: Martin D. Brasier (Oxford), Ivo Chlupac (Prague), Michael A. Fedonkin (Moscow), Raimund Feist (Montpellier), Jens Morten Hansen (Copenhague), Erle G. Kauffman (Boulder), Marcus A. Lamolda (Bilbao), George R. McGhee Jr. (New Brunswick), Michal Szulczewski (Warszawa).

Greatly acknowledged is also the support by IPA, UNESCO, IUGS, and various national science foundations and their representatives - among others Kurt Teichert, Boris S. Sokolov, Arthur J. Boucot and Tony Hallam, presidents of IPA; Eckart v. Braun, Endre Dudich and Vladislav Babuska, IGCP Secretaries at UNESCO; Dieter Maronde, Deutsche Forschungsgemeinschaft; Willi Ziegler, German IGCP Committee. I thank Dr. Wolfgang Engel and his assistants Mrs Theodora Krammer and Mrs Herta Böning of the Springer-Verlag for their cooperation and patience. Finally, I would like to thank Gabriela Meyer, University of Göttingen, for her dedicated help in the handling and final preparation of this volume.

References

Alvarez, L., Alvarez, W., Asaro, F. and Michel, H., 1980. Extraterrestrial Cause for the Cretaceous-Tertiary Extinction. Science 208, 1095-1108.

Davis, M., Hut, P. and Muller, R.A., 1984. Extinction of species by periodic comet showers. Nature 308, 715-717.

Raup, D.M. and Sepkoski, J.J., 1984. Periodicity of Extinctions in the geologic Past. Proc. Natn. Acad. Sci. USA 81, 801-805.

Walliser, O.H., 1985. Natural boundaries and Commission boundaries in the Devonian. Cour.-Forsch.-Inst. Senckenberg 75, 401-408, Frankfurt.

Author´s address:

Otto H. Walliser, Institut und Museum für Geologie und Paläontologie, Goldschmidt-Strasse 3, D - 37077 Göttingen, Germany

Patterns and Causes of Global Events

Otto H. WALLISER

Abstract. Global events are selective with respect to biota, ecosystem and palaeogeography. Global bio-events are mostly connected with a facies change. They give rise to a regular evolutionary pattern, the E-R sequence, which comprises after the extinction event a generative phase followed by a radiation, i.e. a strong diversification. At least most of the global events are proximately caused by changes of environmental conditions. Thereby changes in sea level, oceanic conditions, and climate are prevailing. To a great extent the questions about ultimate causes and periodicity are still unanswered.

Contents

Preface	7
Definitions	8
Patterns of Global Events	10
Occurrence of Global Events	10
Innovation Events	11
Radiation Events	11
Extinction Events and Evolution	11
Causes of Global Events	13
The Complex Biosphere	13
The Non-Actualistic Palaeo-Oceans	15
The Nemesis Affair	15
Periodicity in Global Events	17
The Ultimate Causes	17

Preface

As this chapter is a part of the introduction to a multi-author book on "Global Events and Event Stratigraphy in the Phanerozoic", a general review of the state-of-the-art concerning the understanding of global events is appropriate. Accordingly, all different interpretations, based on a huge amount of observations and data of variable quality, should be reviewed and critically discussed. This is neither possible nor intended.

This chapter presents rather the individual standpoint of the author, who introduced in the late 1970s — under the umbrella of the International Palaeontological Association — an international research programme on global events. The impetus for this programme was the recognition that the answer to the century-old question whether rapid, world-wide simultaneous changes exist or whether they are only simulated by other particular geological phenomena, is crucial to the understanding of the fundamental processes in biotic evolution and in the geologic development of the Earth's crust.

Rapidly growing stratigraphical refinement and international interdisciplinary cooperation provided good chances for a successful re-examination of the global event problem. This subject became enormously popular with the fascinating impact hypothesis of Alvarez et al. (1980), and was followed by a burst of publications. Many of the articles, especially in the early 1980s, were quite speculative. On the other hand, serious worldwide, directed new investigations increased at many sites and time intervals which have been known or were presumed to yield strong changes in fauna and facies. This activity resulted in a great number of different detailed data, thus allowing a new approach to understanding geologic and evolutionary processes.

Definitions

"Science needs a common vocabulary and definitions of the used terms. But a definition must not be necessarily restrictive. Sometimes a relatively wide definition is more adequate or even necessary, e.g. if the term has to describe a large variety of parameters or patterns or, as another example, if the knowledge of the investigated subject is still on a low level" (Walliser, 1990). Since these sentences were written, we have learned that the processes connected with global events are even more complex than recognized at that time. We therefore should retain non-restrictive but clear definitions, as given in Walliser (1990) and partly repeated or modified in the following.

E v e n t . In principle, each change or happening is an event. However, in connection with global biological or geological events we should use this term for short-term exceptional changes and happenings. A few examples may elucidate the term "exceptional": significantly higher extinction or origination rates as compared with the average background extinction or origination rates; intervals with a high mortality or a population burst; and abrupt or short-duration changes of facies or lithology, such as intercalations of black shales in a sequence of limestones with developed benthic communities, or vice versa.

In most cases, an event occupies a much shorter time-span than the interval between two events. Admittedly, this point contains several difficulties which are discussed below under the terms "Global Crisis" and "Global Event versus Global Crisis".

G l o b a l . Extinctions are self-evidently global, even if the last representative of the extinct taxon lived in a restricted area for a longer time. Therefore, the problem is to prove whether the disappearance of a certain taxon, recognized in a certain area, is due to local or regional mortality, to migration, or to actual extinction.

With respect to anorganic events, such as strong changes in facies or geochemical parameters, we should use the term global if the event is traceable world-wide within comparable environments or palaeogeographical situations.

G l o b a l   E v e n t   (G E) . In addition to the definitions given above, the term global event implies the world-wide synchronous appearance of the event. Thus, a global event is a globally synchronous event.

It has to be mentioned that the difficult part in this definition concerns the interpretation of the term synchronous. Synchroneity cannot be established for a time interval which is smaller than our available methods allow. Therefore synchronous means that the event occurs globally within the smallest time unit which we can correlate world-wide. Normally, this is the duration of the smallest biostratigraphic unit, i.e. on the order of several hundred thousand years (in conodont biostratigraphy about 3 to $7 \cdot 10^5$ a). In extraordinary cases the biostratigraphic unit may reach a time resolution of less than a hundred thousand years ($x \cdot 10^4$ a).

With numerous of the recognized Phanerozoic global events we already have reached the necessary preciseness. In other cases the data allow at least the assumption of synchroneity, although they have still to be supplemented by further investigations.

As we know, many different kinds of events may occur globally. Emphasis on a certain kind or prevailing characteristic of an event can be expressed by a prefix: global biological or bio-event, geological or geo-event, lithological or litho-event, etc.

G l o b a l   G e o - E v e n t   (G G E) . The term "geo-event" is used to subsume all abiotic events that happen on Earth. It comprises, e.g., events in sedimentology, volcanism, tectonics, plate movements, geochemistry, sea level, etc. Because not the causation but the documentation of the event in rock sequences is the basis for the recognition of the geo-event, even those events on Earth due to extratelluric (extraterrestrial) influence can also be included. If a geo-event occurs synchronously world-wide, it is a global geo-event.

G l o b a l   B i o - E v e n t   (G B E) . In contrast to geo-event, the term bio-event encompasses events which happen within the biota, i.e. within a certain taxon or ecosystem. Bio-events are, e.g., innovations, radiations, extinctions, etc. If a bio-event occurs synchronously world-wide, it is a global bio-event.

Ranking of Global Bio-Events. The impact of global environmental events on the biota ranges from only mild disturbances to an extensive mass extinction. Accordingly, some kind of ranking, i.e. a distinction of different categories or orders of events, is possible. Such a classification appears to be useful to indicate the degree of importance. Boucot (1990) proposed a certain ordering, based on the taxonomic level of extinct taxa: the highest ranking, i.e. first order event, is given if extinctions at the subclass to phylum level took place, whereas the lowest ranking, i.e. fifth order event, is characterized by the extinction at the species to family level. Other frequently used rankings are based either on the total amount or percentage of extinct taxa.

None of these ranking methods is entirely satisfying. Thus, e.g., the graptolites – a group of high taxonomic rank – underwent a prolonged decline in diversity in the late Silurian and early Devonian, and the last representative disappeared by background extinction, i.e. at a time even when no global event can be recognized. This example shows the weak point of Boucot's method. The other mentioned procedures do not consider the importance of the systagenetic level: as will be discussed below, the extinction rate depends not only on the severity of the environmental changes but also on the diversity of the affected group of organisms. A further point necessary to consider is the ecological valency of an extinct taxon or group: e.g., the extinction of frame-building biohermal organisms during the late Devonian Kellwasser Event has had a much stronger effect on the global ecosystem than the disappearance of highly ranked taxonomic grous, such as graptolites in the early Devonian or placoderms at the Devonian-Carboniferous Boundary Event.

Obviously, a fairly exact ranking of global bio-events appears to be as difficult as the recognition of the complex structure of the events themselves. For the time being we therefore should be content with a relative ordering that considers all relevant parameters, especially such as extinction rate, systagenetic level and ecological valency.

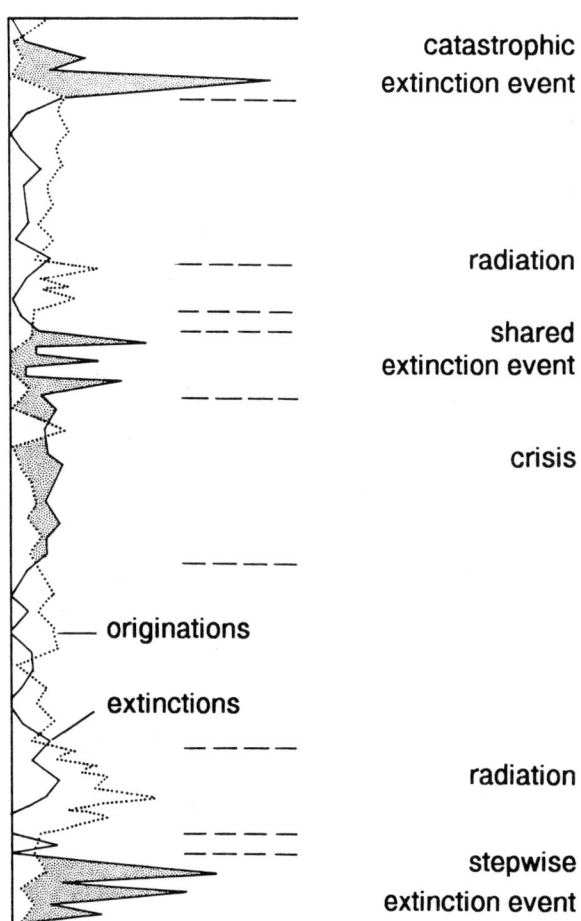

Fig. 1. Characterization of biological events and crises by the ratio between rates of extinction and origination

Global Crisis

In contrast to an event, a crisis spans a relatively long time interval. Among other patterns, this is the case where over a relatively long period the extinction rate exceeds the origination rate. Crises do not only occur in times of long-term environmental changes or fluctuations, but also as the terminal phase of systagenesis (see below, under "Global Events and Systagenesis"), when the diversity of a group of organisms decreases, sometimes towards zero.

Global Event Versus Global Crisis

Fortunately, nature does not follow strict definitions, but rather shows a large variety of patterns with intermediate stages and transitions. Therefore it might be difficult in special cases to assign the observed pattern either to the category of events or of crises (compare Fig. 1). This applies, e.g., to a stepped extinction event, which is relatively short-term but shows a sequence of maxima, as it is examplified in the Kellwasser Event at the Frasnian-Famennian boundary. If a stepped extinction event is less significant and of slightly longer duration than the Fr/Fa Event, it could be difficult to distinguish it from those parts within a crises where the rate of background extinction slightly exceeds the normal quota. Another example is represented in the terminal part of the Cretaceous Preiod, where several separated extinction events follow each other during an interval of about 2 to 3 Ma. In this case the present author is inclined to interpret this pattern as a severe crisis characterized by a series of extinction events.

Patterns of Global Events

As pointed out above, a wide range of events occurs, from only local up to global. Of course, geographically restricted events are also of importance for understanding local and regional geological developments as well as evolution. Later, local and regional events may be influenced by globally effective parameters, fluctuations or changes. In addition, many events originally interpreted as local or regional turned out to be global. That is why the general strategy concerning event research is to investigate whether a locally or regionally recognized event can be traced worldwide. This aspect of partly similar and partly different features of regional and global events is well demonstrated in the chapter on Cretaceous Bio-Events (Kauffman and Hart, this volume). Therein the authors endeavor to establish a superior terminology that is applicable to both global and non-global events, and even for crises or progressive phases, respectively. Because only global events are discussed in the following paragraphs, the global event terminology of Walliser (1986) will be used.

In addition to the patterns listed in Table 1, Walliser (1986) used the term s p r e a d i n g e v e n t, applying that to rapid spreading of new taxa or communities, as well as to the burst of a certain taxon, i.e. population, or community. The contrary could be called a r e s t r i c t i o n e v e n t. In contrast to the innovation, radiation and extinction events, they are not global per se, even if they may sometimes occur globally. Events related to taxon or community burst, spreading, restriction, and migration reflect the kind and magnitude of environmental changes in and between certain regions, which are sometimes even of global extent. Therefore they should be considered in analyzing the pattern of these changes.

Occurrence of Global Events

General Setting

Global events that occur at only a single and sharp time level appear to be relatively rare. These cases are represented, e.g., with the Devonian-Carboniferous (D/C) and the Cretaceous-Tertiary (K/T) Boundary Events. Thereby it is of no importance if one or the other taxon of a largely affected group survives for a certain time. Thus, e.g., at the D/C Event, the previously diverse clymenids became extinct with the exception of only one species, *Cymaclymenia evoluta*. Obviously this species had an advantageous feature which made it less prone to extinction. It might be that the surviving species fed within the free water column, whereas the victims of the event consumed a benthonic food supply that was decimated by the event. In the case of the K/T Event, it would not query the catastrophic impact to dinosaurs if the assumption were verified that one or the other species survived for a short period after the event.

With many global bio-events we recognize that within the short extinction interval the extinctions follow each other in two or more steps. As Schindler (1990) showed, this is the case with the Kellwasser Event at the late Devonian Frasnian-Famennian boundary. The entire event spans an interval on the order of one or a few 100 ka.

Different from the stepwise pattern is if two or more events follow each other closely but are clearly separated. Such a case is described by Kauffman and Hart (this volume) from late Cretaceous time, where

Table 1. General patterns of global events

Occurrence of global geo- and extinction events:
  single and sharp
  single but with differences within the short event interval
  stepwise within a certain time interval, but clearly separated of each other

Types of global bio-events:
  innovation events
  radiation events
    after innovation events
    after extinction events
    after geo-events
  extinction events

Magnitude of global events:
  mild to catastrophic

Selectivity of global events:
  global events are selective with respect to
    palaeogeography
    facies
    ecosystems
    biota

Regular patterns of global extinction events:
  Global extinction events are mostly combined with global facies changes
  Global extinction events trigger a particular sequence of evolutionary phases, here called E-R sequence.

several discrete extinctin events occur within about 2.5 Ma.

Selectivity of Global Events

In principle it is self-evident that global events are selective. Therefore it is unintelligible that again and again the restriction of a global event, e.g. to a certain facies realm or a distinct part of a community, is taken as an argument against the globality of its occurrence. Of course, the event is not global if a worldwide facies realm or community is affected in only a certain region.

For the analysis of a global event in respect to the causes, the selectivity can even be used as a key. Thus it is important whether a global event is affecting both the aquatic biota and the terrestrial communities or only one of them.

Innovation Events

Biological innovation means the evolutionary development of such a Bauplan (constructional pattern) or distinctive feature that can serve as a basis for further successful evolution and diversification. A few obvious examples may be mentioned: At the turn from the Cambrian to Ordovician, the change from benthonic to planktonic life-style led to a rapid diversification and successful evolution of pelagic graptolites; the innovation of coiling in certain cephalopods was the basis for the evolution of ammonites and their subsequent dominance for more than 300 Ma; and the innovation of an additional egg membrane was the precondition for the successful conquest of land areas by amniots.

The importance of innovations for global bio-events is even more evident in the evolution of early life on Earth. Therefore, innovations with subsequent events played a prevailing role in Precambrian and early Cambrian times. These included the creation of protein and the construction of DNA as preconditions for life and reproduction; the appearance of sexuality which enormously enlarged the potency of evolutionary changes; the development of metazoan structures; and the innovation of skeletal mineralization, causing the great revolution around the turning from Precambrian to Phanerozoic.

Radiation Events

Radiations, i.e. a strong diversification within a relatively short time, often occur following innovation and extinction events. In the latter case, the radiation occurs normally after a short time interval. This pattern will be discussed below in the paragraph on "The Extinction-Radiation Sequence".

The interval between an innovation and the subsequent radiation can range from immediately after the innovation up to millions of years later. The latter case is exemplified in the evolution of mammals. There the main innovations, concerning the cranial and other skeletal constructions, were already developed in the terminal Triassic. But the rapid diversification of the mammals started only about 180 Ma later, at the time when the dinosaurs became victims of the terminal Cretaceous extinction events.

Sometimes radiation events are not preceded by innovation or extinction events, but instead by environmental changes. These provided the pre-existing taxa and communities with new and/or altered biotopes and niches.

Extinction Events and Evolution

Principle aspects of extinction events, such as different modes, different magnitude, combination with facies changes, etc., have been mentioned above. In this paragraph the evolutionary pattern caused by an extinction event will be described (Fig. 2).

The Extinction-Radiation (E-R) Sequence

In most cases of extinction events the following evolutionary sequence can be observed: extinction event — interval — radiation — nomismogenesis. According to the main phases it is named extinction-radiation (E-R) sequence. The explanation for this sequence is obvious.

1. The extinctions produce open niches (vacuums, according to Boucot, 1990) within the affected and strongly disturbed ecosystems or biotopes, respectively. In addition, the event-causing environmental changes themselves may create new ecological niches and/or the alteration of pre-existing biotopes by partitioning or by narrowing or widening with respect to controlling and limiting parameters.

2. The existence of niches lowers the selectional stress.

3. Even if the mutation rate does not increase, more of the mutations can be used for adaptive evolution, i.e. the evolution rate increases. With that the diversification starts, at first at the species level and at the same time enlarging the variability. Therefore, the interval between extinction and radiation events is not

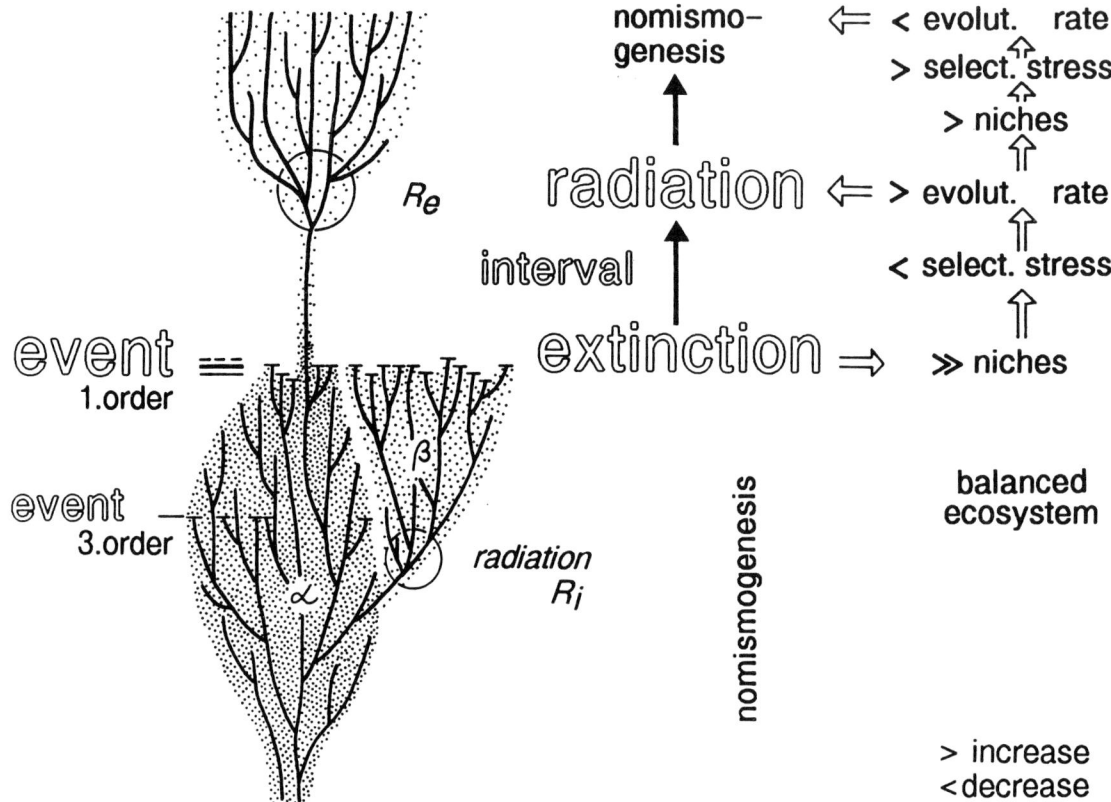

Fig. 2. General pattern of evolutionary changes caused by an extinction event, based on the patterns of goniatites (close dotting) and clymenids (wide dotting) at the Devonian-Carboniferous Boundary Event (after Walliser, 1986, 1990). R_i: radiation after an innovation event; R_e: radiation after an extinction event

at all an evolutionary stillstand as Becker (1986) claimed.

4. After a certain time - which varies within different survivor groups - a strong diversification, called radiation, starts.

5. The adaptive radiation finally leads to the occupation of the niches and thus to a recovery from the extinction event. Of course, this recovery is not manifested simply as a restitution of the pre-event ecosystem but rather the creation of comparable but often quite different ecosystems.

6. With growing completion of the ecosystems, the selection stress increases and the evolution rate decreases.

7. As soon as the biotopes and the ecosystems are established, the phase of nomismogenesis begins, i.e. the "normal" evolutionary mechanisms become dominant.

One general aspect of this regular pattern should be emphasized: With various groups of organisms extinction events ultimately cause an increase in the evolutionary rate, with a subsequent diversification. Thus an appearant paradox pattern is evident: e x - t i n c t i o n s   a c c e l e r a t e   e v o l u t i o n .

The Innovation-Radiation (I-R) Sequence

The pattern of the Extinction-Radiation Sequence also applies to the evolutionary pattern after an innovation event. There, the radiation or strong diversification starts after a certain time interval following the biological innovation. In the Innovation-Radiation sequence, of course, the lowering of the selectional stress is not caused by the appearance of real niches, but by the fact that the newly acquired trait makes this taxon superior to the unchanged competing taxa. In addition we have to consider the possibility that the innovation enables a taxon to evolve in the direction of a pre-existing but as yet unoccupied niche.

Global Events and Systagenesis

Systagenesis is defined as the evolution of a certain taxon group, independent of the taxonomic level, e.g. family, order, class (Walliser, 1994). In many of the fossil groups we observe a common pattern, starting with a generative phase that proceeds a phase of diversification (compare Fig. 2). This latter progressive phase leads to a phase of maximal diversification that is in turn followed by a regressive phase in which the diversity becomes reduced, often ending with the extinction of the group, i.e. with the systagenetic death. The regressive phase therefore could also be called a systagenetic crisis.

The increase of diversity implies an increase in specialized or stenotopic taxa within the whole group. In terms of evolutionary progress, this means that at a low systagenetic level the generalists prevail, in contrast to high systagenetic levels (i.e. high diversity) where most taxa are specialized. Stenotopic and specialized taxa, of course, are prone to extinction to a much higher degree than eurytopic and unspecialized (generalist) taxa. Higher susceptibility means that taxa easily become victims even of minor environmental changes. This correlation appears to explain the background extinctions within a certain group, and thus the phase of systagenetic diversity reduction. It also makes clear an apparently paradox regularity, namely that by the success of a group, i.e. by their diversification, the systagenetic death already is preprogrammed.

With respect to global events the systagenetic level of the affected groups is of paramount importance. If at a time of high systagenetic level an event is mild, i.e. if the event triggers only minor environmental changes, it can nevertheless result in a high extinction rate. By contrast, a very strong event may cause only a small extinction rate if the affected group is still at a low systagenetic level. Therefore the extinction rate (R_{EX}) is the product of the magnitude or intensity of the event, i.e. of the environmental changes (I_{EC}), and of the evolutionary level within the systagenesis (E_{SL}):

$$R_{EX} = I_{EC} \cdot E_{SL}.$$

As mentioned above, another important aspect is the ecologic valency of a certain taxon (V_{EC}) such as, e.g., its position and significance within the food chain or its relevance fo synecologic interactions. Taking this into consideration we have to extend the above equation as follows:

$$R_{EX} = I_{EC} \cdot (E_{SL} + V_{EC}).$$

Causes of Global Events

Discussing proximate as well as ultimate causes of global events, we have to consider all hypothetic possibilities. The most important of those are summarized in Table 2.

Table 2. Possible causes of global events

A) Abiotic causes
  Cosmic   bolide impact
                single
                shower
           orbit (periodicity)
  Telluric sea-level
           ocean physics
           ocean chemistry
           atmosphere physics
           atmosphere chemistry
           climate
           volcanism
           plate movements

B) Biotic causes

C) Combination of two or more abiotic and/or biotic causes

The Complex Biosphere

All processes in Earth's geo- and biosphere are complex, with numerous networked and complicated actions, reactions, interactions and feed-backs. Possible connections of this kind are indicated in Fig. 3, without claiming entirety with regard to involved parameters and interactions. In the search for the causes of global events, this complexity has to be considered.

Global bio-events are caused by the elimination or disturbance of at least one biotope. The term biotope encompasses all synecological, i.e. abiotic and biotic conditions, patterns and parameters that determine a certain biocoenosis and/or ecosystem. Figure 3 therefore is centered on the word "biotope". In many cases the proximate cause (or causes) for the disturbance of the biotope and thus for the event is recognizable. These causes mainly are changes in climate, sea level, and chemo-physical conditions of the ocean and atmosphere. In addition, biological innovations can also be responsible.

In contrast to the proximal causes, the ultimate or

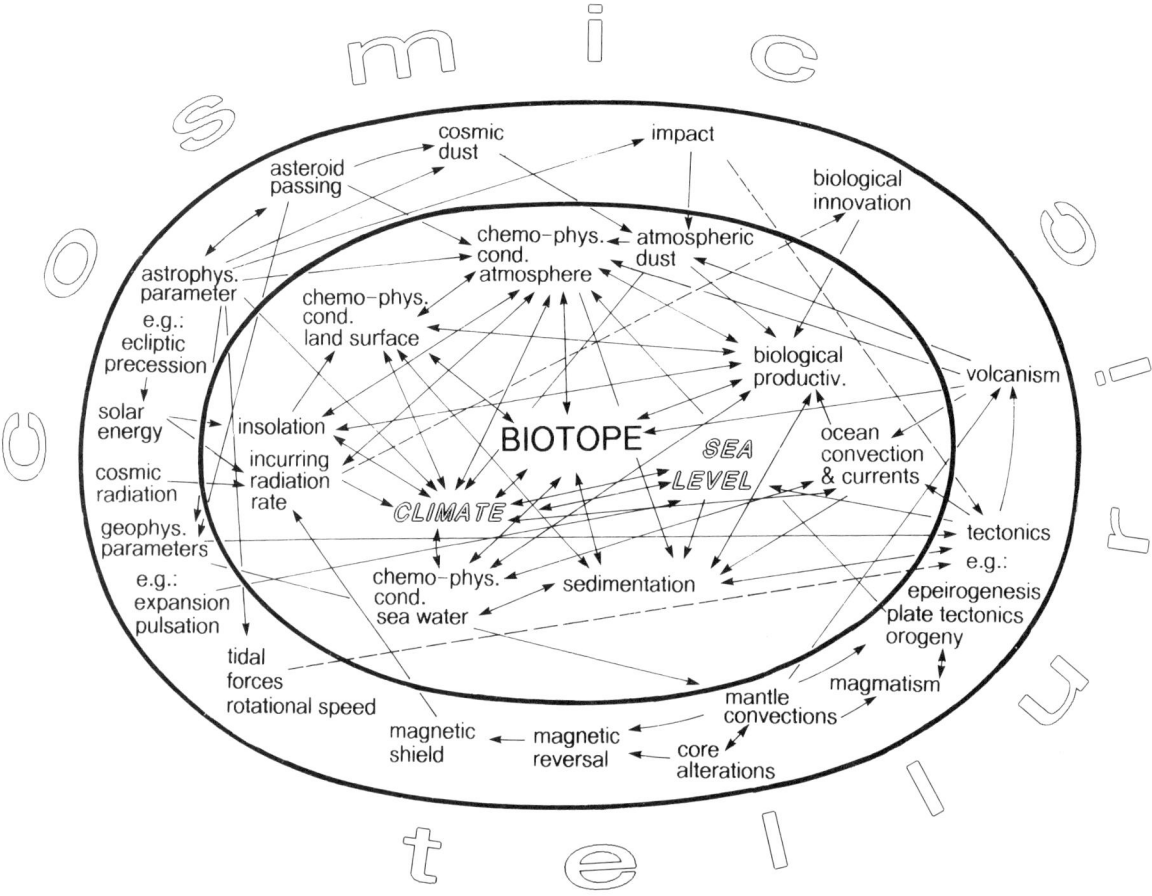

Fig. 3. Flow chart, indicating possible causes and interactions of processes in the complex geo-/bio-sphere that finally may lead to global events

initial causes and the processes connecting both are almost completely unknown. As an example we may take a t r a n s g r e s s i o n as the proximate cause. Then we have to ask two questions: (1) by which process does a global transgression result in a global bio-event?, and (2) what has caused the transgression?

With respect to question (1) the answer could be quite different depending on the kind of process or processes involved. Thus, a biotope can be eliminated if the transgression is rapid enough to prevent a shifting of the biotope. Another possibility is that with the sea-level rise anoxic water sweeps over the area of the biotope. A transgression also influences the supply and transportation of terrigenous clastic sediments, thus eliminating neritic biocoenoses that are adapted to a particular clastic sediment type.

Concerning the second question, i.e. the causation of global transgression, again several answers are possible. Either the available space for the oceanic water masses becomes reduced or the water volume increases. The former possibility could be triggered by plate tectonics, e.g. the development of mid-ocean ridges. The increase in volume needs an increase in the water temperature, which in turn can be caused by various processes, such as a general climatic warming towards a greenhouse state or, on the other hand, by a widening of warm water circulation due to a changing plate or continent configuration. A further well known reason for transgressions is the melting of polar ice sheets. Even the hypothesis of an undulating expansion of the Earth cannot yet be excluded from playing a role in sea-level changes. Without considering further more or less well-founded hypotheses, nor the various possible combinations of causes, we have enumerated already 6 different causations for a transgression.

For the change of c l i m a t e   numerous

causes, each of them separately or in combination, can also be responsible. Here only a few of them will be mentioned, but only for one of numerous climatic factors, namely increasing temperature: increase in insolation that in turn can be caused either by an increase in solar activity or by an increase in atmosphere's permeability; decrease in albedo by decrease of the polar ice sheets or by decrease of epicontinental seas; increase in CO_2 content of the atmosphere by magmatic or metamorphic degasing; by decrease of vegetation or of anorganic CO_2 reduction; melting of the polar ice caps as a consequence of continental movement; or increase in Earth's heat flow.

The Non-Actualistic Palaeo-Oceans

Uniformitarianism may be an excellent tool to decipher processes and patterns in the past. At least this is valid as far as it concerns unchangeable physical and chemical laws. If the same specific sedimentological features occur in the present and in the past, comparable conditions and processes can be assumed. In so far, and in this special case, uniformitarianism is valid. In contrast to that, the superior conditions, e.g. the oceanographic parameters, certainly have changed during the past.

Our recent oceanic system obviously represents an exceptional state due to the Pleistocene glaciation. The temperature gradient between polar and equatorial regions is much steeper than at times without extended polar ice shields. This latter case surely was prevailing in Phanerozoic time. With a relatively shallow temperature gradient, the circulation of ocean currents will be less intensive as at present. This necessarily will lead to a stronger stratification of the oceanic water masses, implying an extension and intensification of the oxygen depleated zone. This extension may affect not only deeper parts of the water column, but may also rise to shallow depths, thus sometimes flooding over large areas of the episialic sea, even reaching the outer shelf areas.

An upwards extended oxygen depleted zone may well explain the widespread occurrence of long-lasting dark and black shale depositions, e.g. in upper Cambrian to Silurian times, as well as short-term black-shale episodes. In this connection it must also be considered that in contrast to the Quaternary, episialic geosynclinal and epicontinental seas prevailed in the preceding Phanerozoic time. At least, almost all Phanerozoic sediments preserved on the continents were deposited in an episialic sea.

Another aspect of the extended oxygen depleted zone concerns biotic evolution. If, e.g., in a certain area, slightly disaerobic water masses remain stable for million of years the neighboring organisms will have the opportunity to encroach into this restricted environment by evolution.

The Nemesis Affair

This is the title of a book written by David M. Raup in 1986. There he excellently describes the dramatic development of the impact hypothesis. Although already in 1970 Digby McLaren proposed the impact of an extratelluric body on Earth as causation for an important global event (the late Devonian Kellwasser Event), it was Alvarez et al. (1980) who demonstrated an unusual high iridium enrichment at the Cretaceous-Tertiary (K/T) boundary for the first time, which provided strong argument for an impact. Within a year these results were corroborated by equivalent findings at numerous K/T Boundary sites, where additional impact signals, such as shocked quartz, were also found.

Alvarez's impact hypothesis was then combined with the 26-Ma periodicity of extinction events proposed by Raup and Sepkoski (1984). Thus the Nemesis hypothesis was born in Nature articles by Whitmire and Jackson (1984). These authors proposed the existence of a companion star of the Sun - the black dwarf Nemesis - which has an excentric orbit. At its closest approach Nemesis passes the "Oort cloud" of comets, thereby triggering a periodic comet shower, which finally leads to at least one impact on Earth. In the same Nature issue of April 1984, other authors (Schwartz and James, 1984, Rampino and Stothers, 1984) proposed the Sun's motion perpendicular to the galactic plain as a cause for the periodicity. In addition Rampino and Stothers recognized a 31 Ma dominant cyclicity in the observed age distribution of impact craters on Earth, i.e. a periodicity similar to the 32 Ma that Fischer and Arthur had proposed already in 1977. Alvarez and Muller (1984) calculated a 28.4 Ma cycle of impacting.

The Nemesis hypothesis received further support by the consideration of observed impact craters on Earth (Grieve and Robertson, 1987; see Morrow et al., this volume), the stratigraphical position of tectites and theoretical calculations of impact rates).

The fascinating Alvarez and Nemesis hypotheses immediately found a host of devotees, as documented by a burst of publications, most of them in Nature or

Science, but also in other periodicals and many special editions. In addition to the K/T Event, extraordinary Ir concentrations were reported from numerous other Phanerozic events. Of course, there have also been articles that critically analysed or refuted the impact hypothesis or even the existence of global events. Although some of them brought forward good arguments, others overshot the mark, as many of the impact disciples also had. To review even the important ones among them is not possible within the framework of this chapter. Therefore only certain aspects of these hypotheses will be discussed.

Iridium enrichment is the original and most important argument for the impact hypothesis. In most published cases this concentration, as well as other geochemical anomalies, is connected with the occurrence of black shales. Yet this fact should evoke a critical consideration, as such a facies change cannot be easily explained by an impact, and the potential for enrichment of certain elements due to an often extremely reduced sedimentation rate within the black shale environment has been known for a long time (e.g. Buggisch, 1972), even if in cases not connected with a global bio-event. Another possibility for the enrichment of iridium and other heavy metals is through biological acticity. A possible example of this process is the co-occurrence of an iridium anomaly with the cyanobacteria *Frutexites* in the early Famennian of the Canning Basin, Western Australia. This extraordinarily high amount of iridium was interpreted as an indication for an impact at the Frasnian-Famennian Kellwasser Event (Playford et al., 1984) although the concerned horizon appears to be markedly younger.

Microtektite enrichment or concentration is also taken as evidence for an impact. With respect to this the experience of the present author is that microtektites in various amounts can be observed within the residues of many conodont samples from time intervals showing no indication of a global event. This random occurrence of microtektites indicates rather a relatively frequent background input of cosmic material. Later we should consider that not each impact which produces microtektites is large enough in size to produce a global bio-event. Therefore, e.g., it is highly improbable that a series of global bio-events such as appear within a few million years in Late Cretaceous time, were caused by "comet showers" (Hut et al., 1987) consisting of interplanetary bodies, each of which was of the extraordinary size needed to produce a global bio-event.

In connection with the 26-Ma periodicity comparable arguments should be considered. It is most improbable that at each closest approach of the assumed Nemesis the earth will be hidden by an interplanetary body of relevant size. Furthermore, it cannot be excluded that a sufficient large impact happens also at other times of Nemesis' orbit, i.e. independent of this artificially constructed dark dwarf.

The scenario connected with an impact at the K/T boundary implies a rapid cooling, producing an "impact or planetary winter" – comparable to a "nuclear winter" – as proximate cause for the mass extinctions. According to several authors (e.g. Besse et al., 1986) a similar abrupt temperature fall could also be caused by extreme volcanic activity, such as connected with the Late Cretaceous Deccan Trap volcanism in India. In the opinion of the present author, the disadvantage of this as well as of the impact hypothesis is that it does not consider the aspect of slow but enormous environmental changes which occurred towards the end of the Mesozoic Era.

An excellent assessment and discussion of the K/T Bio-Event was published by the late Hermann Jaeger in 1986. As this paper is written in German, it is unfortunately not widely known nor cited. In the summary Jaeger wrote: "The alleged sudden mass mortality at the C/T-boundary is an illusion. The diversity of the many animal groups that became extinct, declined stepwise long before the end of the Cretaceous. The impact hypothesis is not necessary to account for the extinctions at the end of the period. The rise and decline of important fossils groups in the Cretaceous were closely linked to earth history, particularly the exceptional geological situation that was characteristic of the whole Upper Cretaceous. The Upper Cretaceous transgression with its resultant unusually equable warm climate and optimum conditions for organic evolution was succeeded by one of the largest regressions. This fundamental transformation of the face of the earth in turn caused widespread changes in the formerly favourable life conditions".

Jaeger's detailed argumentation coincides largely with the opinion of the present author, as summarized in 1984. Differences exist only insofar as I accepted the possibility of an impact that "would have caused most probably only an acceleration of the last phase of extinction. ... This extinction would have taken place a short time later anyway".

Of course, the end-Cretaceous regression and associated loss of epicontinental sea areas and biotopes were sufficient in themselves to cause the enormous loss in marine organisms. Thereby also certain terrestric biotopes, e.g. many of those inhabited by dinosaurs, suffered strong reductions or alterations, as Erben (1969) showed for Western Europe. After Erben (1969, and subsequent papers, until Erben et al., 1995), this ecological stress caused malformations in dinosaur eggs. The present author assumes that if this contributed to the extinction of dinosaurs, it was nothing more than one of the consequences which resulted from the environmental overturn during the terminal Cretaceous. In this connection it should be emphasized that a probable survival of some dinosaur taxa is not at all an argument against the end-Cretaceous catastrophe, even if that had been caused by an impact. It rather would underline the high diversity of this group in respect to the adaptation to manifold environmental conditions.

Besides of the loss of epicontinental seas, the great regression during the terminal Cretaceous caused additional environmental changes: change in the albedo and thus in the climate; at the same time the ocean current system could have changed due to the widening of the Atlantic Ocean; by the regression new ecological niches were formed; and this, in connection with the triggered climatic fluctuations, enhanced the adaptive radiation of angiosperms and thus a strong change with respect to the food supply of all terrestrial phyla.

In summarizing the mentioned aspects and interpreting the contribution of Kauffman and Hart (this volume), we can state that the large Late Cretaceous regression caused a crisis within the terrestrial and marine ecosystems which was accelerated by several mass-extinction events and finally by the K/T Boundary Event. Considering all available data, the present writer is inclined to see the impact of a large extratelluric body as causation of this final boundary event, and thus to accept the impact hypotheses in this unique case. In contrast, the Nemesis Affair appears to me to be a short-lived but nevertheless interesting and thought-provoking scientific affair.

Periodicity in Global Events

As mentioned before, most boundaries of higher-ordered stratigraphic units coincide with global bio-events. Because these boundaries are distributed relatively regularly over the Phanerozoic time-scale, the assumption of a certain periodicity was reasonable. Thus, Fischer and Arthur (1977) supposed a periodicity of 32 Ma, which coincides approximately with the assumed oscillation of the solar system about the galactic plane (Rampino and Stothers, 1984: 33 ± 3 Ma; Schwartz and James, 1984: 26 Ma). By a numerical analysis of data given in the Treatise of Invertebrate Palaeontology, Sepkoski (1983) came to the assumption of a 26-Ma periodicity in post-Palaeozoic time. For the Palaeozoic Era such a regular occurrence of global bio-events could not be recognized.

Sepkoski's results have been enormously stimulating, leading both to further relevant studies and considerations as well as to controversial and sometimes vehement discussions. Of course, the connection of this periodicity with the impact hypothesis can be regarded as obsolete, as discussed above. On the other hand, the meanwhile refined data-base (see Sepkoski, 1995, this volume) confirms the former results and increases the probability of a certain periodicity for larger global bio-events, even if the deviation from the 26-Ma average significantly varies. Thus the question of the cause(s) for the periodicity remains.

Because most global bio-events are connected with facies changes, i.e. obviously with environmental changes, the question of the causation of the periodic pattern leads necessarily to the recognition of a certain periodicity in global geo-events and thus shifts the problem to the question of ultimate causes of global events.

The Ultimate Causes

As pointed out before, even the apparently proximate causes such as changes in oceanographic or atmospheric changes are parts of a complex system, and themselves trigger further complex processes. This has to be assumed also for the ultimate causes. Most probably, many processes and changes are induced by mantle convections: plate tectonics (ocean-floor spreading, subduction, orogenesis), plutonism, volcanism, and heat-flow.

The effect on the biosphere depends on the geophysis, i.e. the respective state of the earth, whereby the position of continents and their geomorphologic features are of decisive importance.

We also have to consider the influence of planetary or galactic parameters. Thus, e.g., the two Phanerozoic megacycles mentioned in Morrow et al. (this volume) may possibly coincide with the revolution of the solar system around the galactic center. These still very hypothetical considerations are summarized in Fig. 4.

In any case, global events can serve as an indication of hitherto unrecognized but important episodic or periodic geological changes. To follow back the chain of causes, from proximate to ultimate, will be one of the great interdisciplinary tasks which will contribute to a better understanding of the development and recent state of our earth.

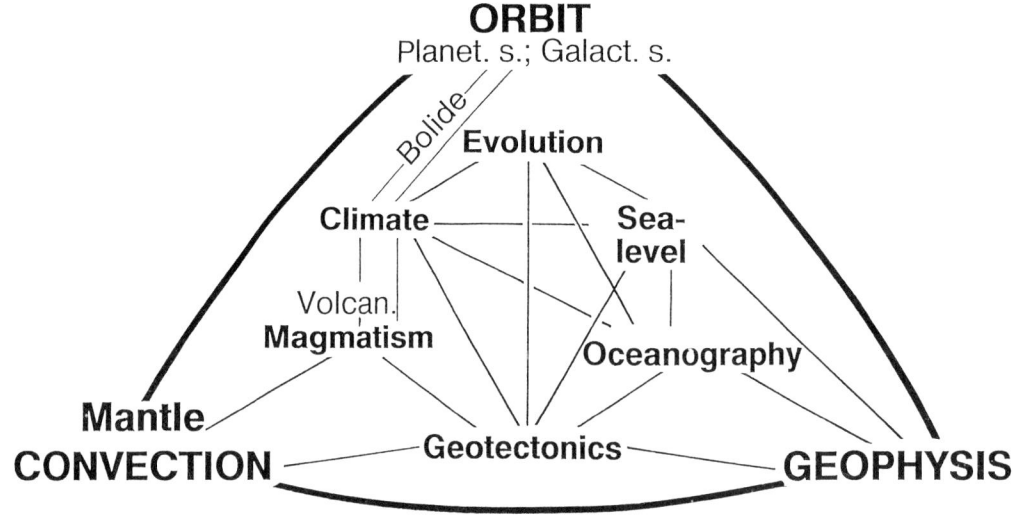

Fig. 4. Possible ultimate causes (outer triangle) and their connections to proximate causes

Acknowledgements

I am grateful for the assistance of Gabriela Meyer for typing and proof-reading and in addition thanks are given to Jared Morrow who assisted in improving linguistically the English text and to Cornelia Kaubisch for her help with graphics.

References

Alvarez, W. and Muller, R.A., 1984. Evidence from crater ages for periodic impacts on Earth. Nature 308, 718-720.

Alvarez, L., Alvarez, W., Asaro, F. and Michel, H., 1980. Extraterrestrial Cause for the Cretaceous-Tertiary Extinction. Science 208, 1095-1108.

Becker, R.Th., 1986. Ammonoid evolution before, during and after the "Kellwasser-Event" – Review and preliminary new results. In: Walliser, O.H. (ed.), Global Bio-Events. A critical Approach. Lecture Notes in Earth Sciences 8, VII+422 pp., Springer, Berlin Heidelberg New York.

Besse, J., Buffetaut, E., Cappeta, H., Courtillot, V., Jaeger, J.-J., Montigny, R., Rana, R., Sahni, R., Vandamme, D. and Vianey-Liaud, M., 1986. The Deccan Traps (India) and Cretaceous-Tertiary boundary events. In: Walliser, O.H. (ed.), Global Bio-Events. A critical Approach. Lecture Notes in Earth Sciences 8, 365-370, Springer, Berlin Heidelberg New York.

Boucot, A.J., 1990. Phanerozoic extinctions: How similar are they to each other? Lecture Notes in Earth Sciences 30, 5-30, Springer, Berlin Heidelberg New York.

Buggisch, W., 1972. Zur Geologie und Geochemie der Kellwasserkalke und ihrer begleitenden Sedimente (Unteres Oberdevon). Abh. hess. L.-Amt Bodenforsch. 62, 1-68, Wiesbaden.

Davis, M., Hut, P. and Muller, R.A., 1984. Extinction of species by periodic comet showers. Nature 308, 715-717.

Erben, H.K., 1969. Dinosaurier: Pathologische Strukturen ihrer Eischale als Lethalfaktor. Umschau Wiss. Techn. 17, 552-553, Frankfurt.

Erben, H.K., Ashraf, A.R., Böhm, H., Hahn, G., Hambach, U., Krumsiek, K., Stets, J., Thein, J. and Wurster, P., 1995. The Cretaceous/Tertiary boundary in the Nanxiong-Basin (Continental Facies, SE-China). Erdwissensch. Forschung, Akad. Wissensch. Literatur Mainz XXXII, 245 pp., Franz Steiner Verlag, Stuttgart.

Fischer, A.G. and Arthur, M.A., 1977. Secular variations in the pelagic realm. In: Cook, H.E. and

Enos, P. (eds.), Deep-Water Carbonate Environments. Soc. Econ. Paleont. Miner., Spec. Publ. 28, 19-50.

Grieve, R.A.F. and Robertson, P.B., 1987. Terrestrial impact structures. Geol. Surv. Canada, Map 1658A; Suppl. to Episodes 10 (2).

Hut, P., Alvarez, W., Elder, W.P., Hansen, T.A., Kauffman, E.G., Keller, G., Shoemaker, E.M. and Weissman, P.A., 1987. Comet showers as a possible cause of stepwise extinctions. Nature 329, 118-126.

Jaeger, H., 1986. Die Faunenwende Mesozoikum/Känozoikum – nüchtern betrachtet. Z. geol. Wiss. Berlin 14, 629-656.

Kauffman, E.G. and Hart, M.B., 1995. Cretaceous Bio-Events. In: Walliser, O.H. (ed.), Global Events and Event Stratigraphy in the Phanerozoic, 285-311, Springer, Berlin Heidelberg New York.

McLaren, D.J., 1970. Time, Life and Boundaries. Journal of Paleontology 44, 801-815.

Morrow, J., Schindler, E. and Walliser, O.H., 1995. Phanerozoic Development of Selected Global Environmental Features. In: Walliser, O.H. (ed.), Global Events and Event Stratigraphy in the Phanerozoic, 53-62, Springer, Berlin Heidelberg New York.

Playford, P.E., McLaren, D.J., Orth, C.J., Gilmore, J.S. and Goodfellow, W.D., 1984. Iridium Anomaly in the Upper Devonian of the Canning Basin, Western Australia. Science 226, 437-439.

Raup, D.M., 1986. The Nemesis Affair. A Story of the Death of Dinosaurs and the Ways of Science. 220 pp. W.W. Norton & Co, New York, London.

Raup, D.M. and Sepkoski, J.J., 1984. Periodicity of Extinctions in the Geologic Past. Proc. natn. Acad. Sci. USA 81, 801-805.

Rampino, M.R. and Stothers, R.B., 1984. Terrestrial mass extinctions, cometary impact and the Sun's motion perpendicular to the galactic plain. Nature 308, 709-712.

Schindler, E., 1990. Die Kellwasser-Krise (hohe Frasne-Stufe, Oberdevon). Göttinger Arb. Geol. Paläont. 46, IV+115 pp., Göttingen.

Schwartz, R.D. and James, P.B., 1984. Periodic mass extinctions and the Sun's oscillation about the galactic plane. Nature 308, 712-713.

Sepkoski, J.J., 1995. Patterns of Phanerozoic Extinction: A Perspective from Global Data Bases. In: Walliser, O.H. (ed.), Global Events and Event Stratigraphy in the Phanerozoic, 35-52, Springer, Berlin Heidelberg New York.

Wallace, M.W., Keays, R.R. and Gostin, V.A., 1991. Stromatolitic iron oxides: Evidence that sea-level changes can cause sedimentary iridium anomalies. Geology 19, 551-554, Boulder.

Walliser, O.H., 1986. Towards a more critical approach to bio-events. In: Walliser, O.H. (ed.), Global Bio-Events. Lecture Notes in Earth Sciences 8, 5-16. Springer, Berlin Heidelberg New York.

Walliser, O.H., 1990. How to define "Global Bio-Events". In: Kauffman, E.G. and Walliser, O.H. (eds.), Extinction Events in Earth History. Lecture Notes in Earth Sciences 30, 1-3. Springer, Berlin Heidelberg New York.

Walliser, O.H., 1994. Regelhaftigkeiten in der Evolution fossiler Organismen-Gruppen. In: Gutmann, W.F. et al. (eds.), Morphologie & Evolution. Senckenberg-Buch 70, 281-294. Verlag Waldemar Kramer, Frankfurt/Main.

Whitmire, D.P. and Jackson IV, A.A., 1984. Are periodic mass extinctions driven by a distant solar companion? Nature 308, 713-715.

Author´s address:

Otto H. Walliser, Institut und Museum für Geologie und Paläontologie, Goldschmidt-Strasse 3, D - 37077 Göttingen, Germany

Evaluating Paleontologic Data Relating to Bio-Events

J. John SEPKOSKI, Jr. and Carl F. KOCH

Abstract. Measurements of diversity and extinction intensity around bio-events in local stratigraphic sections and global taxonomic data bases include some degree of uncertainty, or potential error. Sources of uncertainty involve failure to sample rarer species in all collections and imprecise age correlations and species counts, among others. This chapter presents several methods for evaluating this paleontologic uncertainty. For fossil ranges in local sections, methods are reviewed that might help to discriminate between true gradual (or stepwise) extinction and apparent gradual decline resulting from sampling effects. For tabulated extinctions in regional or global data bases, methods are evaluated for measuring extinction intensity and procedures are presented for calculating relative uncertainty.

Contents

Introduction	21
Fossils in Local Stratigraphic Sections	22
Extinction in Taxonomic Data Bases	26
Definitions	26
Measuring Extinction Intensities	27
Confidence Intervals on Extinction Metrics	29
Conclusions	32

Introduction

An important goal in the study of global bio-events is producing accurate measurements of diversity patterns and extinction intensities below, at, and above event horizons. Such measurements are essential for characterizing events, but they are difficult to make. The fossil record is notoriously incomplete (e.g. Darwin, 1859; Newell, 1959; Simpson, 1960; Durham, 1967), and taxa known from individual fossil collections or global data compilations may have imprecise or even inaccurate information about taxonomy, age range, geography, etc.

The purpose of this chapter is to provide a brief discussion of some sampling problems that may produce imprecise information about diversity and extinction in data from local stratigraphic sections and from global taxonomic compilations. Our purpose is also to review a few techniques for assessing potential problems. First, we discuss some universal sampling problems that are encountered in collecting fossils from local sections for studies of the abruptness of extinction at and below bio-event horizons. We include some techniques that can be used to test paleontologic patterns to determine their robustness to sampling effects. Then we turn to regional and global compilations, or taxonomic data bases, on ages of first and last appearances of taxa, which frequently have been used to measure relative magnitudes of extinction over long time scales. Here we discuss appropriate measures of extinction intensity and estimates of

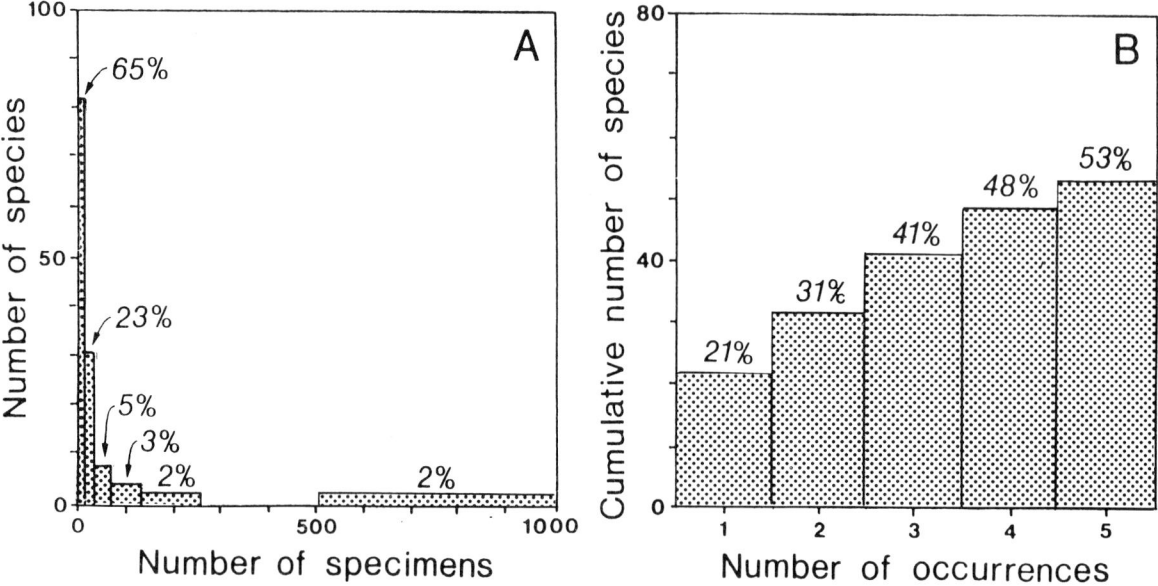

Fig. 1. Typical observed numbers of rare species in large data sets. A, Species abundance curve showing that the majority of species are represented by few specimens even in a very large collection (data from Koch, 1978). The widths of the histogram bars increase geometrically with abundance. B, Cumulative histogram of occurrences of rare species among 578 collections of mid-Maastrichtian molluscs from the Atlantic and Gulf Coastal Plains of North America (data set 9 of Koch, 1987). More than half of the total species occur in fewer than 1% (5 of 578) of the collections

uncertainty associated with them.

Our goal is not to pronounce what kinds of data are good or bad for the study of bio-events. Rather, it is to highlight problems encountered in all kinds of paleontologic data and to review some tools for measuring the magnitudes of these problems so that descriptions of bio-events can be made more accurate.

Fossils in Local Stratigraphic Sections

Basic Data. Studies of local stratigraphic sections (or composite sections) provide detailed information about local manifestations of bio-events and about ranges of species (or higher taxa) below, through, and above event horizons. This is information that is essential for asking how gradual or abrupt an event was, or what the internal pattern of extinction and diversity loss was (e.g. stepwise or smooth).

The general experimental procedure is to locate fossiliferous horizons in a section (outcrop or core), collect fossils from each horizon (or at constant intervals if sections are continuously fossiliferous), and then remove collections to the laboratory. There, fossil identification, counts of abundance, paleoecologic and taphonomic analysis, etc. are conducted. One outcome of this procedure is a local "biostratigraphic range chart" which illustrates, in addition to other information, first and last occurrences of species in the stratigraphic section. These range charts frequently are used to locate bio-events, measure diversity around them, and summarize patterns of extinction below them. These data can provide invaluable information about stratigraphic durations of bio-events, and they have a feel of "ground truth" absent from data compiled solely from the paleontologic literature. However, last occurrences of species and higher taxa in local sections still are last collected occurrences no matter how diligent the collection effort. Below, we discuss general relationships between last collected occurrence and true last occurrence.

Species Abundances. In nearly every ecological setting, living and fossil, there are a few abundant species, a few more common species, and a large number of rare species (Fig. 1A). In fact, generally more than half of species occur in fewer than 1% of collections (i.e. fossil samples) (Fig. 1B). A variety of explanations and mathematical models have been offered for this nearly universal pattern. May (1975), Whittaker (1975), and Maggurran (1988) provide excellent reviews. At higher taxonomic levels, the pattern also holds, although sometimes not so strongly; still, most genera, families, etc. are rare, and only a few are abundant in any given setting (see also Boucot, 1986, appendix).

Implications for Paleontologic Sampling. The consequence of very uneven species abundances for paleontologic sampling is that large numbers of species actually preserved in most fossiliferous horizons will be missing from any collection; this will be true even if the collection contains hundreds of individuals (Koch, 1978, 1991a). Furthermore, multiple collections of hundreds of individuals from the same horizon, each randomly sampled by bulk collecting or the like, will share the same common species but fewer rare species (or rare genera, families, etc.). Buzas et al. (1982) and Koch (1987, 1991a) have considered this problem in some detail for very large data sets (thousands of individuals), analyzing frequencies of occurrences of species among collections rather than abundances of species within collections (which they show to be correlated; see also Rabinowitz et al., 1986).

Buzas et al. (1982) present a very simple probability calculation for determining how likely a species is to occur in a series of new collections. Suppose that a species is present in n previous collections out of a total of N from a given horizon. Then the probability, p, of finding that species in a single new collection is approximately

$$p = n/N. \qquad (1)$$

The probability, P[2], of finding it at least once in two new collections is roughly

$$P[2] = 1 - (1-p)^2, \qquad (2)$$

in three new collections

$$P[3] = 1 - (1-p)^3, \qquad (3)$$

and in N collections

$$P[N] = 1 - (1-p)^N. \qquad (4)$$

Eq. (4) can be used to determine the probability that a given species will be found in a set of N new collections. The equation can also be used to calculate how many new collections must be made in order to resample a species at some given level of confidence. Fig. 2 gives N values for 95% confidence at various values of p.

Comparison of Figs. 1 and 2 illustrates the sampling problem. More than half of the species occur in fewer than 1% of collections (Fig. 1B), and thus

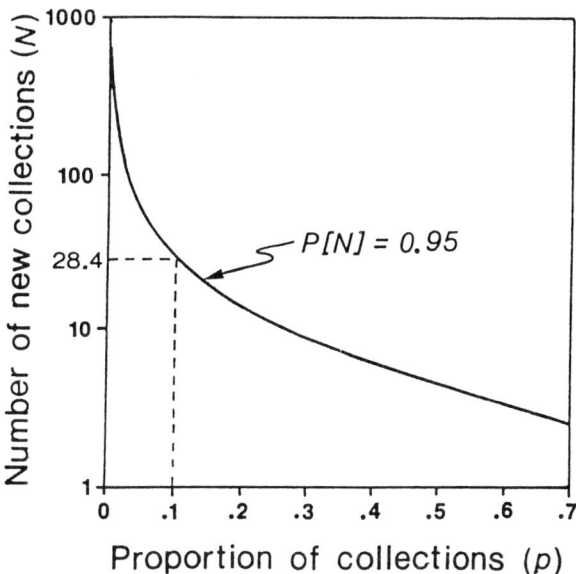

Fig. 2. Relationship between the number of collections (N) needed for 95% confidence (P[N] = 0.95) of sampling a species at least once to the proportion of collections in which it occurs. For example, if the species occurs in 10% (p = 0.1) of previous collections, then 29 (= 28.4 rounded up) new collections are needed to have a 95% confidence of seeing the species again

each requires 298 or more new collections in order to have 95% confidence that it is sampled at least once (Fig. 2). About 1 in 5 species in Fig. 1B occur less often than once in 500 collections and thus require about 1500 collections for 95% confidence of being resampled.

The importance of this to the analysis of bio-events is that collections immediately below an event horizon will contain only a few of the rarer species found lower in any stratigraphic section, e v e n i f a l l s u r v i v e d u p t o t h e b i o - e v e n t. The last sampled occurrences of some rarer species will occur in lower collections, even if collection sizes range in the hundreds or thousands. This problem was first analyzed by Signor and Lipps (1982) and has become known as the "Signor-Lipps effect" (Raup, 1986). Signor and Lipps demonstrated that if sampling is incomplete (as it nearly always is), then last observed occurrences of species will begin below a horizon of abrupt extinction, and diversity will appear to decline at accelerating rate upward to the event (Fig. 3; see also Springer, 1990). Koch and Morgan (1988) provide a more detailed mathematical treatment of this

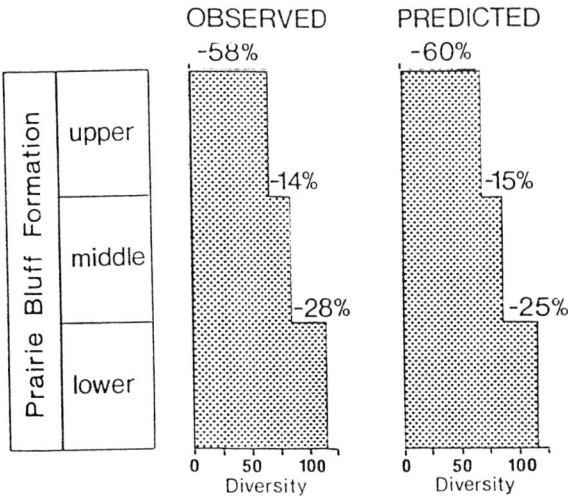

Fig. 3. Comparison of observed and predicted loss of species beneath the Cretaceous-Tertiary boundary (= top of Praire Bluff Formation) at Braggs Cut, Alabama. The observed pattern appears as if the extinction were gradual, or perhaps stepwise. However, the "predicted" pattern, based on calculations assuming abrupt extinction and incomplete sampling, yields a virtually identical pattern. Modified from Koch (1991b, based on equations in Koch and Morgan, 1988)

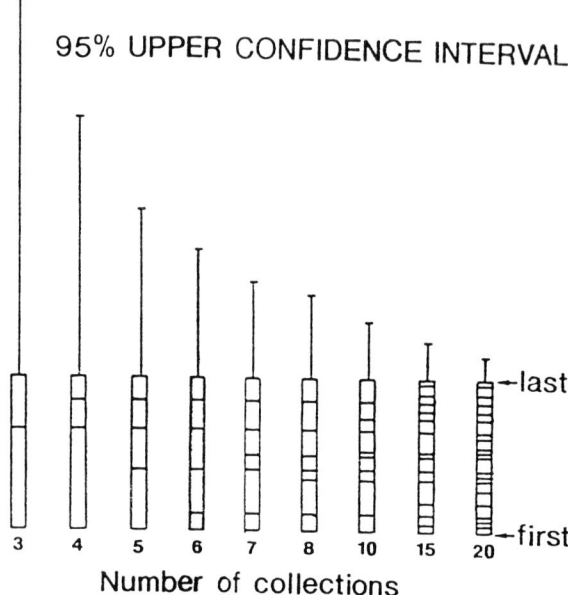

Fig. 4. Upper confidence intervals on the observed stratigraphic ranges of species, based on the number of collections in which they have been sampled. The first (lowest) and last (highest) sampled occurrences of the species are shown along with other sampled occurrences (horizontal tics), which occur randomly between the first and last. The error bar above the last sampled occurrence indicates the range in which there is a 95% probability that the true extinction of the species occurred. This range is always higher than the last sampled occurrence of a species, but decreases exponentially as the number of sampled occurrences increases. Confidence intervals were calculated from Eq. (5); figure modified from Marshall (1990)

sampling problem, considering also what happens when sample sizes vary up a stratigraphic section (see also Koch, 1991a, b). Furthermore, they demonstrate that sampling not only can make abrupt extinctions appear gradual but also can make them appear stepwise (Fig. 3).

Assessing Sampling Problems. Several procedures are available for estimating sampling effects on apparent extinction patterns below bio-event horizons. One is simply to eliminate the less frequently occurring species to determine if apparent patterns change. This approach is recommended implicitly by Buzas et al. (1982) and Koch (1987), but this only lessens the effects of sampling. Even the most abundant species may not occur in every collection, and therefore its last occurrence may be below the last sample in a section.

Another approach is to compute diversity patterns expected in a section assuming that all species become extinct abruptly at the bio-event horizon. This was done by Koch (1991b) to test the apparent stepwise pattern of extinction below the Cretaceous-Tertiary boundary at Braggs, Alabama (Fig. 3). The computed pattern was compared to the observed pattern, and the difference was small. Thus, a sudden mass extinction combined with sampling effects could account for the observed pattern of diversity decline.

Other estimates of sampling effects below bio-event horizons can be obtained from the works of Strauss and Sadler (1989) and Marshall (1990, 1991) (see also Springer and Lilje, 1988; Springer, 1990). These authors are concerned with placing "confidence intervals" on stratigraphic ranges of taxa so that sampled ranges can be used to estimate true ranges (Fig. 4). The two important parameters here are

R = stratigraphic range, or distance (e.g. in meters), between the first and last sampled occurrence of a species;

n = number of sampled occurrences of the species within this range (i.e. number of horizons from which it was collected).

If there is a gap of stratigraphic distance r between the last sampled occurrence of a species and a bio-event horizon, the probability, C_R, that this could be due to sampling failure alone can be estimated as

$$C_R = (1 + r/R)^{-(n-1)} \qquad (5)$$

Marshall (1991) provides a very readable discussion of the logic behind this equation, and Springer (1990) presents a very interesting analysis testing the local abruptness of the end-Cretaceous mass extinction. Notice that Eq. (5) gives a probability of sampling failure equal to unity when a species occurs in only a single collection: one occurrence provides no statistical information about a species' true range.

In the comments above, we have used terms like "estimates of sampling effects" because the equations make specific assumptions about how sampling has proceeded. For example, Eq. (5) assumes that a species has some constant probability of being collected at random horizons throughout its true range (Fig. 4), independent of the occurrence of other species. These conditions are only approximated in true situations: stratigraphic sections often contain intervals lacking in fossils (e.g. Fig. 5A), and fossil-bearing horizons usually vary in fossiliferousness, making equally-sized collections impossible.

In such situations (or if one is uncomfortable with calculations), there are still some qualitative tools that can be used to ask questions about sampling in local sections. One is a modification of the "gap analysis" pioneered by Paul (1982) (see also McKinney, 1986; Springer and Lilje, 1988). If the concern is whether a gap between a species' last occurrence and a bio-event horizon is real, simply look down the stratigraphic range of the species: if gaps of equivalent or larger length exist between known occurrences, then the gap at the top could very well reflect sampling or fossiliferousness, and not extinction prior to the event. This kind of analysis could help to resolve the famous "two-meter gap" problem between the highest dinosaur bones and the Cretaceous-Tertiary boundary in the Hell Creek Formation of eastern Montana (e.g. Alvarez, 1983).

If the concern is decline in overall diversity below an event in a local section, another qualitative tool is available, introduced by Raup (1989). As a working hypothesis, imagine that the bio-event was abrupt, and species could not see it coming. In such a case, the event could have happened anywhere in the section. Now, move down the section to below where the apparent decline in diversity begins, and drop an

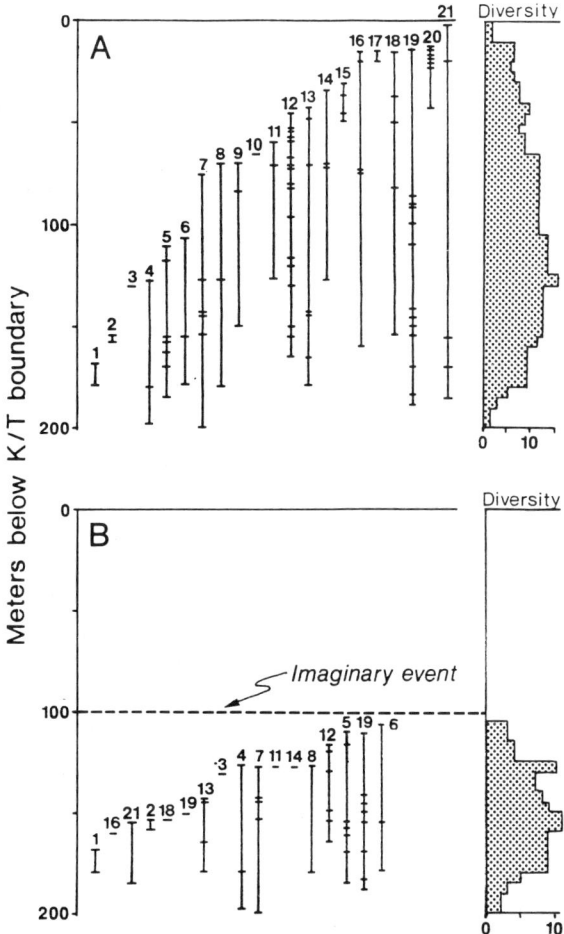

Fig. 5. Range charts for ammonite species (numbers) below the Cretaceous-Tertiary boundary (0 meter) at Zumaya, Spain. Ranges are drawn between first and last sampled occurrences in the local section, with horizontal tics indicating all collections in which the species occur. A, Complete range chart, with apparent local diversity indicated by the histogram on the right. B, Same data (rearranged) with an imaginary bio-event placed at 100 meters. Local diversity, compiled by ignoring all collections above the imaginary event, exhibits a gradual upward decline, with a step at 125 meters. Modified from Raup (1989, based upon data from P.D. Ward)

imaginary event at random (Fig. 5B). Ignoring all species occurrences above the imaginary extinction event, compile a new curve of apparent diversity. If this curve declines similarly to the curve below the real event, then sampling could very well be creating the pattern. If the pattern is different, however, there are two possibilities: extinction was gradual or the random imaginary event is not characteristic. Therefore, try again with another imaginary event, and again and again. Do this, in fact, whether or not the first try looked like the real event; this, too, could have been uncharacteristic. If most imaginary events produce apparent diversity declines similar to or more prolonged than the actual pattern, that pattern probably reflects sampling. (A more quantitative variant on this procedure involves calculating mean diversity and standard deviations within regular intervals below a set of randomly placed imaginary bio-events.)

Intensive Sampling. These statistical considerations have the straight-jacket assumptions that beds through any section (or composite section) are equally fossiliferous (i.e., produce approximately equal numbers of fossils per unit of effort) and that similar effort has been exerted on each sampled bed in order to obtain fossils. These idealistic assumptions are hardly realistic for all fossil collecting and may be violated in the extreme during the study of bio-events. This is the case when beds adjacent to an event are hammered, sifted, and/or dissolved with special attention to find last occurrences of taxa (e.g., Walliser et al., 1988; Schindler, 1990; Harries and Kauffman, 1990). Nonrandom effort greatly weakens the statistical tests reviewed here for asking questions about gradual, stepwise, or abrupt extinction. Still, simply because a bed has been sampled extensively, the absence of rare species may not indicate their prior extinction (cf. Koch, 1987, 1991a), and statistical exercises can be performed to evaluate the data at hand:

1. The statistical tests reviewed above can be performed only on the intensively sampled beds, and probabilities of collecting rare species or confidence intervals on stratigraphic ranges can be calculated for the few focused beds. However, the statistics will lose power because presumably only a few beds have received special scrutiny, making the n-values in Eqs. (4) and (5) uncomfortably small.

2. Alternatively, the sample sizes of intensively sampled beds might be "rarefied" if numbers of collected individuals within species have been counted. This process involves asking how many species should have been found if the number of fossils collected had been smaller. The procedure for doing this is fairly standard in ecology (e.g., Magurran, 1988; see also Foote, 1992) and is used to compare species richnesses in samples of unequal numbers of collected individuals. The expectation is that numbers of identified species should decline as numbers of collected individuals are statistically reduced. With respect to bio-events, the question is how quickly species decline as a function of number of individuals. If species numbers initially remain steady as numbers of individual fossils are statistically rarefied, then one might have some confidence that species disappearing earlier really did become extinct prior to the analyzed horizon (assuming, of course, constant facies). On the other hand, if species numbers decline steadily as sample size is rarefied, then there can be no certainty that rarer species seen last in lower beds really did become extinct around the horizon of their last sampled occurrence. This uncertainty applies not only to macrofossils within collections of hundreds of specimens, but also to microfossils within collections of thousands or even tens of thousands of individuals. Still, how rare a species must be in order to make the judgment that its disappearance is indeed only a matter of sampling defies the idealistic precepts of statistics. Ultimately, conclusions of the collector can only be intelligently informed, but not strictly dictated, by quantitative analyses.

Necessary Conventions. The kinds of sampling considerations discussed here will be of concern to readers whether or not an author is interested in the abruptness of bio-events or the theory of sampling. Therefore, we urge that two conventions be adopted when publishing range charts for taxa in local or composite stratigraphic sections:

1. Indicate all horizons in which each species was collected (e.g. Fig. 5A) and not simply bottom and top of the range;

2. If possible, indicate the number of specimens of each species collected at each horizon.

Without this information, the labors of field paleontology are in vain when asking questions about whether extinction events were gradual, stepwise, or abrupt -- questions necessary for sorting out the causes of any given bio-event.

Extinction in Taxonomic Data Bases

Definitions

Taxonomic data bases are regional or global compilations of information about large numbers of taxa (Sepkoski, this volume). The kind we will be concerned with lists taxa along with their first and last documented occurrences in the fossil record. (With the ready availability of electronic data management systems, many workers are commendably compiling far wider varieties of information.) Taxonomic data bases usually have fossil occurrences documented in terms of some sort of regional or global standard time scale, such as international stratigraphic stages (e.g. Harland et al., 1967; Sepkoski, 1982) or, more recently, substages and even chronozones. Few large data bases have been assembled at the taxonomic level of the species (e.g. Sweet, 1985; Schopf, 1992), and most use higher taxa, such as genera and families. Several large taxonomic data bases have been compiled from workers' personal study of taxa (e.g. Culver et al., 1987; Loeblich and Tappan, 1988), but the majority have been based on compilations of information from the published literature (e.g. Doescher, 1981).

The problems and potentials for studying diversification and extinction with large taxonomic data bases are numerous (Sepkoski, this volume), and only a few will be discussed here. First, we will be concerned with how to measure extinction intensities with imperfect data. Then we will discuss how to measure uncertainty associated with these imperfect data.

Measuring Extinction Intensities

Distributing Data. Information in taxonomic data bases varies in quality. Not only are there variations in systematic quality of taxa and reliability of stratigraphic ages of first and last occurrences, but there can also be differences in the precision of these ages. These can result from two sources:

1. Correlations from certain localities to regional or global standards are uncertain due to lack of suitable fossils or of modern investigation;

2. Available information on age of first and/or last occurrence is from coarser compilations, such as sometimes found in older works, and more modern age determinations have not been found in literature searches.

In either case, these data may be assigned to longer intervals in the hierarchical stratigraphic framework than the units desired (e.g. Table 1).

There are several procedures for dealing with this problem. One is simply to ignore imprecise records, eliminating all affected taxa before counting extinctions. This has been recommended by Patterson and Smith (1987, 1989) and implemented by some authors (e.g. Raup and Boyajian, 1988; Raup, 1991a). Eliminating such data can lead to problems, however, as illustrated by the example in Table 1. This table lists numbers of genus extinctions for stratigraphic intervals of the Silurian. Most of the data are given at the series level, and only a small number of extinctions are located to system level. However, a modest number are located only to "Upper Silurian." If these imprecise data were eliminated, the measured magnitudes of extinction in the Ludlow and Pridoli would be biased downward relative to the Llandovery and Wenlock, for which no comparable imprecise data exist.

The alternative to eliminating imprecise data is to estimate how they might be distributed among stratigraphic intervals. In the absence of any other information, the most probable indicator for this distribution is the precise data. Thus, since 65% of well-documented Ludlovian-Pridolian extinctions fall within the Ludlow, then 65% of the 53 "Upper Silurian" extinctions (= 34.5) should be added there.

Table 1. Worked example showing low-precision extinction data distributed among finer stratigraphic intervals, using marine animal genera and intervals of the Silurian

Stratigraphic interval[1]	Raw numbers	Distributed "super-series" data[2]	New numbers	Distributed system-level data[3]	Final numbers[4]
Silurian	19	--	19	--	0
Upper Sil.	53	--	0	--	0
Pridoli	141	53 x (141/404) = 18.5	159.5	19 x (159.5/834) = 3.6	163
Ludlow	263	53 x (263/404) = 34.5	297.5	19 x (297.5/834) = 6.8	304
Wenlock	207	--	207	19 x (207/834) = 4.7	212
Llandovery	170	--	170	19 x (170/834) = 3.9	174

[1] "Upper Silurian" is undifferentiated Ludlow + Pridoli.   [2] Denominator equals sum of Ludlow + Pridoli extinctions.   [3] Denominator equals sum of Llandovery to Pridoli extinctions.   [4] Rounded to nearest integer

Table 1 provides a worked example for the distribution of all imprecise extinctions recorded for the Silurian. Note that the calculations proceed from smallest coarse interval ("Upper Silurian") to largest (entire Silurian).

This manner of distributing imprecise data is based on a "best guess", but it adds uncertainty to the extinction metrics. This problem will be dealt with further below when dealing with confidence intervals.

Calculating Diversity. A count of diversity within stratigraphic intervals is needed for some measures of extinction intensity. Diversity is very easy to calculate from numbers of extinctions and originations, even with distributed data. For a stratigraphic interval, i, the diversity, D_i, is

$$D_i = D_{i-1} - E_{i-1} + O_i \quad (6)$$

where D_{i-1} is the diversity in the previous interval, E_{i-1} is the number of extinctions in that interval, and O_i is the number of originations in the ith interval. Table 2 provides a worked example with data for Silurian genera.

Table 2. Data for calculating diversity [Eq. (6)] and percentage of extinction, using Silurian genera as an example

Strati-graphic interval	Number of extinc-tions	Number of origi-nations	Calcula-ted diver-sity	Percent extinc-tion
Pridoli	163	104	1067	15.3
Ludlow	304	239	1267	24.0
Wenlock	212	325	1240	17.1
Llandovery	174	369	1089	16.0
Ashgill	688	---	1408	---

Other measures of taxonomic diversity within stratigraphic intervals have been suggested (e.g. Harper, 1975; Gilinsky and Bambach, 1987). But for purposes of calculating some measures of extinction intensity, diversity calculated by Eq. (6) is most appropriate.

Extinction Metrics. There are four basic measures of the intensity of extinction that can be calculated from taxonomic data bases (Sepkoski and Raup, 1986; Raup and Boyajian, 1988; Gilinsky, 1991a):

1. simple number of extinctions (i.e. total number of last documented occurrences);

2. proportion or percentage of extinction (i.e. number of extinctions divided by diversity);

3. total rate of extinction (i.e. number of extinctions divided by estimated duration of the stratigraphic interval);

4. per-taxon rate of extinction, also called the per-capita rate or probability of extinction (i.e. the proportion of extinction divided by estimated duration of the interval).

In addition, extinction intensity could be measured by either absolute or proportional decline in diversity. These are either identical to measures (1) and (2) above or can be modified to take into account originations in the succeeding stratigraphic interval. (However, this second option may underestimate magnitudes of bio-events since episodes of extinction seem often to be followed by intervals of intense evolutionary rebound with high numbers of originations; e.g. Table 2.) Some other measures of extinction intensity in taxonomic data bases have also been presented (e.g. Gilinsky and Good, 1991).

Each of the four basic measures of extinction intensity has merits and drawbacks. Total number of extinctions is the simplest measure and therefore has the simplest kind of statistical uncertainty associated with it. However, it does not account for number of taxa at risk: numbers of extinctions can be much greater when there are 1000 taxa than when there are 10. Percentage of extinction corrects for this by dividing number of extinctions by number of taxa potentially at risk. However, since neither number is known exactly, this ratio has greater relative statistical error than total number of extinctions (Gilinsky, 1991a).

Metrics (3) and (4) incorporate temporal durations into measured extinction intensities. This seems commendable in an intuitive way because longer geologic intervals can accumulate more extinctions. However, in practical terms, we are not sure this should be done (Sepkoski, 1991). The geologic time scale is poorly known over many intervals, and true durations of shorter stratigraphic units such as some stages, substages, and chronozones may be more than twice, or less than half, estimated values.

This problem is illustrated in Table 3, which lists four relatively recent estimates of the durations of Devonian stages. Standard deviations of the listed durations average 2.2 Ma, which ranges from 26 to 33% of mean stage length, depending upon the time scale. Matters become worse, though, because this

Table 3. Estimates of durations (in Ma) of stages of the Devonian from four recent time scales, showing effects on calculated extinction percentages. Time scales: 1 = Harland et al. (1982) and Palmer (1983); 2 = Bayer and McGhee (1986); 3 = Menning (1989); 4 = Harland et al. (1990). Minimum and maximum extinction probabilities computed by adding and subtracting, respectively, one standard deviation from the estimated duration in the denominator, using time scale 2

Stratigraphic	Time scale				Standard	Extinction probability x 100		
interval	1	2	3	4	deviation	actual	minimum	maximum
Famennian	7	13	9	4	3.8	2.1	1.7 (-19%)	3.0 (+43%)
Frasnian	7	9	8	10	1.3	4.0	3.5 (-12%)	4.6 (+15%)
Givetian	6	12	6	4	3.5	2.5	1.9 (-24%)	3.5 (+40%)
Eifelian	7	7	6	5	1.0	3.5	3.0 (-14%)	4.0 (+14%)
Emsian	7	6	8	4	1.7	3.4	2.6 (-24%)	4.7 (+38%)
Pragian	7	4	5	6	1.3	2.2	1.7 (-23%)	3.3 (+50%)
Lochkovian	7	9	6	13	3.1	1.5	1.1 (-27%)	2.2 (+47%)
Mean					2.2		(-20%)	(+35%)

error ends up in the denominator of rate metrics. Table 3 also lists extinction probabilities calculated with the durations of Bayer and McGhee (1986), which have the smallest ratio of standard deviation to duration. Extinction probabilities were calculated by adding and subtracting one standard deviation from the estimated duration in the denominator. This leads to average errors of 20% below and 35% above the simple extinction probability, and this is for only one standard deviation. Furthermore, potential error in the counts of extinction (see below) still must be added, which can lead to estimated uncertainty nearly equal to calculated values! (Errors in duration of longer geologic intervals, such as series, are proportionally smaller, but this stratigraphic resolution is unacceptable in most studies of extinction.)

If potential error in estimated durations of stratigraphic intervals adds so much uncertainty to extinction metrics, how do we circumvent the problem that longer intervals may accumulate more extinction? The only approach we can think of is to attempt making the stratigraphic intervals used as even in duration as possible. This involves amalgamating adjacent short intervals and subdividing long intervals. Examples of this approach can be found in Sepkoski (1986, 1989, 1990, this volume) and Raup and Boyajian (1988).

Confidence Intervals on Extinction Metrics

Just as confidence limits or the like should be calculated for species' ranges in local sections, some sort of measure of uncertainty should be assessed for extinction metrics calculated from taxonomic data bases (see Raup, 1991a). The problem here is that no one knows quite how to do this. Taxonomic data bases are not random samples of the fossil record, and therefore their formal statistical properties are virtually unknown. Procedures for distributing imprecise data hardly help the situation. However, rough measures of uncertainty for one point relative to others can be calculated with some faith that they approximate true uncertainty.

Binomial Confidence Intervals. Because sampling of the fossil record is incomplete, we might think about the known diversity, D, in any stratigraphic interval as being a sample of the true original diversity. Using known diversity and known number of extinctions, a proportion (not percentage), p_e, of the known diversity that becomes extinct before the next stratigraphic interval can be calculated (metric 2, above). This sample proportion will approximate the true proportion of extinction but in all likelihood will not be precisely the same. If the sample represented by the taxonomic data base were random, the uncertainty around p_e could be expressed as

$$s(p_e) = p(1 - p_e)/D \qquad (7)$$

where $s(p_e)$ is the standard deviation of p_e: the standard deviation that would result if multiple random samples of size D could be taken from the true diversity. Buzas (1990) and Raup (1991a) strongly recommend $s(p_e)$ as the best measure of uncertainty for any proportion where the denominator (e.g. D) is greater than 100. For smaller denominators, Raup (1991a) provides a simple computer program that calculates exact binomial confidence intervals.

The binomial standard deviation in Eq. (7) can be used like most other standard deviations for computing confidence intervals. With random samples, there is a 95% probability that the true proportion of extinction lies within $p_e + 1.96\ s(p_e)$, where 1.96 is the usual normal deviate ("z-value") from a Gaussian distribution. If the 95% confidence interval for one value of p_e overlaps another value, there is little reason to believe they are really different.

Time series of extinction percentages should be graphed with the binomial standard deviation multiplied by 100 (Table 4). The usual convention is to display one standard deviation below the measured value of p_e and one above. For other metrics of extinction, $s(p_e)$ could variously be multiplied by diversity, D, for time series of simple numbers of extinction, or be divided by interval duration for time series of per-taxon extinction rates (cf. Sepkoski and Raup, 1986). Both modifications, however, produce only an approximate measure of uncertainty, since there is also uncertainty in the variable used to multiply or divide $s(p_e)$.

Table 4. Binomial confidence intervals ["St. dev.," or $s(p_e)$ from Eq. (7)] on percentage of extinction

Strati-graphic interval	Diver-sity	Number of extinctions[1]	Proportion of extinction	St. dev. x 100
Pridoli	1067	141	0.132	1.04
Ludlow	1267	263	0.208	1.14
Wenlock	1240	207	0.167	1.06
Llandovery	1089	170	0.156	1.10

[1] Only extinctions known definitely to the series level.

We have continuously interjected the term "random" in this discussion because binomial theory is based upon assumptions of random sampling. As stated, there is no reason to believe that diversity or extinction numbers from taxonomic data bases are random samples of true diversity or extinctions; for example, the rare taxa often missed in sampling may be more prone to extinction than common taxa (e.g. Boucot, 1975; Jablonski, 1986; Erwin, 1989). Thus, the binomial standard deviation only approximates the true uncertainty, by a degree diligent statisticians will have to determine. Nevertheless, $s(p_e)$ can be used as a measure of relative uncertainty, and if confidence intervals for adjacent extinction percentages overlap there can be little certainty that the percentages are significantly different.

Uncertainty from Distributed Data. The procedure for distributing imprecise extinction data (Table 1) adds more uncertainty to extinction metrics, and this uncertainty needs to be estimated. This can be done in a rough-and-ready fashion, similar to what was used by Sepkoski and Raup (1986). In Table 1, there are 19 extinctions documented only as "Silurian": approximately 4 of these were distributed to the Llandovery. Now, it is uncertain that 4 is the actual number that really occurred during the Llandovery, but it is intuitively unlikely that the real number was 0 or especially 19. If we allow that there is less than one chance in 100 that either might be the case, then the resulting proportions of extinction (Tables 1 and 2: 170/1089 and 189/1089) would each be around three standard deviations from the measured value of 174/1089. Thus, one standard deviation on each side of the measured value with distributed data could be calculated as the difference with all or no distributed extinctions divided by 3. Thus, the upper deviation for the Llandovery is

$$100 \times [(189/1089) - (174/1089)]/3 = 0.47\%. \quad (8)$$

Worked examples for the entire Silurian are presented in Table 5. Notice that there is no reason why the uncertainty below and above a measured value should be equal, and indeed calculated "standard deviations" for each series are different.

Table 5 illustrates that there is a complication in this exercise: as imprecise extinctions are redistributed to calculate minimum and maximum possible extinctions, apparent diversity will change. For example, when considering the case of no distributed extinctions in the Wenlock (Table 5A), some of the extinctions initially distributed there would have occurred earlier, and this would decrease the apparent Wenlockian diversity. The best guess as to the proportion of extinctions occurring before is the known number of extinctions in lower intervals divided by known numbers below and above. (For the Wenlock this includes extinctions listed as "Upper Silurian.") This proportion multiplied by the number of extinctions distributed to the Wenlock must be subtracted from

Table 5. Worked example showing computation of confidence intervals where potential error results from distributed extinctions

A. Minimum values

Stratigraphic interval	Number of definite extinctions	Additional extinctions distributed to earlier intervals[1]	Recalculated diversity[2]	% extinction[3]	One st. dev.[4]
Pridoli	141	3.6 + 18.5 = 22.1	1045	13.5	-0.60
Ludlow	263	$6.8 \times \frac{170+207}{170+2-7+159.5} = 4.8$	1262	20.8	-1.07
Wenlock	207	$4.7 \times \frac{170}{170+263+141+53} = 1.3$	1239	16.7	-0.13
Llandovery	170	0	1089	15.6	-0.13

[1] These numbers are subtracted from diversity in Table 2.   [2] Resulting diversity rounded to nearest integer.
[3] Number of definite extinctions divided by recalculated diversity.
[4] One-third of the difference between this percent extinction and the corresponding value in Table 2.

B. Maximum values

Stratigraphic interval	Number of definite extinctions	Added extinctions[5]	Addition to diversity[6]	Recalculated diversity[7]	% extinction[8]	One st. dev.[9]
Pridoli	141	19+53=72	9+(304-263)=50	1117	19.1	+1.27
Ludlow	263	19+53=72	4+(212-207)=9	1276	26.3	+0.77
Wenlock	207	19	174-170=4	1244	18.2	+0.37
Llandovery	170	19	0	1089	17.4	+0.47

[5] Total numbers of relevant distributed extinctions, from Table 1.
[6] Differences between total number and definite number of extinctions in all previous series of the system.
[7] Results of the previous column added to diversity in Table 2.
[8] Definite plus added extinctions divided by recalculated diversity.
[9] One-third of the difference between this percent extinction and the corresponding value in Table 2.

estimated Wenlockian diversity, as shown in Table 5A. (Notice that for the uppermost unit, the Pridoli, only its initially distributed extinctions need be subtracted from apparent diversity.)

Calculations for maximum possible extinction percentage are slightly easier (Table 5B): only distributed extinctions assigned to earlier intervals need to be a d d e d to apparent diversity, since none of these extinctions would have occurred previously if all happened during the interval under consideration. Thus, for the Wenlock, its apparent diversity must be incremented by approx. 4, the number of imprecise extinctions initially distributed to the earlier Llandovery.

These calculations become slightly more tedious when there is a hierarchy of imprecise data, as in the case of the Ludlow and Pridoli (Table 1). The worked example in Table 5 should make it reasonably clear how to proceed. Note, still, that the final estimates of standard deviations, or "error," are approximations that measure relative error. Not only is the statistical logic behind these calculations very rough, but they assume no distributed data among originations. Distributed data among both originations and extinctions makes approximation of confidence intervals far more complex. Bootstrapping or randomization procedures (Gilinsky, 1991b) might provide more precise estimates of potential error, but we believe the approximations presented here are good enough to express

Table 6. Total "error," or confidence interval, calculated by adding the binomial confidence interval (Table 4) and the distributional confidence interval (Table 5)

Stratigraphic interval	Standard deviations	
	Below estimated percent	Above estimated percent
Pridoli	- (1.04 + 0.60) = -1.6	(1.04 + 1.27) = +2.3
Ludlow	- (1.14 + 1.07) = -2.2	(1.14 + 0.77) = +1.9
Wenlock	- (1.06 + 0.13) = -1.2	(1.06 + 0.37) = +1.4
Llandovery	- (1.10 + 0.13) = -1.2	(1.10 + 0.47) = +1.6

relative amounts of uncertainty in time series. These approximations are used by Sepkoski (this volume).

Total Confidence Intervals. We have discussed two sources of uncertainty in extinction metrics: uncertainty resulting from binomial sampling and uncertainty resulting from distributed data. The best procedure for combining them probably is simply to add the two together (Table 6). Note that binomial standard deviations here were calculated only for the precise data on extinctions. The rounding of final uncertainties in Table 6 reflects the approximate nature of all the calculations. Note, still, that the resulting uncertainties for the Ludlow and Pridoli are larger than for the Llandovery and Wenlock; this meets intuition since there are more distributed data due to the "Upper Silurian" category of extinction data.

Conclusions

"Sampling effects are so considerable that paleontologists must estimate the magnitude of these effects before exploring species distribution patterns" (Koch, 1991a, p. 17). This admonition applies equally well to local stratigraphic sections and to global taxonomic data bases. It applies whether one is studying the distribution of extinctions below a local bio-event horizon or the distribution of extinctions within a global stratigraphic interval.

We have discussed a few kinds of sampling effects in paleontologic data relating to bio-events and have presented a few tools for measuring the magnitudes of these effects. Paleontologic data relating to bio-events continue to improve, and we think that descriptions of these events continue to become more accurate. However, until estimates of how accurate -- or uncertain -- these are become regularly published with paleontologic analyses, we will not know for sure how good the descriptions really have become.

Acknowledgements

This work received support from the National Aeronautics and Space Administration (U.S.A.) under grant NAGW-1693 to JJS and the National Science Foundation (U.S.A.) under grant EAR-8617291 to CFK.

References

Alvarez, L., 1983. Experimental evidence that an asteroid impact led to the extinction of many species 65 million years ago. Proceedings of the National Academy of Sciences, USA 80, 627-642.

Bayer, U. and McGhee, G.R., 1986. Cyclic patterns in the Paleozoic and Mesozoic: implications for time scale calibrations. Paleoceanography 1, 383-402.

Boucot, A.J., 1975. Evolution and Extinction Rate Controls. Elsevier, Amsterdam.

Boucot, A.J., 1986. Ecostratigraphic criteria for evaluating the magnitude, character and duration of bio-events. In: Walliser, O.H. (ed.), Global Bio-Events. pp. 25-45. Springer, Berlin Heidelberg New York.

Buzas, M.A., 1990. Another look at confidence limits for species proportions. J. of Paleont. 64, 842-843.

Buzas, M.A., Koch, C.F., Culver, S.J. and Sohl, N.F., 1982. On the distribution of species occurrence. Paleobiology 8, 142-150.

Culver, S.J., Buzas, M.A. and Collins, L.S., 1987. On the value of taxonomic standardization in evolutionary studies. Paleobiology 13, 169-176.

Darwin, C., 1859. On the Origin of Species by Means of Natural Selection. John Murray, London.

Doescher, R.A., 1981. Living and fossil brachiopod genera 1775-1979. Lists and bibliography. Smithsonian Contributions to Paleobiology, 42.

Durham, J.W., 1967. The incompleteness of our knowledge of the fossil record. J. of Paleont. 51,

559-564.

Erwin, D.H., 1989. The end-Permian mass extinction: what really happened and did it matter? Trends in Ecology and Evolution 4, 225-229.

Foote, M., 1992. Rarefaction analysis of morphological and taxonomic diversity. Paleobiology 18, 1-16.

Gilinsky, N.L., 1991a. The pace of taxonomic evolution. In: Gilinsky, N.L. and Signor, P.W. (eds.), Analytical Paleobiology, Short Courses in Paleontology Number 4. pp. 157-174. The Paleontological Society, Univ. of Tennessee, Knoxville, Tennessee.

Gilinsky, N.L., 1991b. Bootstrapping and the fossil record. In: Gilinsky, N.L. and Signor, P.W. (eds.), Analytical Paleobiology, Short Courses in Paleontology No 4. pp. 185-206. The Paleontological Society, Univ. of Tennessee, Knoxville, Tennessee.

Gilinsky, N.L. and Bambach, R.K., 1987. Asymmetrical patterns of origination and extinction in higher taxa. Paleobiology 13, 427-445.

Gilinsky, N.L. and Good, I.J., 1991. Probabilities of origination, persistence, and extinction of families of marine invertebrate life. Paleobiology 17, 145-166.

Harland, W.B., Holland, C.H., House, M.R., Hughes, N.F., Reynolds, A.B., Rudwick, M.J.S., Satterthwaite, G.E., Tarlo, L.B.H. and Wiley, E.C., (eds.), 1967. The Fossil Record. Geological Society of London, London.

Harland, W.B., Cox, A.V., Llewellyn, P.G., Pickton, C.A.G., Smith, A.G. and Walters, R., 1982. A Geologic Time Scale. Cambridge University Press, Cambridge.

Harland, W.B., Armstrong, R.L., Cox, A.V., Craig, L.E., Smith, A.G. and Smith, D.G., 1990. A Geologic Time Scale 1989. Cambridge University Press, Cambridge.

Harper, C.W., Jr., 1975. Standing diversity of fossil groups in successive intervals of geologic time: a new measure. J. of Paleont. 49, 752-757.

Harries, P.J. and Kauffman, E.G., 1990. Patterns of survival and recovery following the Cenomanian-Turonian (Late Cretaceous) mass extinction in the Western Interior Basin, United States. In: Kauffman, E.G. and Walliser, O.H. (eds.), Extinction Events in Earth History. pp. 277-298. Springer, Berlin Heidelberg New York.

Jablonski, D., 1986. Background and mass extinctions: the alternation of macroevolutionary regimes. Science 231, 129-133.

Koch, C.F., 1978. Bias in the published fossil record. Paleobiology 4, 367-377.

Koch, C.F., 1987. Prediction of sample size effects on the measured temporal and geographic distribution patterns of species. Paleobiology 13, 100-107.

Koch, C.F., 1991a. Sampling from the fossil record. In: Gilinsky, N.L. and Signor, P.W. (eds.), Analytical Paleobiology, Short Courses in Paleontology No 4. pp. 4-18. The Paleontological Society, Univ. of Tennessee, Knoxville, Tennessee.

Koch, C.F., 1991b. Species extinctions across the Cretaceous-Tertiary boundary: observed patterns versus predicted sampling effects, stepwise or otherwise? Historical Biology 5, 355-361.

Koch, C.F. and Morgan, J.P., 1988. On the expected distribution of species ranges. Paleobiology 14, 126-138.

Loeblich, A. and Tappan, H., 1988. Foraminiferal Genera and Their Classification. Van Nostrand Reinhold, New York.

Magurran, A.E., 1988. Ecological Diversity and Its Measurement. Princeton University Press, Princeton, New Jersey.

Marshall, C.R., 1990. Confidence intervals on stratigraphic ranges. Paleobiology 16, 1-10.

Marshall, C.R., 1991. Estimation of taxonomic ranges from the fossil record. In: Gilinsky, N.L. and Signor, P.W. (eds.), Analytical Paleobiology, Short Courses in Paleontology No 4. pp. 19-38. The Paleontological Society, Univ. of Tennessee, Knoxville, Tennessee.

May, R.M., 1975. Patterns of species abundance and diversity. In: Cody, M.L. and Diamond, J.M. (eds.), Ecology and Evolution of Communities. pp. 81-120. Belknap Press, Cambridge, Massach.

McKinney, M.L., 1986. Biostratigraphic gap analysis. Geology 14, 36-38.

Menning, M., 1989. A synopsis of numerical time scales 1917-1986. Episodes 12, 3-5.

Newell, N.D., 1959. Adequacy of the fossil record. J. of Paleont. 33, 488-499.

Palmer, A.R., 1983. The Decade of North American Geology 1983 geologic time scale. Geology 11, 503-504.

Patterson, C. and Smith, A.B., 1987. Is the periodicity of extinction a taxonomic artefact? Nature 320, 148-150.

Patterson, C. and Smith, A.B., 1989. Periodicity in extinction: the role of systematics. Ecology 7, 802-811.

Paul, C.R.C., 1982. The adequacy of the fossil record. In: Joysey, K.A. and Friday, A.E. (eds.), Problems of Phylogenic Reconstruction. pp. 75-117. Aca-

demic Press, London.
Rabinowitz, D., Cairns, S. and Dillon, T., 1986. Seven forms of rarity and their frequency in the flora of the British Isles. In: Soulé, M.E. (ed.), Conservation Biology: The Science of Scarcity and Diversity. pp. 182-204. Sinauer, Sunderland, Massachusetts.
Raup, D.M., 1986. Biological extinction in earth history. Science 231, 1528-1535.
Raup, D.M., 1989. The case for extraterrestrial causes of extinction. Philosophical Transactions of the Royal Society of London B 325, 421-435.
Raup, D.M., 1991a. The future of analytical paleobiology. In: Gilinsky, N.L. and Signor, P.W. (eds.), Analytical Paleobiology, Short Courses in Paleontology No 4. pp. 207-216. The Paleontological Society, Univ. of Tennessee, Knoxville, Tennessee.
Raup, D.M., 1991b. A kill curve for Phanerozoic marine species. Paleobiology 17, 37-48.
Raup, D.M. and Boyajian, G.E., 1988. Patterns of generic extinction in the fossil record. Paleobiology 14, 109-125.
Schindler, E., 1990. The late Frasnian (Upper Devonian) Kellwasser Crisis. In: Kauffman, E.G. and Walliser, O.H. (eds.), Extinction Events in Earth History. pp. 151-159. Springer, Berlin Heidelberg New York.
Schopf, J.W., 1992. Informal revised classification of Proterozoic microfossils. In: Schopf, J.W. and Klein, C. (eds.), The Proterozoic Biosphere: A Multidisciplinary Study. pp. 1119-1168. Cambridge University Press, Cambridge.
Sepkoski, J.J., Jr., 1982. A compendium of fossil marine families. Milwaukee Public Museum Contributions in Biology and Geology 51, 1-125.
Sepkoski, J.J., Jr., 1986. Global bioevents and the question of periodicity. In: Walliser, O.H. (ed.), Global Bio-Events. pp. 47-61. Springer, Berlin Heidelberg New York.
Sepkoski, J.J., Jr., 1989. Periodicity in extinction and the problem of catastrophism in the history of life. Journal of the Geol. Society London 146, 7-19.
Sepkoski, J.J., Jr., 1990. The taxonomic structure of periodic extinction. In: Sharpton, V.L. and Ward, P.D. (eds.), Global Catastrophes in Earth History; An Interdisciplinary Conference on Impacts, Volcanism, and Mass Mortality. Geological Society of America Special Paper 247, 33-44.
Sepkoski, J.J., Jr., 1991. Population biology models in macroevolution. In: Gilinsky, N.L. and Signor, P.W. (eds.), Analytical Paleobiology, Short Courses in Paleontology No 4. pp. 136-156. The Paleontological Society, Univ. of Tennessee, Knoxville, Tennessee.
Sepkoski, J.J., Jr., and Raup, D.M., 1986. Periodicity of marine extinction events. In: Elliot, D.K. (ed.), Dynamics of Extinction. pp. 3-36. Wiley, New York.
Signor, P.W. and Lipps, J.H., 1982. Sampling bias, gradual extinction patterns, and catastrophes in the fossil record. In: Silver, L.T. and Schultz, P.H. (eds.), Geological Implications of Impacts of Large Asteroids and Comets on the Earth. Geological Society of America Special Paper 190, 291-296.
Simpson, G.G., 1960. The history of life. In: Tax, S. (ed.), Evolution after Darwin, vol. 1. pp. 117-180. University of Chicago Press, Chicago.
Springer, M.S., 1990. The effect of random range truncations on patterns of evolution in the fossil record. Paleobiology 16, 517-520.
Springer, M.S. and Lilje, A., 1988. Biostratigraphy and gap analysis: the expected sequence of biostratigraphic events. Journal of Geology 96, 228-236.
Strauss, D. and Sadler, P.M., 1989. Classical confidence intervals and Bayesian probability estimates for ends of local taxon ranges. Mathematical Geology 21, 411-427.
Sweet, W.C., 1985. Conodonts: those fascinating little whatzits. J. of Paleont. 59, 485-494.
Walliser, O.H., Lottmann, J. and Schindler, E., 1988. Global events in the Devonian of the Kellerwald and Harz Mountains. Courier Forschungs-Institut Senckenberg 102, 190-193, Frankfurt/Main..
Whittaker, R.H., 1975. Communities and Ecosystems, 2nd edition. MacMillan, New York.

Manuscript received October 1993
Revision received July 1994

Authors' addresses:

J. John Sepkoski, Jr., Department of the Geophysical Sciences, University of Chicago, Chicago, Illinois 60637, U.S.A.

Carl F. Koch, Department of Geological Sciences, Old Dominion University, Norfolk, Virginia 23529, U.S.A.

Patterns of Phanerozoic Extinction: a Perspective from Global Data Bases

J. John SEPKOSKI, Jr.

Abstract: Time series of global diversity and extinction intensity measured from data on stratigraphic ranges of marine animal genera show the impact of bio-events on the fauna of the world ocean. Measured extinction intensities vary greatly, from major mass extinctions that eradicated 39 to 82% of generic diversity to smaller events that had substantially less impact on the global fauna. Many of the smaller extinction events are clearly visible only after a series of filters are applied to the data. Still, most of these extinction events are also visible in a smaller set of data on marine families. Although many of the episodes of extinction seen in the global data are well known from detailed biostratigraphic investigations, some are unstudied and require focused attention for confirmation or refutation.

Contents

Introduction	35
Taxonomic Data Bases	35
Data	37
Diversity	37
Time Series of Phanerozoic Marine Extinction	39
Conclusions	46

Introduction

There are a variety of means for recognizing episodes of extinction: disappearance of numerous species in local stratigraphic sections, sudden drop in indicators of preserved biomass, geochemical or other anomalies coinciding with concurrent species extinctions, and drop in diversity or increase in extinction measured in taxonomic compilations or data bases. This chapter will be concerned with the latter. I present curves for diversity and extinction intensity of marine animal genera measured at the level of stratigraphic stage for the whole of the Phanerozoic. These curves show that the global fauna has suffered numerous episodes of extinction that varied from the massive end-Permian event to much more modest bio-events that may have been regional in scale but still sufficiently severe to be seen in global data.

This work is essentially an update of previously published time series of Phanerozoic genus extinction (Sepkoski, 1986a, b). The data are now more refined (but hardly perfect), as are the methods for analyzing them. Below, I begin with some cautionary notes about problems in taxonomic data bases. I then briefly discuss the data base and the diversity patterns derived from it. Following that, I present a series of three graphs for Phanerozoic extinction, progressively attempting to limit noise in the data. Finally, I conclude with a comparison to family-level data on extinction.

Taxonomic Data Bases

In their simplest form, taxonomic data bases are compilations of times of first and last known occurrences of taxa in the fossil record. Such compilations can be at any taxonomic level and temporal resolution;

they can be for a single higher taxon over a limited geographic area, or for all higher taxa over the whole world. The data bases can also include myriads of other information for each taxon, such as classification, geographic range, morphology, paleoecology, etc.

Information in simple taxonomic data bases can be used to study patterns of diversity, origination, and extinction, and therefore can be used to identify bio-events and to measure their timing and magnitudes. This kind of work traces back to Phillips (1860), who compiled a data base on fossil species to argue for the natural division of Phanerozoic life into Paleozoic, Mesozoic, and Cenozoic components, separated by what are now called mass extinctions (see Erwin, 1993). The principal strength in using modern data bases for studying extinction is the ability to construct long time series that allow comparisons of bio-events measured with a consistent type of data; these data can be voluminous and mostly free of local problems of covered section, varying sample size, small number of taxa, facies change, hiatuses, etc. The principal weakness in using data bases is the coarse temporal frameworks used to record first and last occurrences (often international stages or substages). Coarse chronostratigraphic units preclude determination of how gradual, stepwise, or abrupt events are and masks multiple events within single temporal intervals or in contiguous intervals. Still, there is no question about whether bio-events are better studied in taxonomic data bases or local stratigraphic sections: the two provide different kinds of paleontologic information that are complementary, not competitive.

Nonetheless, the use of taxonomic data bases for studying extinction has been variously criticized (e.g. McLaren, 1983; Hoffman, 1985, 1988; Patterson and Smith, 1987, 1989; Ager, 1988; Boucot, 1990). Some of the criticisms are claims that there are better, more accurate methods for studying extinction, which instead I would label complementary. Other criticisms highlight real problems in taxonomic data bases that add noise to any signal of biological extinction and potentially can induce a false signal. Sepkoski and Raup (1986) and Raup and Boyajian (1988) consider some of these problems, several of which are summarized below:

1. Taxonomic Resolution. Most taxonomic data bases contain information on genera, families, etc. and not on species, the true units of evolution and extinction. Because higher taxa in paleontology often are arbitrarily defined and highly variable in size within and among major taxonomic groups, it is sometimes argued that they reveal little about species extinction. However, computer simulations by Sepkoski and Kendrick (1993) indicate that arbitrariness and variability in themselves do not produce false patterns uncharacteristic of underlying species extinction; in fact, arbitrary higher taxa can be truer to extinction patterns when sampling is poor. What multi-species taxa do, however, is damp the magnitude of lineage extinction so that peaks, or local maxima, are lower and less distinct than for species (Raup, 1979a).

2. Taxonomic Practice. Many higher taxa erected in paleontology are monotypic (contain a single species), paraphyletic (contain members ancestral to other taxa), or even polyphyletic. Patterson and Smith (1987, 1989) and Smith and Patterson (1988) have argued that only multi-species, monophyletic (i.e. holophyletic) taxa are valid, and only these provide true information about extinction (see also Briggs et al., 1988). However, the analysis of Sepkoski and Kendrick (1993) indicates this is not the case (see also Sepkoski, 1978, 1989).

3. Pseudoextinction. The last disappearance of a genus or family is generally treated as the termination of one or more species lineages. However, some disappearances may represent simple name changes as the last lineages undergo phyletic evolution and are placed in new genera, etc. It is not certain how large a proportion of fossil genera and families this affects, but perusal of large phylogenies (e.g. House, 1985, 1989) suggests the proportion is not large. "Taxonomic pseudoextinction," in which any member of a paraphyletic taxon is ancestral to another taxon, has been conflated with phyletic pseudoextinction by Patterson and Smith (1987) and Smith and Patterson (1988), but, again, Sepkoski and Kendrick (1993) show that taxonomic pseudoextinction is not a problem.

4. Incomplete Sampling. Knowledge of the fossil record is far from complete, and temporal ranges of taxa are often poorly known. Last occurrences of some taxa in data bases may be listed in intervals below their points of true extinction. This smears extinctions backward from true bio-events: a phenomenon known as the "Signor-Lipps effect" (Signor and Lipps, 1982; Sepkoski and Koch, this volume); thus, extinction intensities will appear to climb upward toward an event. This is a true bias in taxonomic data bases but will not preclude recognition of an event unless the taxa are extemely·poorly sampled.

5. Inaccurate Correlations and Other Bad Data. These are sources of noise in data bases (e.g. Smith and Patterson, 1988; Patterson and Smith, 1989) that make real patterns more difficult to discern. However, bad data do not seem so prevalent as to make the task

impossible. Sepkoski (1993a) compared patterns of extinction in his familial data base (Sepkoski, 1982) over a ten-year period as families were added and deleted, ranges were changed, and bad data were corrected. Although 50% of the data on last occurrences had changed, temporal patterns of extinction remained remarkably consistent. Thus, biological signal from the fossil record seemed to be overwhelming data base noise.

Data

The principal data base analyzed here encompasses extinctions of marine animal genera, including invertebrates, vertebrates, and animal-like protists (foraminiferans and radiolarians). These data have been assembled in an ongoing compilation of published information on the temporal ranges of fossil marine animals, which has been described previously in Sepkoski (1986a, 1989, 1990, 1994). The data base, as of this writing, holds information about first and last geologic occurrences of 33,180 animal genera plus another 1,445 molluscan subgenera, compiled from 1,040 literature sources and 10 unpublished data bases kindly shared by colleagues. Although the data base was founded upon the Treatise on Invertebrate Paleontology (Moore et al., 1953-1992), that information has been continuously updated and refined. At present, approximately 40% of the invertebrate genera in the data base do not appear in any volume of the Treatise.

The stratigraphic framework for locating first and last occurrences is the international stratigraphic time scale (Harland et al., 1990), with 79 "stages". (Series actually are used for the Ordovician, Silurian, and Carboniferous.) Sixty-eight "stages" are subdivided into two or three well-defined substages, so that the data base has 165 temporal intervals at its finest resolution (cf. Sepkoski, 1992).

Not all first and last occurrences have yet been located to the finest resolution. At present, only 55% of generic extinctions are resolved to this resolution, and another 29% are resolved to subdivided "stages". Series-level extinction data compose 14% of the data base, and system-level data 2%. Thus, time series of extinction are presented at the stage, rather than substage, level.

Because there is so much uncertainty about stage durations, particularly in the Paleozoic, extinction intensity is measured by the percentage of extinction (Sepkoski and Koch, this volume). However, because of considerable variation in estimated stage durations (1.6 to 21 Ma in the time scale of Harland et al., 1990), stages estimated to be 15 Ma or longer are divided into two "substages"; these are the Albian, Norian (here including the "Rhaetian"), and Visean. The extremely long Caradocian is subdivided into three substages. Also subdivided into two intervals is the Botomian, which appears to be the second longest stage of the Cambrian; this is done because the data contain very large numbers of extinctions in each substage, and treatment of the Botomian as a single unit would amalgamate too many nonsynchronous extinctions (e.g. Debrenne et al., 1990; Debrenne and Zhuravlev, 1992).

At the other end of durations, two pairs of very short stages are combined: the Turonian and Coniacian in the Cretaceous, and the Induan and Olenekian in the Triassic. All of these manipulations reduce the standard deviation of durations by 20%, from 4.1 Ma for stages to 3.3 Ma for the intervals used here.

Extinction data located to series and system are distributed by the method described by Sepkoski and Koch (this volume). Their method is also used to calculate error bars ("confidence intervals") on measured percentages of extinction.

Diversity

Figure 1 illustrates the pattern of Phanerozoic marine diversity derived from the genus-level data. The fundamental pattern of change through time is basically similar to what is seen among families (Raup and Sepkoski, 1982): rapid rise in the Early Cambrian; a plateau in the later Cambrian; much more increase during the Ordovician; fluctuating diversity through the remainder of the Paleozoic; low diversity again in the Early Triassic; and then generally increasing diversity through the Mesozoic and Cenozoic.

There are two interesting second-order differences between genus and family diversity. One is that the Mesozoic-Cenozoic increase in genera is considerably greater than among families: families are two times more diverse in the late Cenozoic than during the Paleozoic, whereas genera are four times more diverse. This could be a real evolutionary pattern in which post-Paleozoic diversification becomes increasingly damped as one moves up the Linnean hierarchy from species to families and beyond (Valentine, 1969; Signor, 1985). On the other hand, the accentuated increase in genera could be a result of the "pull of the Recent" (Raup, 1979b) in which greater outcrop area of fossiliferous sediment, more preserved provinces,

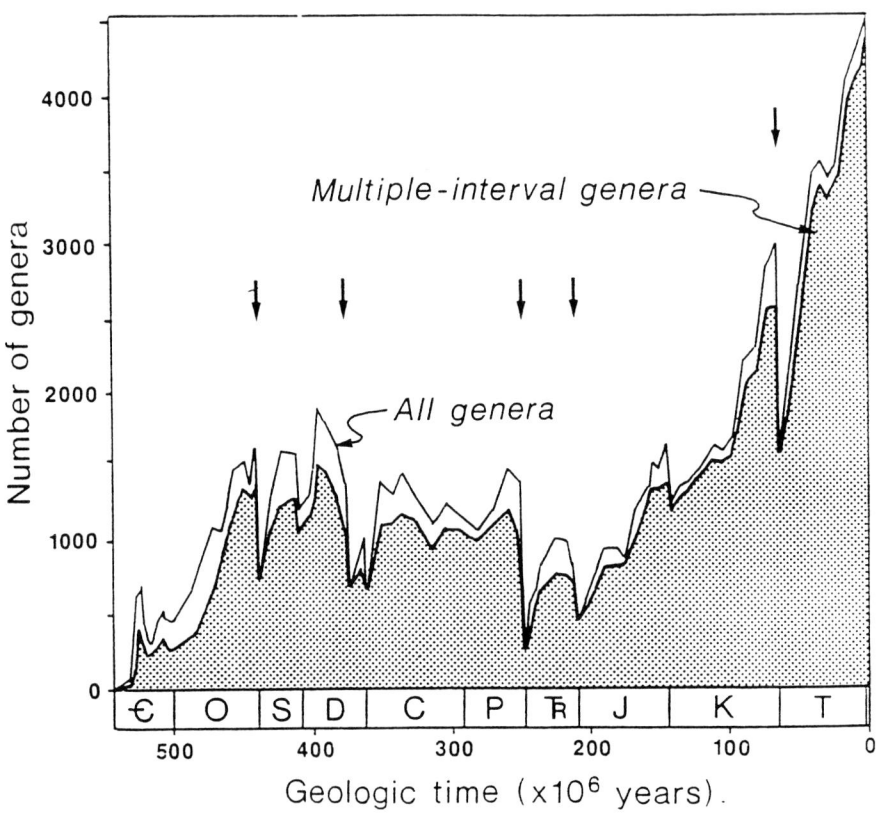

Fig. 1. Diversity of marine animal genera, compiled through 84 stages and substages of the Phanerozoic. Upper curve is for all 33,180 genera in the data base; lower curve is for the 18,310 genera that span two or more time intervals in the data base. Time scale in this and following figures is from Harland et al. (1990) with the Cambrian modified following Bowring et al. (1993). Arrows point to the five great mass extinctions.

Table 1. Percentages of decline in marine animal diversity associated with the five great mass extinctions of the Phanerozoic. Numbers are listed for genera (all genera in the data base as well as "filtered", or multiple-interval, genera) and for families (data modified from Sepkoski, 1992). Declines in diversity were calculated from the first listed stage to the beginning of the second stage.

Mass extinction	Percent decline in diversity		
	all genera	filtered genera	families
End-Ordovician (Ashgill to Llandovery)	-60% +4.4%	-49% +2.7%	-26% +1.9%
Late Devonian (Givetian to Famennian)	-57% +3.3%	-47% +2.0%	-22% +1.7%
End-Permian (Guadalupian to Induan)	-82% +3.8%	-76% +2.1%	-51% +2.3%
End-Triassic (Norian to Hettangian)	-53% +4.4%	-40% +2.4%	-22% +2.2%
End-Cretaceous (Maastrichtian to Danian)	-47% +4.1%	-39% +2.2%	-16% +1.5%

greater taxonomic resolution from the study of living animals, and the sample of the living biota itself all bias the Cenozoic upward relative to the Paleozoic (see also Raup, 1978, 1991).

The other second-order difference is more interesting for the purposes here: the major mass extinctions of the Phanerozoic appear much more pronounced at the genus level than at the familial. These events are the end-Ordovician (Ashgillian), Late Devonian (Frasnian), end-Permian (Tatarian), end-Triassic (Norian), and end-Cretaceous (Maastrichtian) mass extinctions (Flessa et al., 1986). They appear in Fig. 1 as rapid and deep drops in diversity, followed by rapid evolutionary rebounds. Table 1 presents magnitudes of the drops for genera, with familial declines for comparison.

Two diversity curves are actually illustrated in Fig. 1. One is for all genera in the data base. This ensemble includes genera with greatly varying taxonomic study, local abundance, and geographic and temporal range. Approximately 37% of the genera in the data base are confined to a single stratigraphic interval (substage, stage, series, etc., depending upon resolution). There are several reasons to believe that these genera contribute more noise than signal to time series of diversity and extinction:

1. Single-interval genera often are rare, and their presence may reflect sampling intensity (e.g. "monographic bursts") more than true patterns of diversity or extinction (e.g. Boucot, 1978; Koch, 1978; Sepkoski and Koch, this volume).

2. Monte Carlo computer simulations (Sepkoski, 1993b) suggest that numbers of "single-interval" genera are highly dependent upon stage duration, and therefore their apparent diversity and extinction intensity reflect more the irregular durations of stages than any biology.

3. Young supraspecific taxa are subject to effects of taxon aging (Boyajian, 1986, 1991, 1992), and their extinction intensity can be more a function of their species richness than of bio-events affecting the entire fauna.

4. A number of single-interval genera belong to taxa rarely preserved as fossils and come from extraordinary Lagerstätten (e.g. Solnhofen Limestone); thus, their distribution reflects varying preservation through time and not diversity or extinction (cf. Smith and Patterson, 1988).

5. Some single-interval genera may be synonyms, names either published once or used only in a single formation or locality. (Synonyms have been removed from the data base whenever encountered, but some undoubtedly persist.)

For these reasons, single-interval genera were eliminated to produce the lower diversity curve in Fig. 1. Their elimination induces only minor change in the magnitude and pattern of diversity; most of the minor fluctuations (i.e. potential bio-events) persist. However, when time series of extinction are examined, exclusion of single-interval genera can make some difference, as shown below.

Time Series of Phanerozoic Marine Extinction

A l l G e n e r a . There are a number of minor fluctuations in the diversity curves for genera that reflect extinction events smaller than the five great mass extinctions. These smaller bio-events are best examined with time series of extinction intensity. Fig. 2 illustrates a time series of percent extinction for all genera in the data base. A total of 28,856 extinctions are represented, of which 82% are resolved to one of the 84 time intervals. Several patterns in Fig. 2 require comment:

1. There is a strong secular decline in extinction intensity, as noted previously for families (Raup and Sepkoski, 1982; Quinn, 1983; Kitchell and Pena, 1984; Van Valen, 1984). This decline appears both in average intensity and in heights of extinction "peaks" (i.e. local maxima) other than the highest. The decline is not an artifact of longer average stage durations in the Paleozoic; the Cambrian, estimated here to be 44 Ma long, is subdivided into 11 "stages", giving an average duration shorter than in the Mesozoic and Cenozoic. The Cambrian appears to be a time of extraordinary high extinction (Bowring et al., 1993), with several measured percentages nearly equaling the end-Permian (Tatarian).

2. Most peaks of extinction are rather indistinct, with confidence intervals overlapping those of adjacent points on one or both sides. Exceptions include four of the five great mass extinctions: the Ashgillian, Tatarian, upper Norian, and Maastrichtian. The Frasnian also forms a high local maximum, but its confidence interval overlaps the superjacent Famennian (but see below). Several smaller peaks also stand out as distinct: the Tithonian and the Cenomanian, both of which contain extinction events (this volume). The Tommotian also forms a peak significantly higher than adjacent stages, but no decline or inflection in diversity is associated (Fig. 1).

M u l t i p l e I n t e r v a l G e n e r a . As argued above, single-interval genera probably add more noise than signal to the extinction pattern in

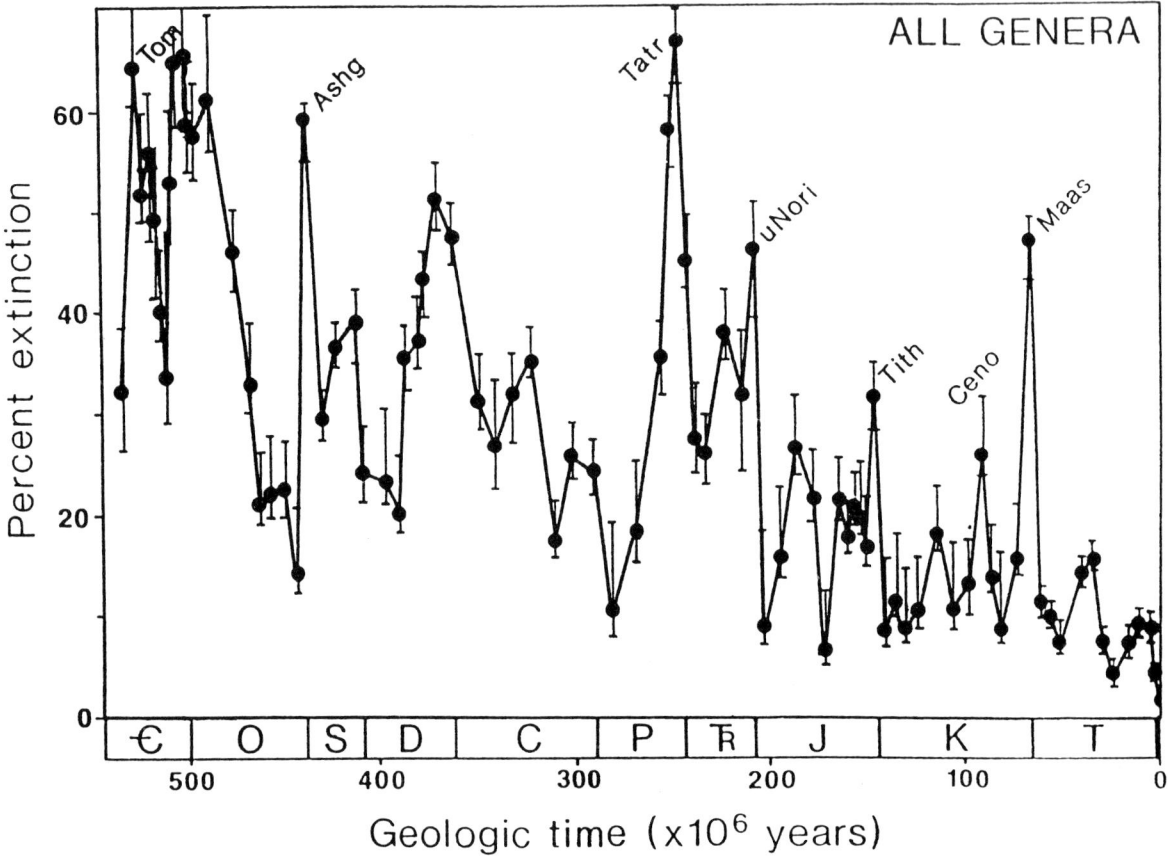

Fig. 2. Percent extinction of all marine animal genera through 84 time intervals of the Phanerozoic. Error bars cover two standard deviations, one on each side of the measured percentage. Only maxima significantly higher than points on each side are labeled (Tom = Tommotian; Ashg = Ashgillian; Tatr = Tatarian; uNori = upper Norian; Tith = Tithonian; Ceno = Cenomanian; Maas = Maastrichtian).

Fig. 2 (Sepkoski, 1986b, 1989, 1990). Fig. 3 illustrates the time series for percent extinction with these genera removed. Extinctions of 13,987 "multiple-interval" genera are included, which is slightly less than half the number in Fig. 2. However, 89% of these extinctions are resolved to one of the 84 time intervals, providing generally smaller error bars. Very importantly, extinctions without single-interval genera exhibit a much lower correlation with numbers of originations: the product-moment correlation (r) is 0.192 as opposed to 0.605 for the entire data base. The much higher correlation for the entire data base is consistent with the hypothesis that frequencies of single-interval genera largely reflect some combination of interval duration, variable research effort, and irregular preservation.

The time series of extinction for the "filtered" data exhibits a secular decline as in the entire data base, but now peaks of extinction are more pronounced. The Ashgillian, Tatarian, upper Norian, and Maastrichtian stand well above surrounding extinction intensities. The Tatarian is significantly higher than all other points; however, the subjacent Guadalupian is also very high, appearing to exceed the intensity of the Maastrichtian. It is not clear whether this reflects real extinction intensity or Signor-Lipps smearing of extinctions backward from the Tatarian. Data compilation over recent years has moved many last occurrences from the Guadalupian to the Tatarian (compare with Sepkoski, 1986a; see also Maxwell, 1989). On the other hand, Erwin (1993) concluded that many disappearances in the upper Guadalupian could represent real extinctions.

The filtered data show the Frasnian in the Devonian as containing a significantly greater extinction intensity than surrounding stages, which is

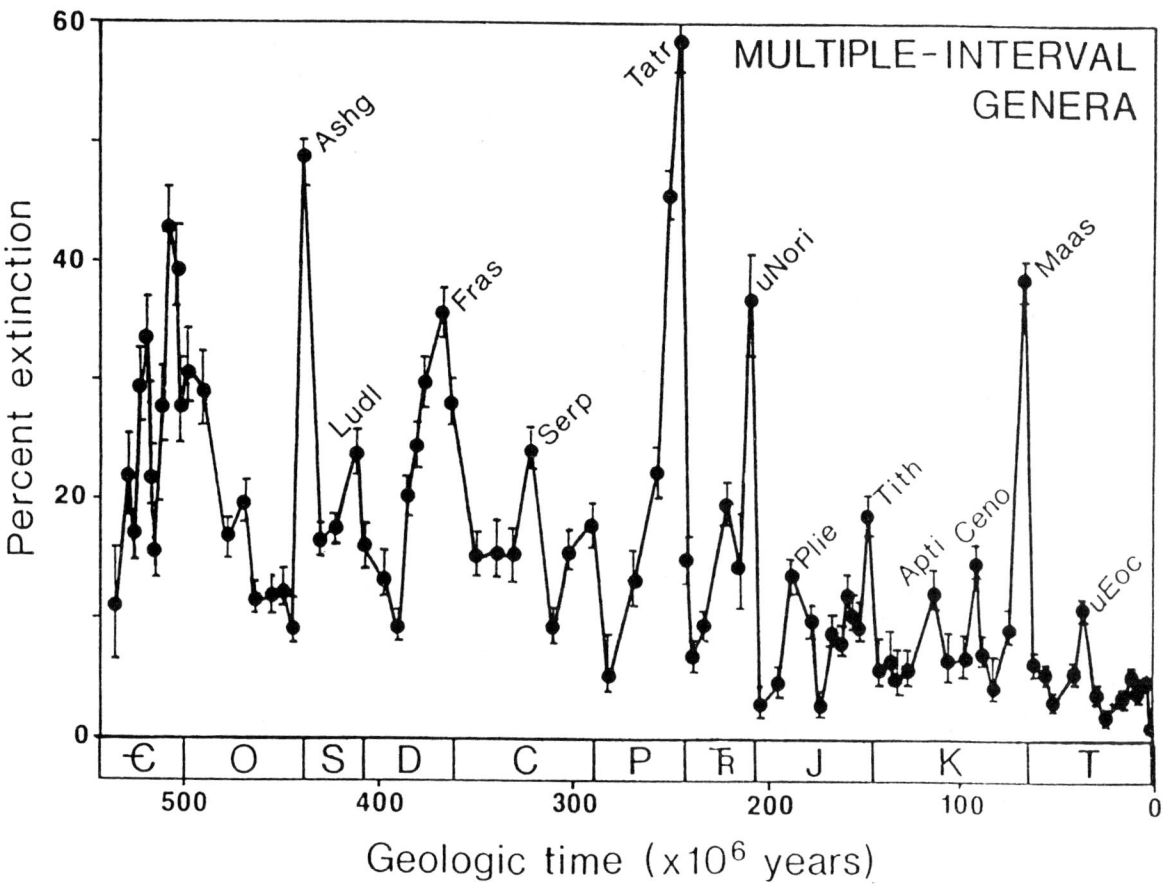

Fig. 3. Percent extinction of multiple-interval marine genera through 84 intervals of the Phanerozoic. Plotting conventions as in Figure 2. Additional abbreviations are Ludl = Ludlovian; Fras = Frasnian; Serp = Serpukhovian; Plie = Pliensbachian; Apti = Aptian; uEoc = Upper Eocene (Priabonian).

consistent with perception that a major mass extinction occurred near the Frasnian-Famennian boundary. Still, the Frasnian is surrounded by high measured percentages of extinction in the Givetian and Famennian, which exceed average values of the Paleozoic. This will be discussed more below.

The elimination of single-interval genera considerably reduces apparent extinction intensities in the Cambrian. Because of rapid faunal turnover during this period, the full data base combines non-synchronous extinctions of short-lived genera. The portrayal of extinction intensities in Fig. 3 is probably more accurate. Notice that the Tommotian no longer forms a significant peak of extinction (although see below); its magnitude in the full data base largely reflects genera described only from the Meishucunian of South China (see Bengtson, 1992).

A number of secondary peaks of extinction appear more pronounced in the filtered data. Most of these are evident in Fig. 2 but are statistically indistinguishable from surrounding points:

1. Llanvirnian: a minor peak associated with a small drop in diversity of all genera (Fig. 1); this event has been emphasized by Boucot (1990) but perhaps represents a regional (Laurentian shallow platform) extinction event.

2. Ludlovian: a more prominent peak of global extinction but poorly characterized in local sections (perhaps a composite of several bio-events affecting different taxa; e.g. Kaljo and Märss, 1991); this peak corresponds to a drop in global diversity in Fig. 1 (see also Boucot, 1975).

3. Serpukhovian: a peak like the Lidlovian with an associated drop in genus diversity. This was noted as a significant extinction event among ammonoids by Ramsbottom (1981) but needs much more local

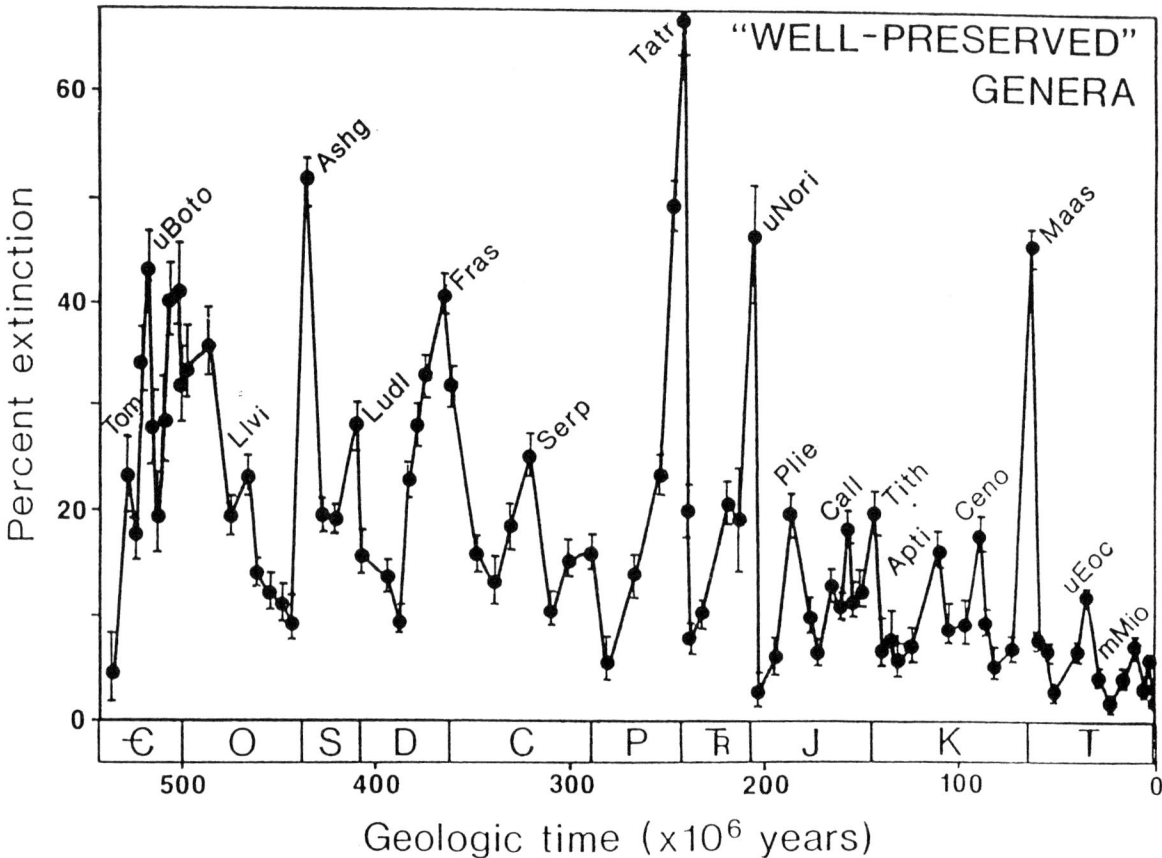

Fig. 4. Percent extinction of "well-preserved" marine genera (see text) through 84 intervals of the Phanerozoic. Plotting conventions as in Figs. 2 and 3. Additional abbreviations are uBoto = upper Botomian; Llvi = Llanvirnian; Call = Callovian; mMio - Middle Miocene.

study (although see Ross and Ross, 1985; Raymond et al., 1990).

4. Pliensbachian: a putatively regional extinction event documented for bivalves by Hallam (1976, 1977, 1986, 1987) but attributed to the lowermost Toarcian; the comparatively high percentage of extinction measured for the superjacent Toarcian may reflect this, with the Pliensbachian intensity resulting from Signor-Lipps backward smearing (although see Fig. 4).

5. Tithonian: another peak associated with a minor decline in global diversity; this event was discussed by Hallam (1976, 1986) but needs better local description (although see Roth, 1986; Bardhan et al., 1988).

6. Aptian: recognized by Leckie (1989) by a decline in the diversity of early planktonic foraminifera but in need of more study in other groups and in local areas.

7. Cenomanian: a substantial event located in the upper part of the stage (Harries and Kauffman, 1990; Hart and Kauffman, this volume).

8. Upper Eocene (Priabonian): associated with a small drop in global diversity; evidently a composite of several bio-events best documented for planktonic groups (e.g. Keller, 1986; Corliss et al., 1984) and in need of better documentation for benthic invertebrates (although see Hansen, 1987).

Several other local maxima of extinction that do not appear significantly different from surrounding extinction intensities need comment:

1. Stephanian (Kasimovian + Gzelian): claimed to be a possible extinction event by Sepkoski (1986a, b) but ridiculed by Boucot (1990). The Stephanian's prominence in Fig. 3 is emphasized by low values of extinction in the Bashkirian and Asselian, and the Stephanian is not significantly different from the subjacent Moscovian. However, the two series are associated with a decline in global diversity in Fig. 1. It is

possible, as "substage" data suggest, that there may be separate events in the upper Moscovian (Desmoinesian) (Schutter and Heckel, 1985) and upper Stephanian (Virgilian or Gzelian); alternatively, the maximum of extinction in the Stephanian could result from artificial name changes at a systemic boundary. More local study is needed.

2. Carnian: event here argued by Benton (1986, 1991), Johnson and Simms (1989), and Simms and Ruffell (1989, 1990). The Carnian has a local maximum in Fig. 3, but it is statistically indistinguishable from the superjacent lower + middle Norian, partially because distributed data within the Norian lead to large error bars. More study is needed.

"Well-Preserved" Genera. The time series in Fig. 3 might be improved somewhat by eliminating genera in groups that are poorly studied (e.g. sponges) or best preserved in Lagerstätten (e.g. marine vertebrates). Determining which higher taxa fall into these categories is more subjective than the mechanical culling of single-interval genera, but I have performed one experiment. The following groups with uncertain taxonomy or irregular preservation (cf. Sepkoski, 1990) were eliminated from the data base:

1. Radiolarians: systematics poorly studied except for biostratigraphically important genera.

2. Sponges: poorly studied and/or infrequently preserved; stromatoporoids and "sphinctozoans" (Senowbari-Daryan, 1990) retained.

3. Coelenterates: only conulariids and "corals" (Tabulata, Rugosa, and Scleractinia) retained.

4. Problematica: only genera recently monographed and assigned to orders retained (see Sepkoski, 1992).

5. Polychaetes: all eliminated.

6. Arthropods: only trilobites retained. (Ostracodes, which are common fossils, have undergone tremendous taxonomic revision over recent years, and the data base poorly reflects first and last occurrences.)

7. Echinoderms: eliminated, since usually only relatively complete, uncommon specimens can be identified to the level of genus.

8. Hemichordates: only graptolites retained.

9. Vertebrates: eliminated in entirety.

The wholescale elimination of these groups excludes some genera with widespread and well-documented fossil records and retains some genera that are inadequately studied. Still, this represents a first attempt to rarify the record toward its best studied and most consistently preserved components.

The culling leaves 10,119 extinct genera that range over two or more time intervals. Approximately 91% of these extinctions are located to one of the 84 stratigraphic intervals, which is slightly better than the previous data. The correlation between numbers of extinctions and originations drops to 0.132, suggesting further removal of the effects of Lagerstätten and monographic bursts.

Figure 4 illustrates the time series for percent extinction calculated from the rarified data set. Despite 23% fewer extinctions, the time series differs little from that for all multiple-interval genera (Fig. 3). Interesting differences are concentrated mostly in the Mesozoic and Cenozoic:

1. Callovian: contains a significant maximum of extinction similar in magnitude to the Pliensbachian and Tithonian. This was recognized by Sepkoski (1990) but has not been verified in any local or regional studies (although see Brochwicz-Lewinski et al., 1984, 1986). More work is needed.

2. Middle Miocene (Langhian and Serravallian): a significant if low peak, claimed to be part of a periodic array by Raup and Sepkoski (1984, 1986; Sepkoski, 1986b, 1989, 1990). More work is needed (although see Ujiie, 1984; Barron, 1986; Petuch, 1993).

3. Pliocene: documented as a regional extinction event by Stanley (1982, 1986). Its intensity appears to be lower than the poorly characterized Middle Miocene event.

Ten peaks of extinction in the interval from Permian to Recent are the prediction of the 26 Ma periodicity of Raup and Sepkoski (1984). Their refined analysis of fossil families (Raup and Sepkoski, 1986; Sepkoski and Raup, 1986) identified only eight of the expected ten periodic events, with peaks missing in or near the Jurassic Callovian and Cretaceous Aptian. The time series for genus-level extinction in Figs. 3 and 4 display all ten expected maxima of periodic extinction, along with the two smaller peaks in the Carnian and Pliocene.

However, the time scale on which these are plotted has changed, degrading the apparent 26 Ma periodicity. The timing of the younger putatively periodic events, from Tithonian to Middle Miocene, remains largely unchanged. However, the time scale of Harland et al. (1990) eliminates apparent periodicity in earlier Jurassic events. Essentially there has been a phase shift, taking 7 Ma from perviously uncertain ages in the Late Jurassic and adding them to the Early Jurassic.

This may or may not be correct. Jurassic and Triassic chronometric time scales are still rather poorly constrained, and Harland et al. (1990) indicate uncertainties of up to 15 Ma in this interval. Other

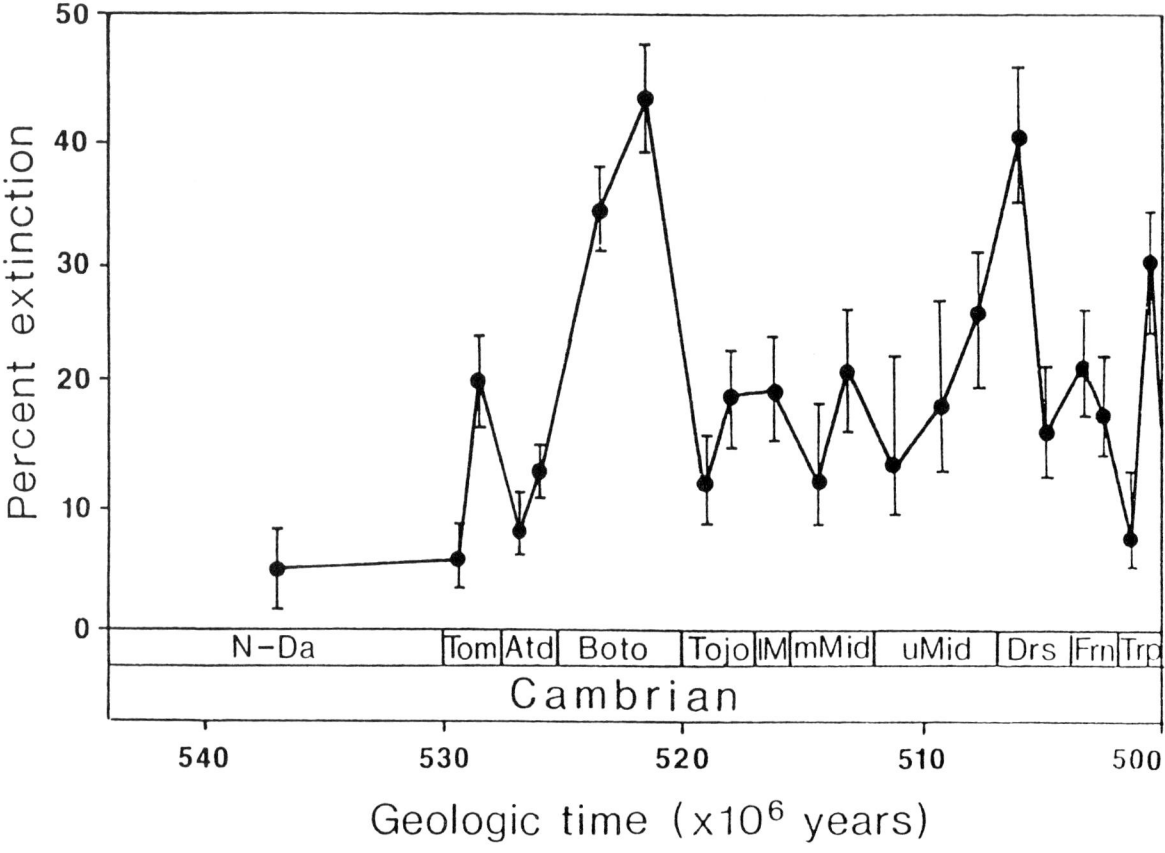

Fig. 5. Percent extinction of "well-preserved" genera through 21 substages of the Cambrian. Time scale modified from Bowring et al. (1993) with placement of the Cambrian-Ordovician boundary at 500 Ma and subdivision of the Botomian into Botomian + Tojonian; units in the Middle and Upper Cambrian are scaled from Sepkoski (1979). Abbreviations are Tom = Tommotian; Atd = Atdabanian; Boto = Botomian; Tojo = Tojonian; lM = lower Middle Cambrian; mMid = middle Middle Cambrian; uMid = upper Middle Cambrian; Drs = Dresbachian; Frn = Franconian; Trp = Trempealeauan.

time scales, such as that of Westermann (1984) which was based on numbers of biostratigraphic zones rather than a small number of radiometric tie points, maintain the periodicity claimed by Raup and Sepkoski (1984). Thus, until the Jurassic and Triassic chronometric time scales are more accurately determined, the question of periodicity must remain open. Nevertheless, even the most recent estimates of the ages of the events imply a non-random recurrence of extinction in the Mesozoic and Cenozoic, and perhaps the Paleozoic, and this needs explanation (although see McKinney, 1989; Stanley, 1990).

Cambrian and Devonian Substage Patterns. The time series in Figs. 3 and 4 exhibit continuously high extinction over much of the Cambrian and Devonian. To determine if these patterns might resolve into individual bio-events recoverable in the taxonomic data base, I have stretched the data shown in Fig. 4 to the level of substage.

1. Cambrian extinctions: Fig. 5 illustrates percent extinction calculated for 21 intervals of the Cambrian. The data comprise 926 genera with 779 extinctions. Approximately 80% of these extinctions are located to substage, and another 17% are located to stages that have been subdivided. Consequently, the error bars about calculated percentages are reasonably small.

Several peaks of extinction now stand out from the smear of high extinction. The upper Tommotian seems to have a small peak; this corresponds to the disappearance of many early "small shelly fossils" (Bengtson, 1992), although it could reflect more con-

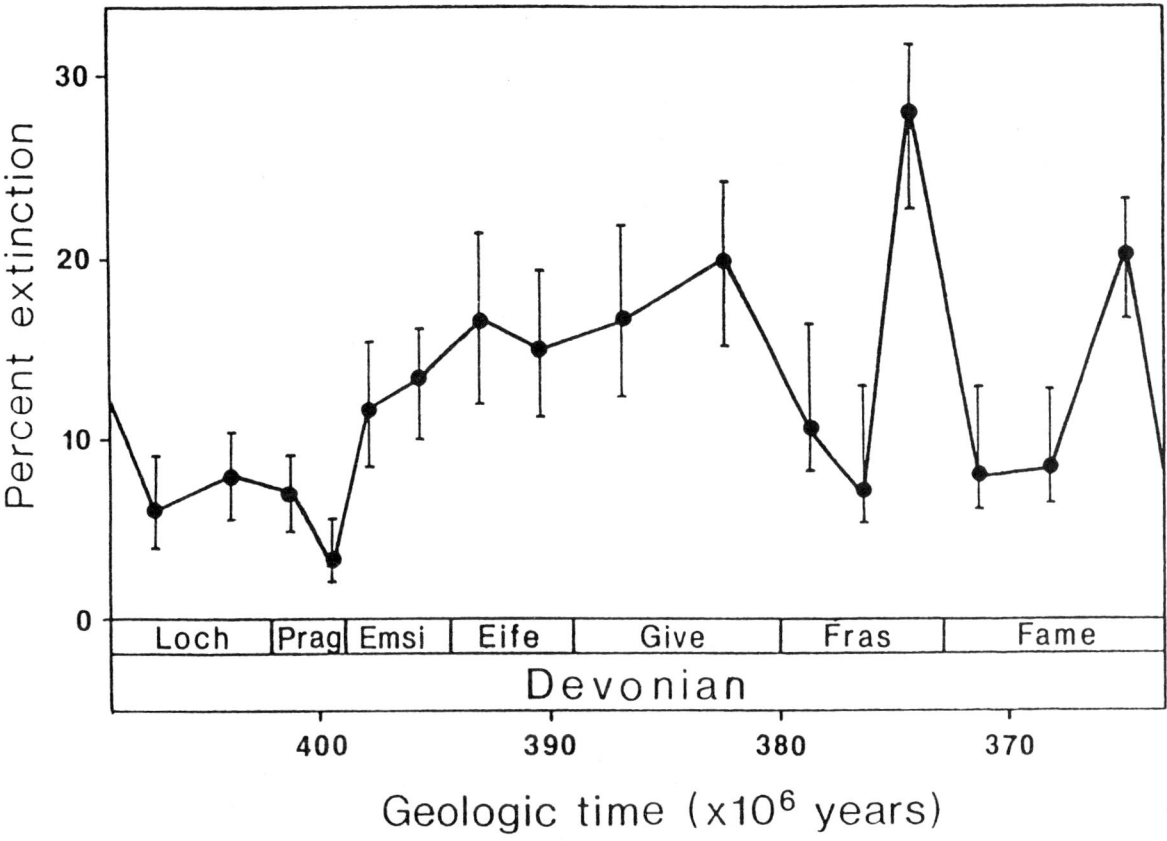

Fig. 6. Percent extinction of "well-preserved" marine genera through 16 substages of the Devonian. Time scale from Bayer and McGhee (1986), rescaled to the period boundary ages of Harland et al. (1990). Abbreviations: Loch = Lochkovian; Prag = Pragian; Emsi = Emsian; Eife = Eifelian; Give = Givetian; Fras = Frasnian; Fame = Famennian.

centration of workers' efforts on earliest Cambrian invertebrate microfossils. The succeding broad Botomian peak remains the highest of the Cambrian and is associated with a decline in global diversity (Fig. 1). Signor (1992) recognized the Botomian as a major mass extinction, which seems to be true to some degree. However, much of the magnitude appears to result from extended rapid turnover, especially among archaeocyathids, which do lose considerable diversity by the end of the stage (Debrenne, 1991).

Following the Botomian, lower intensities of extinction persist through the Tojonian and Middle Cambrian; none of the measured percentages are significantly different from others. Discrete peaks of extinction reappear in the Upper Cambrian: the second highest peak of the Cambrian is over the lower Dresbachian, corresponding to the event of trilobite extinction at the top of Palmer's (1965, 1984) Marjumid biomere. Another biomere event is represented by the extinction peak in the upper Trempealeauan near the Cambrian-Ordovician boundary, which corresponds to the top of the Ptychaspid biomere (Westrop, 1989). Note that this peak is significantly smaller than the lower Dresbachian maximum. Thus, there is no major mass extinction at the Cambrian-Ordovician, despite persistent references in the literature. (This mythic mass extinction seems to have come from older analyses that summed all extinction over the Upper Cambrian in coarse, series-level data bases; see Sepkoski, 1979, 1994; also Fortey, 1989; Edgecombe, 1992.) The third biomere event in the Upper Cambrian, at the top of the Pterocephalid biomere, is reflected by the insignificant peak over the lower Franconian and is evidently less severe or less widespread than the other two biomere events.

2. Devonian extinctions: Fig. 6 illustrates percent extinction for 16 intervals in the Devonian. There are 1943 genera and 1507 extinctions represented. Only

42% of the extinctions are located to substage, with another 53% located to stages that are subdivided. Thus, this exercise can only partially resolve the pattern of global extinction during the Devonian.

The best resolved portion of the time series is for the Upper Devonian, and two separate peaks of extinction are prominent. The higher is in the upper Frasnian, corresponding to the well-known Kellwasser event or Frasnian/Famennian mass extinction (Walliser et al., 1988, this volume; Schindler, 1990). Notice, however, that the peak in the upper Famennian, corresponding to the Hangenberg event at the Devonian-Carboniferous boundary, has 70% of the magnitude of the upper Frasnian maximum; in the unculled and the multiple-interval data, it has 80% of the magnitude of the upper Frasnian peak. This indicates that the upper Famennian event had major impact on the global fauna and corroborates its classification as a second-order bio-event (Walliser, this volume).

Figure 6 gives the impression of relatively high intensities of extinction throughout the Middle Devonian. The data here, however, have only half the stratigraphic resolution of the Upper Devonian and thus poorly capture the nature of global extinction through this interval. Walliser (this volume) document a high frequency of third- and fourth-order bio-events in the Emsian to Givetian, which probably cause the measured high intensities of global extinction. Two of the bio-events (including the Taghanic event) are concentrated in the upper Givetian, which may contribute the insignificant maximum there (see also House, 1985, 1989; Bayer and McGhee, 1986; Scrutton, 1988; Feist, 1991; McGhee, 1991). More highly resolved data, probably at the level of chronozone (House, 1985, 1989), are necessary to measure the effects of these bio-events on the global fauna.

Family-Level Extinction. The various manipulations of the data for genera were designed to reduce noise but were not entirely nonarbitrary. Furthermore, the genera themselves might be very arbitrary taxonomic units without the evolutionary and ecological coherency of taxa of other ranks, such as families.

Figure 7 presents a time series of percent extinction for marine animal families for comparison to the genera. The data are from Sepkoski (1992) with 110 corrections and additions. Single-interval families (15% of the total data) have been eliminated, leaving 3461 families and 2349 extinctions. The time scale, including subdivided and amalgamated stages, is the same as for the genera, and 95% of the familial extinctions have been resolved to the time intervals (with another 2% resolved to stages that have been subdivided). Large error bars, especially through the Paleozoic, reflect the much lower diversity of families relative to genera, and thus the larger binomial standard deviations (see Sepkoski and Koch, this volume).

The extinction pattern for families in Fig. 7 is very similar to that for genera in Fig. 3, even down to many minor peaks. The principal difference is that absolute magnitudes of extinction are lower for families, which is expected since the average family contains more species than the average genus. There are some large differences in pattern over the Cambrian, where diversity is low and error bars large. The upper Botomian, for example, appears as the second highest peak of extinction among families but is more minor among genera. (This may reflect traditional oversplitting of archaeocyathid families; cf. Sepkoski, 1979.) Other more minor differences include lower relative magnitude of the Givetian point, significance of the Stephanian maximum, disappearance of the Carnian peak, lower prominence of the Aptian peak, lower relative magnitude of the Maastrichtian maximum, and appearance of a very low peak in the Upper Miocene.

Conclusions

Patterns of extinction derived from a global taxonomic data base corroborate many of the bio-events recognized in careful studies of local stratigraphic sections and detailed syntheses of regional biostratigraphic information. The data base also provides some measure of the importance of these events for the global marine fauna relative to major mass extinctions and other, more limited extinction events. In some intervals, such as the Devonian, the data base has insufficient biostratigraphic resolution, and the results of detailed local studies of bio-events provide interpretation of the measured pattern of global extinction. On the other hand, the global data base shows some undocumented or weakly characterized peaks of extinction that look like well-known lower-order bio-events and thereby suggests intervals in need of detailed study. These include the Ludlovian, Moscovian and Stephanian, Aptian, and Middle Miocene.

Diversification following major extinction events appears very rapid in the curves of global generic diversity. This suggests that whatever the inimical

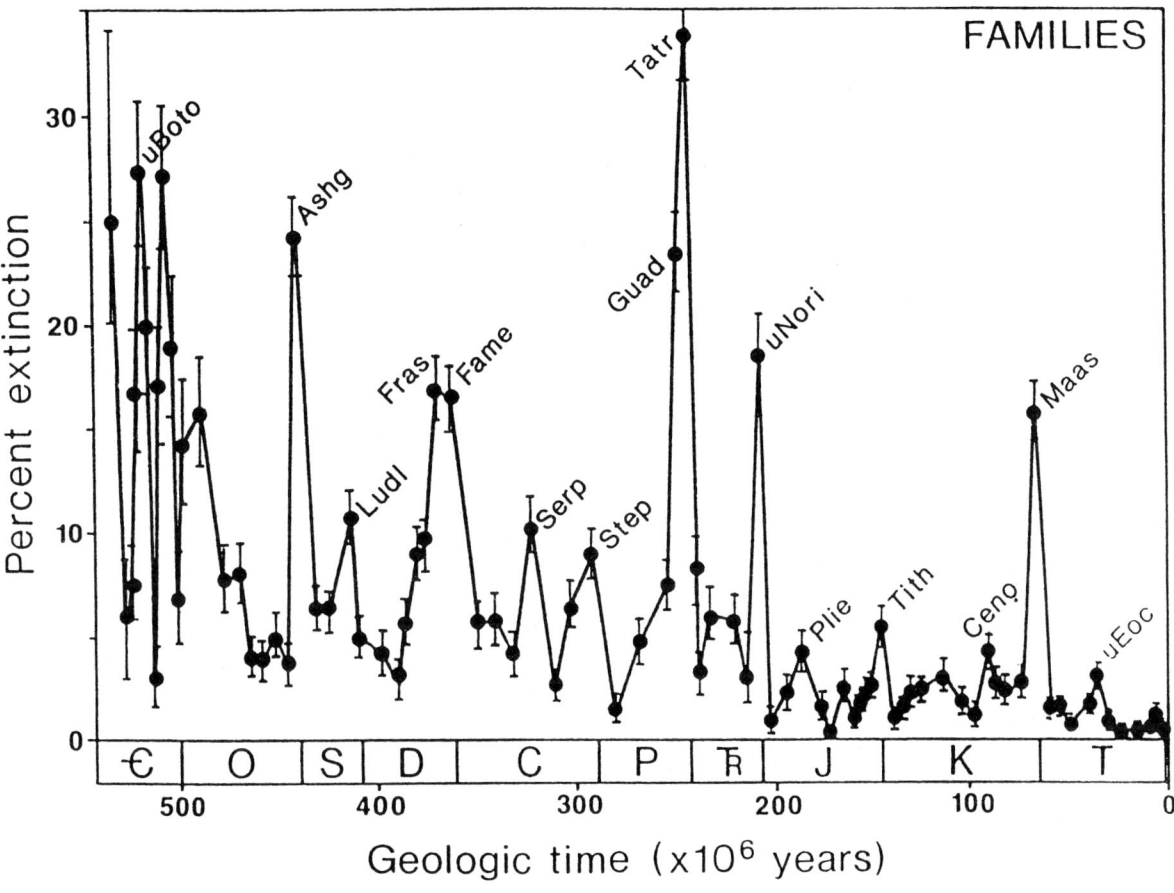

Fig. 7. Percent extinction for multiple-interval families of marine animals through 84 intervals of the Phanerozoic. Data from Sepkoski (1992) with modification. Plotting conventions and abbreviations as in Figs. 2-4. Step = Stephanian; Guad = Guadalupian.

conditions were that decimated the taxa, they quickly abated, allowing the biota to return to its full evolutionary potential and repopulate the world ocean.

Acknowledgments

Compilation and analysis of the taxonomic data base has received support from the National Aeronautics and Space Administration (USA) under grants NAG 2-282 and NAGW-1693.

References

Ager, D.V., 1988. Extinctions and survivals in the Brachiopoda and the dangers of data bases. In: Larwood, G.P. (ed.), Extinction and Survival in the Fossil Record. pp. 89-97. Oxford University Press, Oxford.

Bardhan, S., Shome, S., Bose, P.K. and Ghosh, G., 1988. Faunal crisis and marine regression across the Jurassic-Cretaceous boundary in Kutch, India. Mesozoic Research 2, 1-10.

Barron, J.A., 1986. The end of the Middle Miocene: hiatuses and polar cooling and their consequences. Geological Society of America Abstracts with Program 18, 535-536.

Bayer, U. and McGhee, G.R., 1986. Cyclic patterns in the Paleozoic and Mesozoic: implications for time scale calibrations. Paleoceanography 1, 383-402.

Bengtson, S., 1992. Proterozoic and earliest Cambrian skeletal metazoans. In: Schopf, J.W. and Klein, C. (eds.), The Proterozoic Biosphere. A Multidisciplinary Study. pp. 1017-1033. Cambridge University Press, Cambridge.

Benton, M.J., 1986. More than one event in the Late Triassic mass extinction. Nature 321, 857-861.

Benton, M.J., 1991. What really happened in the Late Triassic? Historical Biology 5, 263-278.

Boucot, A.J., 1975. Evolution and Extinction Rate Controls. Elsevier, Amsterdam.

Boucot, A.J., 1978. Community evolution and rates of cladogenesis. Evolutionary Biology 11, 545-654.

Boucot, A.J., 1990. Phanerozoic extinctions: how similar are they to each other? In: Kauffman, E.G. and Walliser, O.H. (eds.), Extinction Events in Earth History. pp. 5-30. Springer, Berlin Heidelberg New York.

Bowring, S.A., Grotzinger, J.P., Isachsen, C.E., Knoll, A.H., Pelechaty, S.M. and Kolosov, P., 1993. Calibrating rates of Early Cambrian evolution. Science 261, 1293-1298.

Boyajian, G.E., 1986. Phanerozoic trends in background extinction: consequence of an aging fauna. Geology 14, 955-958.

Boyajian, G.E., 1991. Taxon age and selectivity of extinction. Paleobiology 17, 49-57.

Boyajian, G.E., 1992. Taxon age, origination, and extinction through geologic time. Historical Biology 6, 281-291.

Briggs, D.E.G., Fortey, R.A. and Clarkson, E.N.K., 1988. Extinction and the fossil record of arthropods. In: Larwood, G.P. (ed.), Extinction and Survival in the Fossil Record. Systematics Association Special Volume 34, 171-209.

Brochwicz-Lewinski, W., Gasiewicz, A., Suffczynski, S., Szatkowski, K. and Zbik, M., 1984. Luki i kondensacje na pograniczu jury srodkowej i gornej. Przeglad Geologiczny 31, 285-290.

Brochwicz-Lewinski, W., Gasiewicz, A., Krumbein, W.E., Melendez, G., Sequeiros, L., Suffczynski, S., Szatkowski, K., Tarkowski, R. and Zbik, M., 1986. Anomalia irydowa na granicy jury srodkowej i gorney. Przeglad Geologiczny 2 (394), 83-88.

Corliss, B.H., Aubry, M.-P., Berggren, W.A., Fenner, J.M., Keigwin, L.D., Jr. and Keller, G., 1984. The Eocene/Oligocene boundary event in the deep sea. Science 226, 806-810.

Debrenne, F., 1991. Extinction of the Archaeocyatha. Historical Biology 5, 95-106.

Debrenne, F. and Zhuravlev, A., 1992. Irregular Archaeocyaths. Morphology, Ontogeny, Systematics, Biostratigraphy, Palaeoecology. Cahiers de Paléontologie, CNRS Editions, Paris.

Debrenne, F., Rozanov, A.Yu. and Zhuravlev, A., 1990. Regular Archaeocyaths. Morphology, Systematics, Biostratigraphy, Palaeogeography, Biological Affinity. Cahiers de Paléontologie, CNRS Editions, Paris.

Edgecombe, G.D., 1992. Trilobite phylogeny and the Cambrian-Ordovician "event": cladistic reappraisal. In: Novacek, M.J. and Wheeler, Q.E. (eds.), Extinction and Phylogeny. pp. 144-177. Columbia University Press, New York.

Erwin, D.H., 1993. The Great Paleozoic Crisis. Columbia University Press, New York.

Feist, R., 1991. The Late Devonian trilobite crisis. Historical Biology 5, 197-214.

Flessa, K.W., Erben, H.K., Hallam, A., Hsü, K.J., Hüssner, H.M., Jablonski, D., Raup, D.M., Sepkoski, J.J., Jr., Soulé, M.E., Sousa, W., Stinnesbeck, W. and Vermeij, G.J., 1986. Causes and consequences of extinction. In: Raup, D.M. and Jablonski, D. (eds.), Patterns and Processes in the History of Life. pp. 235-257. Springer, Berlin Heidelberg New York.

Fortey, R.A., 1989. There are extinctions and extinctions: examples from the lower Palaeozoic. Philosophical Transactions of the Royal Society of London B 325, 327-355.

Hallam, A., 1976. Stratigraphic distribution and ecology of European Jurassic Bivalves. Lethaia 9, 245-259.

Hallam, A., 1977. Jurassic bivalve biogeography. Paleobiology 3, 58-73.

Hallam, A., 1986. The Pliensbachian and Tithonian extinction events. Nature 319, 765-768.

Hallam, A., 1987. Radiations and extinctions in relation to environmental change in the marine Lower Jurassic of northwest Europe. Paleobiology 13, 152-168.

Hansen, T.A., 1987. Extinction of Late Eocene to Oligocene molluscs: relationship to shelf area, temperature changes, and impact events. Palaios 2, 69-75.

Harland, W.B., Armstrong, R., Cox, A.V., Craig, L.E., Smith, A.G. and Smith, D.G., 1990. A Geologic Time Scale 1989. Cambridge University Press, Cambridge.

Harries, P.J. and Kauffman, E., 1990. Patterns of survival and recovery following the Cenomanian-Turonian (Late Cretaceous) mass extinction in the Western Interior Basin, United States. In: Kauffman, E.G. and Walliser, O.H. (eds.), Extinction Events in Earth History. pp. 277-298. Springer, Berlin Heidelberg New York.

Hoffman, A., 1985. Patterns of family extinction: dependence on definition and geologic time scale. Nature 315, 659-662.

Hoffman, A., 1988. Arguments on Evolution. Oxford University Press, New York.

House, M.R., 1985. Correlation of mid-Palaeozoic ammonoid evolutionary events with global sedimentary perturbations. Nature 313, 17-22.

House, M.R., 1989. Ammonoid extinction events. Philosophical Transactions of the Royal Society of London B 325, 307-325.

Johnson, A.L.A. and Simms, M.J., 1989. The timing and cause of the Late Triassic marine invertebrate extinctions: evidence from scallops and crinoids. In: Donovan, S.K. (ed.), Mass Extinctions: Processes and Evidence. pp. 174-194. Columbia University Press, New York.

Kaljo, D. and Märss, T., 1991. Patterns of some Silurian bioevents. Historical Biology 5, 145-152.

Keller, G., 1986. Stepwise mass extinctions and impact events: Late Eocene to Early Oligocene. Marine Micropaleontology 10, 267-294.

Kitchell, J.A. and Pena, D., 1984. Periodicity of extinction in the geologic past: deterministic versus stochastic explanations. Science 226, 689-692.

Koch, C.F., 1978. Bias in the published fossil record. Paleobiology 4, 367-372.

Leckie, R.M., 1989. A paleoceanographic model for the early evolutionary history of planktonic foraminifera. Palaeogeography, Palaeoclimatology, Palaeoecology 73, 107-138.

Maxwell, W.D., 1989. The end Permian mass extinction. In: Donovan, S.K. (ed.), Mass Extinctions: Processes and Evidence. pp. 152-173. Columbia University Press, New York.

McGhee, G.R., 1991. Extinction and diversification in the Devonian Brachiopoda of New York State: no correlation with sea level? Historical Biology 5, 215-227.

McKinney, M.L., 1989. Periodic mass extinctions; product of biosphere growth dynamics? Historical Biology 2, 273-287.

McLaren, D.J., 1983. Bolides and biostratigraphy. Geological Society of America Bulletin 94, 313-324.

Moore, R.C., Teichert, C., Robison, R.A. and Kaesler, R.L. (eds.), 1953-1992. Treatise on Invertebrate Paleontology. Geological Society of America and University of Kansas Press, Lawrence, Kansas.

Palmer, A.R., 1965. Biomere -- a new kind of biostratigraphic unit. Journal of Paleontology 39, 149-153.

Palmer, A.R., 1984. The biomere problem: evolution of an idea. Journal of Paleontology 58, 599-611.

Patterson, C. and Smith, A.B., 1987. Is the periodicity of extinctions a taxonomic artefact? Nature 330, 248-251.

Patterson, C. and Smith, A.B., 1989. Periodicity in extinction; the role of systematics. Ecology 70, 802-811.

Petuch, E.J., 1993. Patterns of diversity and extinction in Transmarian muricacean, buccinacean, and conacean gastropods. The Nautilus 106, 155-173.

Phillips, J., 1860. Life on Earth: Its Origin and Succession. Macmillan, Cambridge.

Quinn, J.F., 1983. Mass extinctions in the fossil record [discussion]. Science 219, 1239-1240.

Ramsbottom, W.H.C., 1981. Eustatic control in Carboniferous ammonoid biostratigraphy. In: House, M.R. and Senior, J.R. (eds.), The Ammonoidea. pp. 369-388. Academic Press, New York.

Raup, D.M., 1978. Cohort analysis of generic survivorship. Paleobiology 4, 1-15.

Raup, D.M., 1979a; Size of the Permo-Triassic bottleneck and its evolutionary implications. Science 206, 217-218.

Raup, D.M., 1979b. Biases in the fossil record of species and genera. Carnegie Museum of Natural History Bulletin 13, 85-91.

Raup, D.M., 1991. A kill curve for Phanerozoic marine species. Paleobiology 17, 37-48.

Raup, D.M. and Boyajian, G.E., 1988. Patterns of generic extinction in the fossil record. Paleobiology 14, 109-125.

Raup, D.M. and Sepkoski, J.J., Jr., 1982. Mass extinctions in the marine fossil record. Science 215, 1501-1503.

Raup, D.M. and Sepkoski, J.J., Jr., 1984. Periodicity of extinction in the geologic past. Proceedings of the National Academy of Sciences USA 81, 801-805.

Raup, D.M. and Sepkoski, J.J., Jr., 1986. Periodic extinction of families and genera. Science 231, 833-836.

Raymond, A., Kelly, P.H. and Lutken, C.B., 1990. Dead by degrees: articulate brachiopods, paleoclimate and the mid-Carboniferous extinction event. Palaios 5, 111-123.

Ross, C.A. and Ross, J.P., 1985. Carboniferous and Early Permian biogeography. Geology 13, 27-30.

Roth, P.H., 1986. Mesozoic paleoceanography of the North Atlantic and Tethys Oceans. In: Summerhayes, C.P. and Shackleton, N.J. (eds.), North Atlantic Paleoceanography. Geological Society of America Special Publication 21, 299-320.

Schindler, E., 1990. The late Frasnian (Upper Devonian) Kellwasser crisis. In: Kauffman, E.G. and Walliser, O.H. (eds.), Extinction Events in Earth History. pp. 151-160. Springer, Berlin Heidelberg New York.

Schutter, S.R. and Heckel, P.H., 1985. Missourian (early late Pennsylvanian) climate in Midcontinent North America. International Journal of Coal Geology 5, 111-140.

Scrutton, C.T., 1988. Patterns of extinction and survival in Palaeozoic corals. In: Larwood, G.P. (ed.), Extinction and Survival in the Fossil Record. Systematics Association Special Volume 34, 65-88.

Senowbari-Daryan, B., 1990. Die systematische Stellung der thalamiden Schwämme und ihre Bedeutung in der Erdgeschichte. Münchner Geowissenschaftliche Abhandlungen Reihe A, Geologie und Paläontologie 21, 1-326.

Sepkoski, J.J., Jr., 1978. A kinetic model of Phanerozoic taxonomic diversity. I. Analysis of marine orders. Paleobiology 4, 223-251.

Sepkoski, J.J., Jr., 1979. A kinetic model of Phanerozoic taxonomic diversity. II. Early Phanerozoic families and multiple equilibria. Paleobiology 5, 222-251.

Sepkoski, J.J., Jr., 1982. A compendium of fossil marine families. Milwaukee Public Museum Contributions in Biology and Geology 51, 1-125.

Sepkoski, J.J., Jr., 1986a. Phanerozoic overview of mass extinction. In: Raup, D.M. and Jablonski, D. (eds.), Patterns and Processes in the History of Life. pp. 277-295. Springer, Berlin Heidelberg New York.

Sepkoski, J.J., Jr., 1986b. Global bioevents and the question of periodicity. In: Walliser, O.H. (ed.), Global Bio-Events. pp. 47-61. Springer, Berlin Heidelberg New York.

Sepkoski, J.J., Jr., 1989. Periodicity in extinction and the problem of catastrophism in the history of life. Journal of the Geological Society London 146, 7-19.

Sepkoski, J.J., Jr., 1990. The taxonomic structure of periodic extinction. In: Sharpton, V. and Ward, P. (eds.), Global Catastrophes in Earth History. Geological Society of America Special Paper 247, 33-44.

Sepkoski, J.J., Jr., 1992. A compendium of fossil marine animal families, 2nd edition. Milwaukee Public Museum Contributions in Biology and Geology 83, 1-156.

Sepkoski, J.J., Jr., 1993a. Ten years in the library: new data confirm paleontological patterns. Paleobiology 19, 43-51.

Sepkoski, J.J., Jr., 1993b. Phanerozoic diversity at the genus level: problems and prospects. Geological Society of America Abstracts with Program 25.

Sepkoski, J.J., Jr., 1994. What I did with my research career: or how research on biodiversity yielded data on extinction. In: Glen, W. (ed.), Mass Extinction Debates: How Science Works in a Crisis. pp. 132-144. Stanford University Press, Stanford, California.

Sepkoski, J.J., Jr. and Kendrick, D.C., 1993. Numerical experiments with model monophyletic and paraphyletic taxa. Paleobiology 19, 168-184.

Sepkoski, J.J., Jr. and Raup, D.M., 1986. Periodicity in marine extinction events. In: Elliott, D.K. (ed.), Dynamics of Extinction. pp. 3-36. Wiley, New York.

Signor, P.W., 1985. Real and apparent trends in speies richness through time. In: Valentine, J.W. (ed.), Phanerozoic Diversity Patterns: Profiles in Macroevolution. pp. 129-150. Princeton University Press, Princeton, New Jersey.

Signor, P.W., 1992. Taxonomic diversity and faunal turnover in the Early Cambrian: did the most severe mass extinction of the Phanerozoic occur in the Botomian Stage? In: Lidgard, S. and Crane, P.R. (eds.), Fifth North American Paleontological Convention Abstracts and Program. Paleontological Society Special Publication 6, p. 272. University of Tennessee, Knoxville, Tennessee.

Signor, P.W. and Lipps, J.H., 1982. Sampling bias, gradual extinction patterns, and catastrophes in the fossil record. In: Silver, L.T. and Schultz, P.H. (eds.), Geological Implications of Impacts of Large Asteroids and Comets on the Earth. Geological Society of America Special Paper 190, 291-296.

Simms, M.J. and Ruffell, A.H., 1989. Synchroneity of climate change and extinctions in the Late Triassic. Geology 17, 265-268.

Simms, M.J. and Ruffell, A.H., 1990. Climatic and biotic change in the Late Triassic. Journal of the Geological Society London 147, 321-327.

Smith, A.B. and Patterson, C., 1988. The influence of taxonomic method on the perception of patterns of evolution. Evolutionary Biology 23, 127-216.

Stanley, S.M., 1982. Glacial refrigeration and Neogene regional mass extinction of marine bivalves. In: Gallitelli, E.M. (ed.), Paleontology, Essential of Historical Geology. pp. 179-191. S.T.E.M.

Mucchi, Modena, Italy.

Stanley, S.M., 1986. Anatomy of a regional mass extinction: Plio-Pleistocene decimation of the Western Atlantic bivalve fauna. Palaios 1, 17-36.

Stanley, S.M., 1990. Delayed recovery and the spacing of major extinctions. Paleobiology 16, 401-414.

Ujiie, H., 1984. A Middle Miocene hiatus in the Pacific region: its stratigraphic and paleoceanographic significance. Palaeogeography, Palaeoclimatology, Palaeoecology 46, 143-164.

Valentine, J.W., 1969. Patterns of taxonomic and ecological structure of the shelf benthos during Phanerozoic time. Palaeontology 12, 684-709.

Van Valen, L.M., 1984. A resetting of Phanerozoic community evolution. Nature 307, 50-52.

Walliser, O.H., Lottman, J. and Schindler, E., 1988. Global events in the Devonian of Kellerwald and Harz Mountains. Courier Forschungs-Institut Senckenberg 102, 190-193.

Westermann, G., 1984. Gauging the duration of stages: a new appraisal for the Jurassic. Episodes 7, 26-28.

Westrop, S.R., 1989. Trilobite mass extinction near the Cambrian-Ordovician boundary in North America. In: Donovan, S.K. (ed.), Mass Extinctions: Processes and Evidence. pp. 89-103. Columbia University Press, New York.

Manuscript received October 1993
Revision received July 1994

Author's address:

J. John Sepkoski, Jr., Department of the Geophysical Sciences, University of Chicago, 5734 South Ellis Avenue, Chicago, Illinois 60637, U.S.A.

Phanerozoic Development of Selected Global Environmental Features

Jared R. MORROW, Eberhard SCHINDLER, and Otto H. WALLISER

Abstract. Data on diversity and extinction as well as the development of certain environment-controlling parameters for the entire Phanerozoic are widely scattered in the literature. Therefore, selected Phanerozoic physical, chemical, and biological data are compiled herein in order to allow a comparison and the testing of their relevance for the causation of global events.

Contents

Introduction	53
Comments on Curves and on Additional Data	58
Interpretations	59

Introduction

Global events are mostly caused by short-term changes of the environmental conditions. These changes are initiated either by extratelluric (extraterrestrial) or telluric (terrestrial, earth-bound) forces (we give preference to the terms extratelluric and telluric because the term 'terrestrial' is often restricted to describing continental features).

Extratelluric forces derive from earth-impacting meteorites or planetoids, the more or less steady input of cosmic dust, ongoing insolation, the input of hard cosmic radiation, passing comets, etc. In some cases, the planetary parameters of the Earth (e.g. excentricity, obliquity, precession, ecliptic, orbit) have to be included in the extratelluric forces. We must also consider astrophysical forces which may change in their intensity during the revolution of our solar system around the galactic center.

Telluric forces, on the other hand, imply all parameters which are earth-bound, such as mantle convection, plate tectonics, volcanism, orogeneses, or those factors which concern the state of oceans, continents, atmosphere, or magnetism.

With respect to the manifestation of these forces, three categories can be recognized:

1. Processes which directly cause strong, global-scale changes in the Earth´s environment. Examples of such changes include the drastic lowering of average annual temperature through extreme volcanic activity accompanied by high material output into the atmosphere; or major paleogeographic changes caused by short-term, widespread transgression.

2. Global-scale changes that initially have no marked effect on the environment, but instead induce long-term processes which only later cause direct and serious environmental changes leading to an event. An example of this category would be gradual temperature change which must reach a critical threshold value before drastic environmental effects result.

3. Two or more long-term processes that alone produce no dramatic effect, but which interact and can potentially reinforce each other at critical intervals to cause drastic, short-term environmental changes.

Within the biota, these process-based changes may initiate not only short-term extinction events but also long-lasting crises that cause an acceleration of back-

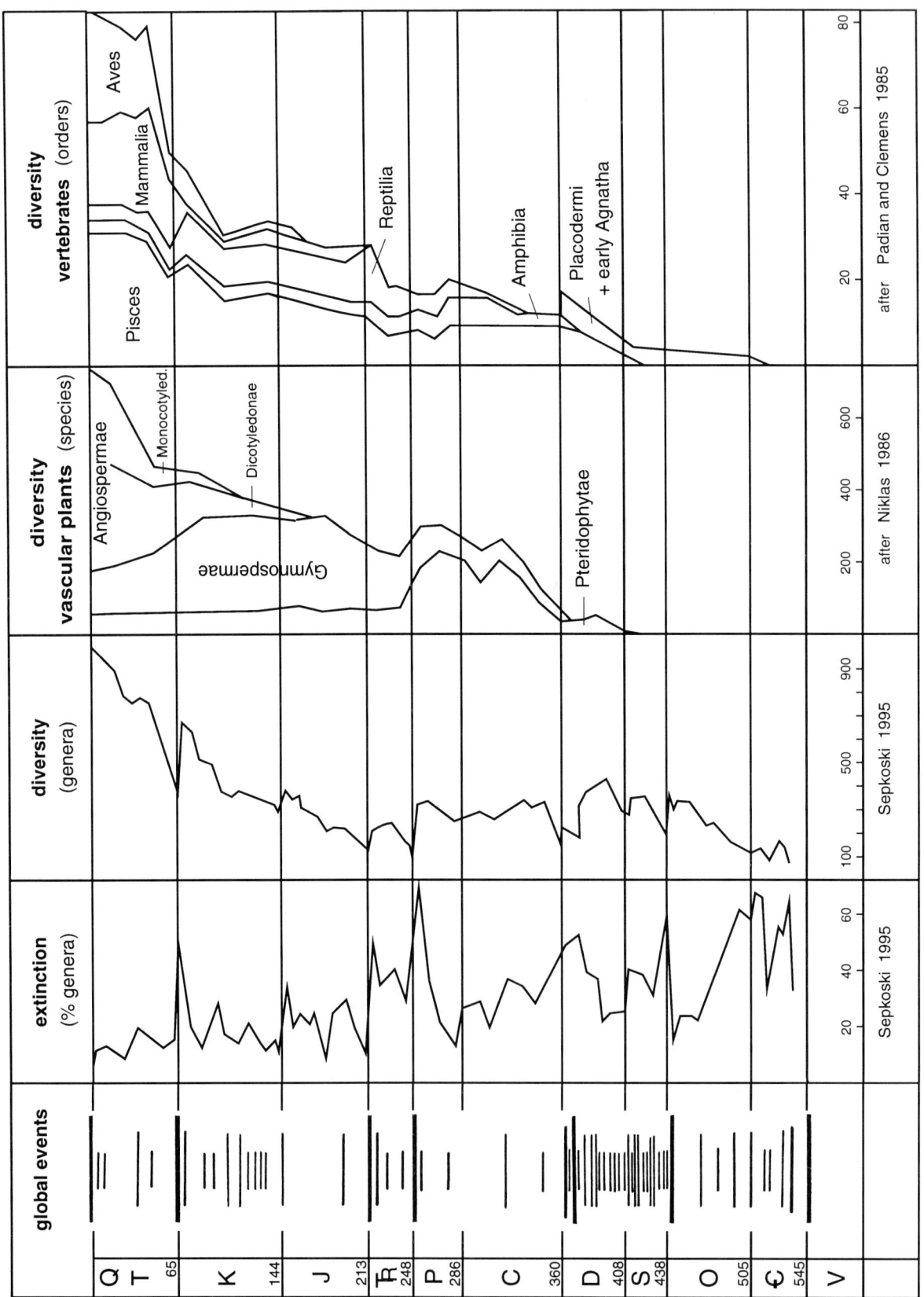

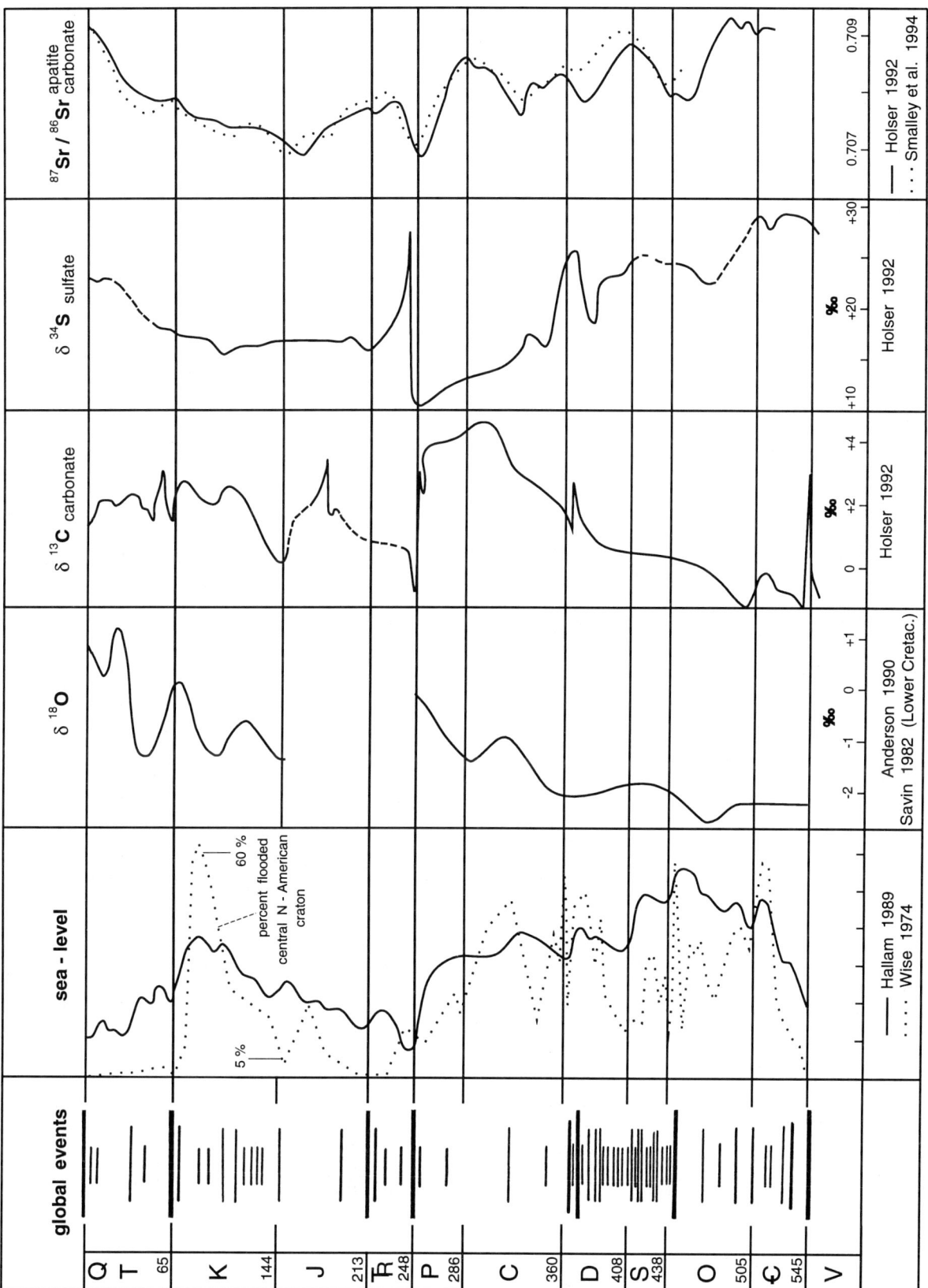

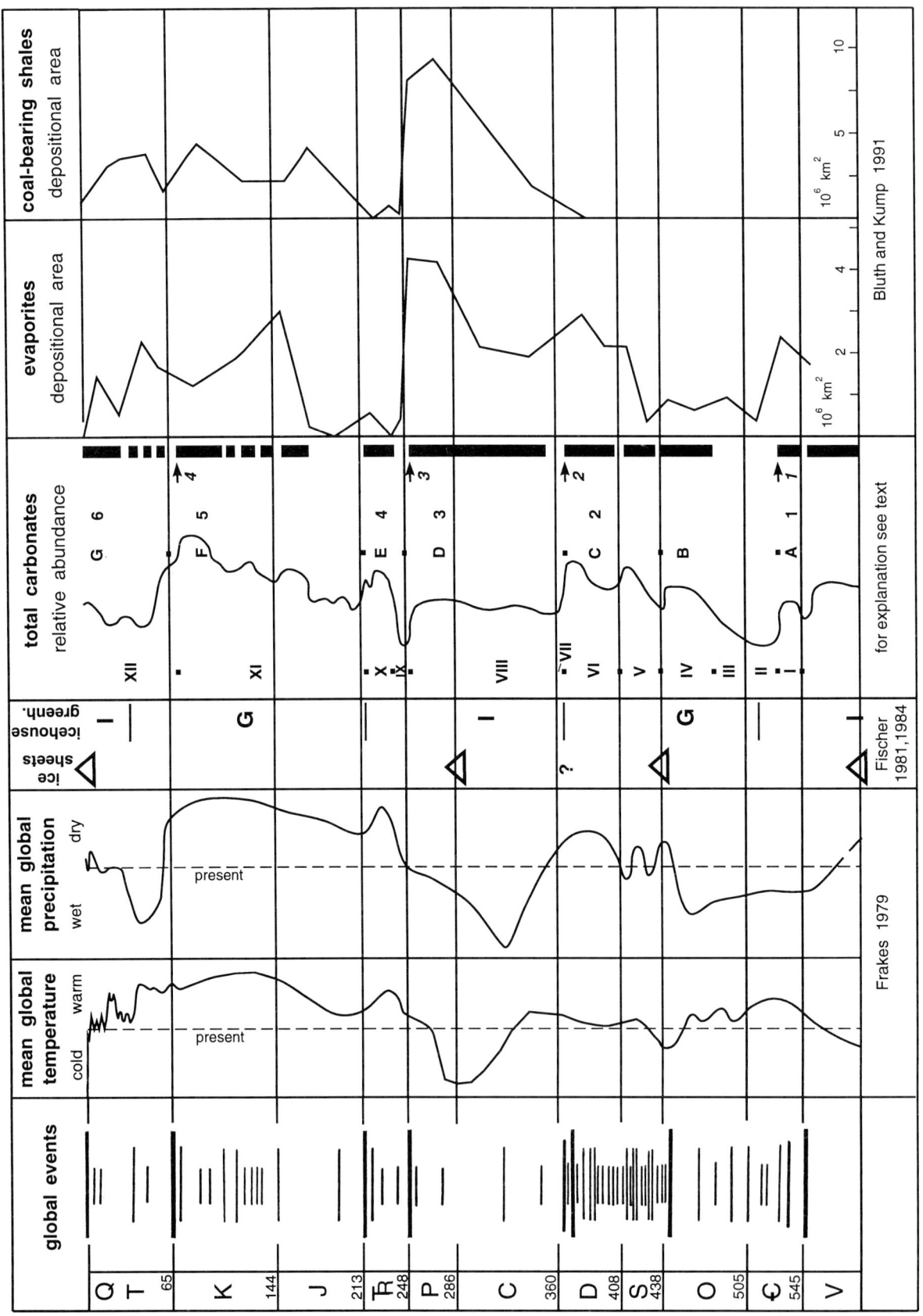

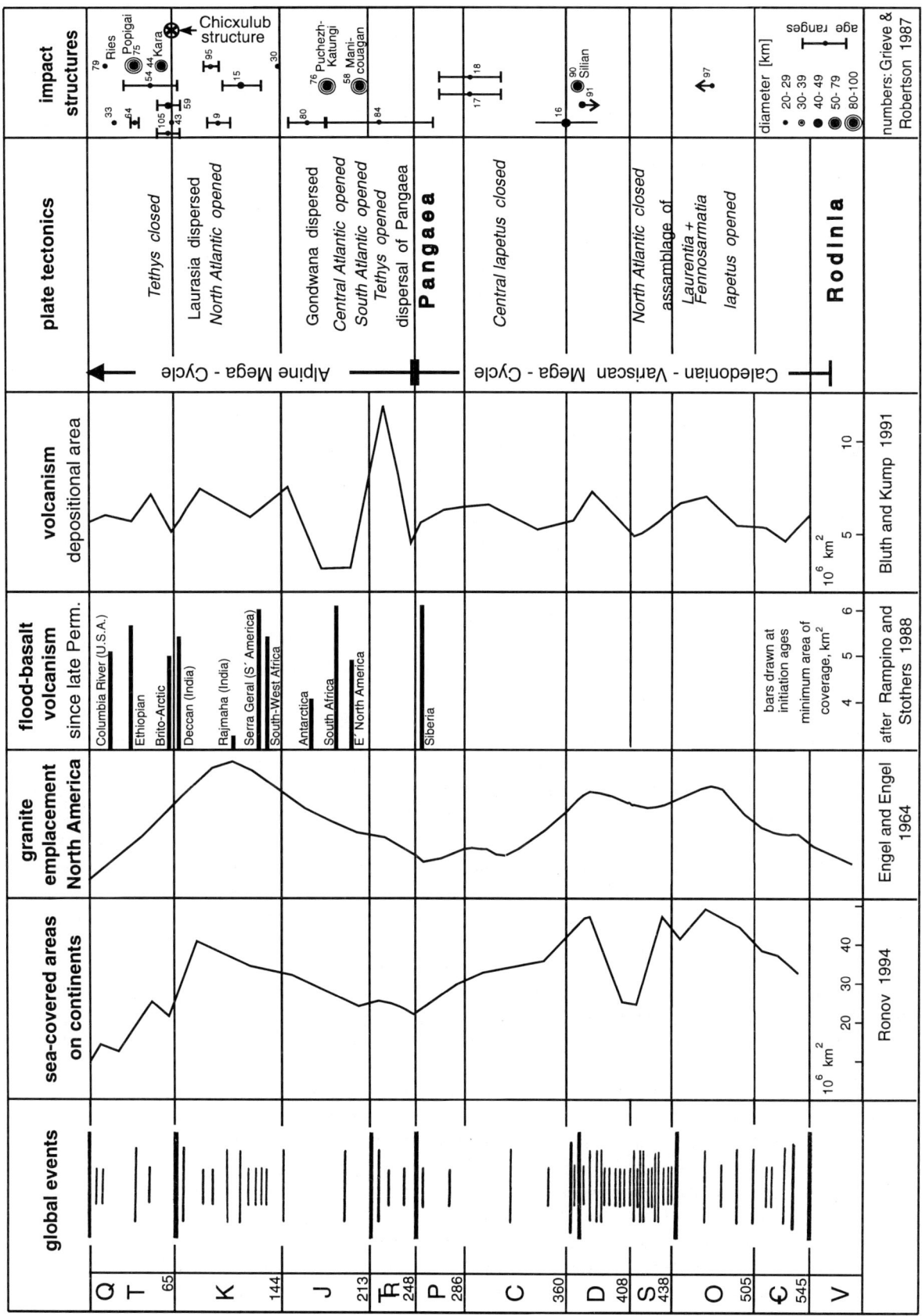

ground extinctions and possibly a final mass extinction. Alternately, these processes could lead to such a weakened or stressed ecosystem that only a small-scale environmental change is required to cause a major extinction event. Therefore, long-term processes must be considered when analyzing global events. In this chapter, selected data summarizing long-term global chemical, physical, and biological changes or fluctuations, and other important features have been compiled from the literature. In the compilation process various difficulties arose, primarily concerning the geologic time scale. In numerous publications a correlation of the geochronologic time scale with System or Series boundaries is not given. In addition, within the published curves there are often incomplete references regarding specific sources of the data used. As a result, when translating and scaling these previously published curves to the time-scale of Harland et al. (1990) (used in this volume), small deviations from the original authors' published works may occur.

The curves used in this chapter are, of course, of different quality and resolution. As a result, they are intended to show only general trends, and may omit minor excursions. Compilations which are similar in some respect have been published several times, as e.g. by Fischer (1984) and Kauffman and Fagerstrom (1993). For some of the curves there is more detailed information available within the chapters of this volume.

In most cases, we present the curves or data as they are published. In a few specific cases, we made some modifications, which are explained in the following paragraphs.

The curves in Fig. 1 which are not discussed below are reproduced without changes except for adjusting them to a uniform time-scale.

Comments on Curves and on Additional Data

Sea-Level Curve. The trends of the Hallam curve agree in general with the strongly generalized sea-level curve of Vail et al. (1991). The older curve of Vail et al. (1977) is more detailed; however, it was not favored as it depicts slow transgressions and rapid regressions which is just the opposite to most cases known by us. In the reproduced Hallam curve we avoid modifications, although probably needed at some specific levels, as can be seen in the detailed curves of other chapters within this volume and other publications such as assembled in the volume edited by Wilgus et al. (1988). As an example of a more developed curve in a well-known region (North American Craton), that of Wise (1974) is added. For the purpose of this chapter it is most important to recognize the overall development of major features, such as the two obvious first-order super-cycles, and also severe regressions in the late Permian and late Cretaceous preceding two of the biggest Phanerozoic mass extinction events.

Sr Curve. For the $^{86}Sr/^{87}Sr$ ratio we figure two curves which cover most of the Phanerozoic. Although the curve of Smalley et al. (1994; dotted) is based on a larger data set, it confirms in principle the earlier published curve of Holser (1984, 1992; solid line). Nevertheless, both curves are presented because of some minor differences which are possibly important (e.g. the stratigraphic position of the Upper Jurassic minimum).

Icehouse/Greenhouse Episodes. We reproduce the intervals of Fischer (1981, 1984) although it is obvious that this plot shows only the prevailing global icehouse or greenhouse state. Obviously, there occur fluctuations contrary to the prevailing trend. Thus, e.g., the latest Ordovician glaciation is intercalated within the early Palaeozoic greenhouse episode.

Total Carbonates and Reefs. There are several published curves depicting total carbonates, reefs, and ramp carbonates. In this chapter we present only a curve for total carbonate abundance, supplemented by vertical bars indicating times of major reef development. Concerning the development of reefs, there exist numerous published, qualitative portrayals. However, well-based quantitative analyses for the whole Phanerozoic are lacking or still rather tentative (e.g. Burchette and Wright, 1992, who present a quantitative analysis separately for ramp and reef carbonates). A very detailed summary of reef development for the Phanerozoic is given by Talent (1988). The curve presented herein is mainly based on the literature cited in this paragraph and on our own experience, and is therefore also only tentative. The rough base for our total carbonate curve is that given by Kazmierczak et al. (1985). In contrast, our curve deviates in several intervals for the following reasons. In the Vendian, with the exception of the topmost strata, we recognize a greater contribution from globally distributed stromatolite carbonates. Within the Cambrian, the occurrence of archaeocyathid limestones preceding the mid-Cambrian transgression is emphasized. For the Silurian and Devonian our curve depicts the maxima of reef construction. Although stromatoporoid- and coral-dominated bioherms disappear at the end of the Frasnian, widespread carbonate platforms exist into the Carboniferous and Permian.

In addition to the curve, we indicate the interpretation of various colleagues with respect to main intervals of reef building and connected ecosystems:

A - G: Reef intervals with complex framework (Sheehan, 1985).
1 - 6: Erathemic successions (Copper, 1988).
1 - 4: Reef collapses (Newell, 1972).
I - XII: Level bottom communities (Boucot, 1983, 1990).

Evaporites, Coal-Bearing Shales and Volcanics (from Bluth and Kump, 1991). The data and the methods on which these curves are based are rather problematic as in the sense discussed by the authors themselves. On the other hand, it is a first and helpful attempt and therefore included herein.

Flood Basalt Volcanism. The data from Rampino and Stothers (1988), given in their Table 1 and in their text are graphically presented herein.

Impact Structures. In the compilation of Grieve and Robertson (1987) 116 impact structures of all sizes are listed. For our purpose, we choose only structures with diameters ≥ 20 km. Impacts creating smaller structures are not thought to cause effects on a global scale. The projectile that made the 25 km diameter Ries crater in southern Germany, which caused a huge amount of dislocated material and additional extensive impact melt rock (Suevit), caused no more than a regionally limited disaster. In addition to the data of Grieve and Robertson (1987) the Chicxulub structure (Yucatan peninsula, Mexico, close to the K/T boundary) is included herein, although not unanimously considered of impact origin by all workers (e.g. Meyerhoff et al., 1994).

Interpretations

The given curves and data were assembled in order to detect major Phanerozoic trends, rather than short-term changes. The latter can be better seen in the more detailed curves within other chapters of this volume. Unfortunately, there are presently many intervals of Phanerozoic time where detailed data are still lacking.

The most obvious trend, discernible in most of the general curves, is a super-cyclicity of which the older super-cycle spans the interval from the uppermost Precambrian to the end-Permian, i.e. at least the entire Palaeozoic. This older super-cycle was recognized more than a hundred years ago and gave rise to the - as we know today - well-founded establishment of the Palaeozoic Era. Similarly, more recent geochemical (i.e. stable isotope), palaeontological, and relative sea-level data all tend to support the recognition of this older super-cycle. However, one must be aware that the biota characteristic of the Phanerozoic Era did not start at the beginning of this super-cycle, but evolved later, i.e. in post-Tommotian time. For post-Permian times interpretation of super-cyclicity is even more difficult. From a biological-evolutionary point of view the further subdivision of the younger Phanerozoic super-cycle into the Mesozoic and Cenozoic is evident. Therefore, the Phanerozoic is made up of three large-scale evolutionary cycles. The second, of course, coincides with the Mesozoic, i.e., it is terminated at the close of the Cretaceous. Thus, the second evolutionary cycle spanning about 180 Ma is significantly shorter than the Palaeozoic one of about 290 Ma. Besides the impossibility to preview the geological processes responsible for the formation of these evolutionary cycles, all predictions could be upset by the disastrous impact of man on the biosphere, which by many estimates will accelerate the ending of the third evolutionary cycle.

Considering only some abiotic parameters, one could also take the whole post-Palaeozoic as a single geological super-cycle, following, e.g., Fischer (1981, 1984). This is well seen in the sea-level curve. Starting from a lowstand in the late Precambrian, sea level reached its Phanerozoic maximum in the Ordovician. During the Silurian to early Permian, sea level remained relatively high, although characterized by minor but significant fluctuations. It is only in the later part of the Permian that a strong regressive trend leads to the Phanerozoic minimum at the end of the Permian which terminates this older, 300 Ma-long super-cycle. The second super-cycle, spanning the Mesozoic and Cenozoic, has a duration of only about 250 Ma up to the present. However, this super-cycle has not ended and it is not known what the final duration will be. The highest sea level of this super-cycle is reached in the late Cretaceous. Following this maximum, a strong regression occurs which lasts until about the end of the Cretaceous. It can be seen that in the case of both super-cycles, the sea level maxima are followed by regressions which are associated with severe biotic turnovers.

The Permian sea-level lowstand coincides with the final assemblage of Pangaea. However, for the same time interval also other curves show strong deviations from the general trend. When analyzing this pattern, one should consider that with the fomation of a super-continent the coastline as well as the shelf area both have their shortest possible length. This could be one of the main causes for the great end-Palaeozoic biotic turnover. However, one should be aware that in most cases not only one single cause is responsible for a major change, but rather it is the potential interaction of different processes and their mutual amplification. It is hoped that the present compilation may help to identify some of these complex connections.

Acknowledgements

Assistance with drawings (Cornelia Kaubisch), computer graphics (Dr. Angelika Wedel) and careful handling of the text (Gabriela Meyer) is gratefully appreciated; we owe fruitful discussions to Profs. Joachim Reitner and Walter Riegel (all Göttingen).

References

Anderson, T.F., 1990. Temperature from oxygen isotope ratios. In: Briggs, D.E.G. and Crowther, P.R. (eds.), Palaeobiology - a synthesis. pp. 403-406. Blackwell Sci. Publ., Oxford-London-Edinburgh-Boston-Melbourne.

Bluth, G.J.S. and Kump, L.R., 1991. Phanerozoic paleogeology. Amer. J. Sci. 291, 284-308.

Boucot, A.J., 1983. Does evolution take place in an ecological vacuum. J. Paleont. 57, 1-30.

Boucot, A.J., 1990. Phanerozoic extinctions: How similar are they to each other? Lecture Notes Earth Sci. 30, 5-30.

Burchette, T.P. and Wright, V.P., 1992. Carbonate ramp depositional systems. Sediment. Geol. 79, 3-57.

Copper, P., 1988. Ecological succession in Phanerozoic reef ecosystems: Is it real? Palaios 3, 136-151.

Engel, A.E.J. and Engel, C.G., 1964. Continental accretion and the evolution of North America. In: Subramaniam, A.P. and Balakrishna, S. (eds.), Advancing frontiers in geology and geophysics. pp. 17-37. Indian Geophys. Union, Hyderabad.

Fischer, A.G., 1981. Climatic oscillations in the biosphere. In: Nitecki, M.H. (ed.), Biotic crises in ecological and evolutionary time. pp. 103-131. Acad. Press, New York-London-Toronto-Sydney-San Francisco.

Fischer, A.G., 1984. The two Phanerozoic supercycles. In: Berggren, W.A. and Van Couvering, J.A. (eds.), Catastrophes and Earth history. pp. 129-150. Princeton Univ. Press, Princeton.

Frakes, L.A., 1979. Climates throughout geologic time. 310 pp. Elsevier, Amsterdam-Oxford-New York.

Grieve, R.A.F. and Robertson, P.B., 1987. Terrestrial impact structures. Geol. Surv. Canada, Map 1658 A, scale 1:63 000 000, Suppl. to Episodes 10 (2).

Hallam, A., 1989. The case for sea-level change as a dominant causal factor in mass extinction of marine invertebrates. Phil. Trans. R. Soc. London B 325, 437-455.

Harland, W.B., Armstrong, R.L., Cox, A.V., Craig, L.E., Smith, A.G. and Smith, D.G., 1990. A geologic time scale 1989. 263 pp. Cambridge Univ. Press, Cambridge-New York-Port Chester-Melbourne-Sydney.

Holser, W.T., 1984. Gradual and abrupt shifts in ocean chemistry during Phanerozoic time. In: Holland, H.D. and Trendall, A.F. (eds.), Patterns of change in Earth evolution. Dahlem Workshop Reports, Phys. Chem. Earth Sci. Res. Rep. 5, 123-143. Springer, Berlin Heidelberg New York.

Holser, W.T., 1992. Stable isotope geochemistry of sulfate and chloride rocks. Lecture Notes Earth Sci. 43, 153-176.

Kauffman, E.G. and Fagerstrom, J.A., 1993. The Phanerozoic evolution of reef diversity. In: Ricklefs, R. and Schluter, D. (eds.), Species diversity in ecological communities. pp. 315-329. Univ. Chicago Press, Chicago-London.

Kazmierczak, J., Ittekkot, V. and Degens, E.T., 1985. Biocalcification through time: Environmental challenge and cellular response. Paläont. Z. 59, 15-33.

Meyerhoff, A.A., Lyons, J.B. and Officer, C.B., 1994. Chicxulub structure: A volcanic sequence of Late Cretaceous age. Geology 22, 3-4.

Newell, N.D., 1972. The evolution of reefs. Sci. Amer. 226, 54-65.

Niklas, K.J., 1986. Large-scale changes in animal and plant terrestrial communities. In: Raup, D.M. and Jablonski, D. (eds.), Patterns and processes in the history of life. Dahlem Workshop Reports, Life Sci. Res. Rep. 36, 383-405. Springer, Berlin Heidelberg New York.

Padian, K. and Clemens, W.A., 1985. Terrestrial vertebrate diversity: Episodes and insights. In: Valentine, J.W. (ed.), Phanerozoic diversity patterns. pp. 41-96. Princeton Univ. Press, Princeton-San Francisco.

Rampino, M.R. and Stothers, R.B., 1988. Flood basalt volcanism during the last 250 million years. Science 241, 663-668.

Ronov, A.B., 1994. Phanerozoic transgressions and regressions on the continents: A quantitative approach based on areas flooded by the sea and areas of marine and continental deposition. Amer.

J. Sci. 294, 777-801.

Savin, S.M., 1982. Stable isotopes in climatic reconstructions. Stud. in Geophys., Climate in Earth Hist. pp. 164-171. National Acad. Press, Washington, D.C.

Sepkoski, J.J., Jr., 1995. Patterns of Phanerozoic Extinction: a Perspective from Global Data Bases. In: Walliser, O.H. (ed.), Global Events and Event Stratigraphy in the Phanerozoic, 35-51, Springer, Berlin Heidelberg New York.

Sheehan, P.M., 1985. Reefs are not so different -- They follow the evolutionary pattern of level-bottom communities. Geology 13, 46-49.

Smalley, P.C., Higgins, A.C., Howarth, R.J., Nicholson, H., Jones, C.E., Swinburne, N.H.M. and Bessa, J., 1994. Seawater Sr isotope variations through time: A procedure for constructing a reference curve to date and correlate marine sedimentary rocks. Geology 22, 431-434.

Talent, J.A., 1988. Organic reef-building: Episodes of extinction and symbiosis? Senckenbergiana lethaea 69, 315-368.

Vail, P.R., Audemard, F., Bowman, S.A., Eisner, P.N. and Perez-Cruz, C., 1991. The stratigraphic signatures of tectonics, eustasy and sedimentology - an overview. In: Einsele, G., Ricken, W. and Seilacher, A. (eds.), Cycles and events in stratigraphy. pp. 617-659. Springer, Berlin Heidelberg New York.

Vail, P.R., Mitchum, R.M., Jr. and Thompson, S. III, 1977. Seismic stratigraphy and global changes of sea level, part 4: Global cycles of relative changes of sea level. Amer. Assoc. Petrol. Geol. Mem. 26, 83-97.

Wilgus, C.K., Hastings, B.S., Kendall, C.G.S.C., Posamentier, H.W., Ross, C.A. and Van Wagoner, J.C. (eds.), 1988. Sea-level changes: An integrated approach. Soc. Econ. Paleont. Mineral. Spec. Publ. 42, 407 pp.

Wise, D.U., 1974. Continental margins, freeboard and the volumes of continents and oceans through time. In: Burke, K. and Drake, C.L. (eds.), The geology of continental margins. pp. 45-58. Springer, Berlin Heidelberg New York.

Manuscript received December 1994

Authors' addresses:

Jared R. Morrow, Department of Geological Sciences, University of Colorado, Boulder, CO 80309, U.S.A.

Eberhard Schindler, Forschungsinstitut Senckenberg, Senckenberganlage 25,
D - 60325 Frankfurt am Main, Germany

Otto H. Walliser, Institut und Museum für Geologie und Paläontologie, Goldschmidt-Strasse 3,
D - 37077 Göttingen, Germany

Global Isotopic Events

William T. HOLSER, Mordeckai MAGARITZ† and Robert L. RIPPERDAN

Abstract: More than 60 global "events" have been identified in the isotopic records of $\partial^{13}C$, $\partial^{18}O$, $\partial^{34}S$ and $^{87}Sr/^{86}Sr$. Over half of these are carbon isotopic events, about equally divided between positive excursions (or a simple rise) and negative excursions (or a fall). The positive excursions generally have been ascribed to "oceanic anoxic events" or similar incidents of gross storage of organic carbon, and negative isotopic events have been related to a catastrophic reduction of primary productivity, but many records suggest a complex origin. Likewise strontium isotopic shifts are generally related to changes in the balance of inputs to the ocean of light strontium from reaction with MOR basalts and of heavy strontium from uplift and erosion of old cratonic terranes. Despite this variety of origins, most of the isotopic events of these elements have potential for worldwide stratigraphic correlation, with resolution that may equal or exceed that of biostratigraphy. Diagenetic distortion of isotopic profiles should be minimized by screening samples with appropriate textural and trace element criteria.

Contents

Introduction	63
A Catalog of Isotopic Events	70
Isotopic Events in Carbonate Carbon ($\partial^{13}C_{carb}$)	71
Isotopic Events in Carbonate Oxygen ($\partial^{18}O_{Carb}$)	75
Isotopic Events in Sulfate Sulfur ($\partial^{34}S$)	77
Isotopic Events in Strontium ($^{87}Sr/^{86}Sr$)	78
Conclusions	79

Introduction

The record in marine sediments of changes in the oceanic isotopic ratios of C, O, S, Sr and possibly other elements has been interrogated with questions about the history of the sedimentary cycle and about the secular variations of biological evolution, marine and atmospheric composition, oxidation/reduction, climate, erosion and oceanic circulation. Typically these effects have been studied by geochemical modelling, using the long-term isotopic variations as control.

More recently short-term events in the isotope record have been studied not only for the insight they provide into processes such as bolide impact, volcanic eruption, oceanic overturn and mass extinction, but also for an entirely empirical application to chronostratigraphy. This chapter outlines the characteristics of marine isotopic events, and catalogs the events that can be recognized at this time.

The nature of the marine geochemical cycles of these elements have been reviewed repeatedly, and is summarily illustrated in Fig. 1 (Holser, Magaritz and Wright, 1986; Holser et al., 1988). The dominant mechanisms for changes in the recorded isotopic ratios of C and S are both biologically mediated, by photosynthetic reduction and anaerobic bacterial reduction, respectively. Levels of oxygen isotopic ratio are, on the other hand, a function of the temperature of crystallization of the carbonate mineral, and of the isotopic ratio of the oxygen in sea water. For Sr, changes in the marine ratio depend on an imbalance of inputs from light (young, oceanic) and heavy (old,

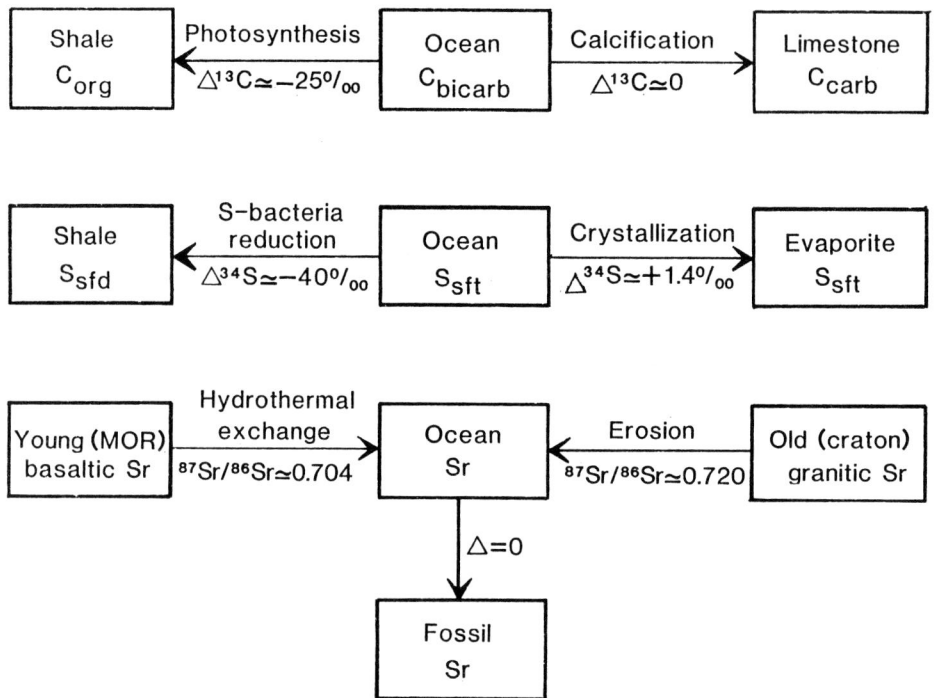

Fig. 1. Important sources of secular variations of the isotopes of C, S and Sr in the exogenic cycle. After Holser et al. (1984)

continental) sources. To a first approximation the fluxes in Fig. 1 may be considered as a steady state--inputs equal outputs in each "reservoir".

A change in an isotopic ratio--an isotopic event--represents a shift of input or output to set up a new steady state. The rate of this sort of change will depend on the "residence time" of the particular element in the particular reservoir. The residence time in the ocean reservoir--the ratio of the mass of the element in the ocean to the rate of input of the element, e.g., from rivers, serves as a rough measure of how easily the steady state isotope level of a reservoir can be changed. If an element has a long residence time in the ocean it will take a long time to come to a new equilibrium steady state; if the residence time is short the new steady state can be established rapidly. A long residence time makes it more likely that the isotopic age curve will pick up transient excursions, as explained below in the section on carbon isotopes. Column 2 of Table 1 lists nominal residence times for elements that are useful in marine isotope geochemistry, for the whole ocean reservoir. Column 3 of Table 1 splits off the first hundred meters or so of the surface ocean (above the thermocline) to give a distinctly shorter residence time, because: (1) this zone is mixed faster, by waves, currents, and tides than is the ocean deep; (2) this zone is the scene of primary productivity in the carbon cycle, causing local changes in carbon isotope ratio (see below); and (3) this zone is the site of deposition and sampling of most (not all) marine carbon isotopes.

Figure 2 (Holser and Magaritz, 1989) gives an overview of gross changes in the marine isotopic ratios of C, S and Sr through Phanerozoic time. This set of curves includes transient as well as steady-state levels of isotopic ratio, and furthermore has not been updated to include the details of some of the isotopic events that have been recognized recently.

Marine waters are well mixed with respect to each of these isotopic ratios because the residence times of C, S and Sr in the marine system are so long (Table 1) relative to the mixing time of the surface ocean (a few hundreds years) or even the deep ocean (a few thousand years). Therefore the pattern of change-events that is generated in any part of the world ocean may be expected to be recorded in marine rocks worldwide with stratigraphic resolution that is effectively instantaneous geologically speaking. In this respect isotopic chronostratigraphy, as practically applied, shares much of the character of paleomagnetic reversal stratigraphy,

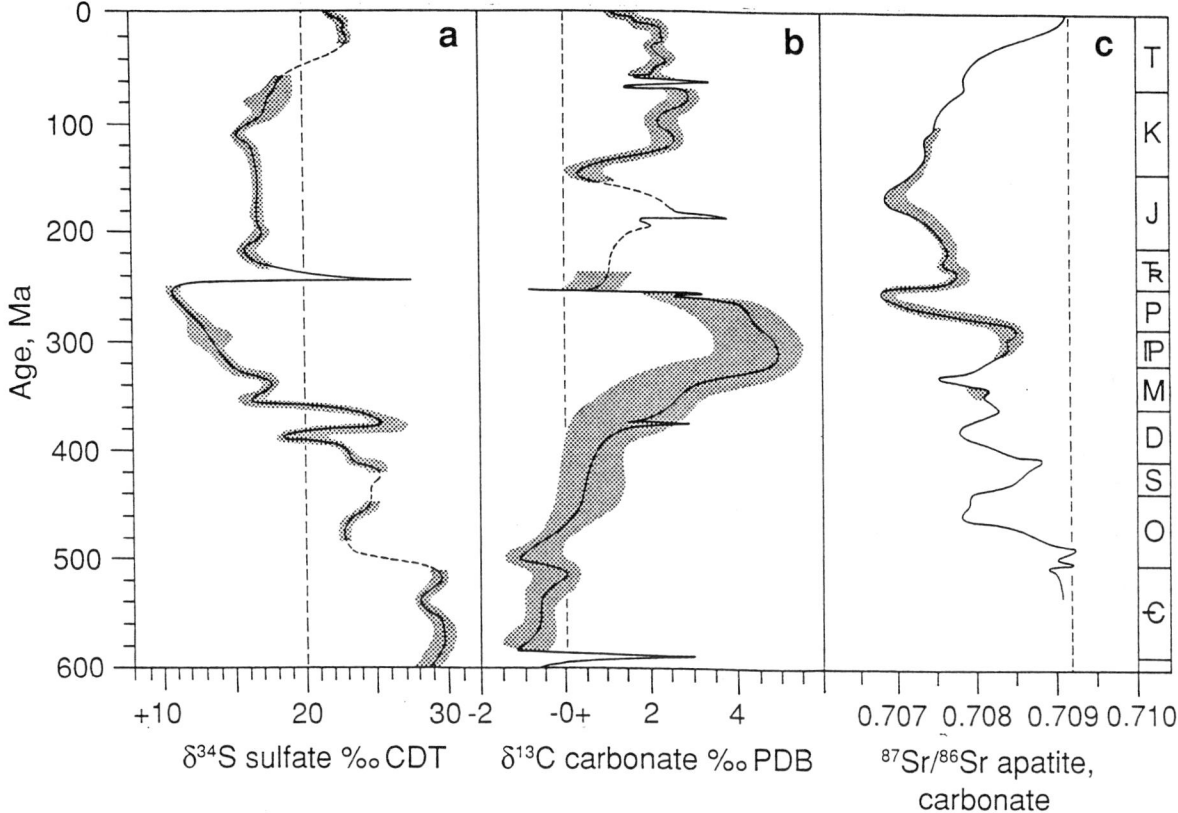

Fig. 2. Age curves for isotopes of S, C and Sr through Phanerozoic time. Both long-term steady state and short-term transient changes are included. The reference levels (dashed lines) are modern sea water. After Holser and Magaritz (1989); not updated with more recent information as listed in Table 2

which has a characteristic response time of 10^4 yr. Both paleomagnetic and isotopic stratigraphy provide only multivalued age correlations. Isotopic events thus add an additional dimension to stratigraphic correlation, particularly if the conventional biostratigraphic information can be compounded with both paleomagnetic and (one or more) isotopic parameters. Both methods also suffer in similar ways from degradation of the primary signal by subsequent overprinting through diagenesis.

Most applications of both isotopic and paleomagnetic reversals to stratigraphy depend only on the recognition of a "pattern" of changes in depositional sequences. But an isotopic shift that is cleanly developed in unaltered sediments allows a substantially refined resolution in chronostratigraphy. In particular, the monotonic rise in $^{87}Sr/^{86}Sr$ through much of the Cenozoic can be calibrated to give stratigraphic dates of high resolution, better than ± 0.5 Ma in many cases (Capo and DePaolo, 1990; Hodell et al., 1990; Hodell et al., 1991; McArthur, 1991). Lack of precision in the data base for carbon and sulfur has not yet allowed this level of refinement in those systems.

Table 1. Residence Times of Elements in the Modern Ocean [a]

Element	Whole Ocean	Surface Ocean
S	45 Ma	1.5 Ma
Sr [b]	4.1 Ma	140 ka
C_{carb}	55 ka	1.8 ka
C_{org} [c]	1000 a	33 a
H_2O [d]	1000 a	100 a
Nd [e]	300 a	10 a

[a] Preanthropogenic. Holser et al. (1988)
[b] Data from Veizer (1989)
[c] New productivity: "biological pump" (Kump, 1991)
[d] Mixing time
[e] Elderfield and Greaves (1982)

Table 2. Catalog of isotopic events

EVENT	AGE	CHAR	SIZE	BASINS	CORR	SAMP	RESOL	BOUN	NOTES AND REFERENCES
87Sr Pli	2.5	rise	2	3	4	5	5	0	1-3
87Sr Mes	5.5	rise	1	3	4	5	5	0	1-3 "Chron 6 Event" (9) Non-marine rise in $\delta^{13}C$ (4)
13C Mes	6	fall	1	5	5	5	4	0	4-13
18O Srv	12	rise	1	3	3	2	3	0	13-17 "Event Mi4" (14) Glaciation (15)
18O Lan$_2$	14	rise	2	4	5	4	5	1	8, 13-19 "Event Mi3" (14) Glaciation (15)
13C Lan	16	fall	2	4	5	5	3	2	8, 9, 12, 13, 18-21 "Chron 16 Event" (9) Numerous minor events (8, 20)
18O Lan$_1$	16	+ exc	1	3	3	2	3	2	14 "Event Mi2" (14)
13C Bur	17	rise	2	3	5	5	3	0	8, 10, 12, 13, 22-24 "Monterey Excursion" (22)
18O Cha$_2$	24	+ exc	1	2	3	4	3	3	14, 16, 25 "Event Mi1" (14) "Event E" (25)
18O Cha$_1$	28	fall	1	2	2	1	3	0	14 "Event Oi2a"(?) (14)
18O Rup$_2$	32	rise	1	4	4	4	4	0	14 "Event Oi2" Not observed in low-latitude planktonic records (14)
18O Rup$_1$	35	fall	1	5	5	5	4	0	14, 26 "Event Oi1" (14)
18O Eoc/Oli	36	rise	2	5	5	5	5	3	13, 15-17, 25, 27, 28 "Events C, D" (25) Glaciation (15) Maximum ice-rafting (27)
13C Eoc/Oli	36	+ exc	2	3	3	4	5	3	11, 25, 29
87Sr Cz	36	rise	5	5	5	5	1	0	3, 26, 29-32
18O Ypr	55	- exc	1	4	5	5	2	3	13, 16, 17, 25, 33 "Event A" (25)
13C Pal/Eoc$_2$	56	- exc	3	3	5	5	3	3	11, 34-37 Duration 150 ka (34); also continental (35)
13C Pal/Eoc$_1$	56	- exc	4	4	5	3	2	3	13, 17, 25, 37-40 "Event X" (25); duration 6 Ma
18O Pal/Eoc	56	- exc	2	5	5	3	5	3	34, 36, 41
13C Tha	57	+ exc	3	2	5	5	3	3	11, 13, 17, 25 "Event Y" (25)

^{18}O Tha	57	− exc	2	3	1	3	5	3	36, 41	Antarctic Basin only (*41*)
^{13}C K/T	66	− exc	2	5	5	5	2	5	11, 13, 17, 25, 38, 40-45	
^{87}Sr K/T	66	+ exc	2	4	4	4	4	5	46-51	
^{13}C Maa$_2$	67	rise	1	3	5	4	4	5	44, 45, 52, 53	
^{13}C Maa$_1$	68	− exc	2	1	1	2	4	1	44	
^{13}C Cen/Tur	90	+ exc	2	5	5	4	4	2	11, 13, 54-65	"Oceanic Anoxic Event (OAE)"
^{13}C Apt/Alb	112	+ exc	3	2	5	5	2	3	11, 13, 65, 66	
^{13}C J/K	146	− exc	3	3	4	2	4	4	65	
^{87}Sr Dog/Mlm	157	− exc	5	5	3	3	1	3	3, 67, 68	
^{13}C Tor	186	+ exc	3	2	5	2	4	0	69	
^{13}C Plb	188	+ exc	2	1	5	2	3	0	69	
^{87}Sr Nor	215	− exc	2	2	2	2	2	0	68, 70	
^{34}S Spa	241	+ exc	5	5	3	5	2	0	71-75	"Röt Event" (*71*)
^{13}C P/Tr	245	− exc	5	5	5	5	3	5	76-83	
^{34}S P/Tr	245	+ exc	5	3	2	4	2	5	73, 84, 85	
^{87}Sr P/Tr	245	− exc	5	5	5	3	2	5	3, 47, 85-90	
^{13}C Cap	251	rise	5	5	3	5	5	0	91-93	
^{13}C Bsk/Mos	323	− exc	3	2	4	3	3	4	94	
^{87}Sr Vis	340	− exc	5	5	2	3	1	0	3, 88, 89, 95	Glaciation
^{18}O D/C	363	+ exc	3	5	3	3	1	4	96-99	"Hangenberg Event"
^{87}Sr D/C	363	+ exc	3	1	3	5	2	4	68, 100, 101	"Lower *P crepida* Zone Event"
^{13}C Fam	365	− exc	3	2	5	4	4	0	102-104	
^{13}C Fra/Fam	367	rise (?)	3	1	5	4	3	2	102, 105-111	"Upper Kellwasser Event"
^{13}C Fra	369	+ exc	3	1	5	4	3	1	102, 110, 111	"Lower Kellwasser Event"

Event	Age	Direction							Refs	Notes
^{34}S Fra	370	+ exc	5	3	2	3	2	0	71, 73, 102, 106, 112	Sulfate and sulfide
^{87}Sr Eif$_2$	384	+ exc	1	1	2	2	4	0	101	
^{87}Sr Eif$_1$	385	+ exc	1	1	4	2	4	0	101	
^{87}Sr Ems	389	rise	3	1	4	2	4	0	101	
^{13}C Wen/Lud	424	fall	4	2	3	5	3	2	113, 114	
^{87}Sr Lly	431	rise	5	3	3	4	1	0	114a	
^{18}O O/S	439	rise	2	5	2	5	1	0	120	
^{13}C Hir	440	+ exc	4	3	3	4	3	2	115-120	Glaciation
^{13}C Tre	508	fall	2	3	5	3	5	4	121-122	
^{13}C Dol$_2$	509	rise	2	3	5	3	5	4	121-122	
^{13}C Dol$_1$	511	- exc	4	3	5	3	5	4	121-122	
^{13}C Mnt	515	+ exc	5	1	3	1	2	0	123	
^{13}C Atd	558 (520)	+ exc	3	2	2	4	2	1	124-129, 140	
^{87}Sr Atd	560 (522)	+ exc	5	5	3	5	1	0	130	
^{13}C Tom$_2$	565 (533)	- exc	2	2	4	5	4	2	124-129, 140	"Badaowan $\delta^{13}C$ Minimum"; "Late Tommotian Minimum" (*140*)
^{13}C Tom$_1$	575 (544)	rise	5+	4	4	5	4	0	124-132, 131, 140	"Pre-Tommotian Rise" (*140*)
^{13}C Edi$_1$	590	+ exc	5	3	2	4	2	0	130-132	Relation to ^{13}C Tom$_1$ unclear
^{13}C Var	600	- exc	5	3	2	5	2	4	130, 133, 134	Glaciation
^{87}Sr Var	600	rise	5	4	2	5	2	4	130, 133, 135-138	
^{13}C Stu	780	- exc	2	2	2	5	2	4	130, 133	Glaciation
^{87}Sr Rif	830	- exc	5	2	2	4	1	4	130, 133, 135-139	

Footnotes to Table 2.

[a] Event named by isotope and stratigraphic age (designated by standard abbreviation according to Harland et al., 1990, p. 10, 186)

[b] Nominal age of the event according to the scale of Harland et al. (1990) except for ^{13}C Atd - ^{13}C Tom$_1$, which are based on currently accepted radiometric dates

[c] Columns 3-9 are descriptors of the event: CHAR--character, as a rise, a fall, or a positive or negative excursion. Each of the following on a scale from 1 (worst) to 5 (best): SIZE--in permil for $\partial^{13}C$, $\partial^{18}O$, $\partial^{34}S$, and steps of 0.0001 for strontium events; BASINS--the number of sedimentary basins in which the event has been verified; CORR--the quality of correlations for coeval timing of the events in these basins; SAMP--the density of sampling that defines the event; RESOL--the level of time-resolution of the event; and BOUN--the importance of a stratigraphic boundary with which the event is associated, if any. See text for details

[d] References: (1) Hodell et al., 1989; (2) Hodell et al., 1990; (3) Burke et al., 1982; (4) Cerling et al., 1993 (see also Morgan et al., 1994); (5) Cita and McKenzie, 1986; (6) Haq et al., 1980; (7) Hodell et al., 1989; (8) Loutit, Pisias and Kennett, 1983; (9) Margolis et al., 1975; (10) Miller and Fairbanks, 1985; (11) Renard, 1986; (12) Savin et al., 1981; (13) Williams et al., 1988; (14) Miller et al., 1991; (15) Miller et al., 1987; (16) Raymo and Ruddiman, 1992; (17) Shackleton, 1987; (18) Woodruff and Savin, 1991; (19) Woodruff et al., 1981; (20) Loutit, Kennett and Savin, 1983; (21) Berger and Vincent, 1986; (22) Vincent and Berger, 1985; (23) Vincent et al., 1985; (24) Shackleton and Kennett, 1975; (25) Shackleton, 1986; (26) Miller et al., 1988; (27) Zachos, Breza et al., 1992; (28) Wei, 1991; (29) Zachos, Berggren et al., 1992; (30) Elderfield, 1986; (31) Hodell et al., 1991; (32) Richter et al., 1992; (33) Kennett and Stott, 1990; (34) Thomas, 1992; (35) Koch et al., 1992; (36) Kennett and Stott, 1991; (37) Pak and Miller, 1992; (38) Shackleton, 1985; (39) Oberhänsli and von Salis Perch-Nielsen, 1990; (40) Zachos, Rea et al., 1992; (41) Stott, 1992; (42) Alcala-Herrera et al., 1992; (43) Magaritz, Benjamin et al., 1992; (44) Margolis et al., 1987; (45) Stott and Kennett, 1989; (46) Javoy and Courtillot, 1989; (47) Koepnick et al., 1985; (48) Macdougall, 1988; (49) Martin and Macdougall, 1991; (50) McArthur et al., 1992; (51) Nelson et al., 1991; (52) Shackleton et al., 1984; (53) Zachos et al., 1989; (54) Corfield et al., 1991; (55) Hilbrecht et al., 1986; (56) Hilbrecht and Hoefs, 1986; (57) Hilbrecht et al., 1992; (58) Jarvis, Carson, Hart et al., 1988; (59) Jarvis, Carson, Cooper et al., 1988; (60) Kaufman, 1986; (61) Leary et al., 1989; (62) Peryt and Wyrwicka, 1991; (63) Peryt et al., 1992; (64) Schlanger et al., 1987; (65) Scholle and Arthur, 1980; (66) Weissert, 1989; (67) Fischer and Gygi, 1989; (68) Koepnick et al., 1990; (69) Jenkins and Clayton, 1986; (70) Faure et al., 1978; (71) Holser, 1977; (72) Chen et al., 1986; (73) Claypool et al., 1980; (74) Holser et al., 1988; (75) Huang and Liu, 1989; (76) Baud et al., 1989; (77) Chen et al., 1991; (78) Chen et al., 1984; (79) Holser et al., 1989; (80) Magaritz et al., 1988; (81) Magaritz, Krishnamurthy and Holser, 1992 (see also Morante, 1993, and Morante et al., 1994); (82) Oberhänsli et al., 1989; (83) Scholle et al., 1991; (84) Cortecci et al., 1981; (85) Kramm and Wedepohl, 1991; (86) Brookins, 1988; (87) Holser and Magaritz, 1987; (88) Nishioka et al., 1991; (89) Popp, Podosek et al., 1986; (90) Scholle et al., 1990; (91) Holser, Magaritz and Clark, 1986; (92) Magaritz et al., 1983; (93) Magaritz and Turner, 1982; (94) Magaritz and Holser, 1990; (95) Brand, 1991 (see also Denison, Koepnick, Burke et al., 1994); (96) Brand (1989); (97) Hudson and Anderson, 1989; (98) Popp, Anderson et al, 1986; (99) Veizer et al., 1986 (see also Lohmann and Walker, 1989; Railsback, 1990); (100) Ebneth et al., 1991; (101) Kürschner, Ebneth et al., 1992; (102) Halas et al., 1992; (103) Playford et al., 1984; (104) Wang and Geldsetzer, 1992; (105) Wang et al., 1991; (106) Geldsetzer et al., 1987; (107) Hurley and Lohmann, 1990; (108) Buggisch, 1991 (see also Joachimski and Buggisch, 1993); (109) Goodfellow et al., 1989; (110) McGhee et al., 1986; (111) Schindler, 1990; (112) Goodfellow and Jonasson, 1984; (113) Corfield et al., 1991; (114) Jux and Steuber, 1992; Bertram et al., 1992; (115) Long, 1993; (116) Marshall and Middleton, 1990; (117) Middleton et al., 1991; (118) Wang, Orth et al., 1993; (119) Wang, Chatterton et al., 1993; (120) Wadleigh and Veizer, 1982; (121) Ripperdan et al., 1992; (122) Ripperdan et al., 1993; (123) Brasier, 1990; (124) Brasier, 1991; (125) Brasier, 1992; (126) Brasier, Anderson et al., 1992; (127) Brasier, Magaritz et al., 1990; (128) Kirschvink et al., 1991; (129) Magaritz et al., 1991; (130) Donnelly et al., 1990; (131) Kaufman et al., 1991; (132) Magaritz et al., 1986; (133) Derry et al., 1992; (134) Knoll et al., 1986; (135) Asmeron et al., 1991; (136) Burns et al., 1994; (137) Derry and Jacobsen, 1988; (138) Derry et al., 1989; (139) Veizer et al., 1983; (140) Ripperdan, 1994.

A Catalog of Isotopic Events

A catalog of isotopic events is given in Table 2. This preliminary catalog of isotopic events is restricted to the marine record before the Quaternary and after mid-Proterozoic time. A beginning for such a catalog, listing 18 carbon-isotope "excursions" during Phanerozoic time, was recently published by Schidlowski and Aharon (1992). Our listing is more than 60 events, including 34 of carbon isotope events.

Our time scale is chronostratigraphic, with each event dated as closely as possible by biostratigraphic or lithostratigraphic correlation with a geological period, stage, or biozone. For convenience, a correlative radiometric date (rounded to a million years or more) is also tabulated (Column 2) for each event, based on the scale of Harland et al. (1990). But it should be clearly understood that most of these isotopic events occur in marine sediments and are timed by the geological (biostratigraphic) time scale, which of course is subject to refinement of the radiometric scale, and to more precise definitions of chronostratigraphic units. For example, both kinds of adjustment have recently improved the radiometric time scale in the Cambrian, with the new age assignments given in parentheses in Table 2. But until the whole time scale has undergone a complete overhaul, the Harland scale provides an internally consistent listing.

The descriptive name given to an event (Table 2, column 1) is consequently related to the geological time scale, with a prefix that designates the isotope involved. The name designation is abbreviated using the codes as standardized by Harland et al. (1990, p. 10). The time of the event is taken as the peak of an excursion, or as the time of steepest rate of a rise or fall. This arbitrary definition does not take account of an earlier onset of an event. Events closely associated with a period or stage boundary are given a boundary designation, in deference to current usage, e.g., ^{13}CK/T, and ^{13}CFra/Fam.

In contrast to paleomagnetic reversal events, isotopic events may have a complex geometry, for which at least some simple descriptors will be useful. The CHARacter (Table 2, column 3) of a typical event may be a "fall" (or negative shift), a "rise" (positive shift), a "negative excursion" (falls sharply and quickly returns to substantially its former level), or (analogously) a "positive excursion". Where events follow closely, these designations are somewhat arbitrary: thus when a rise closely follows a fall, the rise and fall may be designated as a single negative excursion event. To simplify the descriptive categories, a negative excursion following continuously on a positive excursion is arbitrarily designated as those two excursions, although it might equally well be regarded as a single "cycle".

The SIZE (column 4) of an excursion or shift is a useful descriptor. In the catalog tables we have designated the magnitude of shifts in ∂ values of C, O and S isotopes on a scale of 5: $1 = \leq 1‰$, $2 = >1$ and $\leq 2‰$, ..., $5 = \geq 5‰$; and similarly in steps of 0.0001 for Sr isotope ratios (shifts in $\in$ units). Note that "5" includes all larger values.

We have made a preliminary attempt to evaluate the quality of an event, in columns 5 to 9 of Table 2. The level of confidence in our knowledge about the event can be ranked on appropriate scales, ranging from "1" for the worst to "5" for the best.

BASINS. Most important is the global character of an event, which we have rated on a scale of 1 to 5 (or more) separate sedimentary basins in which the event has so far been verified. For the most part we have not listed "events" that have yet to be confirmed for at least a second site, e.g., Xu et al. (1986): negative excursion of ∂^{13}C in the Tournaisian; Magaritz and Holser (1990): two negative excursions of ∂^{13}C in the Desmoinesian; Pawellek and Veizer (1992): positive excursion of ∂^{13}C across the Eifelian/Givetian boundary; and Gao and Land (1992) excursions of ∂^{13}C in the early Ordovician.

CORRelation. The quality of biostratigraphic correlation by which the events at the various sites have been determined as isochronous: 1 = single site, no correlation; 2 = poorly correlated, less precise than a stage ... 5 = very well correlated (to a biozone or better).

SAMPling. The quality of mapping a rise or fall is ranked as: 1 = defined in the best section by only three samples; 2 = 5 samples; 3 = 7 samples; 4 = 10 samples; $5 = \geq 15$ samples--and double these numbers of samples for an excursion.

RESOLution. We rank the time resolution of the event, that is, the sharpness of rise, fall or excursion on a scale of 1 to 5, as follows: 1 = spread over much of a geological period, 2 = confined to a stratigraphic stage; 3 = confined to a single biozone; 4 = confined to half a biozone, or, say, about a meter of section; and 5 = a step function within the limits of bioturbation.

BOUNdary. We note the association of the event with the stratigraphic hierarchy, with a scale of 0 = not associated with any boundary, 1 = at a biozone boundary, 2 = at a stage boundary, 3 = at an epoch boundary, 4 = at a period boundary, and 5 = at an era boundary. Our criterion for association with a boundary is rather liberal.

The final column of Table 2 lists a few notes, mainly earlier names of the events, and numbered references.

Isotopic Events in Carbonate Carbon ($\partial^{13}C_{carb}$)

Steady-State Levels of $\partial^{13}C$

The dominant mechanism that determines the general level of $\partial^{13}C$ is photosynthetic reduction of oxidized carbon to organic carbon, either in the photic zone of the ocean or on land. Explicitly the $\partial^{13}C$ of the remaining marine bicarbonate depends on the (net) fraction of carbon reduced to organic carbon and buried in sediments, and on how much this buried carbon differs in its isotopic composition from the oxidized carbon--the isotopic fractionation, $\Delta^{13}C = \partial^{13}C_{org} - \partial^{13}C_{carb}$, of the photosynthetic reduction process. The isotopic fractionation depends on the photosynthetic "pathway" that is operative, on whether the photosynthesis takes place in the marine or non-marine (air) milieu, on the concentration of dissolved inorganic carbon (or the partial pressure of CO_2), and on the temperature (e.g. Rau et al., 1989). Because of these factors, global Δ can shift with time. The marine bicarbonate is "sampled" by $CaCO_3$ crystallization in marine fossils and limestones, with a fractionation of only 1 or 2‰, virtually independent of temperature. Most photosynthesized organic carbon is relatively light ($\partial^{13}C$ -20 to -30‰); a nominal mean value is $\partial^{13}C = -25$‰. Much of the photosynthesis takes place in the surface waters of the ocean (100 m), and to the extent that the generated organic carbon is buried in the sediment (in either the deep or the shallow ocean) $\partial^{13}C_{carb}$ of the surface zone becomes enriched in ^{13}C--it becomes slightly positive. This process, diagrammed in the outer loop of Fig. 3, might be called the "biological dump". The level recorded in the main shallow-water carbonates under long-term steady-state conditions is thus a measure of (net) worldwide carbon burial. It will be at differing levels at different geological times because of differing oceanic conditions that favor burial (anoxic bottom conditions, fast sedimentation, high productivity) instead of re-oxidation by dissolved oxygen (Fig. 3). It is buried in the sedimentary section, to complete its cycle only after uplift and erosion or re-mobilization by metamorphism or volcanism (including re-oxidation). Secular

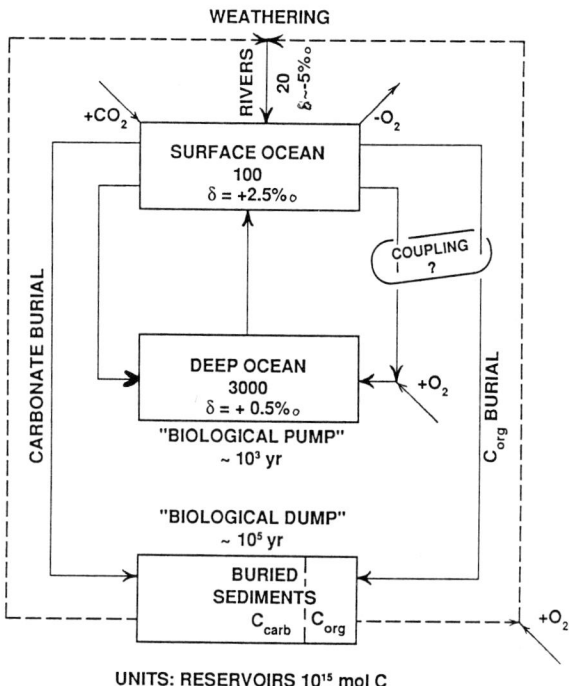

Fig. 3. The exogenic cycles of carbon. The steady-state level of $\partial^{13}C$ in the surface ocean is maintained by the relative burial rates of carbonate and organic carbon--the "biological dump" shown in the outer circuit. Most of the new productivity of C_{org} in the surface ocean is circulated without burial in the "biological pump" of the inner circuit. The pump lowers the isotope level of the deep ocean relative to the surface, maintaining a gradient between surface and deep in proportion to the new productivity. Modified after Kump, 1991

change in this cycle is a consequence of shifts in the proportions of net fluxes of C_{org} and C_{carb}: inputs by rivers less outputs to burial.

The Surface-To-Deep Gradient, and Transient Events

The above description of marine carbon isotope events assumes, as stated, that for any time-point on the isotopic age curve, the carbon isotope system is at steady state, that is, that the inputs and outputs to the oceanic system are balanced. Consequently the level of $\partial^{13}C$ is at a steady value depending on the proportion of carbon buried as organic carbon. To reach this steady state takes several lifetimes of carbon flowing through the oceanic system; a single residence time of carbon in the present (whole) ocean is about 55 ka

(Table 1). In most carbonate sections such a time lapse would not correspond to more than a few tens of centimeters of stratigraphic thickness, and might ordinarily be overlooked or ignored. However, transient changes on a short time scale may contain useful information about the internal workings of the marine carbon cycle.

Consider the inner loop of Fig. 3, called the "biological pump" (Berger and Vincent, 1986; Kump, 1991). This subsystem is independent of the biological dump that dumps organic carbon into the sediment and so sets the steady-state level of $\partial^{13}C$ in the surface waters and in the shelf limestones derived from them. In the biological pump a much larger flux of organic carbon falls out of the photic zone into the deep sea, but instead of being buried it is metabolized or otherwise oxidized by dissolved oxygen, and returned to the surface, to be recycled again and again (Fig. 3). This fast processing generates a surface-to-deep gradient of $\partial^{13}C$, with the $\partial^{13}C$ of the deep-sea waters depressed relative to that of the surface ocean by the addition of recently oxidized organic carbon. The separation of surface and deep $\partial^{13}C$, typically 1 or 2‰, is a measure of photosynthetic "new productivity"--that is, organic production that escaped re-oxidation in the surface layer but was re-oxidized in the deep. In steady state the $\partial^{13}C$ of the surface waters is established by the river input and burial of organic carbon, hence the level of $\partial^{13}C$ of the deep water is depressed below that of the surface water by the amount of the surface-to-deep gradient, as shown in the lower part of Fig. 4 ("earlier steady state"). This can be detected by comparing $\partial^{13}C$ of pelagic (surface, floating) and benthic (deep, sedimentary) fossils picked from the same samples, if present. A minimum measure of the relative activity of the biological pump is also revealed by transients in the surface isotope age curve, as explained in a later paragraph.

Now consider the transient return of both surface and deep profiles to new steady-state levels, caused by sharp changes in either the biological dump or the biological pump, or both. The middle part of Fig. 4a displays a transient consequent to decreasing the biological dump. The change to the new lower steady state level of $\partial^{13}C$ is throttled by the limited change of input or output flux of organic carbon, so the decrease drops gradually along the asymptotic exponential curve illustrated, with a characteristic time of carbon throughput, say 55 ka. If the biological pump is unchanged through this same time interval, its fast circulation and short time constant (residence time of advecting re-oxidized carbon about 35 years) allow it to keep the surface-to-deep gradient of $\partial^{13}C$ at a constant difference, tracking the surface change (Fig. 4a). If, on the other hand, the biological pump partially or fully stops (while the feed to the dump remains constant), the gradient is decreased, bringing the surface $\partial^{13}C$ temporarily down, until the slow inflow gradually (time constant 55 ka, Table 1) is allowed to catch up with the new situation (central part of Fig. 4d). If both biological dump and biological pump are decreased at the same time (perhaps owing to connected causes), typical results are illustrated in Fig. 4b (change in dump greater than in pump) and Fig. 4c (change in pump greater than in dump). Curves for stepped <u>increases</u> in either circuit are analogous.

If transients can be detected by close sampling, and recognized by their characteristic form and estimated time constant, they have potential for separating the effects of changing burial of organic carbon, from those of changing productivity. This ploy helped untangle the complex relations in the isotope age curve at the Cretaceous/Tertiary boundary (Zachos et al., 1989; Kump, 1991). It helps to have both surface and deep $\partial^{13}C$ age curves; but even without the pelagic data, note that the transient drop on the surface age curve (Figs. 4b, 4c and 4d) is still a measure of the drop in productivity. Thus the two or three transient minima in the earliest Triassic of Austria (Holser et al., 1989) bear witness to corresponding catastrophic drops in productivity, superimposed on a more radical long-term drop in rate of organic burial. The potential of analysis and separation of steady-state and transient events in isotope age curves is only beginning to be recognized.

Although the two circuits conjointly operate to govern the isotope levels, they participate quite differently in the geochemical cycles of CO_2 and O_2. The biological pump circuit is substantially balanced in O_2, taking up as much in the deep as it produces in the surface (Fig. 3). In contrast, any changes in the proportion of C_{org} storage by the biological dump will have to add or lose O_2 to or from the atmosphere or elsewhere in the oxygen cycle. Conversely, the level of CO_2 that is distributed between the atmosphere plus the oceans, is also dependent on the biological dump (Fig. 3). But a full realization of the transient variations of pCO_2 and pO_2 requires more complex dynamic modelling (Caldeira et al., 1990; Caldeira, 1991).

One may make a plausible thesis that the two circuits act together, that generally a more or less constant proportion of C_{org} reaching the deep sea actually escapes re-oxidation and is buried in the

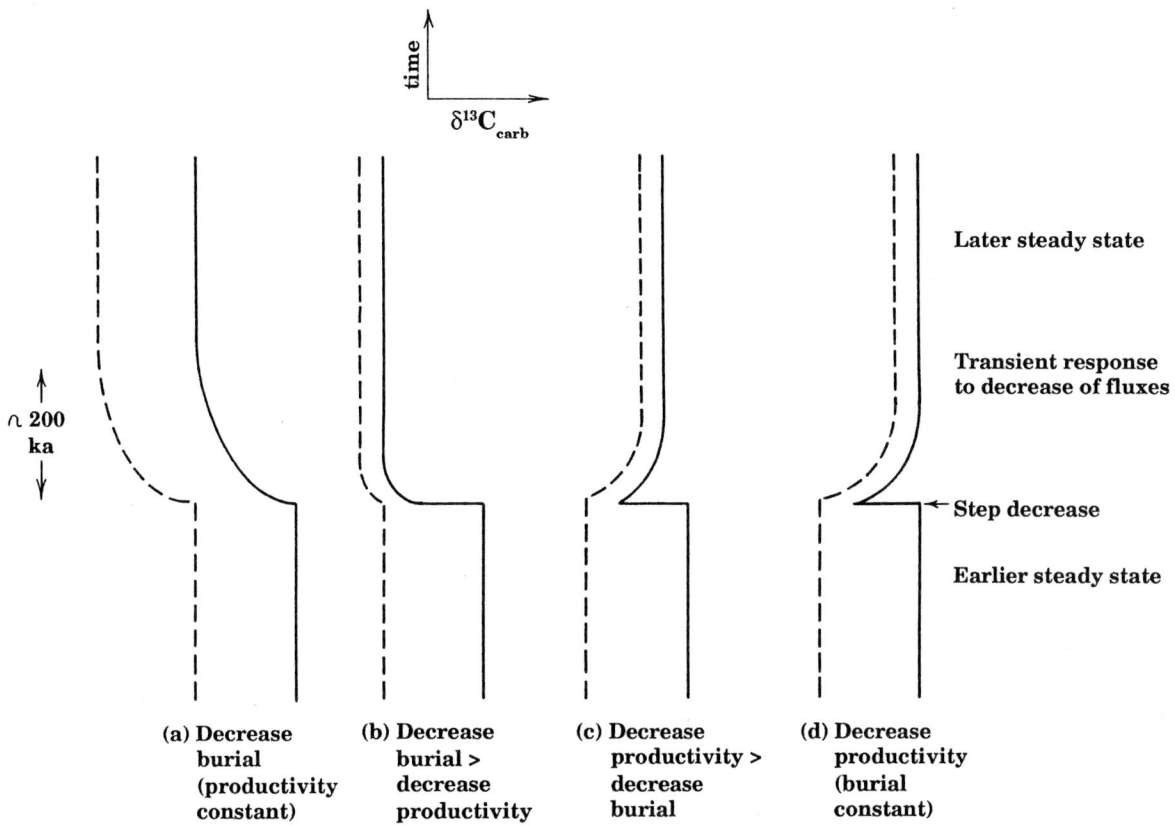

Fig. 4. Transients in the carbon isotope system, illustrated for four differing relations of relative importance of the biological pump and the biological dump. The solid line indicates signals seen in the surface ocean upon imposition of a step decrease of burial and/or productivity. The dashed line represents the consequent track of the deep ocean. Step increases of the carbon fluxes will generate comparable transients but of opposite sign (not shown)

sediments. If carbon burial were thus coupled to carbon production (coupled loops suggested in Fig. 3), we would expect the carbon isotope events to partake proportionally of both mechanisms. However, the record from the K/T boundary event suggests that at least in that case the two parameters are substantially decoupled (Kump, 1991). Involvement of terrestrial organic carbon in the burial cycle but not in the biological pump cycle (Caldeira, 1991), may contribute to its lack of coordination with deep-sea burial, particularly in the forest-rich post-Silurian times. There is need for more deep-sea records of complex carbon-isotope events to be studied and modelled in detail.

Departures from Primary Equilibrium Values

The above models of variations in $\partial^{13}C$ were simplified by assuming equivalence of the precipitated carbonate carbon and the sea-water bicarbonate from which it was derived. A variety of particular situations may distort this simple relation (e.g., Rush and Chafetz, 1990; Grant, 1992; Marshall, 1992; Grossman, 1994): (1) "Vital effects"--a particular fauna may precipitate its carbonate exoskeleton either out of equilibrium, or in equilibrium with a local microenvironment of differing isotope ratio that it has established at the crystallization site. (2) Instead of the calcite specified by equilibrium, a particular fauna may precipitate aragonite or magnesian calcite. Either of these is essentially unstable, and in the process of stabilization to low-Mg calcite, isotope ratios may be altered, sometimes in carbon but especially in oxygen. (3) Diagenesis of the original limestone may alter isotope ratios, by replacement of primary mineralogy or by crystallization of late carbonate as cement filling porosity, shell cavities, or vein fractures. (4) Diagenetic alteration of isotope ratios may be exacerbated at horizons subject to long-term circulation of meteoric water (typically lower in $\partial^{13}C$ derived from soil gas, and lower in $\partial^{18}O$ derived from precipitation), such as soil horizons, or vadose-phreatic or fresh-salt water boundaries, or marine hard grounds (e.g., Allan and Matthews, 1982; Rush and Chafetz, 1990). But these effects (4) ordinarily do not penetrate more than a meter or two into the subjacent formation.

These distortions of the sea-water signal may be factored out by several tactics, as discussed in the following section.

Using Carbon-Isotope Age Curves

The distortions, discussed above, of the steady-state equilibrium isotope age curves, are of more or less consequence depending on the use to which the curves are to be put. If the purpose is chemostratigraphic correlation, then it does not matter whether the curve represents a steady state of the carbon cycle, or also includes transient responses to variations in productivity in the surface ocean, because both signals should be global. In fact, the sharp transient drop in $\partial^{13}C$ that characterizes most mass extinction events (Magaritz, 1989), owing to a drop in new productivity as explained in a previous section, adds an additional characteristic shift to the correlation data set.

Vital effects, caused by disequilibrium or kinetic variations in shell deposition of different genera or species, may shift the isotope age curve by one or more per mil. It will be prudent to sample the same taxon when practical to do so, or to calibrate by comparing different species from the same sample. But vital effects generally do not change the shape of the curve, so useful correlations can often be made with undifferentiated shelly material or with whole-rock samples.

Diagenesis, metamorphism or other post-depositional alteration can be a problem for stratigraphic correlation with carbon-isotope data, even with the low water/rock ratio that generally prevails in carbonate terranes. By their very nature, diagenetic signals are usually local, and consequently can confound long-distance correlations. The following are some bases for screening out samples that may have been affected strongly by diagenesis (Kaufman and Knoll, in press).

1. Avoid sampling evident veins, vugs, paleosols, hardgrounds, caliche or reefal facies.
2. Be wary of samples with less than 50% carbonates, or more than 1% organic carbon.
3. Avoid materials or fauna with low preservation potential, as outlined in Fig. 5 (Marshall, 1992).
4. Test for diagenesis by chemical analysis of a sample aliquot dissolved in acetic acid: most diagenetic recrystallization will raise the traces of Mn (and Fe) and decrease Sr (and Na) content of the calcite (Brand and Veizer, 1980, 1981; Veizer, 1983; Banner and Hanson, 1990). A cutoff at Mn/Sr > 2 (by weight) is suggested (Veizer, 1983).
5. Alternatively test for diagenesis by cathodoluminescence, which is activated by Mn^{2+} in carbonate. The Mn cathodoluminescence is quenched by Fe^{2+}, which can be checked for by chemical analysis or ferrocyanide stain.
6. Sections of shell free of diagenesis may be recognized (and microsampled) by the presence of nacreous surfaces and primary crystalline fabric, by freedom from cathodoluminescence, or by low Mn/Sr determined with microdrill or laser spot sampling.
7. It is also becoming evident by experience that real global shifts in carbon-isotope levels in marine carbonate are approximately paralleled in primary organic matter (e.g., kerogen) in coeval rocks (Knoll et al., 1986; Magaritz, Krishnamurthy and Holser, 1992; Morante et al., 1994; Kaufman and Knoll, in press). We know of no secondary mechanism, including all forms of diagenesis, that will alter carbon isotopic levels in carbonate and inorganic carbon by the same amount or even in the same direction. Although the isotopic analysis of organic carbon is somewhat more difficult than the routine analysis of carbonate, it is a potentially powerful tool for the verification of isotope shifts.
8. Whatever material has been sampled, the ultimate test of global character of the isotopic trends and events is a comparison of stratigraphically equivalent sections. These may be simply sampling of neighboring outcrops along equivalent bedding planes, sampling of well-correlated sections in the same sedimentary basin, or stacking of interbasin profiles (Baud et al., 1989).

If the objective of the carbon isotope profiling is an evaluation or modelling of variations in global burial of organic carbon, all of the above constraints and limitations still apply, but in addition it will be useful to minimize the effects of productivity transients in the carbon isotope record, because they do not represent burial of carbon. As described in a previous section (see also Berger and Vincent, 1986; Kump, 1991), a drop (or rise) in productivity of the surface ocean will show up within decades as a transient drop (or rise) in the surface isotopic record,

ISOTOPE SIGNAL PRESERVATION POTENTIAL		Components		Bulk Sediments
		Skeletal	Non skeletal	
HIGH Good chance of preservation of Carbon and Oxygen values	Pristine aragonite	Molluscs	Marine cements	Pelagic sediments particularly Coccolith Oozes
	Pristine Low magnesium calcite fossils, grains and cements	Brachiopods Belemnites Foraminifera Bivalves	Marine cements LMC Ooids	
MODERATE Carbon values may be preserved, oxygen values commonly altered.	Secondary calcites (stabilised in relatively closed system with low W/R ratio)	Molluscs, Foraminifera Corals, Echinoderms Calcareous algae	Marine cements Ooids, peloids intraclasts	Some Micrites Some shallow water limestones Some dolomites
LOW Carbon and oxygen values likely to have been altered	Secondary calcites (stabilised or cemented in relatively open systems with high W/R ratio)	Limestones and components altered by near surface meteoric diagnosis or intensive cementation or recrystallisation during burial. Many dolomites		

Fig. 5. Preservation potential of various geological materials for carbon and oxygen isotopes (Marshall, 1992)

by an amount up to the isotopic surface-to-deep gradient previously (or subsequently) established. The transient will be relaxed, and the desired long-term relative level of the isotope record re-established, when the carbon cycle has more slowly circulated incoming river water through the deep ocean. The time constant for this process, of about 55 ka, still represents a fast transient relative to the longer term changes in organic carbon burial. The record can be cleaned of transients by a low-pass filter of appropriate cutoff (say 100 ka) (Williams et al., 1988, chapter 13). Processing with a high-pass filter has potential to measure changes in productivity, although this possibility has yet to be tested.

Isotopic Events in Carbonate Oxygen ($\partial^{18}O_{carb}$)

Steady-State Levels of $\partial^{18}O$

The controls on oxygen isotopes in the carbonate system are categorically different from those for carbon (Fallick and Hamilton, 1989; Hudson and Anderson, 1989). The most important fractionation of oxygen occurs during crystallization of the carbonate, and the resulting $\partial^{18}O$ of the carbonate crystal is strongly dependent on the temperature, as well as on the isotopic composition of the aqueous medium in which the crystallization takes place. If the carbonate is calcite, experiment and theory relate the temperature of equilibrium crystallization, t, as a function of the fractionation $\Delta_c = \partial^{18}O_{calc} - \partial^{18}O_{aq}$, where the reference standards are PDB for calcite and SMOW for water:

$$t°C = 16.0 - 4.14\,\Delta_c + 0.13\,\Delta_c^2 \qquad (1)$$

The curvature of this equation is small, and as a good approximation a rise in $\partial^{18}O_{calc}$ of 1‰ corresponds linearly to a drop in temperature of 4°C. Similar equations with essentially the same slope apply to the crystallization of other minerals, such as aragonite or apatite (e.g., Hudson and Anderson, 1989). Because of the inverse relation of the change of $\partial^{18}O$ to change of temperature, some authors in graphing profiles of $\partial^{18}O$ reverse the direction of its axis so that the equivalent temperature increases to the right or upward, but there is no consistency of practice.

In Equation (1) the value of $^{18}O_c$ (PDB) is related not only to the temperature but also to $\partial^{18}O_w$ (SMOW) of the water. By definition present sea water $\partial^{18}O_w = 0‰$ (SMOW), but in geological perspective this value has been temporarily raised by the crystallization of an appreciable fraction of the water mass as glacial ice that is very light in $\partial^{18}O$ owing to evaporative cycling in the atmosphere as it moved upward and poleward. Consequently, for an ice-free world (most of pre-Miocene Phanerozoic time) the oceanic level of $\partial^{18}O_w$ is lighter, and for a fully glacial period (Pleistocene, Carboniferous, or late Ordovician) is heavier than the present reference ocean, with a glacial to interglacial shift ranging from -1 to -2‰. Note that the sign of the glacial-interglacial shift is the same whether it is ascribed to a lighter water (melted glacial ice) or to a warmer climate. Although it has been difficult to quantitatively untangle these two effects, comparisons of coeval benthic vs. pelagic and low- vs. high-latitude profiles are consistent with a conclusion that most of the glacial shift is due to ice storage rather than to a colder climate.

Other processes of evaporation-condensation also can distort the ideal SMOW, and consequently can also distort the temperatures that have been derived by assuming a reference $\partial^{18}O = 0$ for conditions similar to the present, or a suitably negative level for ice-free conditions. For example, lakes and rivers are generally appreciably negative, say $\partial^{18}O \approx -10‰$ SMOW, for the same reason that glacial ice is negative. Consequently brackish mixtures, such as those that may result from river input in a coastal zone, are reflected in $\partial^{18}O$ a few per mil light. Conversely, marginal seas with excess evaporation may be a few per mil heavy because they store the liquid residue of the evaporative process. These two related shifts resulting from evaporation lead to a general correlation between salinity and $\partial^{18}O_w$, and consequently between salinity and $\partial^{18}O_c$ crystallized from such water; the slope of the correlation depends on $\partial^{18}O_w$.

The mean $\partial^{18}O_w$ of all water in the hydrological cycle is apparently set by a dynamic isotope exchange with the solid Earth's crust, and this is a baseline for all of the possible shifts discussed above. Apparently the principal exchange takes place by hydrothermal convection of sea water beneath mid-ocean ridges, which raises the circulating sea water to about 350°C, and exchanges oxygen isotopes with silicate minerals $\partial^{18}O_{silicate} \approx +6‰$: Muehlenbachs, 1986; Cathles, 1990). Gradual shifts in the baseline value for sea water that prevailed during the Paleozoic, have been controversial as to whether they represent positive shifts in the general level of $\partial^{18}O_w$ established by this hydrothermal circulation, or a long-term decrease in ocean temperature (Hudson and Anderson, 1989; Holland, 1984). But for application to stratigraphic correlation this controversy can be safely ignored, because at the most any such changes in the mean level of $\partial^{18}O_w$ should be very slow (50-250 Ma: Muehlenbachs, 1986).

Transient Events

In the oxygen isotope system all of the main controls on steady-state $\partial^{18}O_c$--temperature, ice storage, salinity--are each constitutionally capable of fast short-term shifts to new steady states. The present mixing time of the world ocean--a thousand years--is short enough that changes of ice storage or mean surface sea temperature will appear in carbonates worldwide essentially simultaneously. Some of these events will be global: isotope shifts due to ice storage will certainly be global; temperature shifts will have a mean global component with local changes superimposed; isotope salinity shifts due to local river input are likely to be limited to a basin.

Departures from Primary Equilibrium Values

The oxygen isotope system is subject to all of the alterations described above for the carbon system: vital effects, mineralogy, and diagenesis. These changes in $\partial^{18}O_c$ are generally greater than those seen previously for $\partial^{13}C_c$. This is because the concentration of oxygen (as liquid water) in altering solutions is generally a couple of orders of magnitude greater than that of carbon (as bicarbonate ions in solution). Consequently the altering water/rock ratio required to attain a given degree of alteration is very much lower for oxygen isotopes than for carbon isotopes (Banner and Hanson, 1990). Furthermore, there is a tendency in many natural situations for the carbon system to be locally dominated by the surrounding lime mud or limestone having (a nearly) primary $\partial^{13}C$, whereas with time the water in the porosity is likely to have been displaced by a solution of differing $\partial^{18}O$ (Marshall, 1992; Banner and Hansen, 1990).

Using Oxygen-Isotope Age Curves

Oxygen isotope records have been most successfully applied to the stratigraphic correlation of lime muds, particularly in the Quaternary, when a strong (1 to 2‰) ice-storage signal is developed (Crowley and

North, 1991). During non-glacial (most) times, however, characteristic event signals are less often recognized than those from the carbon record (Table 2).

Oxygen isotope data is also applied to determining the temperature of sedimentary deposition, and as an additional refinement the comparison of $\partial^{18}O$ in planktonic and benthic fauna gives the surface-to-bottom temperature gradient, which in turn is applied to modelling ocean (Railsback et al., 1989) and atmospheric circulation.

Isotopic Events in Sulfate Sulfur ($\partial^{34}S$)

Although the sulfur system was the first in which time-variations were verified (Holser and Kaplan, 1966), and these variations were very large (Fig. 2), the age curve for $\partial^{34}S$ has not had the detailed development that we see in carbon or even in strontium. This lack may be ascribed mainly to the fact that the main source of measurements has been sulfate minerals of evaporite facies (Fig. 1). The evaporites in which these sulfate minerals occur have a spotty stratigraphic record, and when they do occur a lack of fossils makes stratigraphic calibration of the age curve imprecise. On the other hand, this lack of precision in biostratigraphy makes isotope stratigraphy more uniquely useful.

The main control on the sedimentary sulfur cycle is the reduction of sulfate to sulfide by sulfur-reducing bacteria, which occurs mainly in reducing muds on the sea floor (Holser et al., 1988). The fractionation factor from reduction of the sulfur is about -45‰. Organic carbon is consumed in the proccess, producing CO_2. The fraction of sulfur that is finally fixed in the formation (eventually as pyrite) modifies the isotope level in the remaining sulfate of the ocean, and the ocean is sampled by inflow and crystallization as calcium sulfate in a bordering evaporite basin. Whereas the marine sulfate is globally mixed and therefore gives a $\partial^{34}S$ characteristic of its age, the corresponding sedimentary pyrite generally displays a high variability of its primary $\partial^{34}S$. Consequently isotopic analysis of sedimentary sulfide has not been widely applied to isotope stratigraphy (Strauss et al., 1992; Bottomley et al., 1992). $\partial^{18}O$ in sulfate constitutes, rather surprisingly, an isotope system completely independent of $\partial^{18}O$ in carbonate or in water, and of $\partial^{34}S$ in sulfate (Holser et al., 1979), and the fundamental processes underlying the secular variations of $\partial^{18}O_{sulfate}$ are not understood (Richardson and Hanson, 1991). The following discussion will be concerned only with $\partial^{34}S$ of sulfate.

The residence time of sulfate in the ocean is long, perhaps 45 Ma for the whole ocean (Table 1), and consequently the system is very resistant to isotopic change. Sharper changes are possible if only the surface ocean is involved, in an initial transient. That must account for the sharp positive excursion of about 15‰ in the $^{34}SSmi$ event (Table 2), which is found worldwide. A similar rise occurs in the Neoproterozoic record of $\partial^{34}S$ (Strauss, 1993), but is not yet clearly enough defined to be included in Table 2.

Marine evaporites, where they do occur, are particularly suited to sampling the world surface ocean. Evaporite rocks, that are of sufficient thickness (typically tens of meters) to gain the attention of the field geologist, require more than a casual input of sea water for their formation--indeed the supply of sulfate-bearing sea water must have filled the basin many times over, either continuously or intermittently (Holser, 1979a, p. 264). Consequently the sulfate rock is likely to be a good solid sample of the oceanic sulfate. However, the site of evaporite accumulation requires near isolation, typically marginal to a continent. It requires high aridity to attain crystallization concentration, typically in medium latitudes. While marine input dominates such a paleogeographic milieu, the landward side of the basin is susceptible to mixing with surface or ground waters of continental origin. These waters draining the continental hinterland may have gained sulfate from the erosion of older evaporites, which would have a $\partial^{34}S$ dependent on their age (Fig. 3). Or their sulfate may have been oxidized from pyrite or other sulfide minerals of shaley, igneous or hydrothermal rocks, any of which would be likely to have light $\partial^{34}S$, in the range -20 to +5‰.

Several criteria are available to verify the marine character of an evaporite rock section: the mineralogy and its sequence should be typical of marine origin:

calcite/dolomite → gypsum/anhydrite → halite → potash minerals

The bromide content of halite should be in the typical marine range of 50-150 ppm (Holser, 1979b).

Evaporite rocks are subject to alteration: calcium sulfate may go through one or more cycles gypsum ⇆ anhydrite, and salt rock may dissolve to leave a residue of its primary anhydrite. But the situation is so dominated by the original sulfate and its $\partial^{34}S$, that the primary marine value usually prevails. Even secondary solution of marine evaporite sulfate, and its subsequent re-deposition as cement in a later sandstone, has been shown to retain its original isotopic signature (Holser, 1979b, p. 333).

Isotopic Events in Strontium ($^{87}Sr/^{86}Sr$)

Nomenclature

The nomenclature for strontium isotopes has not been stabilized (Veizer, 1989). The most common statement of analyses (and the one used here) is a direct ratio $^{87}Sr/^{86}Sr$, usually normalized to an assumed value (which must be stated for interlaboratory comparisons) for modern sea water (0.70920), NBS/SRM-987 (0.71022), or Eimer and Amend $SrCO_3$ (0.70800) (Elderfield, 1986). Alternatively some authors use a "del" notation, analogous to that used (e.g., above) for H, C, N, O and S, expressing the deviation from a standard such as sea water, or modern foraminifera (0.709168: Dia et al., 1992), for example:

$$\partial^{87}Sr = [\{(^{87}Sr/^{86}Sr)_{sample}/(^{87}Sr/^{86}Sr)_{standard}\} -1] \times 10^3 \text{ (or } 10^5, \text{ or } 10^6)$$

In another alternative, designated $\epsilon^{87}Sr/^{86}Sr$, the deviation of the sample is relative to a chondritic "uniform reservoir" (0.7045), x 10^4. Mass spectrometry of the strontium ratio attains a precision ranging in various dates and laboratories from 5 to 20 x 10^{-6}.

Steady-State Levels of $^{87}Sr/^{86}Sr$

The geochemistry of strontium in the exogenic cycle has been reviewed by Veizer (1989), Palmer and Edmond (1989) and McArthur (1991). The system is fundamentally radiogenic, from the decay of ^{87}Rb through its long half-life of 48.8 Ga. But the half-life is so long, and the relative concentration of Rb/Sr so low, that new radiogenic ^{87}Sr is added to the sedimentary cycle very slowly. Consequently Sr acts much like a stable isotope system in the exogenic cycle. Furthermore, the mass ratio of radiogenic to non-radiogenic isotopes, 87/86 = 1.012 is so low (cf 34/32 = 1.063 for S) that Sr is not visibly fractionated by either equilibrium or kinetic processes.

The residence time of strontium in modern sea water is long enough (Table 1) that Sr in sea water is expected to be constant in both concentration of 7.85 ppm, and in isotope ratio of 0.70920. The main fluxes that establish these levels are (Fig. 1): (1) river inflow, especially those dominated by erosion of old metamorphic and igneous rocks, with heavy $^{87}Sr/^{86}Sr$, and (2) "exchange" of Sr with young basalts, with light $^{87}Sr/^{86}Sr$, in sea-water hydrothermal systems beneath mid-ocean ridges [MOR]. Much of the river inflow is loaded with re-eroded carbonate sediments, and to that extent the sea-water isotope ratio is buffered against potentially shifting its isotopic level by varying the dynamic balance of the riverine isotopic level and basalt interaction.

Strontium is taken up from sea water in the crystallization of carbonate (and phosphate) minerals. The Sr^{2+} ion in solution proportionates by a distribution coefficient ($\underline{D}$ <1) into a firmly held solid solution in calcite, so it requires complete dissolution to be altered in either concentration or isotope ratio (Pingitore et al., 1992). There is no fractionation in this crystallization, so the strontian calcite is a direct sample of the isotopic ratio of the sea-water solution. However, if the original precipitate or shell material is aragonite or high-Mg calcite, both the concentration and isotopic ratio of the strontium are at risk for change during re-crystallization, as described below.

Transient Events

One might expect that abrupt shifts in the steady-state balance of the strontium system (Fig. 1) would generate a transient response in the marine record, analogous to that for carbon (Fig. 4). However, in view of the long residence time of Sr in the whole ocean (Table 1), its response will be considerably diminished. Richter and Turekian (1993) use a simple model to demonstrate that the 800 ka cycles detected in the Quaternary deep sea by Capo and DePaolo (1990) could be actual global signals, but that the glacial-interglacial (100 ka) cycles claimed by Dia et al. (1992) are an unlikely event for any reasonable forcing (Henderson et al., 1993). The observed Pleistocene oscillations are of the same order of magnitude as the controversial positive excursion of $^{87}Sr/^{86}Sr$ at the K/T boundary (Table 2: event $^{87}SrK/T$) (e.g., Martin and Macdougall, 1991; Javoy and Courtillot, 1989). A very large and sharp positive excursion of $^{87}Sr/^{86}Sr$ was also tentatively identified in the Pennsylvanian (Brand, 1991). If the observed variations are not artifacts, they may be limited to the surface ocean or some other semi-isolated reservoir. Transient events may appear first and most strongly in the surface ocean, from a shift of riverine input, before being diluted in the deep ocean, while the deep ocean, with its greater mass, can only shift slowly in its isotopic level.

Thus, the interpretation of transient variations of the strontium-isotope age curve remains ambiguous.

Departures from Primary Equilibrium Values

Although strontium is firmly incorporated in marine calcite without isotopic fractionation, both the level of concentration and the isotopic ratio are at risk for several kinds of alteration by solution and recrystallization (Veizer, 1989; Denison, Koepnick, Fletcher et al., 1994). The strontium isotope record in carbonates is subject to some of the same diagenetic effects as the carbon system, described above. Replacement of original shell material, that may have been aragonite or high-Mg calcite, by stable low-Mg calcite, will shift $^{87}Sr/^{86}Sr$ in the direction of the permeating solutions. Most such solutions are of lower Sr^{2+} content than the primary crystallizing sea water, so its concentration level in the diagenetically altered shell or matrix will be lowered, along with a corresponding rise of Mn^{2+} described in the section on $\partial^{13}C$. The shift in isotope ratio $^{87}Sr/^{86}Sr$ that accompanies this diagenesis depends on the source of the strontium in the altering solution. There is some tendency for the strontium in pore waters to be dominated by the original permeating sea water, as was the case for $\partial^{13}C_{carb}$ discussed above (Banner and Hanson, 1990; Veizer, 1989). Common sources of non-primary solutions are micas and clay minerals, which often are rich in Rb and hence in radioactive ^{87}Rb that would have generated ^{87}Sr while resident in the rock or solution. As a consequence there is a tendency for altered Sr levels to be more radiogenic, that is, higher in $^{87}Sr/^{86}Sr$ (Brand, 1991). For this reason Burke et al. (1982) picked the levels for their standard strontium-isotope age curve from among the lower values at each date. However, permeating solutions moving upward through deep-sea sediments may bear a light ratio of $^{87}Sr/^{86}Sr$ (≈ 0.705) from young basalts underlying the sediments. Also, the altering solutions would generally be in contact with newly sedimented carbonate muds, and would therefore likely be charged with primary or re-dissolved Sr of a slightly earlier age, which would have $^{87}Sr/^{86}Sr$ that was either lighter, if the time was one of increasing $^{87}Sr/^{86}Sr$ along the sea water age curve (as in the Cenozoic), or vice-versa (Brand, 1991).

Using Strontium-Isotope Age Curves

The strategies for selecting samples and evaluating analytical results so as to avoid diagenesis are mostly parallel to those described above for carbon: avoid mineralogy or fauna with a low preservation potential (Fig. 5), and test for diagenesis using indicators such as decreased Sr and increased Mn (or cathodoluminescence) levels (Denison, Koepnick, Fletcher et al., 1994). Brand (1991), and others, emphasize the usefulness of the scanning electron microscope for detecting primary textures of aragonite in molluscs and of low-Mg calcite in brachiopods. Strontium in solid solution in apatite of conodonts is also a useful measure of sea water isotope ratio (Burke et al., 1982), especially in those with low color alteration index (CAI 1 to 3: Bertram et al., 1992). The age curve in Fig. 2 is based in part on conodont data. However, recent research by Kürschner, Becker et al. (1992) and Kürschner, Ebneth et al. (1992) finds $^{87}Sr/^{86}Sr$ in conodont apatite shifted to more radiogenic Sr than coeval unaltered brachiopod calcite. Surprisingly, this alteration is a consistent increase of about 0.0001 in the material studied, so that strontium isotopic events could be recognized in spite of the alteration; we do not expect this to always be the case. Deep-sea authigenic barite also has potential for a reliable record of global $^{87}Sr/^{86}Sr$ (Paytan et al., 1993).

An important additional strategy in the selection of strontium isotope samples, beyond those that already apply for carbon isotopes, is to minimize the chance of contamination by radiogenic strontium from local sources of rubidium. Any rubidium is likely to be contained in silicate minerals of the acid-insoluble residue, not in the carbonate itself. As a first step, analyses are confined to carbonate-rich samples. Then the sample is treated gently with acetic acid to dissolve the carbonate and any rubidium loosely held on silicates. If the ratio of Rb^+/Sr^{2+} in the leach is high enough to account for any significant increase of ^{87}Sr, the sample is screened out.

Conclusions

1. More than 60 distinct shifts in the marine isotopic record of carbon, oxygen, sulfur and strontium during the Phanerozoic and Proterozoic are potential markers for chemostratigraphic correlation.
2. The precision and resolution of chemostratigraphic correlations are of the same order as for biostratigraphic correlations, and sometimes better (< 0.5 Ma).
3. Carbon isotope events are the most numerous events presently recognized, and they include both the long-term effects of burial of organic carbon, and short-term effects of varying productivity. An analogous dichotomy may be present in the isotopic records of other elements, but has not yet been clearly recognized.
4. A true marine isotopic record is often distorted, particularly by diagenesis.
5. A variety of petrographic, chemical and taxonomic criteria are available to assist in screening out such unsuitable samples.

6. The ultimate test for the real global nature of an isotopic event is its verification in distinct and distant sedimentary basins, controlled by the best available biostratigraphic correlation.

Acknowledgements

The authors are indebted to many individuals for data or other information furnished to us before formal publication, especially: Werner Buggisch, Chen Jinshi, Lou Derry, Stefan Ebneth, Helmut Geldsetzer, E.L. Grossman, E.A. Hetherington, M.M. Joachimski, Alan Kaufman, R.B. Koepnick, W. Kürschner, R. Morante, Danuta Peryt, Ellen Thomas, Jan Veizer, Ken Wang and Fay Woodruff. We also thank Lou Derry for his helpful comments on an earlier version of the manuscript.

References

Alcala-Herrera, J.A., Grossman, E.L. and Gartner, S., 1992. Nannofossil diversity and equibility and fine-fraction $\partial^{13}C$ across the Cretaceous/Tertiary boundary at Walvis Ridge Leg 74, South Atlantic. Marine Micropaleo. 20, 77-88.

Allan, J.R. and Matthews, R.K., 1982. Isotope signatures associated with early meteoric diagenesis. Sedimentology 29, 797-817.

Asmeron, Y., Jacobsen, S.B., Knoll, A.H., Butterfield, J.J. and Swett, K., 1991. Strontium isotopic variations of Neoproterozoic seawater: Implications for crustal evolution. Geochim. Cosmochim. Acta 55, 2883-2894.

Baker, A.J. and Fallick, A.E., 1989. Evidence from Lewisian limestones for isotopically heavy carbon in two-thousand-million-year-old sea water. Nature 337, 352-354.

Banner, J.L. and Hanson, G.N., 1990. Calculation of simultaneous isotopic and trace element variations during water-rock interaction with applications to carbonate diagenesis. Geochim. Cosmochim. Acta 54, 3123-3127.

Baud, A., Magaritz, M. and Holser, W.T., 1989. Permian-Triassic of the Tethys: Carbon isotope studies. Geol. Rundsch. 78, 649-677.

Berger, W.H. and Vincent, E., 1986. Deep-sea carbonates: Reading the carbon-isotope signal. Geol. Rundsch. 75, 249-269.

Bertram, C.J., Elderfield, H., Aldridge, R.J. and Morris, S.C., 1992. $^{87}Sr/^{86}Sr$, $^{143}Nd/^{144}Nd$ and REEs in Silurian phosphatic fossils. Earth Planet. Sci. Lett. 113, 239-249.

Bottomley, D.J., Veizer, J., Nielsen, H. and Moczyclowska, J., 1992. Isotopic composition of disseminated sulfur in Precambrian sedimentary rocks. Geochim. Cosmochim. Acta 56, 3311-3322.

Bowring, S.A., Grozinger, J.P., Isachsen, C.E., Knoll, A.H., Pelechaty, S.M. and Kolosov, P., 1993. Calibrating rates of Early Cambrian evolution. Science 261, 1293-1298.

Brand, U., 1982. The oxygen and carbon isotope composition of Carboniferous fossil components: Sea-water effects. Sedimentology 29, 139-147.

Brand, U., 1989; Biogeochemistry of late Paleozoic North American brachiopods and secular variation of seawater composition. Biogeochemistry 7, 159-193.

Brand, U., 1991. Strontium isotope diagenesis of biogenic aragonite and low-Mg calcite. Geochim. Cosmochim. Acta 55, 505-513.

Brand, U. and Veizer, J., 1980. Chemical diagenesis of a multicomponent carbonate system-1: Trace elements. Jour. Sediment. Petrol. 50, 1219-1236.

Brand, U. and Veizer, J., 1981. Chemical diagenesis of a multicomponent carbonate system-2: Stable isotopes. Jour. Sediment. Petrol. 51, 987-997.

Brasier, M.D., 1990. Towards a carbon isotope stratigraphy of the Cambrian System: Potential of the Great Basin succession. Geol. Soc. [London] Spec. Publ. 70, 341-350.

Brasier, M.D., 1991. Nutrient flux and the evolutionary explosion across the Precambrian-Cambrian boundary interval. Hist. Geol. 5, 85-93.

Brasier, M D., 1992. Global ocean-atmosphere change across the Precambrian-Cambrian transition. Geol Mag. 129, 161-168.

Brasier, M.D., Magaritz, M., Corfield, R., Luo H., Wu X., Ouyang L., Jiang Z., Hambdi, B., He T. and Fraser, A.G., 1990. The carbon- and oxygen-isotope record of the Precambrian-Cambrian boundary interval in China and Iran and their correlation. Geol. Mag. 127, 319-332.

Brasier, M.D., Anderson, M.M. and Corfield, R.M., 1992. Oxygen and carbon isotope stratigraphy of early Cambrian carbonates in southeastern Newfoundland and England. Geol. Mag. 129, 265-279.

Brookins, D.G., 1988. Seawater $^{87}Sr/^{86}Sr$ for the Late Permian Delaware Basin evaporites (New Mexico, USA). Chem. Geol. 69, 516-519.

Buggisch, W., 1991. The global Frasnian-Famennian "Kellwasser Event". Geol. Rundsch. 80, 49-72.

Burke, W.H., Dennison, R.E., Hetherington, E.A.,

Koepnick, R.B., Nelson, H.F. and Otto, J.B., 1982. Variation in sea water $^{87}Sr/^{86}Sr$ throughout Phanerozoic time. Geology 10, 516-519.

Burns, S.J., Haudenschild, U. and Matter, A., 1994. The strontium isotopic composition of carbonates from the Late Precambrian carbonates (560-540 Ma) from Oman. Chem. Geol. Isotope Geosci. Sec. 111, 269-282.

Caldeira, C., 1991. Continental-pelagic carbonate partitioning and the global carbonate-silicate cycle. Geology 19, 204-206.

Caldeira, C., Rampino, M.R., Volk, T. and Zachos, J.C., 1990. Biogeochemical modeling at mass extinction boundaries: Atmospheric carbon dioxide and ocean alkalinity at the K/T boundary. In: Kauffman, E.G. and Walliser, O.H. (eds.), Extinction Events in Earth History. pp. 333-345. Springer, Berlin Heidelberg New York.

Capo, R.C. and DePaolo, D.J., 1990. Seawater strontium isotopic variations from 2.5 million years ago to the present. Science 249, 51-55.

Cathles, L.M., III, 1990. Scales and effects of fluid flow in the upper crust. Science 248, 323-329.

Cerling, T.E., Wang Y. and Quade, J., 1993. Expansion of C4 ecosystems as an indicator of global ecological change in the Late Miocene. Nature 361, 344-345.

Chen J., Chu X. and Shao M., 1986. Sulfur isotopes of the Triassic sea. Scient. Geol. Sinica, 1986 (4), 330-338.

Chen J.S., Chu X.-L., Shao M.-R. and Zhong H., 1991. Carbon isotope study of the Permian-Triassic boundary sequences in China. Chem. Geol. Isotope Geosci. Sec. 89, 239-251.

Chen J.S., Shao M.R., Huo W.G. and Yao Y.Y., 1984. Carbon isotopes of carbonate strata at Permian-Triassic boundary in Changxing, Zhejiang. Scient. Geol. Sinica, 1984 (1), 88-93.

Cita, M.B. and McKenzie, J.A., 1986. The terminal Miocene event. Amer. Geophys. Un. Geodynamics Ser. 15, 123-140.

Claypool, G.E., Holser, W.T., Kaplan, I.R., Sakai, H. and Zak, I., 1980. The age curves of sulfur and oxygen isotopes in marine sulfate and their mutual interpretation. Chem. Geol. 28, 199-260.

Corfield, R.M., Cartlidge, J.E., Premoli-Silva, I. and Housley, R.A., 1991. Oxygen and carbon isotope stratigraphy of the Palaeogene and Cretaceous limestones in the Bottaccione Gorge and the Contessa Highway sections, Umbria, Italy. Terra Nova 3, 414-422.

Corfield, R.M., Siveter, D.J., Cartlidge, J.E. and McKerroe, W.S., 1992. Carbon isotope excursion near the Wenlock-Ludlow (Silurian) boundary in the Anglo-Welsh area. Geology 20, 371-374.

Cortecci, B., Reyes, E., Berti, G. and Casati, P., 1981. Sulfur and oxygen isotopes in Italian marine sulfates of Permian and Triassic ages. Chem. Geol. 34, 65-79.

Crowley, T.J. and North, G.R., 1991. Paleoclimatology. 339 pp. Oxford University Press.

Denison, R.E., Koepnick, R.B., Burke, W.H., Hetherington, E.A. and Fletcher, A., 1994. Construction of the Mississippian, Pennsylvanian and Permian seawater $^{87}Sr/^{86}Sr$ curve. Chem. Geol. (Isotope Geosci. Sec.) 112, 145-167.

Denison, R.E., Koepnick, R.B., Fletcher, A., Howell, M.W. and Callaway, W.S., 1994. Criteria for the retention of original seawater $^{87}Sr/^{86}Sr$ in ancient shelf sediments. Chem. Geol. (Isotope Geosci. Sec.) 112, 131-143.

Derry, L.A. and Jacobsen, S.T., 1988. The Nd and Sr isotopic evolution of Proterozoic seawater. Geophys. Res. Lett. 15, 397-400.

Derry, L.A., Keto, L.S., Jacobsen, S.B., Knoll, A.H. and Swett, K., 1989. Strontium isotope variations in Upper Proterozoic carbonates from Svalbard and East Greenland. Geoch. Cosmoch. Acta 53, 2331-2339.

Derry, L.A., Kaufman, A.J. and Jacobsen, S.B., 1992. Sedimentary cycling and environmental change in the Late Proterozoic: Evidence from stable and radiogenic isotopes. Geochim. Cosmochim. Acta 51, 317-1389.

Dia, A.N., Cohen, A.S., O'Nions, R.R. and Shackleton, N.J., 1992. Seawater Sr isotope variation of the past 300 kyr and influence of global climate cycles. Nature 356, 786-788.

Donnelly, T.H., Shergold, J.H., Southgate, P.N. and Barnes, C.J., 1990. Events leading to global phosphogenesis around the Proterozoic/Cambrian boundary. Geol. Soc. [London] Spec. Publ. 52, 273-287.

Ebneth, S., Kürschner, W., Diener, A., Buhl, D. and Veizer, J., 1991. Strontium isotopic evolution of sea water across the Devonian-Carboniferous transition. Geol. Soc. Amer. Abstr. Prog. 23, A111.

Elderfield, H., 1986. Strontium isotope stratigraphy. Palaeogeog. Palaeoclimat. Palaeoecol. 57, 71-90.

Elderfield, H. and Greaves, M.J., 1982. Rare earth elements in seawater. Nature 296, 214-219.

Fallick, A.E. and Hamilton, P.J., 1989. The isotopic geochemistry of ocean waters through time. Trans.

Roy. Soc. Edinburgh Earth Sci. 80, 177-181.

Faure, G., Assereto, R. and Tremba, E.L., 1978. Strontium isotope composition of marine carbonates of Middle Triassic to Early Jurassic age, Lombardic Alps, Italy. Sedimentology 25, 523-543.

Fischer, H. and Gygi, R., 1989. Numerical and biochronological time scales correlated at the ammonite sub-zone level: K-Ar, Rb-Sr ages, and Sr, Nd and Pb seawater isotopes in an Oxfordian (Late Jurassic) succession of northern Switzerland. Geol. Soc. Amer. Bull. 101, 1584-1597.

Gao, G. and Land, L.S., 1992. Geochemistry of Cambro-Ordovician Arbuckle Limestone, Oklahoma: Implications for diagenetic $\partial^{18}O$ alteration and secular $\partial^{13}C$ and $^{87}Sr/^{86}Sr$ variation. Geochim. Cosmochim. Acta 55, 2911-2920.

Geldsetzer, H.H.J., Goodfellow, W.D., McLaren, D.J. and Orchard, M.J., 1987. Sulfur-isotope anomaly associated with the Frasnian-Famenian extinction, Medicine Lake, Alberta, Canada. Geology 15, 393-396.

Goodfellow, W.D. and Jonasson, I.R., 1984. Ocean stagnation and ventilation defined by $\partial^{34}S$ secular trends in pyrite and barite, Selwyn Basin, Yukon. Geology 12, 583-586.

Goodfellow, W.D., Geldsetzer, H.H.J., McLaren, D.J., Orchard, M.J. and Klapper, G., 1989. Geochemical and isotopic anomalies associated with the Frasnian-Famennian extinction. Hist. Biol. 2, 51-72.

Grant, W.F., 1992. Carbon isotope vital effect and organic diagenesis, Lower Cambrian Forteau Formation, northwest Newfoundland: Implications for $\partial^{13}C$ chemostratigraphy. Geology 20, 243-246.

Grossman, E.L., 1994, in press. The carbon and oxygen isotope record during the evolution of Pangea--Carboniferous to Triassic. Geol. Soc. Amer. Spec. Pap. 284.

Halas, S., Balinski, A., Gruszczynski, M., Hoffman, A., Malkowski, K. and Narkiewicz, M., 1992. Stable isotope record at the Frasnian/Famennian boundary in southern Poland. Neues Jahrb. Geol. Paläontol. Monatsh. 1992, 129-138.

Haq, B.U., Worsely, T.R., Burckle, L.M., Douglas, R.G., Keigwin, L.D., Jr., Opdyke, N.D., Savin, S.M., Sommer, M.W., II, Vincent, E. and Woodruff, F., 1980. Late Miocene carbon-isotopic shift and synchroneity of some phytoplanktonic biostratigraphic events. Geology 8, 427-431.

Harland, W.B., Armstrong, R.L., Cox, A.V., Craig, L.E., Smith, A.G. and Smith, D.G., 1990. A Geologic Time Scale. 263 pp. Cambridge University Press.

Henderson, G.M., O'Nions, R.K. and Shackleton, N.J., 1993. Sr-isotopes in Quaternary planktonic foraminifera from the Pacific and Indian oceans. Eos 74, 176.

Hilbrecht, H. and Hoefs, J., 1986. Geochemical and palaeontological studies of the ^{13}C anomaly in boreal and north Tethyan Cenomanian-Turonian sediments in Germany and adjacent areas. Palaeogeog. Palaeoclimat. Palaeoecol. 53, 169-189.

Hilbrecht, H., Arthur, M.A. and Schlanger, S.O., 1986. The Cenomanian-Turonian boundary event: Sedimentary, faunal and geochemical criteria developed from stratigraphic studies in NW-Germany. In: Walliser, O.H. (ed.), Global Bio-Events. pp. 345-351. Springer, Berlin Heidelberg New York.

Hilbrecht, H., Hubberten, W.-W. and Oberhänsli, H., 1992. Biogeography of planktonic foraminifera and regional carbon isotope variations: Productivity and water masses in Late Cretaceous Europe. Palaeogeog. Palaeoclimat. Palaeoecol. 92, 407-421.

Hodell, D.A., Benson, R.H., Kennett, J.P. and El Bied, K.R., 1989. Stable isotope stratigraphy of latest Miocene sequences in northwest Morocco: The Bou Regreg section. Paleoceanography 4, 467-482.

Hodell, D.A., Mueller, P.A., McKenzie, J.A. and Mead, G.A., 1989. Strontium isotope stratigraphy and geochemistry of the Late Neogene ocean. Earth Planet. Sci. Lett. 92, 165-178.

Hodell, D.A., Mead, G.A. and Mueller, P.A., 1990. Variation in the strontium isotopic composition of seawater (8 Ma to present): Implications for chemical weathering rates and dissolved fluxes to the ocean. Chem. Geol. 80, 291-307.

Hodell, D.A., Mueller, P.A. and Garrido, J.R., 1991. Variations in the strontium isotopic composition of seawater during the Neogene. Geology 19, 24-27.

Holland, H.D., 1984. The Chemical Evolution of the Atmosphere and Oceans. 582 pp. Princeton University Press.

Holser, W.T., 1977. Catastrophic chemical events in the history of the ocean. Nature 267, 403-408.

Holser, W.T., 1979a. Mineralogy of evaporites. Mineral. Soc. Amer. Rev. Mineral. 6, 211-294.

Holser, W.T., 1979b. Trace elements and isotopes in evaporites. Mineral. Soc. Amer. Rev. Mineral. 6,

295-346.

Holser, W.T. and Kaplan, I.R., 1966. Isotope geochemistry of sedimentary sulfates. Chem. Geol. 1, 93-135.

Holser, W.T. and Magaritz, M., 1987. Events near the Permian-Triassic boundary. Mod. Geol. 11, 155-180.

Holser, W.T. and Magaritz, M., 1989. Application of isotopes in stratigraphic correlation. Intern. Geol. Cong., 28th, Washington, D.C.

Holser, W.T., Kaplan, I.R., Sakai, H. and Zak, I., 1979. Isotope geochemistry of oxygen in the sedimentary sulfate cycle. Chem. Geol. 25, 1-17.

Holser, W.T., Magaritz, M. and Clark, D.L., 1986. Carbon-isotope stratigraphic correlations in the Late Permian. Amer. Jour. Sci. 286, 390-402.

Holser, W.T., Magaritz, M. and Wright, J., 1986. Chemical and isotopic variations in the world ocean during Phanerozoic time. In: Walliser, O.H. (ed.), Global Bio-Events. pp. 63-74. Springer, Berlin Heidelberg New York.

Holser, W.T., Schidlowski, M., Mackenzie, F.T. and Maynard, J.B., 1988. Geochemical cycles of carbon and sulfur. In: Gregor, C.B., Garrels, R.M., Mackenzie, F.T. and Maynard, J.B. (eds.), Chemical Cycles in the Evolution of the Earth. pp. 105-173. John Wiley and Sons, New York.

Holser, W.T., Schönlaub, H.P., Attrep, M., Jr., Boeckelmann, K., Klein, P., Magaritz, M., Orth, C.J., Fenninger, A., Jenny-Deshusses, D., Kralik, M., Mauritsch, H., Pak, E., Schramm, J.-M., Stattegger, K. and Schmöller, R., 1989. A unique geochemical record at the Permian/Triassic boundary. Nature 337, 39-44.

Huang J. and Liu S., 1989. Sulfur isotope distribution of Triassic evaporite and its geological significance in Sichuan Basin. Acta Sedimentologica Sinica 7 (2), 105-110.

Hudson, J.D. and Anderson, T.F., 1989. Ocean temperatures and isotopic compositions through time. Trans. Roy. Soc. Edinburgh Earth Sci. 80, 183-192.

Hurley, N.F. and Lohmann, K.C., 1990. Diagenesis of Devonian reefal carbonates in the Oscar Range, Canning River Basin, Australia. Jour. Sed. Petrol. 59, 127-146.

Jarvis, I., Carson, G., Hart, M., Leary, P. and Tocher, B., 1988. The Cenomanian-Turonian (Late Cretaceous) anoxic event in SW England: Evidence from Hooken Cliffs near Beer, S. Devon. Newsl. Stratig. 18, 147-164.

Jarvis, I., Carson, G.A., Cooper, M.K.E., Hart, M.B., Leary, P.N., Tocher, B.A., Horne, D. and Rosenfeld, A., 1988. Microfossil assemblages and the Cenomanian-Turonian (Late Cretaceous) Oceanic Anoxic Event. Cretaceous Res. 9, 3-103.

Javoy, M. and Courtillot, V., 1989. Intense acidic volcanism at the Cretaceous-Tertiary boundary. Earth Planet. Sci. Lett. 94, 409-416.

Jenkins, H.C. and Clayton, C.J., 1986. Black shales and carbon isotopes in pelagic sediments from the Tethyan Lower Jurassic. Sedimentology 33, 87-106.

Joachimski, M.M. and Buggisch, W., 1993. Anoxic events in the Late Frasnian--Causes of the Frasnian-Famennian faunal crisis? Geology 21, 675-678.

Jux, U. and Steuber, T., 1992. C_{carb}- und C_{org}-Isotopenverhältnisse in der silurischen Schichtenfolge Gotlands als Hinweise auf Meeresspiegelschwankungen und Krustenbewegungen. Neues Jahrb. Geol. Paläontol. Monatsh. 1992, 385-413.

Kaufman, A.J., Hayes, J.N., Knoll, A.H. and Germs, G.J.B., 1991. Isotopic compositions of carbonates and organic carbon from Upper Proterozoic successions in Namibia: Stratigraphic variation and the effects of diagenesis and metamorphism. Precambrian Res. 49, 301-327.

Kaufman, A.J. and Knoll, A.M., in press. Neoproterozoic variations in the C-isotope composition of seawater: Stratigraphic and biogeochemical implications.

Kaufman, E.E., 1986. High resolution event stratigraphy: Regional and global Cretaceous bioevents. In: Walliser, O.H. (ed.), Global Bio-Events. pp. 279-335. Springer, Berlin Heidelberg New York.

Kennett, J.P. and Stott, L.D., 1990. Proteus and proto-Oceanus: Ancestral Paleogene oceans as revealed from Antarctic stable isotopic results, ODP Leg 113. Proc. Ocean Drilling Prog. Sci. Results 113, 865-880.

Kennett, J.P. and Stott, L.D., 1991. Abrupt deep-sea warming, palaeoceanographic changes and benthic extinctions at the end of the Paleocene. Nature 353, 225-229.

Kirschvink, J.L., Magaritz, M., Ripperdan, R.L., Shuravlev, A.Yu. and Rozanov, A.Yu., 1991. The Precambrian/Cambrian boundary. Magnetostratigraphy and carbon isotopes resolve correlation problems between Siberia, Morocco and South China. GSA Today 1, 69-91.

Knoll, A.H., Hayes, J.M., Kaufman, A.J., Swett, F. and Lambert, T.B., 1986. Secular variations in carbon isotope ratios from Upper Proterozoic successions of Svalbard and East Greenland. Nature 321, 832-838.

Koch, P.L., Zachos, J.C. and Gingerich, P.D., 1992. Correlation between isotope records in marine and continental carbon reservoirs near the Paleocene/Eocene boundary. Nature 358, 319-322.

Koepnick, R.B., Denison, R.E., Burke, W.H., Hetherington, E.A. and Dahl, D.A., 1990. Construction of the Triassic and Jurassic portion of the Phanerozoic curve of seawater $^{87}Sr/^{86}Sr$. Chem. Geol. Isotope Geosc. Sec. 80, 327-349.

Koepnick, R.B., Denison, R.E., Burke, W.H., Hetherington, E.A., Nelson, H.F., Otto, J.B. and Waite, L.E., 1985. Construction of the seawater $^{87}Sr/^{86}Sr$ curve for the Cenozoic and Cretaceous. Chem. Geol. Isotope Geosci. Sec. 58, 55-81.

Kramm, U. and Wedepohl, K.H., 1991. The isotopic composition of strontium and sulfur in seawater of Late Permian (Zechstein) age. Chem. Geol. 90, 253-262.

Kump, L.R., 1991. Interpreting carbon-isotope excursions: Strangelove oceans. Geology 19, 299-302.

Kürschner, W., Becker, R.T., Buhl, D. and Veizer, J., 1992. Strontium isotopes on conodonts: Devonian-Carboniferous transition, the northern Rhenish Slate Mountains, Germany. Ann. Soc. Geol. Belg. 115, 595-621.

Kürschner, W., Ebneth, S., Veizer, J. and Buhl, D., 1992. Variations of strontium isotopes in Paleozoic conodonts. Profil 1, 29.

Leary, P.N., Carson, G.A., Cooper, M.R.E., Hart, M.B., Horne, D., Jarvis, I., Rosenfeld, A. and Tocher, B.A., 1989. The biotic response to the Late Cenomanian oceanic anoxic event: Integrated evidence from Dover, SE England. Jour. Geol. Soc. London 146, 311-317.

Lohmann, K.C. and Walker, J.C.G., 1989. The ^{18}O record of Phanerozoic abiotic marine calcite cements. Geophys. Res. Lett. 16, 319-322.

Long, D.G.F., 1993. Oxygen and carbon isotopes and event stratigraphy near the Ordovician-Silurian boundary, Anticosti Island, Quebec. Palaeogeog. Palaeoclimat. Palaeoecol. 104, 49-59.

Loutit, T.S., Kennett, J.P. and Savin, S.M., 1983. Miocene equatorial and southwest Pacific paleoceanography from stable isotope evidence. Marine Micropaleo. 8, 215-233.

Loutit, T.S., Pisias, N.G. and Kennett, J.P., 1983. Pacific Miocene carbon isotope stratigraphy using benthic foraminifera. Earth Planet. Sci. Lett. 66, 48-62.

Macdougall, J.D., 1988. Seawater strontium isotopes, acid rain, and the Cretaceous-Tertiary boundary. Science 239, 485-487.

Magaritz, M., 1989. $\partial^{13}C$ minima follow extinction events: A clue to faunal radiation. Geology 17, 337-340.

Magaritz, M. and Holser, W.T., 1990. Carbon isotope shifts in Pennsylvanian seas. Amer. Jour. Sci. 290, 977-994.

Magaritz, M. and Turner, P., 1982. Carbon cycle changes of the Zechstein Sea: Isotopic transition zone in the Marl Slate. Nature 297, 389-390.

Magaritz, M., Anderson, R.Y., Holser, W.T., Saltzman, E.S. and Garber, J., 1983. Isotope shifts in the Late Permian of the Delaware Basin, Texas, precisely timed by varved sediments. Earth Planet. Sci. Lett. 66, 111-124.

Magaritz, M., Holser, W.T. and Kirschvink, J.L., 1986. Carbon-isotope events across the Precambrian/Cambrian boundary on the Siberian Platform. Nature 320, 258-259.

Magaritz, M., Bär, R., Baud, A. and Holser, W.T., 1988. The carbon isotope shift at the Permian/Triassic boundary in the southern Alps is gradual. Nature 331, 337-339.

Magaritz, M., Kirschvink, J.L., Latham, A.J., Zhuravlev, A.Yu. and Rozanov, A.Yu., 1991. Precambrian/Cambrian boundary problem: Carbon isotope correlations for Vendian and Tommotian time between Siberia and Morocco. Geology 19, 847-850.

Magaritz, M., Benjamin, C., Keller, G. and Moshkovitz, S., 1992. Early diagenetic isotopic signal at the Cretaceous/Tertiary boundary, Israel. Palaeogeog. Palaeoclimat. Palaeoecol. 91, 291-304.

Magaritz, M., Krishnamurthy, R.V. and Holser, W.T., 1992. Parallel trends in organic and inorganic carbon isotopes across the Permian/Triassic boundary. Amer. Jour. Sci. 292, 727-739.

Margolis, S.V., Kroopnick, P.J., Goodney, D.E., Dudley, W.C. and Mahoney, M.E., 1975. Oxygen and carbon isotopes from calcareous nannofossils as paleoceanographic indicators. Science 189, 555-557.

Margolis, S.V., Mount, J.F., Doehne, E., Showers, W. and Ward, P., 1987. The Cretaceous/Tertiary boundary carbon and oxygen isotope stratigraphy, diagenesis, and paleoceanography at Zumaya, Spain. Paleoceanography 2, 361-377.

Marshall, J.D., 1992. Climatic and oceanographic

isotopic signals from the carbonate rock record and their preservation. Geol. Mag. 129, 143-160.

Marshall, J.D. and Middleton, P.D., 1990. Changes in marine isotopic composition and Late Ordovician glaciation. Jour. Geol. Soc. London 147, 1-4.

Martin, E.E. and Macdougall, J.D., 1991. Seawater Sr isotopes at the Cretaceous/Tertiary boundary. Earth Planet. Sci. Lett. 104, 166-180.

McArthur, J.M., 1991. Strontium-isotope stratigraphy. Geol. Today 7 (6), i-iii.

McArthur, J.M., Burnett, J. and Hancock, J.M., 1992. Strontium isotopes at the K/T boundary. Nature 355, 28.

McGhee, F.R., Jr., Orth, C.J., Quintana, L.R., Gilmore, J.S. and Olson, E. J., 1986. Geochemical analyses of the Late Devonian "Kellwasser Event" stratigraphic horizon at Steinbruch Schmidt. In: Walliser, O.H. (ed.), Global Bio-Events. pp. 219-224. Springer, Berlin Heidelberg New York.

Middleton, P.D., Marshall, J.D. and Brenchley, P.J., 1991. Evidence for isotopic change associated with Late Ordovician glaciation, from brachiopods and marine cements of central Sweden. Geol. Surv. Canada Pap. 90-9, 313-321.

Miller, K.G. and Fairbanks, R.G., 1985. Oligocene and Miocene global carbon isotope cycles and abyssal circulation changes. Amer. Geophys. Un. Geophys. Monogr. 32, 469-486.

Miller, K.G., Fairbanks, R.G. and Mountain, G.S. 1987. Tertiary oxygen isotope synthesis, sea level history and continental margin erosion. Paleoceanography 2, 1-19.

Miller, K.G., Feigenson, C.D., Kent, D.V. and Olsson, R.K., 1988. Upper Eocene to Oligocene isotope ($^{87}Sr/^{86}Sr$, $\partial^{18}O$, $\partial^{13}C$) standard section, Deep Sea Drilling Project Site 522. Paleoceanography 3, 223-233.

Miller, K.G., Wright, J.D. and Fairbanks, R.G., 1991. Unlocking the Ice House: Oligocene-Miocene oxygen isotopes, eustasy, and margin erosion. Jour. Geophys. Res. 96(B4), 6829-6848.

Morante, R., 1993. Determining the Permian/Triassic boundary in Australia through C-isotope chemostratigraphy. In: Flood, P.G. and Aitchison, J.C. (eds.), New England Orogen, Eastern Australia. pp. 293-298. Department of Geology and Geophysics, University of New England, Armidale, Australia.

Morante, R., Veevers, J.J., Andrew, A.S. and Hamilton, P.J., 1994. Determination of the Permian-Triassic boundary in Australia from carbon isotope stratigraphy. APEA Jour. 34, 330-336.

Morgan, M.E., Kingston, J.D. and Marino, B.D., 1994. Carbon isotopic evidence for the emergence of C4 plants in the Neogene from Pakistan and Kenya. Nature 367, 162-165.

Muehlenbachs, K., 1986. Alteration of the ocean crust and the $\partial^{18}O$ history of sea water. Mineral. Soc. Amer. Rev. Mineral. 12, 425-444.

Nelson, B.K., MacLeod, G.K. and Ward, P.D., 1991. Rapid change in strontium isotopic composition of sea water before the Cretaceous/Tertiary boundary. Nature 351, 644-647.

Nishioka, S., Arakawa, Y. and Kobayashi, Y., 1991. Strontium isotope profile of Carboniferous-Permian Akiyoshi Limestone in southwest Japan. Geochem. Jour. 25, 137-146.

Oberhänsli, H., Hsü, K.J., Piasecki, S. and Weissert, H., 1989. Permian-Triassic carbon isotope anomaly in Greenland and in the southern Alps. Hist. Biol. 2, 37-49.

Oberhänsli, H. and von Salis Perch-Nielsen, A.K., 1990. The Paleocene ^{13}C event: Was it due to changes in the storage rate of terrestrial biomass? Veröff. Übersee-Mus. A10, 99-112.

Pak, D.K. and Miller, K.G., 1992. Paleocene to Eocene benthic foraminiferal isotopes and assemblages: Implications for deepwater circulation. Paleoceanography 7, 405-422.

Palmer, M.R. and Edmond, J.M., 1989. The strontium isotope budget of the ocean. Earth Planet. Sci. Lett. 92, 11-26.

Pawellek, F. and Veizer, J., 1992. C- and O-isotopes in Middle Devonian brachiopod shells. Presented at 82nd Annual Meeting of the Geologische Vereinigung, February 26-29, 1992, Stuttgart, Germany.

Paytan, A., Kastner, M., Martin, E.E., Macdougall, J.D. and Herbert, T., 1993. Marine barite as a monitor of seawater strontium isotope composition. Nature 366, 445-449.

Peryt, D. and Wyrwicka, K., 1991. The Cenomanian-Turonian Oceanic Anoxic Event in SE Poland. Cret. Res. 12, 65-80.

Peryt, D., Wyrwicka, K., Orth, C.J., Attrep, M., Jr. and Quintana, L.R., 1992. The Cenomanian-Turonian boundary interval in central Poland. Fifth International Conference of Bio-Events, Abstracts, p. 88.

Pingitore, N.R., Jr., Lytle, F.W., Davies, B.M., Eastman, M.P., Eller, P.G. and Larson, E.M., 1992. Mode of incorporation of Sr^{2+} in calcite:

Determination by X-ray absorption spectroscopy. Geochim. Cosmochim. Acta 58, 1531-1538.

Playford, P.E., McLaren, D.J., Orth, C.J., Gilmore, J.S. and Goodfellow, W.D., 1984. Iridium anomaly in the Upper Devonian of the Canning Basin, Western Australia. Science 226, 437-439.

Popp, B.N., Anderson, T.F. and Sandberg, P.A., 1986. Brachiopods as indicators of original isotopic composition in some Paleozoic limestones. Geol. Soc. Amer. Bull. 97, 1262-1269.

Popp, B.N., Podosek, F.A., Brannon, J.C., Anderson, T.F. and Pier, J., 1986. $^{87}Sr/^{86}Sr$ ratios in Permo-Carboniferous sea water from the analyses of well-preserved brachiopod shells. Geochim. Cosmochim. Acta 50, 1321-1328.

Railsback, L.B., 1990. Influence of changing deep ocean circulation on the Phanerozoic oxygen isotope record. Geochim. Cosmochim. Acta 54, 1501-1509.

Railsback, L.B., Anderson, T.F., Ackerly, S.C. and Cisne, J.L., 1989. Paleoceanographic modeling of temperature-salinity profiles from stable isotopic data. Paleoceanography 4, 585-591.

Rau, G.H., Takahashi, T. and Des Marais, D.J., 1989. Latitudinal variations in plankton $\partial^{13}C$: Implications for CO_2 and productivity in past oceans. Nature 341, 516-518.

Raymo, M.E. and Ruddiman, W.F., 1992. Tectonic forcing of late Cenozoic climate. Nature 359, 117-122.

Renard, M., 1986. Pelagic carbonate chemostratigraphy (Sr, Mg, ^{18}O, ^{13}C). Mar. Micropaleo. 10, 117-164.

Richardson, S.M. and Hansen, K.W., 1991. Stable isotopes in the sulfate evaporites from southeastern Iowa, U.S.A.: Indications of postdepositional change. Chem. Geol. 90, 79-90.

Richter, F.M. and Turekian, K.K., 1993. Simple models for the geochemical response of the ocean to climatic and tectonic forcing. Earth Planet Sci. Lett. 119, 1212-1231.

Richter, F.M., Rowley, D.B. and DePaolo, D.J., 1992. Sr isotope evolution of seawater: The role of tectonics. Earth Planet. Sci. Lett. 109, 11-23.

Ripperdan, R.L., 1994. Global variations in carbon isotope composition during the latest Neoproterozoic and earliest Cambrian. Ann. Rev. Earth Planet. Sci. 22, 385-417.

Ripperdan, R.L., Magaritz, M., Nicoll, R.S. and Shergold, J.H., 1992. Simultaneous changes in carbon isotopes, sealevel and conodont biozones within the Cambrian-Ordovician boundary interval at Black Mountain, Australia. Geology 20, 1039-1042.

Ripperdan, R.L., Magaritz, M. and Kirschvink, J.L. 1993. Carbon isotope and magnetic polarity evidence for non-depositional events within the Cambrian-Ordovician boundary section at Dayangcha, Jilin Province, China. Geol. Mag. 130, 443-452.

Rush, P.F. and Chafetz, H.S., 1990. Fabric-retentive, non-luminescent brachiopods as indicators of original $\partial^{13}C$ and $\partial^{18}O$ composition: A test. J. Sed. Petrol. 60, 968-981.

Savin, S.M., Douglas, R.G., Keller, G., Killingsley, J.S., Shaughnessy, L., Sommer, M.Q., Vincent, D. and Woodruff, F., 1981. Miocene benthic foraminifera isotope records: A synthesis. Marine Micropaleo. 6, 423-450.

Schidlowski, M. and Aharon, P., 1992. Carbon cycle and carbon isotope record: Geochemical impact of life over 3.5 Ga of Earth history. In: Schidlowski, M. et al. (eds.), Early Organic Evolution: Implications for Mineral and Energy Resources. pp. 147-175. Springer, Berlin Heidelberg New York.

Schindler, E., 1990. Die Kellwasser-Krise (hohe Frasne-Stufe, Ober-Devon). Göttinger Arb. Geol. Paläontol. 46, 115 p.

Schlanger, S.O., Arthur, M.A., Jenkyns, H.C. and Scholle, P.A., 1987. The Cenomanian-Turonian Oceanic Anoxic Event, I. Stratigraphy and distribution of organic carbon-rich beds and the marine $\partial^{13}C$ excursion. Geol. Soc. Spec. Publ. 26, 371-399.

Scholle, P.A. and Arthur, M.A., 1980. Carbon isotope fluctuations in Cretaceous pelagic limestones: Potential stratigraphic and petroleum exploration tool. Amer. Ass. Petrol. Geol. Bull. 64, 67-87.

Scholle, P.A., Stemmerick, L. and Harpøth, O., 1990. Origin of major karst-associated celestite mineralization in Karstryggen, central East Greenland. J. Sed. Petrol. 60, 397-410.

Scholle, P.A., Stemmerick, L. and Ulmer, D.S., 1991. Diagenetic history and hydrocarbon potential of Upper Permian carbonate buildups, Wegener Halvø area, Jameson Land Basin, East Greenland. Amer. Assoc. Petrol. Geol. Bull. 75, 701-725.

Shackleton, N.J., 1985. Oceanic carbon isotope constraints on oxygen and carbon dioxide in the Cenozoic atmosphere. Amer. Geophys. Un. Geophys. Monogr. 32, 412-417.

Shackleton, N.J., 1986. Paleogene stable isotope

events. Palaeogeog. Palaeoclimat. Palaeoecol. 57, 91-102.

Shackleton, N.J., 1987. The carbon isotope record of the Cenozoic history of organic carbon burial and of oxygen in the ocean and atmosphere. Geol. Soc. Spec. Publ. 26, 423-444.

Shackleton, N.J., Hall, M.A. and Boersma, A., 1984; Oxygen and carbon isotopic data from Leg 74 foraminifera. Init. Rep. Deep Sea Drill. Proj. 74, 599-612.

Shackleton, N.J. and Kennett, J.P., 1975. Paleotemperature history of the Cenozoic and the initiation of Antarctic glaciation: Oxygen and carbon isotope analyses in DSDP Sites 277, 279, and 281. Init. Rep. Deep Sea Drill. Proj. 29, 743-755.

Stott, L.D., 1992. Higher temperatures and lower oceanic pCO_2: A climatic enigma at the end of the Paleocene Epoch. Paleoceanography 7, 395-404.

Stott, L.D. and Kennett, J.P., 1989. New constraints on early Tertiary paleoproductivity from carbon isotopes in foraminifera. Nature 342, 526-529.

Strauss, H., 1993. The sulfur isotopic record of Precambrian sulfate: New data and a critical evaluation of the existing record. Precamb. Res. 63, 225-246.

Strauss, H., Bengston, F., Myrow, P.M. and Vidal, G., 1992. Stable isotope geochemistry and palynology of the Late Precambrian to Early Cambrian sequence in Newfoundland. Can. J. Earth Sci. 29, 1662-1673.

Thomas, E., 1992. Cenozoic deep-sea circulation: Evidence from deep-sea benthic foraminifera. Amer. Geophys. Union Antarctic Res. Ser. 56, 141-165.

Veizer, J., 1983. Chemical diagenesis of carbonates: Theory and application. In: Arthur, M.A., Anderson, T.F., Kaplan, I.R., Veizer, J. and Land, L.S.. Society of Economic Paleontologists and Mineralogists, Short Course no. 10, p. 3.1-3.100.

Veizer, J. 1989. Strontium isotopes in seawater through time. Ann. Rev. Earth Planet. Sci. 17, 141-167.

Veizer, J., Compston, W., Clauer, N. and Schidlowski, M., 1983. $^{87}Sr/^{86}Sr$ in Late Proterozoic carbonates: Evidence for a "mantle event" at 900 Ma ago. Geochim. Cosmochim. Acta 47, 295-302.

Veizer, J., Fritz, P. and Jones, B., 1986. Geochemistry of brachiopods: Oxygen and carbon isotopic records of Paleozoic oceans. Geochim. Cosmochim. Acta 50, 1679-1696.

Vincent, E. and Berger, W.M., 1985. Carbon dioxide and polar cooling in the Miocene: The Monterey hypothesis. Amer. Geophys. Un. Geophys. Monogr. 32, 455-468.

Vincent, E., Killingsley, K.S. and Berger, W.M., 1985. Miocene oxygen and carbon isotope stratigraphy of the tropical Indian Ocean. Geol. Soc. Amer. Mem. 163, 103-130.

Wadleigh, M.A. and Veizer, J., 1982. $^{18}O/^{16}O$ and $^{13}C/^{12}C$ in lower Paleozoic articulate brachiopods: Implications for the isotopic composition of seawater. Geochim. Cosmochim. Acta 56, 431-443.

Wang, K., Chatterton, B.D.E., Attrep, M., Jr. and Orth, C.J., 1993. Late Ordovician mass extinction event in the Selwyn Basin, northwestern Canada: Geochemical, sedimentological and paleontological evidence. Can. J. Earth Sci. 30, 1870-1880.

Wang, K. and Geldsetzer, H.H.J., 1992. A Late Devonian impact event (about 1.5 Ma after the F/F crisis) in South China and Western Australia and its association with a possible mass extinction event. In: Walliser, O.H. (ed.), Phanerozoic Global Bio-events and Event Stratigraphy. pp. 118-119. Fifth International Conference on Bio-Events, Göttingen, 15-19 February, 1992, Abstracts.

Wang, K., Orth, C.M., Attrep, M., Jr., Chatterton, B.D.E., Hou, H. and Geldsetzer, H.H.J., 1991. Geochemical evidence for a catastrophic biotic event at the Frasnian/Famennian boundary in south China. Geology 19, 776-779.

Wang, K., Orth, C.J., Attrep, M., Jr., Chatterton, B.D.E., Wang X. and Li J.-J., 1993. The great latest Ordovician extinction on the South China Plate: Chemostratigraphic studies of the Ordovician-Silurian boundary interval on the Yangtse Platform. Palaeogeog. Palaeoclimat. Palaeoecol. 104, 61-79.

Wei, W., 1991. Evidence for an earliest Oligocene abrupt cooling in the surface waters of the Southern Ocean. Geology 19, 760-763.

Weissert, H., 1989. C-isotope stratigraphy, a monitor of paleoenvironmental change: A case study from the Early Cretaceous. Surv. Geophys. 10, 1-61.

Williams, D.F., Lerche, I. and Full, W.F., 1988. Isotope Chronostratigraphy: Theory and Methods. 345 pp. Academic Press, San Diego.

Woodruff, F. and Savin, S.M., 1989. Miocene deepwater stratigraphy. Paleoceanography 4, 87-140.

Woodruff, F. and Savin, S.M., 1991. Mid-Miocene isotope stratigraphy in the deep sea: High-resolution correlations, paleoclimatic cycles, and sediment preservation. Paleoceanography 6, 755-806.

Woodruff, F., Savin, S.M. and Douglas, R.G., 1981. Miocene stable isotope record: A detailed deep Pacific Ocean study and its paleoclimatic implications. Science 212, 665-668.

Xu D.-Y., Yan Z., Zhang Q.-W. and Sun Y.Y., 1986. Three main mass extinctions--significant indicators of major natural divisions of geological history in the Phanerozoic. Mod. Geol. 10, 365-375.

Zachos, J.C., Arthur, M.A. and Dean, W.E., 1989. Geochemical evidence for suppression of pelagic marine productivity at the Cretaceous/Tertiary boundary. Nature 337, 61-64.

Zachos, J.C., Berggren, W.A., Aubry, M.-P. and Mackensen, A., 1992. Isotope and trace element geochemistry of Eocene and Oligocene foraminifera from Site 478, Kerguelen Plateau. Proc. Ocean Drill. Prog. Sci. Res. 120, 839-854.

Zachos, J.C., Breza, J.R. and Wise, W.W., 1992. Early Oligocene ice-sheet expansion on Antarctica: Stable isotope and sedimentological evidence from Kerguelen Plateau, southern Indian Ocean. Geology 20, 569-573.

Zachos, J.C., Rea, D.K., Seto, K., Nomura, R. and Niitsuma, N., 1992. Paleogene and Early Neogene deep water paleoceanography of the Indian Ocean as determined from benthic foraminifer stable carbon and oxygen isotope records. Amer. Geophys. Un. Geophys. Monogr. 70, 351-385.

Manuscript received August 1993
Revision received June 1994

Authors' addresses:

William T. Holser, Departments of Geological Sciences, Cornell University, Ithaca, NY 14850, U.S.A. and University of Oregon, Eugene, OR 97403, U.S.A.

Mordeckai Magaritz †, formerly Department of Environmental Sciences and Energy Research, The Weizmann Institute, 76100 Rehovot, Israel

Robert L. Ripperdan, Department of Environmental Sciences and Energy Research, The Weizmann Institute, 76100 Rehovot, Israel. Present address: Department of Geological Sciences, University of California, Santa Barbara, CA 93106, U.S.A.

Geobiological Trends and Events in the Precambrian Biosphere

Mikhail A. FEDONKIN

Abstract. Reconstruction of the global biological events in the Precambrian requires a multidisciplinary approach. Data from the poor fossil record of the prokaryotes has to be supplemented by the interpretation of some sedimentological, geochemical and paleoclimatic phenomena as the biologically controlled processes. Biochemical and ecological conservatism of the prokaryotic ecosystems makes it possible to decode their signals documented in the Precambrian geological history. Expansion of the eukaryotes was the cause of the strong transformation of the fossil record as a whole. In addition to the biological innovation represented by the growth and change in the taxonomic diversity and cell (or body) size of the organisms, there were other global bio-events connected with the restructuring of the ecosystems, colonization of new environments, and rise of new physiologies which strongly affected the sedimentological and taphonomic processes.

Contents

Introduction	89
Tectonic Evolution in the Precambrian	91
Evolution of the Precambrian Ocean	91
Evolution of the Sedimentation and "Extinct" Deposits	94
Geochemical Trends and Events	96
Composition of the Atmosphere and Climate	97
Oxygen Evolution of Atmosphere	98
Precambrian Biological Events	98
Events in the Realm of the Precambrian Microorganisms	99
Stromatolites as By-Products of the Bacterial Ecosystems	101
The Rise of Eukaryotes and the Eukaryotization of Ecosystems	103
Biomarkers	103
Biominerals and Events in Biomineralization	103
Megascopic Fossils	105
Carbonaceous Fossils	105
Metazoan Radiations and Extinctions	105
Conclusions	107

INTRODUCTION

P.S. Laplasse was probably the first to emphasize that the recognition of the immense extent of space and time can provide the key to discovering new classes of phenomena and their particular characteristics.

The study of the great continuum of Precambrian biosphere history seems to support this methodologically important idea. In the Precambrian history we perceive trends and events of a kind which cannot be identified in the Phanerozoic because the latter is too short and oversaturated by other brighter and shorter events. Being closer to us in time these events are more familiar to us by their nature and that is why they are more discernible.

However, the Phanerozoic fossil record does not exhaust the whole diversity of global biological events. This statement is easy to prove taking into account an event of the origin of life; but there were other events which may differ from the Phanerozoic ones qualitatively.

In contrast to the Phanerozoic fossil record, the study of the Precambrian history of life is predominantly oriented to the world of the prokaryotes which dominated in most parts of the aquatic habitats. But the most striking circumstance concerning the prokaryotes is that they seem to stay unchangeable (at least at the level of general cell morphology and ecology) from the Early Archean up to the recent time. The prokaryotic part of the organic world seems not to follow the laws of Darwinian evolution (Zavarzin, 1984).

However, in spite of the conservatism of the bacterial biota, its potential to reveal global biological events is very high. The prokaryotes left distinctive signals in the fossil record as a result of their direct biochemical and reproductive reactions to the external physico-chemical events in the biosphere. These signals seem to be easier to decode because of the relatively simple and distinct character of their biogeochemical ties and reactions, and because in contrast to the Phanerozoic these signals are not overshadowed by the eukaryotic superstructure.

In the Phanerozoic, eukaryotes act as if they wipe off the picture of the prokaryotic world, competing with the bacteria for some nutrients and habitats, forcing them out of numerous biotopes, utilizing some by-products of prokaryotic life activity and just consuming their biomass for food. Symbiogenetic processes and colonization of the host cell by prokaryotes, accompanied by the loss of their individuality and free mode of life, could be attributed to the same class of phenomena. A great portion of the prokaryotes disappeared just because they became the organelles of eukaryotic cells or colonized other organisms as endosymbionts.

Nevertheless, prokaryotes were and still are the important group of primary producers at the base of the trophic pyramid. Thus, the productivity of the prokaryotes controls to a large extent the state of the eukaryotic world.

Due to their biogeochemical activity prokaryotes strongly influence the chemical composition of the hydrosphere, the atmosphere, the upper lithosphere, and the sedimentary processes as well.

Extremely important is the role of the prokaryotes in the colonization of new environments and in the recolonization of the habitats after ecological crises. Last but not least, through the diverse symbiotic connections with eukaryotes, the prokaryotes actually determine the fate of the higher organisms.

Methodologically interesting is the idea that the eukaryotization of the organic world during the Proterozoic may resemble to some extent the process of rehabilitation of geologically younger ecosystems after ecological crises. The long duration of the eukaryotization of the biota in the Precambrian makes it possible to distinguish some details which may escape our attention in the Phanerozoic history of life because of higher speed of the evolutionary processes and because of additional difficulties in decoding the paleontological or biogeochemical signals from the far more complex living systems of the Phanerozoic.

In principle, every ecological catastrophe may be described in terms of the "prokaryotization" of the biota, and every recovery of the ecosystem may be described in terms of the "eukaryotization" of the biota. It is not the intention here to apply the recapitulation law to ecosystem successions and to the Proterozoic history of biosphere, but some similarities between the rise of the eukaryotic world and the rehabilitation of the ecosystems after crises may be noticed.

In spite of an exponential growth of paleobiological information from the Precambrian, we still have a lot of difficulties in identifying real global biological events. Reasons are for example:

1) low biological diversity at the species level prevent detailed biostratigraphy and evolutionary modeling;

2) poor morphological characters and uncertain nature of many fossil groups; imperfect classifications;

3) lack of data on the evolutionary lineages for most of the Precambrian groups of organisms;

4) low precision of the biostratigraphic divisions and their telecorrelation.

The essence of stratigraphic methods strongly changes while we are moving down within the Precambrian. Organisms of uncertain systematic position and of unclear nature (problematics sensu lato) are becoming the major paleontological objects. And even more, the highest places in the hierarchy of the paleontological objects which have stratigraphic significance are gradually shifting to remnants of biogeocoenoses and fossil products of life activity, rather than to fossil assemblages and to remnants of the organisms. Recent stratigraphy of the Precambrian involves more and more the study of biochemical and biogeochemical signals, reflecting the integrated result of an interaction of the biota as a whole with geological and climatic factors (though in many cases the biota is the major factor driving the climate).

Thus, moving down within the Precambrian, the pure paleontological methods must be supplemented or even partially replaced by geobiological methods based on the analysis of an interaction of the biota and the abiotic factors.

Further, the more we go down within the Precambrian, in particular into the Lower Archean, the volume of rocks available for investigations seriously decreases because of long-term erosion, tectonic and metamorphic processes. All this complicates the identification of global events and an adequate comparison of events of different time. In addition, many if not all the former isotopic datings of the oldest rocks made in the first half of this century are not reproducible.

That is why we are able to distinguish in the Precambrian only the most resistant trends and the largest, cardinal global bio-events. For the same reasons we are dealing mainly with biological novelties and to a lesser degree with a change in the biotic diversity and with mass extinctions. One must admit that taphonomic factors play a great role in the appearance of biologic diversity and its evaluation because in general non-skeletal organisms and microscopic forms of life have a low preservation potential and are extremely dependent on the facies.

Nonetheless there is reason to hope that the diversity of Precambrian biological events which can be recognized will grow. Weak differentiation of the Precambrian organic world, low biotic diversity and relatively simple trophic structure of ecosystems made it impossible to utilize all the energy and organic matter without essential loss. The same circumstances did not allow the establishment of efficient self-regulated ecosystems comparable to the Phanerozoic ones. One can suppose that during the major part of Precambrian history the biosphere might not have shown homeostasis. By-products of those not-quite-well-regulated ecosystems seem to comprise an essential part of the Precambrian geological record.

On the other hand, the biological novelties which appeared on the background of the pre-existing prokaryotic ecosystem could introduce new disbalances into the biogeochemical cycles. Thus, some geochemical or sedimentary signals in the Precambrian geological record theoretically could have an indirect connection to biological innovations.

Tectonic Evolution in the Precambrian

There is growing evidence that oceanic lithosphere dominated on the Archean Earth (Veizer and Jansen, 1979, 1985; Lowe, 1992). Major land areas were small fragments of microcontinents which comprised less than 5% of the recent continental crust, as well as unstable volcanic islands which were forming along the spreading zones or areas of intraplate hot spots.

Prior to 3.8 Ga the lithosphere seemed to recycle totally. Accretion complexes which began to form later were converting to the large blocks of the stable continental crust which were not subjected to tectonic recycling.

Analysis of radiometric data indicates that the process of continental growth had three major episodes: A) 3.3-3.1 Ga when about 5% of the present Precambrian crust has been formed, B) 2.7-2.3 Ga, 58%, and C) 2.1-1.6 Ga, 33%. Thus at about 1.6 Ga more than 90% of the present Precambrian crust was formed and Late Proterozoic tectonics were dominated by the continental blocks (Lowe, 1992).

Intensive rifting and accumulation of thick deposits inside the rifts and around the passive edges of the continents took place between 2.5 and 1.7 Ga. Relatively rare occurrence of volcanic rocks in the time interval 1.8-1.2 Ga may indicate a tectonic stagnation at that period.

Diverse evidence indicates that the late Precambrian large supercontinent broke up at the end of the Proterozoic (Kirschvink, 1992a). Thereby, at about 900-800 Ma and 600 Ma, vast extension basins were formed (Knoll, 1992).

Evolution of the Precambrian Ocean

Analysis of the Ar, Xe and He isotopic record allows us to suggest that Early Archean degassing of the planet had a catastrophic character though the maximum rate of water input into the ocean took place at the very beginning of the Proterozoic (Sozokhtin and Ushakov, 1991). The liquid water and primary ocean appeared quite early but this ocean was different from the recent ocean by a number of parameters over a very long time. These differences are as follows (Timofeyev, Kholodov and Zverev, 1986a, b, 1988):

1) the hypsometric contrast of the Precambrian ocean floor was rather weak;

2) the sediment accumulation rate was on the average lower than recent;

3) the lateral differentiation of clastic sediments was insignificant.

According to some models of the early Earth, the proximity of the melting point of the rocks to the surface of the planet and higher velocity of its rotation around its axis resulted in a permanently low relief. The relatively smooth surface of the planet was entirely covered by water of moderate depth.

The Archean ocean, though not as deep as the

Tab. 1. Environmental and biological trends in the Precambrian

Era	Period	Ga b.p.	TECTONICS	OCEAN	SEDIMENTS	ATMOSPHERE	CLIMATE	MAJOR BIOEVENTS	MICROFOSSILS	STROMATOLITES
NEOP.	Vend.	0.5	continental breakup	regression	dolomite decrease. ophyolites. phosphogenic episode	oxygen flux decrease	glaciation	radiation of megascop. metazoans	radiation of acanthomorph acritarchs. extinction of large acritarchs	trombolites
NEOP.	LATE RIPHEAN		extention basins	regression	evaporites. S-Cu ores. Mg/Ca minimum	transparent atmosphere	glaciation	restructuring of global ecosystem	cell size 1-7 mm. maxim. diversity	cosmopolites
MESOPROTEROZOIC	LATE RIPHEAN	1.0	widespread rifting	regression	major Cu sulfides	low CO_2 content	glaciation	major diversificat. of eucaryotes red algae	cell size > 1000 μm	major decline
MESOPROTEROZOIC	MIDDLE RIPHEAN			regression		oxygen increase	glaciation			greatest diversity and abundance
MESOPROTEROZOIC	MIDDLE RIPHEAN		carbonate platforms	water temp. decrease	decrease in cherts		steady cooling	multicellular algae	cell size 200-600 μm	cosmopolites
MESOPROTEROZOIC	EARLY RIPHEAN	1.5	tectonic stagnation	vast shallow basins	shale-hosted Pb-Zn	steady CO_2 decrease		heterotroph. protoz. manganese bacteria		ministromatolite decline
PALEOPROTEROZ.			orogeny		last BIF	ozone screen in place				cosmopolites
PALEOPROTEROZ.			continental crust formation	upwelling increased	first sedimentary Cu			aerobic eucaryots, rise of eucaryotes	cell size 60-100 μm	diversity increase episode
PALEOPROTEROZ.		2.0	intensive rifting		Mn deposits last siderite ores	dramatic growth of O_2 content from 1% PAL to 20% PAL	glaciation	magnetobacteria methane cycle decrease	Gunflint biota	
PALEOPROTEROZ.			carbonate platforms		last detrital uraninite. redbeds and paleosols widespread. sulfates			Cu-Zn enzymes	diverse cyanobacteria	diversity increase episode
PALEOPROTEROZ.		2.5	onset of plate tectonics	alluvial shelves formations				Mn-enzymes sulfate-reducing bacteria	cell size < 60 μm	

Eon/Era	Age (Ga)	Geology/Crust	Oceans/Water	Sediments	Atmosphere	Climate/Solar	Biology/Metabolism	Microfossils	Stromatolites
ARCHEAN — L. ARCH.	2.5	major episode of the continental crust formation; low hypsometric contrast	rise of upwelling; local oxic habitats	abundant BIF; conglomeratic Au-U	slightly oxidizing atmosphere	low solar luminosity (18% below present)	eubacterial methylotrophs		abundant ministromatolites
ARCHEAN — EARLY ARCHEAN	3.0	first continental blocks; oceanic crust dominating	permanent stratification of water	detrital uraninite; growing BIF	slightly reducing atmosphere		cyanobacterial oxygen; methan cycle dominates; methanogens	oldest organic-walled microfossils	growing diversity and abundance; low diversity
ARCHEAN — EARLY ARCHEAN	3.5		low bathymetric contrast; high water temperature	oldest little metamorph. sediments; BIF; oldest sedimentary rocks	high content of CO_2 (> 100 PAL), water steam and other volcanic gases	warm biosphere	photoautotrophic procaryotes; anaerobic bacteria	cyanobacterium filaments; oldest chert microbiota	no columnar and branching forms; oldest stromatolites
ARCHEAN — HADEAN	4.0	total recycling of lithosphere			dense, timid atmosphere		Fe-enzymes anoxigenic photosynthesis; diversification of anaerobic organisms		
ARCHEAN — HADEAN	4.5	accretion of Earth							

recent one, was strongly and permanently stratified (Klein and Beukes, 1989). The ocean body consisted of two layers: a thick and anoxic low stratum of water and a thin upper layer which was mixed by the wind and was better oxidized.

The temperature of ocean water in the Archean 3.5-3.2 Ga was at least about 30-50°C (Lowe, 1992), and if so, even in the polar regions the water could have been warm enough to prevent the formation of marine shelf ice. Thus, the ventilation of the deep ocean by cold water submerging down to the ocean floor from the polar ice caps (Wilde and Berry, 1982) was impossible for a long time. In addition, the small areas of continental blocks in the Archean did not cause upwelling at a scale comparable with the recent one.

Constant stratification of the Archean ocean left the surface water layer extremely poor on metabolites which were rapidly utilized by the phytoplankton. That is why the bioproductivity of the Archean ocean stayed very low, except in those small areas where deep ocean water enriched by the biophile elements was mixing with surface water, e.g. edge zones of the microcontinents and volcanic islands. Life oases could exist also in the vicinity of shallow water hydrothermal vents and volcanic springs.

Local mixing of the stratified ocean water could be produced sporadically by strong earthquakes or by the impacts of large extraterrestrial bodies.

A permanently stratified ocean existed until the Early Proterozoic (Lowe, 1992). Between 2.7 and 2.5 Ga the large blocks of the continental crust were formed, thus developing vast shallow water habitats of the shelf and epiplatform sea. Both the shallow water environments and the increase of upwelling as a phenomenon of global significance promoted the recycling of the major biophile elements in the biosphere.

Evolution of the Sedimentation and "Extinct" Deposits

The most general sedimentological trend of the Precambrian geological history was the increasing rate of accumulation of sediments through the time, especially during the Proterozoic. This persistent trend is connected with the increase of the hypsometric contrast of the planetary relief, as well as with the change of feeding provinces through the time from the alkaline rocks to the acid ones and then to the sedimentary rocks of the platform cover. The resistance to the weathering agents decreased in the same direction.

The change in the composition of the feeding provinces could have been caused by the decrease of phosphorus input to the ocean, as on the average the content of this element decreases in the rock sequence from the ultramafic to the acid ones. However, the acid rocks weather much faster than the alkaline ones; that should compensate for the low phosphorus content in the former ones. Weathering rates of sedimentary rocks are even higher; that makes recycling of the biophile elements more rapid since the middle Proterozoic.

According to Zakrutkin (1993), the organic carbon content in the sedimentary rocks decreases from Archean to Mesozoic (Archean - 11; Proterozoic - 8.7; Paleozoic - 3.3; Mesozoic - 2.1 x 10^{19} gram per 10^6 years). This author explains this trend by the decrease in productivity of the primary producers through the geological time, especially within the photosynthetic organisms. Major cause of the bioproductivity decrease might be the shift of some environmental parameters, in particular the decrease of the CO_2 content in the atmosphere and the dropping of the average temperature through time, especially at the interval from the Archean to the Middle Proterozoic.

Decrease of CO_2 and temperature (as the result of the reduction of some greenhouse gases, in particular carbon dioxide, methane, and water vapour in the atmosphere) seems to happen during Precambrian time. However, the apparent decrease in organic carbon in the sediments through this time can be explained by different causes. In particular, the growth of the sedimentation rates through time may cause a similar effect in the case of more or less constant or even slightly growing productivity of the biota, as carbon was diluted by the sediment.

But an even more important factor may have been the increasing complexity of the trophic structure in the ocean because of the diversification of the heterotrophic consumers and decomposers. This trend was accompanied by the decreasing loss of carbon and energy from the ecosystem. And finally, the progressive oxygenization of the atmosphere and of the sediments (with the participation of bioturbators since the Late Proterozoic) increased the catagenetic loss of the buried organic carbon.

Actually the story of organic carbon might be far more complex. For example, Late Proterozoic (850 Ma and younger) carbonates and organic matter have carbon isotopic ratios which are unusually high relative to those of either younger or older deposits (Knoll et al., 1986; Knoll, 1992). The magnitude and direction of this shift ($\partial^{13}C_{carb}$ = +5 to +7‰ PDB) may be interpreted as a response to a strong increase in the burial ratio of organic carbon to carbonate carbon usually attributed to an absolute increase in the rate of organic carbon burial (Holser, 1984).

Late Neoproterozoic intervals of anomalously high $\partial^{13}C$ are interrupted by negative excursions to values which are close to the recent ones. Some if not all of the negative excursions can be correlated stratigraphically with glaciogenic rocks and iron formations (Kaufman et al., 1991). Rate fluctuations of organic carbon burial could directly influence the CO_2/O_2 balance in the atmosphere and ocean and thus affect the evolution of biota (Knoll, 1991). In particular, a rather rapid growth in oxygen concentration in the atmosphere at the beginning of the Vendian Period is suggested by several independent biogeochemical models.

For example, the Derry et al. (1992) model of organic carbon burial during the Neoproterozoic demonstrates that erosion rate and organic carbon burial were very low through most of Late Riphean (1050-650 Ma). However, after the Varanger glaciation (ca. 600 Ma) organic carbon burial rates may have been 2-4 times the present rates.

This event could be caused by an increasing flux of oxygen into the atmosphere just before the mass appearance of Ediacaran fauna (Kaufman and Knoll, 1992). Thus, the rise of megascopic metazoans at the beginning of the Vendian might be connected with the rise of O_2 content in the atmosphere.

One more sedimentological trend in the Late Proterozoic is the gradual decrease in dolomites and increase in limestones noticed by many authors (Rozanov, 1986). This observation, especially obvious on such carbonate platforms as the Siberian Platform, was challenged recently by Sochava and Podkovyrov (1992, 1993). Analysis of more than 1200 carbonate samples led them to the conclusion that the Mg/Ca ratio is approximately the same for the Vendian and Paleozoic carbonates and that this ratio is even lower in the Pre-Vendian (850-610 Ma) deposits.

However, many authors indicated that Proterozoic dolomites are often early diagenetic and fabric-retentive, while younger counterparts commonly do not preserve the primary sedimentary textures (Tucker, 1983). Thus, the similarity of dolomite/limestone ratios through the Proterozoic-Phanerozoic transition may mask some important differences in the environmental, geochemical or even geobiochemical processes (Knoll, 1992).

Superimposed on the general trend of the growing role of limestone and decreasing role of dolomite in the global carbonate accumulation there is long term cyclicity. One of the minima in the dolomite/limestone ratio occurs in the Late Riphean (900 Ma B.P.). From the Early to Late Riphean the relative portion of the dolomite decreased, then increased in the Vendian, and later decreased again down to the very low value in the Mesozoic and Cenozoic.

On the background of more or less resistant trends in the sedimentological processes during the Precambrian, one can see relatively short episodes of some ore accumulation, including deposits unknown from later geological times ("extinct sediments"). Well known examples are uranium-gold-bearing conglomerates, banded-iron formations and others. Though the banded-iron formations occur in a wide time-interval of the Precambrian geological record from 3.8 to 0.8 Ga this kind of ore deposits is especially abundant between 3.5 and 1.8 Ga with its maximum at about 2.5 Ga (Klein et al., 1992). Global appearance of these mineralizations is connected with important geological and biological events of unknown or intensively debated nature. But the time intervals and the intensity of those mineralizations may well serve as time markers in the Precambrian geological record.

Some of those mineralizations could be directly or indirectly connected with bacterial or even eukaryotic biochemical activity through processes like bio-mineralization, concentration of some elements in the cell membranes during active life or in the organic debris after death, bio-oxidation or bio-reduction, as well as through biologically induced chemical changes of the environments (including microenvironments and sharp biogeochemical gradients).

The role of biota in the sedimentation processes changed during the Precambrian. The major negative trend was the reducing importance of the prokaryotic activity in connection with the global environmental change (exhaustion of some geochemical resources of energy, oxygenization of many habitats, lowering of the surface temperature of the planet etc.) and with the rising importance of the eukaryotes in the global biogeochemical cycles. Good evidence in favour of this statement might be the dynamics of the banded-iron formations' accumulation (see above) and the story of the rise and decline of the stromatolites (see below). Bacterial mats as the major accumulators of organic matter and as the geochemical filters and traps for some mineralizations reduced the area of their active growth after about 1 Ga.

Extremely important bio-events which affected the sedimentation processes were: a) increasing colonization of the shallow-water soft sediments; b) rise of the active filtrators, especially those suspension-feeders who produced fecal pellets, and c) growing diversity of the biomineralization in the eukaryotes. All these events took place during the Vendian and early Cambrian in connection with the major biological innovations.

The increase in diversity, size and depth of bioturbations promoted a better aeration of the sediment, a more efficient oxidation of the buried organic matter, and the recycling of some chemical compounds, including the biophile ones.

The role of active filtrators in the Recent ocean is well known: they clean the sea water by extracting small food particles. But probably more important is that the non-digested remains are packed into pellets which reach the bottom much more rapidly than fine particles. This well-known phenomenon in recent copepods, for example, makes the sea water clean (like the "blue waters" in the recent open ocean). The photic zone expanded downward and led to the expansion of habitats for the photosynthetic organisms both in the pelagic zone and on the deeper shelf bottom, to better aeration of the bottom water and the sediment, and to the expansion of aerobic eukaryotic organisms, including decomposers and bioturbators. Thus, the effect of the active filtrators is not simply a sedimentological one but strongly influences and changes most of the habitat parameters.

Another very important consequence was the decrease in rates of early diagenesis and lithification of the sediments because of its better permeability, lower content of organic matter, aeration and bioturbation (Fedonkin, 1987).

The beginning of active and diverse skeletal biomineralization in the eukaryotes had changed the sedimentation processes in the shallow water zones of low latitude, first of all by the formation of large reef-like build-ups which replaced the stromatolitic landscapes, and by the formation of new kinds of carbonate- and phosphate-bearing deposits.

Geochemical Trends and Events

Among the geochemical signals received from the Precambrian fossil record, the most interesting are those which reflect directly the life activity of the ancient biotas. Of particular importance is the behaviour of such biophile elements as carbon, oxygen, sulfur and phosphorus. However, the behaviour of other elements which are not involved so actively in the biological recycling may be indicative of some abiotic environmental parameters to which the biota could not be indifferent.

For example, the strontium isotope record in the Proterozoic demonstrates a steady change in the isotopic composition of this element in sea water from very "light" ($\Delta\ ^{87}Sr = -500$) up to very heavy ($\Delta\ ^{87}Sr = +30$) numbers. The range of the Proterozoic long-term variations of the strontium isotope composition exceeds the range of strontium variations in the Phanerozoic (Derry et al., 1989).

Extremely low volumes of $\Delta\ ^{87}Sr$ in the Riphean may indicate an essential input of submarine hydrothermal vents which contributed the major portion of the strontium to ocean waters at that time. Long-term growth of $\Delta\ ^{87}Sr$ in the water during the Late Proterozoic can reflect both the change of the main source of the strontium from hydrothermal to continental, as well as the change of the average isotope composition of strontium transported from the continents. This latter change can be related in particular to accretion processes and to the later break-up of the Late Proterozoic supercontinent, as well as to the global period of marine regressions connected partially with a glacio-eustatic dropping of the sea-level during the Late Riphean and Vendian.

The carbon isotope record gives evidence both of the environmental change and of the change in bioproductivity or even composition of the Precambrian biota.

A remarkable contrast is observed between the isotopic characteristics of the Early and Late Archean carbon cycle estimated by analysis of the ^{13}C isotope record. The Early Archean isotopic record of ^{13}C may correspond to the known processes of photoautotrophy based on carbon fixation resembling the one driven by RuBisCo. Rather early participation of metagens in the carbon cycle is very probable. About one third of the organic carbon might be produced from the methane (Hayes et al., 1992a). Recycling of the organic matter with a minimum isotopic effect is interpreted as evidence that carbon dioxide and methane have had approximately equal participation in the global carbon cycle. Decreasing importance of the methane cycle during 2.7-2.2 Ga might reflect the formation of the oxygen enriched atmosphere.

Carbon isotope fluctuations observed in some carbonate sequences of the Late Proterozoic demonstrate strikingly similar behaviour through geological time (Knoll, 1992). Interpreted as a reflection of tectonic and climatic events interacting through the biota, these fluctuations may correspond to episodes of higher carbon burial rates that decreased the greenhouse effect and initiated the four Late Proterozoic glacial periods which took place between 0.85 and 0.57 Ga (Chumakov and Elston, 1989).

The sulfur isotope record may be interpreted in a similar way. The evolution of the $^{34}S/^{32}S$ isotope ratio from almost zero in the Archean to values comparable to recent ones at about 2.4-2.2 Ga indicates that biogeochemical cycles of sulfur resembling the recent ones were formed in the very Early Proterozoic. Isotopic records of carbon and sulfur indicate that biogeochemical cycles of carbon and sulfur became coupled at the end of the Proterozoic (Hayes et al., 1992b).

According to Sochava (1993) the Vendian rocks differ strongly from the underlying Riphean and overlying Cambrian deposits by the extremely high values of ^{34}S in sulfates and $^{87}Sr/^{86}Sr$ in carbonates, as well as by the extremely low values of ^{13}C in carbonates. These geochemical anomalies are interpreted as evidence that during the Vendian the oxygen content in the atmosphere essentially decreased and then increased again at the Vendian/Cambrian boundary.

Composition of the Atmosphere and Climate

According to many geochemists, the concentration of carbon dioxide in the Archean atmosphere exceeded the present day concentration of CO_2 by about 100 times, if not more (Ronov, 1964; Holland, 1984; Walker, 1985). One-dimensional climate models (Kasting, 1987) indicate that concentration of CO_2 in the atmosphere must have decreased from more than 100 PAL (present atmospheric level) to a few PAL or less during the two billion years of the Proterozoic.

High concentrations of CO_2, water vapours and methane in the Archean atmosphere seem to compensate for the low luminosity of the Sun. The standard model of solar evolution predicts that Sun luminosity was 18% below its present value 2.5 Ga and about 5% below today's value at the beginning of the Cambrian (Gough, 1981). Nevertheless, the surface temperature in the Early Archean could have been 80° to 100°C (Kasting and Ackerman, 1986) dropping down to 30-50°C at 3.5-3.2 Ga (Lowe, 1992).

Holland (1992, 1994) assumes that the oldest organisms used H_2 and H_2S to reduce CO_2 down to the organic compounds. The rise of the photosynthetic chlorophyll-bearing algae has led to the disappearance of hydrogen from the atmosphere and to the appearance of O_2 as an essential component of the atmosphere. Gradual change from a slightly reducing to slightly oxidizing atmosphere could have taken place as early as 3.8 Ga and not later than 2.3 Ga.

During the Archean and Early Proterozoic the major part of the biogenic oxygen was spent in the oxidation of iron and other metal ions. After the major episode of the accumulation of the banded iron formations between 2.6 and 1.7 Ga, the oxygen concentration in the atmosphere grew faster.

This increase in atmospheric oxygen was accompanied by a rise in atmospheric ozone and the development of an efficient screen against solar ultraviolet radiation. Photochemical models demonstrate that the ozone UV screen is effective at O_2 levels exceeding 0.01 PAL (Kasting, 1987) and thus the UV screen was fully in place by 1.7 Ga. This was especially more important for the organisms colonizing the land environments than for the marine organisms.

One of the critical factors which might drive the chemical evolution of an atmosphere was the appearance of the vast continental plates (Fedonkin, 1992a, 1993). Emergence and accretion of the continental crust, especially active between 2.7 and 2.5 Ga and 2.1-1.6 Ga (Lowe, 1992), ultimately created a new situation on the globe. Large scale upwelling became a factor of global importance. Extensive shallow water habitats which appeared at that time opened great ecological opportunities because of a huge diversity of microenvironments, in contrast to the pelagic realm which is more ecologically uniform. Both phenomena (upwelling and shallow water habitats) promoted the recycling of the metabolites in the ocean ecosystem, which was impossible in the pre-existing stratified ocean.

There appeared a special kind of shallow water landscape formed by bacterial mats and stromatolite-forming bacterial communities. This landscape was spread over the vast and shallow epiplatformic basins. The Siberian Platform is a good example of such an "amphibian landscape" (terminology by Krylov and Zavarzin, 1988) which existed for over a billion years in the Late Proterozoic.

Greater diversity of the microenvironments in the shallow water benthic habitats and an intensified recycling of the major biophile elements promoted the growth of primary productivity of the phytoplankton and phytobenthos in shallow water environments, and led to the appearance of new groups of consumers, to a greater complexity of the trophic structure, and, as a result, to a decreasing loss of energy by the ocean ecosystem.

The importance of the shallow water habitats for

an Early Proterozoic ecosystem may be underlined by the fact that 83% of the total benthic biomass in the recent ocean is concentrated on 8% of the bottom surface, i.e. on the shelf (Leont'ev, 1982).

Intensification of the metabolite recycling promoted first of all the bioproductivity of the primary producers (phytoplankton and phytobenthos) which both consumed CO_2 through photosynthesis and released O_2 to the atmosphere.

The removal of carbon dioxide from the geochemical cycle, the burial of organic matter, and the formation of carbonate rocks on great areas of the carbonate platform might be the critical factors for the decreasing greenhouse effect during the Proterozoic. Conservation of carbon in the sediments of the platforms decreased the greenhouse effect down to a very sensitive balance at about 900 Ma. Beginning at this moment the glacial periods became more or less regular events of the subsequent geological history.

Cooler climate, in particular the glacial periods, was accompanied by the formation of polar shelf ice and by cold water submerging down to the ocean floor. This initiated ocean mixing, the destruction of the stratified water column and the return of the biophile matter back into the biological recycling.

After the Gowganda (Huronian) Glaciation (2.3 Ga), whose intensity is not quite clear, there was a rather long period of a relatively warm climate on the globe until 900 Ma. From this moment the Earth saw at least four glacial periods during the Late Proterozoic which began 850, 740, 650 and 570 Ma (Chumakov and Elston, 1989). These glaciations were accompanied by regression, decreasing of the shelf areas, growth of the land surface, radical shrinkage of the areas of benthic environments and the shift of the life to the pelagic realm. Geographic isolation of the species might increase because of growing land surface and climatic differentiation. Low content of the biophile elements in the open ocean might be compensated by the input from the eroded land surface and by an elevation of the metabolites from the deeper ocean by the glacial ventilation of the ocean floor and by upwelling.

Oxygen Evolution of Atmosphere

Most of the specialists are in agreement that atmospheric oxygen concentrations prior to about 2.4 Ga B.P. were 10^{-14} PAL or below (Walker, 1987; Holland and Kasting, 1992), though this opinion is challenged by Towe (1988, 1994) who proves that as early as during the Late Archean the oxygen content in the atmosphere should be at least at the Pasteur Point level of 10^{-2} PAL (0.2% O_2).

Holland (1992) suggests that the oldest organisms should use H_2 and H_2S in order to reduce the CO_2 down to the organic compounds. The rise of green photosynthetic algae resulted in the disappearance of hydrogen from the atmosphere and the growth of oxygen concentration as an essential component of the atmosphere. Paucity of redbeds (Holland, 1994) and the relatively narrow range of isotopic composition of Archean sulfides and sulfates (Lambert and Donnelly, 1991) are used as evidence of an extremely low oxygen level prior to 2.4 Ga.

Analysis of data from the study of the oxidation state of paleosols, the nature of the uranium ores, the trace metal content of black shales, the age distribution of the banded iron formations, and the evolution of eukaryotes indicates that the oxygen content of the atmosphere increased dramatically between 2.2 and 1.9 Ga B.P., from less than 1% PAL to about 15-20% PAL (Holland, 1992, 1994; Klein et al., 1992).

Strong variations of the isotopic composition of carbon and strontium during the Late Proterozoic, especially between 850 and 570 Ma B.P., seem to reflect the cardinal environmental change (Kaufman and Knoll, 1992). According to the model of organic carbon burial during the Late Proterozoic (Derry et al., 1992) the rate of erosional weathering and the rate of accumulation of organic carbon were extremely low during most of the Riphean (1.6-0.65 Ga). However, after the Varanger glaciation which took place about 620-650 Ma B.P., the rate of organic carbon burial increased rapidly to volumes 2-4 times greater than the average rate of organic carbon burial in recent marine ecosystems.

This event led to the rapid growth of oxygen input into the atmosphere immediately before the appearance of the megascopic forms of Ediacara fauna in the Vendian. One may not exclude that the Vendian radiation of metazoans was connected directly with the rapid growth of the free oxygen content in the biosphere.

Precambrian Biological Events

Going down within the Precambrian we sooner or later reach a certain moment where the preserved rocks represent such insignificant parts of the past sedimentary basins that it is impossible to judge on global biological events.

But even in those portions of the Precambrian

where the geological record is representative, the global biological events have some peculiarities which make them different from their Phanerozoic counterparts.

By their nature one can distinguish: a) biological events sensu stricto (e.g. biological novelties), b) biologically produced events (e.g. accumulations of biogenic minerals), and c) biologically induced events (e.g. the oxygenation of the atmosphere and the hydrosphere by photosynthetic organisms has had geochemical and biochemical consequences which may be considered as events).

An absence of evolutionary lineages in the Precambrian paleontological record makes it difficult to identify some true biological events. For instance, radiation may be simulated by immigration and extinction may be simulated by the change of some taphonomic factors.

The special character of the Precambrian fossil record requires consideration of such events as the appearance of new body size groups of microfossils, the periods of cosmopolitism and provincialism in stromatolite history, the taphonomic trends and events, and, of course, the taxonomic, ecological and ecosystem novelties. Among the novelties we should mark first of all are the appearance of new ecosystems and the colonization of new environments (including the colonization of the host cell by the endosymbiotic bacteria in the process of the symbiogenetic rise of the eukaryotes).

The majority of data on Precambrian life have been obtained during the last 30 years. Precambrian fossils were collected from more than 3000 localities all over the world. More than 1300 taxa at the genus level have been described up to now, though some critical paleontologists evaluate the number of "real taxa" at somewhere between 500 and 900 (Schopf and Klein, 1992).

The major conclusion from this calculation can be that the biotic diversity of the Precambrian, if compared with that of the Phanerozoic (that is 15% of geological history), is extremely low.

The stratigraphic distribution of the fossil localities is very uneven in the Precambrian. About 1-5 localities per every 100 million years is usual for the Lower Proterozoic, and about 10-25 localities per every 100 million years is typical for the Late Proterozoic. The Archean fossil record appears much poorer. For example, no more than 20 stromatolite localities are known from the whole Archean.

This disproportion in the stratigraphic distribution of the fossil localities can be related to the accessibility of rocks and outcrops where the fossils could be preserved, and the number of sites of this kind decrease with the growing age of the rocks. The second circumstance is that macroscopic organisms appeared relatively late, and among the microfossils which represent the microbial realm we identify only the larger bacterial cells, in particular the cyanobacteria, while all other groups of the prokaryotes lack a fossil record because of their very small cell size (for example, the whole Archaeobacteria kingdom).

Paleontology of the Precambrian involves the study of the following objects: 1) structurally preserved fossils - microfossils, thalii of megascopic algae, imprints, moulds and mineralized parts of animals and bioturbation features, 2) structural biolites - stromatolites, oncolites and other phytolites as well as sediments and ore deposits accumulated due to the direct or indirect activity of the biota, 3) structureless remains of life, e.g. kerogen and diverse organic films, 4) biogenic minerals, and 5) biomarkers and organic chemofossils.

Events in the Realm of the Precambrian Microorganisms

Precambrian microfossils are studied in two kinds of preservation: mummified or organic walled microfossils and mineral pseudomorphs (silica is most common). Being heterogeneous by nature (bacteria, lower eukaryotic algae, lower fungii, protozoans, cysts, eggs and egg cases) and by their ecological specialization, the microfossils demonstrate two important trends through the Precambrian fossil record, namely, the growth of diversity and the increasing individual cell size (Schopf, 1992a, b, c).

Analysis of more than 200 genera of silicified microfossils shows that they first appear as early as Early Archean (3.56 Ga B.P.) but they become especially diverse and abundant after 2.2 Ga B.P., especially in the Late Proterozoic (Riphean and Vendian).

Conservative communities of cyanobacteria inhabiting the extremely shallow water environments do not demonstrate any visible morphological evolution during almost 2.5 Ga. These communities seem to have their morphological and ecological analogues in the recent cyanobacterial communities of sabkha, marshes and lagoons of the arid climatic zones. Extremely rare events in the communities are biological innovations such as the appearance of the spirally-cylindrical *Obruchevella* in the Late Riphean.

Microorganisms of the open marine environments

show more dynamic change through time. For example, an assemblage of morphologically complex prokaryotes which are common from the Gunflint Iron Formation of Southern Ontario, Canada (*Kakabekia, Eoastrion* et al.), are known as early as 2.2 Ga B.P. but do not range later than 1.8 Ga B.P. (Schopf and Klein, 1992).

Large spheroidal microfossils including *Chuaria*, with a smooth wall surface, and *Trachyhystrichosphaera*, a form with processes, appear in the Late Riphean, and the Vendian period saw the rise of true acanthomorph acritarchs (*Micrhystridium, Baltisphaeridium, Skiagia* et al.), reflecting the general process of eukaryotization of the planktic ecosystems.

A similar picture is observed in the fossil record of organic walled microfossils. The oldest organic walled microfossils are known from the Late Archean (2.9 Ga B.P.) but they become very diverse and abundant after the beginning of the Riphean (1.6 Ga B.P.) and in the Vendian. General trends in the history of this group of microfossils are the increase of the upper size limit of the sphaeromorphic and filamentous forms, the growing morphologic complexity, morphological innovations (appearance of acanthomorph and marginate acritarchs, etc.) and the growth of taxonomic diversity.

On the background of the general trends mentioned above one can see at least three global events reflected in the paleontological record of microorganisms.

The first event is connected with the disappearance of the Gunflint-type microbiota in the Early Proterozoic (about 1.8 Ga B.P.) which was replaced by the biota of morphologically complex organic walled sphaeromorph acritarchs in the Lower Riphean (1.6 Ga B.P.).

The second event is related to the rise of acanthomorph acritarchs (*Trachyhystrichosphaera*), polygonal forms (*Octoedryxium*), ball-like (*Tortunema*) and other morphologically complex microfossils in the late Middle Riphean.

And the third event is connected with the Vendian radiation of microorganisms marked by the appearance of *Bavlinella, Micrhystridium, Leiomarginata* and others with complex morphology.

Along with the biological innovations there were some trends and events with respect to the upper size limit of the cell as well as the change in diversity and ecology of the microorganisms. Forms 60-100 µm in diameter appeared no later than 1850 Ma B.P., the forms with diameter of 200-600 µm no later than 1400 Ma B.P., the forms with diameter more than 1000 µm including morphologically complex acritarchs have appeared no later than 1050 Ma ago, and large sphaeromorph microfossils of 1-7 mm in diameter have become widespread about 850 Ma B.P..

Taking into account that the upper size limit of the prokaryotic cell is near 60 µm, the size trend described above may reflect the evolution of eukaryotes in the Precambrian. On the other hand, we know that the lower size limit of the eukaryotic cell is about 20-30 µm, and thus their real history might begin far prior to 1850 Ma B.P..

Size trends in the realm of the eukaryotic microfossils may reflect the growing size of their primary biotopes dependent on the degree and stability of oxidation (Fedonkin, 1993).

Eukaryotes, and in particular microscopic algae, seem to evolve rather slowly from the time of their appearance up to 1200 Ma B.P..

However, the later diversity of eukaryotes had grown very rapidly, being probably connected with the development of sexual reproduction in this group (Schopf, 1992a, c).

A maximum in abundance and taxonomic diversity of microfossils occurred in the time interval between 950 and 850 Ma B.P.. During this period, the megasphaeromorphs (up to a few millimeters in diameter) were dominant among the phytoplankton, although these gigantic microfossils became extinct prior to the end of the Proterozoic. After this 950-850 Ma maximum one can observe a sharp decrease in abundance and diversity of microfossils (Vidal and Knoll, 1982, 1983). Interpreted as the earliest recorded extinction event in the Neoproterozoic protists, this collapse of the phytoplankton can be explained by the negative effect of CO_2 concentration decrease and of the growing O_2 concentration in the atmosphere upon the enzymes which controlled the photosynthesis. This hypothesis has an experimental confirmation by the study of recent populations of microalgae (Schopf, 1992a).

The change in CO_2/O_2 balance might be connected with the glaciations which took place in the interval between 850 and 650 Ma B.P.. Unsuitable conditions of preservation offered by widespread glacial deposits may have influenced the fossil record of the microorganisms as well (Vidal, 1994).

Between 850 Ma B.P. and the beginning of the Cambrian most of the planktic taxa with cell diameter more than 600 mm became extinct as well as the numerous groups of the morphologically complex acritarchs.

After the great Varanger glacial period (650-630 Ma B.P.), a new radiation of the eukaryotic organisms took place but did not reach the former diversity which existed prior in the Vendian.

There were some ecological trends and events in the history of microorganisms, such as an apparent increase in diversity of the planktic eukaryotic forms and a parallel decrease of diversity and the space distribution of benthic prokaryotes in the interval between 1500 and 1000 Ma B.P., followed by some increase in diversity of both prokaryotes and eukaryotes between 1000 and 850 Ma B.P. (Schopf, 1992a).

Large acanthomorph acritarchs, a few hundred microns in diameter and widespread between 950 and 800 Ma B.P., disappeared prior to the end of the Proterozoic. Diverse acanthomorph acritarchs which appeared during the Early Cambrian radiation of the phytoplankton did not exceed 75 µm in diameter (Mendelson and Schopf, 1992).

Size increase in the acritarchs during Late Proterozoic may be interpreted as a strategy of defense from planktotrophic organisms, comparable by their body size to protozoans or even metazoan larvae (Burzin, 1987). However, an increase of the cell size in the benthic microorganisms during the same period of time may indicate different causes.

An opposite trend can be observed during the time after 850 Ma B.P., i.e. the disappearance of most planktic taxa with a cell size more than 600 µm, and an explosive radiation of small acanthomorph acritarchs at the Atdabanian of the Early Cambrian may reflect a natural selection in favour of the forms with high surface/volume ratio. There are two ways to increase this ratio, namely to decrease the cell size or to develop special devices increasing the cell surface (spines, processes, ornamentations, sculptured elements etc.). Lower Cambrian acritarchs, demonstrating both morphological strategies, obviously had increased abilities to extract the biophile matters per a volume unit of the cell (Fedonkin, 1993).

Stromatolites as By-Products of the Bacterial Ecosystems

The most notable evidence of the microbial life activity in the Precambrian are the stromatolites. These bio-sedimentary structures appeared as early as 3.5 Ga ago and became widespread in the Early Proterozoic. The vast expansion of the stromatolite forming bacterial communities over the new born carbonate platform in the Early Proterozoic (2.5-2.2 Ga B.P. and later) may be considered as an important biological event of the Precambrian. Since that time the stromatolites became an essential or even dominant part of the platform sedimentary covers. Thickness of the stromatolite bearing carbonates can be very high; taking into account the extremely low growth rate in recent analogues of stromatolites, this looks very strange.

It is very difficult to judge on the diversity of stromatolites because their morphology might be controlled by a number of heterogeneous factors like the composition of the bacterial communities, intensity of metabolism, and geochemical, sedimentary and hydrodynamic factors of the environment, etc. On the other hand, the formal classification of stromatolites based on the hierarchy of morphological characters of the build-up and microstructure may have no strong correlation with the taxonomic composition and the structure of the bacterial (or bacterial-algal) communities which formed the stromatolites.

Nevertheless the stromatolite fossil record reveals some important temporal trends (Walter, Grotzinger and Schopf, 1992; Walter, 1992). Early and Middle Archean stromatolites are rare, their diversity is low and the morphology is simple (for example, no columnar branching forms, as is common in the Proterozoic, are known).

From about 2.7-2.8 Ga B.P. the stromatolite fossil record became more diverse and abundant, continuing without interruption through the whole Proterozoic. Known from 20-22 localities, Late Archean stromatolites demonstrate a wide variety of stratiform, domical and columnar forms which were growing mostly in marine shallow water environments associated with tectonically active areas with abundant volcanism (Grotzinger, 1989).

Abundance and diversity of stromatolites increased rapidly during Proterozoic and reached the maximum at about 1.0 Ga B.P.. This trend was accompanied by the vast lateral expansion and increasing role of the columnar varieties of stromatolites.

The Late Proterozoic saw a decline of stromatolites after 1.0 Ga B.P., especially at 600-700 Ma B.P.. This decline resulted in the disappearance of stromatolites as the dominant group of the shallow water marine environments (Awramik, 1971; Walter and Heys, 1985).

On the background of those general trends in the stromatolite fossil record one can see more detailed temporal patterns. For instance, ministromatolites with radial fibrous fabrics were abundant in the peritidal environments of the Late Archean and especially in the Early and Middle Proterozoic, and declined rapidly thereafter.

Columnar stromatolites with conical laminae

(*Conophyton* and related forms) characteristic of quiet subtidal environments were abundant in the Early and Middle Proterozoic with a marked decline thereafter.

Rare in the Proterozoic, non-laminated stromatolites (or trombolites) became widespread during the Early Cambrian.

Empirically substantiated, mostly by biostratigraphers of the Russian school (see Krylov, 1975, and references herein), the stromatolite zonation of the Proterozoic may be essentially improved after a period of criticism and even nihilism by the extensive study of the biological, ecological and actuopaleontological aspects of the stromatolites (Schopf and Klein, 1992).

Recent understanding of the nature of stromatolites makes it possible to divide and correlate the Proterozoic deposits at a level of precision within 300-100 Ma. That is a great achievement for the biostratigraphy of the Proterozoic. The Early Middle Proterozoic (or the Early Riphean) is characterized by *Kussiella, Omachtenia, Conophyton cyclindri-cum*, and others. The later Middle Proterozoic (the Middle Riphean) contains diverse forms of *Conophyton, Baicalia* and other groups. The early Late Proterozoic (the Late Riphean) contains less abundant forms of *Conophyton* but demonstrates a high diversity of small columnar branching stromatolites *Gymnosolen, Minjaria* and *Inzeria* which disappeared prior to the lower boundary of the Vendian. Transitional groups of the stromatolites such as *Juruzania, Paniscollenia, Collenia* and *Boxonia* totally change their taxonomic composition at the "species" level while crossing the Riphean-Vendian boundary. All the Vendian groups of stromatolites disappeared at the Precambrian-Cambrian boundary, and the Lower Cambrian strata contain totally new groups of stromatolites and trombolites (Krylov, 1975; Semikhatov, 1976; Semikhatov, Fedonkin et al., 1990; Walter, 1992).

These distinct changes in the stromatolite assemblages through time may reflect important events in the evolution of the benthic microbial communities and their biogeochemical activity, as well as the evolution of external biotic and abiotic factors of the environment. All these aspects of the stromatolite fossil record are to be studied parallel to the revision of the most important taxa, taking into account the recent data on the morphogenesis of recent analogues of stromatolites and diagenetic alterations of the carbonates.

The Late Proterozoic history of stromatolites was marked by the relatively short periods of cosmopolitism 1.6-1.5, 1.2-1.0 and 0.85-0.75 Ga B.P. (Krylov, 1985).

Among the major possible causes of the stromatolite decline after 1.0 Ga B.P. and especially after 700 Ma B.P. there might be:

1) major sea-level low-stands (Gebelein, 1976); 2) negative effect of the newly evolved grazing and burrowing metazoans (Awramik, 1971; Walter and Heys, 1985); 3) appearance of the eukaryotic algae, competing with the cyanobacteria for nutrients and habitats (Monty, 1974); 4) negative affect of the growing concentration of biogenic oxygen upon the bacterial stromatolite-building communities (Krylov, 1985); 5) decreasing carbonate saturation of the sea water during the Late Proterozoic (Grotzinger, 1990); 6) climate change, in particular the African Glacial Era and its paleogeographic and geochemical consequences (Semikhatov and Raaben, 1993, 1994).

These hypotheses do not exclude each other but are rather complementary. One can recognize some additional factors which have negatively influenced the stromatolite communities.

In general, it was a gradual shift in the abiotic environmental parameters from the optimum favorable for the growth and reproduction of bacteria, and blue-green algae in particular. Among these factors we should mention a decrease in the geochemical energy resources connected with the steady lowering of volcanic and hydrothermal activity during the Late Proterozoic. In addition, great volumes of the nutrients important for the prokaryotes were removed from the biogeochemical cycles, being buried in the sedimentary cover on the stable platforms or their passive margins.

Average temperature in most parts of the environments decreased far below the optimum for many groups of the prokaryotic organisms. Growing concentration of free oxygen in the atmosphere might inhibit the life activity of both anaerobic prokaryotes and photosynthesizing cyanobacteria which do not tolerate the high saturation of oxygen (during the Late Proterozoic, newly evolved groups of eukaryotic algae became the active producers of the oxygen in addition to the bacterial producers).

Growing intensity of the solar luminosity and increasing transparency of the atmosphere, especially because of the decreasing amount of water vapour caused by the lowering temperature (in particular, during African glacial era which began about 850 Ma B.P.), might have had a negative affect on the photosynthezing cyanobacteria. Self-shading by the pigments and burial in the sediment are well known in recent benthic microbial communities as an adaptation against the very bright light (Castenholz, Bauld and

Pierson, 1992).

Late Proterozoic glaciations were accompanied by numerous glacio-eustatic sea-level falls and by long periods of regressions. Stromatolite habitats in the vast epiplatformic and shelf basins decreased down to narrow strips along the edge of the continents. Every post-glacial transgression was accompanied by the recolonization of the shallow water habitats, but under the new biological circumstances connected with the growing role of eukaryotes. Thus, glacial eustatic regression could destabilize those relative resistant stromatolite ecosystems, making it easier for the eukaryotes to dominate in the shallow water environment (Fedonkin, Chumakov and Jankauskas, 1987).

Recent detailed analysis of the stromatolite diversity from Europe and Asia carried out by Semikhatov and Raaben (1993) reveals the continuing growth of the stromatolite diversity until 850 Ma B.P. (not 1.0 Ga, according to Walter and Heys, 1985). The decline in the stromatolite diversity and abundance is connected by these authors with the beginning of the African glacial era, which was accompanied as well by long-term regressions, decrease in the CO_2 concentration in the atmosphere, drop in the carbonate saturation and dramatic decrease of the carbonate precipitation rate including the stromatolite build-ups.

Global analysis of the Proterozoic stromatolite diversity accompanied by the revision of the stromatolite taxonomy and their stratigraphic distribution (Semikhatov and Raaben, 1995, in press) demonstrates the trimodal appearance of the diversity curve with maximum at 2.3-2.0, 1.65-1.45 and, the most intensive, at 1.0-0.85 Ga B.P.. Periods of maximum diversity of the pre-Vendian stromatolites are correlated with large subglobal transgressions, and minimum diversity chronologically coincides with the time of intensive orogeny.

The Rise of Eukaryotes and the Eukaryotization of Ecosystems

The origin of the eukaryotic cell as an important event (or a series of events?) remains one of the enigmas and the key seems to occur further and further into the geological past. Quite possibly we shall never recognize this moment because the first eukaryotes could have appeared in very narrow habitats.

Three independent lines of evidence indicate that eukaryotes appeared not later than 1.85 Ga B.P.: 1) paleontological data; 2) phylogenetic interpretation of the molecular genetics, and 3) taxonomic interpretation of the biomarkers.

Biomarkers

The last decades are characterized by success in the separation, identification and taxonomic interpretation of the biologically meaningful organic compounds (or biomarkers) from the Precambrian kerogens and hydrocarbons (Ourisson, 1987, 1990; Summons and Walter, 1990). Some biomarkers, e.g. the terpenoids can serve as the base for a simple molecular "phylogenetic tree" which helps to identify the appearance of new large groups of organisms by the occurrence of corresponding hydrocarbons in the fossil organic matter. For example, pentacyclic triterpane hydrocarbons represent the molecular remnants of the eubacteria, sterols represent eukaryotes, and some acyclic isoprenoids correspond to the archaebacteria.

Special kinds of biomarker hydrocarbons are indicative of the groups of microorganisms which existed in the Late Proterozoic, became extinct later on, and which do not have recent counterparts at even high taxonomic level (Summons, 1988).

The study of the oldest biomarkers from well-preserved Proterozoic sediments gave results which are consistent with the presence of the major terpenoid biosynthetic pathways characteristic of eubacteria, archaebacteria and eukaryotes as far back as 1700 Ma B.P. (Summons, 1988). Biomarker data indicate that microbial ecosystems of the Middle and Late Proterozoic included not only cyanobacteria and phototrophic eukaryotes, but possibly already heterotrophic protozoans. There were organisms among the Late Proterozoic plankton which resemble recent dinoflagellates (Summons and Walter, 1990).

The general trend in the biomarker time-distribution demonstrates their limited diversity in the Early-Middle Proterozoic and increasing diversity towards the end of the Proterozoic.

Biominerals and Events in Biomineralization

Biomineralization is a rather usual phenomenon in the organic world. About 60 minerals are produced by the representatives of the five kingdoms of organisms and this diversity is far from being exhausted (Lowenstam and Weiner, 1989). Biominerals which have a unique crystal shape as well as other physico-chemical properties are easy to distinguish. That is why the biominerals can be used as markers of certain groups of organisms and as time-markers.

At least a third of the known biominerals is produced by the prokaryotes and in principle the bio-

Table 2. Vendian Bio-Events

Ga b.p.				
0.54	active filtrators	new feeding strategies	archeos-algal reefs	radiation of acanthomorph acritarchs
0.55	radiation of skeletal fauna	radiation and size increase of bioturbators	colonization of subsurface sediment by invertebrates	
0.56				
0.57	taphonomic threshold			
0.58	extinction and/or miniaturization of metazoans	ichnofossil diversity decrease		
0.59				
0.60		radiation of vendotaenian algae		
0.61	expansion of Ediacaran organisms			
0.62		metazoan colonization of sediment in shallow water environments		
0.63	radiation of soft-bodied metazoans			
0.64				
0.65			Varanger Glaciation	

chemical possibilities of the prokaryotes are wider than is the geochemical diversity of the recent natural environments. We can hope to discover a higher diversity of biominerals than is known in present day eukaryotes. But up to now, only the first steps in this area have been made.

Thus, the isotopic composition of the sulphide minerals may indicate the presence of biologically induced biomineralization as far back as 2.7 Ga B.P. (Monster et al., 1979).

Direct evidence of the biologically produced minerals is given by the discoveries of biogenic modification of magnetite, precipitated by the magnetobacteria in the Early Proterozoic about 2.0 Ga B.P. or earlier (Kirschvink, 1992b). Magnetotactic bacteria are considered as the most suitable candidates for the host cell group which united different prokaryotic cells in the process of the symbiogenetic origin of the eukaryotes.

Incrusting manganese bacteria have left the products of their life activity in deposits 1.6 Ma old (Muir, 1978). First evidence of the carbonate biomineralization represented by the slightly calcified cyanobacteria occurs as early as 1.0 Ga B.P., continuing through the Vendian (Riding and Voronova, 1982, 1984).

The earliest known example of metazoan biomineralization is the multi-layered tubular shell of *Cloudina* formed of a rigid calcite-impregnated organic-rich material (Germs, 1972; Grant, 1990). This geographically widespread fossil form is interpreted as a filter-feeding metazoan of at least a coelenterate grade of organisation.

Diverse organisms which appeared at the Vendian-Cambrian boundary were secreting carbonates, phosphate and silica to produce their skeletal elements. Phosphate biomineralization, appearing first in the Vendian, reached its maximum during the Cambrian Period and became more common in the vertebrates since the Middle Ordovician. Numerous groups of the phosphate-producing problematic invertebrates apparently went extinct prior to the end of the Cambrian Period and the carbonate-producing organisms became the dominant forms among the skeletal organisms.

Megascopic Fossils

Increasing abundance and diversity of megascopic fossils is observed through the Proterozoic geological record. These fossils are preserved as carbonaceous remains, casts and moulds, and as mineral remains and pseudomorphs.

The large size of these fossils may reflect the rise of multicellularity which could appear many times in all kingdoms of organisms, depending primarily on the environmental conditions (Stebbins, 1974). However, the observed long-term trend in the growing abundance and diversity of megascopic fossils may be interpreted as the result of selective advantages obtained due to the large size and individual biomass as well as due to the primary cell specialization. On the other hand, one can suppose the growing influence of the external environmental factors (for example, the growing oxygen concentration in the sea water or the increasing stress on heterotrophic organisms) which might be "neutralized" by the development of large size and biomass, cell specialization and tissue grade of organization.

Carbonaceous Fossils

Carbonaceous fossils of diverse morphology occur in many Proterozoic sequences all over the globe. The oldest of them are the moranids which may represent the flattened remains of microbial colonies which were living about 2.0 Ga B.P.. Also present are megascopic band-like fossils similar to the vendotaenids and spiraliform filaments (similar to *Crypania*, 1.4 Ga B.P.), and round and angulate compressions of moraniaceans and beltinaceans, dating from approximately 2 Ga B.P. (Hofmann, 1992, 1994). These fossils may also represent the flattened remains of the microbial colonies or bacterial mat fragments.

More complex fossils with possible anatomical details appear in the rocks at about 1.4 Ga B.P., although far more diverse and complex morphologies are observed in the carbonaceous fossils 0.9-0.8 Ga B.P. (Chuariaceae, Ellipsophysceae, Tawuiaceae, Longfengshaniaceae etc.). The first annulated organisms resembling annelid worms (Sinosabelliditida) occur in the rocks 0.7 Ga B.P.. The Vendian Period saw the radiation of the Vendotaeniaceae and Eoholyniaceae.

According to Gnilovskaya (1990) at least three different floras may be distinguished during the Vendian Period. The earliest one containing *Eoholynia, Caudina, Orbisiana, Leotrichoides* and some other forms is typical for the Redkino Stage (V_2). The second flora with *Vendotaenia, Aataenia, Sarmenta* and *Primoflagella* occur in the Kotlin Stage of the Vendian (V_3).

The uppermost Vendian, including the top of the Kotlin and the Rovno Stage (V_4), is characterized by the *Tyrasotaenia* and *Dvinia*. Though this succession of algal assemblages is recognized on the analysis of extremely abundant fossil material from the Vendian of the Russian Platform and partially from the Siberian Platform, the global character of this change is still to be confirmed.

Metazoan Radiations and Extinctions

The Varanger glaciation (0.65-0.62 Ga) was the most intensive one in the series of the glacial episodes that took place in the Late Proterozoic (Chumakov, 1978). The Post-Varanger glacioeustatic transgression over the continents saw a rapid radiation of megascopic soft-bodied invertebrates known as the Ediacaran fauna. Although the most vivid paleozoological element of the Vendian Period (Fedonkin, 1990a, b), this fauna does not so far demonstrate any prominent sequence in the appearance of the major fossil groups. This suggests that either the post-Varanger radiation of metazoans was very rapid or invertebrates had a cryptic period of earlier history.

The pre-Vendian origin of the metazoans is confirmed indirectly by, or is consistent with, a number of heterogeneous data including 1) extrapolation of the rate of evolution (Durham, 1970); 2) molecular clock interpretation of the rates of change of globine molecules (Runnegar, 1982a), as well as cytochrome c (McLaughlin and Dayhoff, 1973), 18S r RNA (Field et al., 1988), 28S r RNA (Christen et al., 1991) and other molecules; 3) analysis of diversity of the metazoan cell types through time (Valentine, 1991); 4) high differentiation of the Vendian animal world already present in the first half of this period (Fedonkin, 1992b) as well as the discovery of medusae-like fossils in the intertillite beds, e.g. close to the lower boundary of the Vendian in northwestern Canada (Hofman, Narbonne and Aitken, 1990); and 5) the Proterozoic stromatolite decline interpreted as the result of ecological restriction by grazing and burrowing metazoans (Awramik, 1971).

Though, according to these data, the moment of the origin of the metazoans can be estimated in a very wide time range (from the maximum 1.75 Ga b.p. to 645-680 Ma B.P.), one may speculate on the earlier pre-Vendian (or pre-Ediacaran) radiation(s) of the invertebrates and on their probable mass extinction during the Varanger glacial episode. Thus, along with the newborn high rank taxa which might have appeared during the early Vendian radiation of the metazoans (and some of them continued to the Phanerozoic) there could well be among the Ediacaran fauna some relict forms of the pre-Vendian groups.

The interpretation of Vendian metazoan body fossils known as the Ediacaran fauna has met some methodological difficulties for a number of reasons, e.g. unusual taphonomic circumstances, low diversity of the Vendian organisms, poor fossil record of the Phanerozoic soft-bodied animals (few forms to compare), and absence of data on the pre-Vendian metazoans. Comparison of Ediacara organisms with recent soft-bodied invertebrates is not easy because of the great evolutionary distance between them.

Three major approaches have been developed in respect to the Vendian fauna. The first one stressed the similarity of Ediacaran organisms with recent invertebrates (Glaessner, 1984). A chronologically second approach pointed out the dissimilarity of the Vendian animals with their recent counterparts (Fedonkin, 1983, 1987). The third approach underlined an unusual taphonomy and architecture of the Vendian body fossils and the necessity to interpret these fossils as the representatives of the non-metazoan Vendobionta kingdom (Seilacher, 1984, 1989, 1992). This last approach puts the Ediacara fauna in a position of apparent but false phylogenetic isolation.

Key arguments in favour of the metazoan nature of the Vendian body fossils are being received from taphonomic and paleoecological study (Gehling, 1991), from the body plan analysis of these fossils (Fedonkin, 1990b) and from the study of the Ediacara-like fossils of the later periods (Conway Morris, 1993).

Mass preservation of the soft-bodied organisms in the Vendian deposits was a relatively short, globally observed episode. This kind of preservation, which was observed as a global phenomenon in a variety of the siliclastics and carbonates (Fedonkin, 1990a, b), seems to disappear well prior to the end of the Vendian. The biotic and abiotic factors which could promote the preservation of the soft-bodied animals are discussed in detail elsewhere (Fedonkin, 1985, 1992a, b). To some extent the appearance of the Ediacaran fauna in the fossil record could be considered as the result of the body size increase, if the pre-Vendian animals were indeed of a small size because of lower concentration of oxygen in the sea water (Runnegar, 1982b).

Metazoan diversification seems to be very rapid in the Vendian (Redkino Stage). A period of relatively slow evolution or even stasis in the middle part of the Vendian was followed by mass extinctions in many metazoan groups, while others seem to decrease their body size during Kotlin time of the Vendian. The fall of the taxonomic diversity of the phytoplankton and an increase of the buried vendotaenian algae in sediments of the Kotlin basins (Sokolov and Iwanowski, 1990) may be indicative of eutrophication of the shallow marine environments which might act as a selective factor in favour of the forms having small body size by an analogy with recent eutrophic basins.

The hypothesis on the miniaturization in some metazoan groups close to the end of the Vendian (Fedonkin, 1987) can be indirectly confirmed by the fact that the first skeletal metazoans which showed explosive radiation in the early Cambrian were indeed very small creatures (Rozanov and Zhuravlev, 1992; Bengtson, 1992).

The second increase of the body size of invertebrates took place at the very end of the Vendian (Rovno Stage according to Russian nomenclature). This event, observed in the paleoichnological fossil record, coincides with the increasing colonization by invertebrates of the bottom sediments of shallow water environments and was immediately followed by the rise of small shelly organisms (Fedonkin, 1990c). Trace fossils of the Rovno Stage of the Vendian are more diverse, larger and deeper than those which are known from the earlier Redkino and Kotlin Stages. Active colonization of the sediment by the vagile benthos as well as new bioturbation strategies facilitated the aeration and decomposition of the buried organic matter (Fedonkin, 1987; Droser and Bottjer, 1988).

Actually the Vendian-Cambrian transition was the time of a rapid vertical expansion of the habitats of the benthic metazoans due to the colonization of the deeper sediment below the bottom surface and because of the appearance of skeletons and reef-like structures which supported the organisms to rise above the bottom. An expansion of the benthic habitat increased the diversity of the microenvironments both inside the sediment and on the greatly enlarged surface of such biogenic structures like reefs, exoskeletons etc. Growing environmental diversity in the benthic realm might give additional opportunities for the newborn Early Cambrian groups of invertebrates.

It is important to understand that an explosive radiation of invertebrates was accompanied by an appearance of new physiologies that could affect strongly the environments in the Late Proterozoic and

Early Cambrian. Along with the growing diversity of bioturbation and biomineralization, it was active suspension feeding or filtering that could radically change the properties of the sediment and water habitats (Fedonkin, 1987, 1992b).

The rise of active filtrators such as sponges, brachiopods, molluscs and some arthropods in the Early Cambrian should have made the ocean water clear and the photic zone deeper (see above), thus providing additional opportunities for the photosynthesizing organisms to occupy lower levels of the water column and deeper benthic environments. The expansion of the photic zone therefore could have resulted in better oxygenation of the pelagic and bottom habitats via the activity of the chlorophyll-bearing organisms.

Removal of the fine particles from the sea water and packing them into pellets should have increased the permeability of the sediment which could lead to a better aeration and colonization of the subsurface bottom environments and to more rapid oxidation of the buried organic carbon.

The growing length of the trophic chains during the Vendian and in particular in the Cambrian has decreased the loss of the major biophile elements and of energy from the ecosystems because of more efficient biological recycling. That could lead to the global oligotrophication of ocean waters. This hypothesis is consistent with the general decrease in buried organic carbon during the Early Cambrian (Knoll, 1992) as well as by the radiation of the Early Cambrian phytoplankton having very large surface-volume ratio due to external processes, spines, ornamentation and very small cell size (Yankauskas, 1989). Noteworthy is that 70% of the biomass and 80% of the chlorophyll belong to the picoplankton in the oligotrophic waters of the Recent ocean (Krupatkina, Berlan and Maestrini, 1985).

Conclusions

The history of the biosphere may be subdivided into two parts. The earlier one was marked by conservative bacterial biota and by radically changing physico-chemical parameters. In contrast to this first eon, the later and shorter one was characterized by the rapidly evolving eukaryotic biota and a relatively stable abiotic environment.

In spite of the incredible individual and populational tolerance of the prokaryotes and their ability to restore rapidly to an optimum population density, they were forced out as dominant groups from most habitats by the eukaryotes.

What do the eukaryotes possess to oppose the enormous population resistance of the prokaryotes? The relatively low tolerance of the eukaryotic organisms was compensated by a more active metabolism and by a larger individual biomass. But the most important "argument" against the prokaryotic ecosystems was the stupefying diversity of the eukaryotes. Due to their growing diversity the eukaryotes were able to give rise to forms competing with the prokaryotes, and they created longer trophic chains between the primary producers and the decomposers. More complex ecosystems turned out to be more economical energetically. That is why the later periods in the history of the biosphere represent mainly the history of the eukaryotic world.

References

Awramik, S.M., 1971. Precambrian columnar stromatolite diversity: reflection of metazoan appearance. Science 174, 825-827.

Bengtson, S., 1992. Proterozoic and Early Cambrian skeletal metazoans. In: Schopf, J.W. and Klein, C. (eds.), The Proterozoic Biosphere. A Multidisciplinary Study. pp. 397-411. Cambridge University Press, Cambridge.

Burzin, M.B., 1987. Strategy of defence from predation, trophic structure and body size distribution in the pelagic zone, and the stages of morphological evolution of acritarchs in the Late Precambrian and Early Cambrian. In: Sokolov, B.S. (ed.). 3rd All-Union Symp. Paleontol. Precambrian and Early Cambrian, Petrozavodsk, Abstr., 12-14 [In Russian].

Castenholz, R.W., Bauld, J. and Pierson, B.K., 1992. Photosynthetic activity in modern microbial mat-building communities. In: Schopf, J.W. and Klein, C. (eds.), The Proterozoic Biosphere. A Multidisciplinary Approach. pp. 279-285. Cambridge University Press, New York.

Christen, R., Ratto, A., Baroin, A., Perasso, R., Grell, K.G. and Adoutte, A., 1991. Origin of metazoans. A Phylogeny deduced from sequence of the 28S ribosomal RNA. In: Simonetta, A.M. and Conway Morris, S. (eds.), The Early Evolution of Metazoa and the Significance of Problematic Taxa. pp. 1-9; Cambridge University Press, Cambridge.

Chumakov, N.M., 1978. Precambrian Tillites and Tilloids. pp. 1-201; Nauka, Moscow [in Russian].

Chumakov, N.M. and Elston, D.P., 1989. The paradox of Late Proterozoic glaciations at low latitudes.

Episodes 12(2), 115-120.

Conway Morris, S., 1993. Ediacaran-like fossils in Cambrian Burgess Shale-type faunas of North America. Paleontology 36(3), 593-635.

Derry, L.A., Keto, L.S., Jacobsen, S.B., Knoll, A.H. and Sweet, K., 1989. Sr isotopic variations in Upper Proterozoic carbonates from Svalbard and East Greenland. Geochimica et Cosmochimica Acta 53, 2331-2339.

Derry, L.A., Kaufman, A.J. and Jacobsen, S.B., 1992. Sedimentary cycling and environmental change in the Late Proterozoic: evidence from stable and radiogenic isotopes. Geochimica et Cosmochimica Acta 56, 1317-1329.

Droser, M.L. and Bottjer, D.J., 1988. Trends in extent and depth of Early Paleozoic bioturbation in the Great Basin (California, Nevada and Utah). In: This Extended Land, Geological Journey in the Cordilleran Section, Las Vegas, Nevada. pp. 123-135.

Durham, J.W., 1970. The fossil record and the origin of the Deuterostomia. Proc. North Amer. Paleontological Conv., 1969 H., 1104-1132.

Fedonkin, M.A., 1983. Organic world of the Vendian. Itogi nauki i techniki, stratigrafia, paleontologia 12, 1-128 [in Russian].

Fedonkin, M.A., 1985. Precambrian metazoans: the problems of preservation, systematics and evolution. Philosophical Trans. Roy. Soc. London Part B 311, 27-45.

Fedonkin, M.A., 1987. Non-skeletal fauna of the Vendian and its place in the evolution of metazoans. Trans. Paleontol. Inst. N226, 1-175; Nauka, Moscow [In Russian].

Fedonkin, M.A., 1990a. Precambrian metazoans. In: Briggs, D.E.G. and Crowther, P.R. (eds.), Palaeobiology. A Synthesis. pp. 17-24; Blackwell Scientific Publ. Ltd.

Fedonkin, M.A., 1990b. Systematic description of the Vendian Metazoa. In: Sokolov, B.S. and Iwanowski, A.B. (eds.), The Vendian System. Vol. 1. Paleontology, 71-120; Springer, Berlin Heidelberg New York.

Fedonkin, M.A., 1990c. Paleoichnology of Vendian Metazoa. In: Sokolov, B.S. and Iwanowski, A.B. (eds.), The Vendian System. Vol. 1. Paleontology, 132-137; Springer, Berlin Heidelberg New York.

Fedonkin, M.A., 1992a. Neoproterozoic ecosystem restructuring: from net to pyramid. Vth Internat. Conf. Global Bio-Events, Göttingen, February 16-19, 1992, Abstr., 33-34; Göttingen.

Fedonkin, M.A., 1992b. Vendian faunas in the early evolution of Metazoa. In: Lipps, J.H. and Signor, P.W. (eds.), Origin and Early Evolution of the Metazoa. pp. 87-129; Plenum Press, New York.

Fedonkin, M.A., 1993. Paleobiology of the Precambrian: on the way to the synthesis. In: Sokolov, B.S. and Iwanowski, A.B. (eds.), Faunas and Ecosystems of Geological Past. pp. 7-21; Moscow, Nauka.

Fedonkin, M.A., 1995, in press. Vendian body fossils and trace fossils. In: Bengtson, S. (ed.), Early Life on Earth. Nobel Symp. N 84, Björkborn, Karlskoa, 297-316.

Fedonkin, M.A., Chumakov, N.M. and Jankauskas, T.V., 1987. The problem of the global biotic and abiotic events in the Late Precambrian. 3rd All-Union Symp. Paleontol. Precambrian and Early Cambrian, Petrozavodsk, Abstr., 99-101 [in Russian].

Field, K.G., Olson, G.J., Lane, D.J., Giovannoni, S.J., Ghiselin, M.G., Raff, E.C., Pace, N.R. and Raff, R.A., 1988. Molecular phylogeny of the animal kingdom. Science 239, 748-753.

Gebelein, C.D., 1976. The effect of the physical, chemical and biological evolution of the Earth. In: Walter, M.R. (ed.), Stromatolites: Developments in Sedimentology 20, 499-516; Elsevier, Amsterdam.

Gehling, J.G. 1991. The case for Ediacara fossils roots to the metazoan tree. Memoirs of the Geological Society of India 20, 181-224.

Germs, J.G.B., 1972. New shelly fossils from the Nama Group, South West Africa. American Journal of Science 272, 752-761.

Glaessner, M.F., 1984. The Dawn of Animal Life. A Biohistorical study. pp. 1-244; Cambridge University Press, Cambridge.

Gnilovskaya, M.B., 1990. Vendian actinomycetes and organisms of uncertain systematic position. In: Sokolov, B.S. and Iwanowski, A.B. (eds.), The Vendian System. Vol. 1. Paleontology. pp. 148-153; Springer, Berlin Heidelberg New York.

Gough, D.O., 1981. Solar interior structure and luminosity variations. Solar Physics 74, 21-34.

Grant, S.W.F., 1990. Shell structure and distribution of Cloudina, a potential index fossil for the terminal Proterozoic. American Journal of Science 290-A, 261-294.

Grotzinger, J.P., 1989. Facies and evolution of Precambrian carbonate depositional systems: emergence of the modern platform achetype. In: Crevello, P.D., Wilson, J.L., Sarg, J.F. and Read, J.F. (eds.), Controls on Carbonate Platform and Basin Development. Soc. Economic Paleontol.

Mineralog., Spec. Publ. 44, 79-106.

Grotzinger, J.P., 1990. Geochemical model for Proterozoic stromatolite decline. Amer. J. Sci. 290-A, 80-103.

Hayes, J.M., Des Marais, D.J., Lambert, I.B., Strauss, H. and Summons, R., 1992a. Proterozoic biogeochemistry. In: Schopf, J.W. and Klein, C. (eds.), The Proterozoic Biosphere. A Multidisciplinary Study. pp. 81-134; Cambridge University Press, Cambridge.

Hayes, J.M., Lambert, I.B. and Strauss, H., 1992b. The sulfur isotopic record. In: Schopf, J.W. and Klein, C. (eds.), The Proterozoic Biosphere. A Multidisciplinary Study. pp. 129-132; Cambridge University Press, Cambridge.

Hofmann, H.J., 1992. Proterozoic carbonaceous films. In: Schopf, J.W. and Klein, C. (eds.), The Proterozoic Biosphere. A Multidisciplinary Study. pp. 349-357; Cambridge University Press, New York.

Hofmann, H.J., 1994, in press. Proterozoic carbonaceous compressions ("metaphytes" and "worms"). In: Bengtson, S. (ed.), Early Life on Earth. Nobel Symp. N 84, Björkborn, Karlskoga, 271-286; Columbia University Press.

Hofmann, H.J., Narbonne, G.M. and Aitken, J.D., 1990. Ediacaran remains from intertillite beds in northwestern Canada. Geology 18, 1199-1202.

Holland, H.D., 1984. The Chemical Evolution of the Atmosphere and Oceans. pp. 1-582; Princeton University Press, Princeton, N.J.

Holland, H.D., 1992. Major aspects of atmospheric evolution during the Precambrian. 29th Int. Geol. Congr., Kyoto, Abstr. 1, p. 170.

Holland, H.D., 1994, in press. Early Proterozoic atmospheric change. In: Bengtson, S. (ed.), Early Life on Earth. Nobel Symp. N 84, Björkborn, Karlskoga, 185-193; Columbia University Press.

Holland, H.D. and Kasting, J.F., 1992. The environment of the Archean Earth. In: Schopf, J.W. and Klein, C. (eds.), The Proterozoic Biosphere. A Multidisciplinary Study. pp. 21-24; Cambridge University Press, New York.

Holser, W.T., 1984. Gradual and abrupt shifts in ocean chemistry during Phanerozoic time. In: Holland, H.D. and Trendall, A.F. (eds.), Patterns of Change in Earth Evolution. pp. 123-144; Springer, Berlin Heidelberg New York.

Kasting, J.F., 1987. Theoretical constraints on oxygen and carbon dioxide concentrations in the Precambrian atmosphere. Precambrian Research 34, 205-229.

Kasting, J.F. and Ackerman, T.P., 1986. Climatic consequences of very high CO_2 levels in earth's early atmosphere. Science 234, 1383-1385.

Kaufman, A.J. and Knoll, A.H., 1992. Neoproterozoic variations in the isotopic composition of sea-water: stratigraphic and biogeochemical implications. 29th Int. Geol. Congr., Kyoto, Abstr. 1, p. 239.

Kaufman, A.J., Hayes, J.M., Knoll, A.H. and Germs, G.J.B., 1991. Isotopic compositions of carbonates and organic carbon from Upper Proterozoic successions in Namibia: stratigraphic variation and the effect of diagenesis and metamorphism. Precambrian Research 49, 301-327.

Kirschvink, J.L., 1992a. A paleogeographic model for Vendian and Cambrian time. In: Schopf, J.W. and Klein, C. (eds.), The Proterozoic Biosphere. A Multidisciplinary Study. pp. 567-582; Cambridge University Press, Cambridge.

Kirschvink, J.L., 1992b. Magnetite biomineralization and the evolution of eucaryotic cell. 29th Int. Geol. Congr., Kyoto, Abstr. 2, p. 340.

Klein, C. and Beukes, N.J., 1989. Geochemistry and sedimentology of a facies transition from limestone to iron-formation deposition in the early Proterozoic Transvaal Supergroup, South Africa. Economic Geol. 84, 1733-1774.

Klein, C., Beukes, N.J., Holland, H.D., Kasting, J.F., Kump, L.R. and Lowe, D.R., 1992. Proterozoic atmosphere and ocean. In: Schopf, J.W. and Klein, C. (eds.), The Proterozoic Biosphere. A Multidisciplinary Study. pp. 135-174; Cambridge University Press, Cambridge.

Knoll, A.H., 1991. End of Proterozoic Eon; Scientific American 256 (4), 64-73.

Knoll, A.H., 1992. Neoproterozoic evolution and environmental change. In: Bengtson, S. (ed.), Early Life on Earth. Nobel Symp. N 84, Björkborn, Karlskoga, 261-270; Columbia University Press.

Knoll, A.H., Hayes, J.M., Kaufman, A.J., Sweet, K. and Lambert, I., 1986. Secular variations in carbon isotope ratios from Upper Proterozoic successions of Svalbard and East Greenland. Nature 321, 832-838.

Krupatkina, D.K., Berlan, B. and Maestrini, C., 1985. Leader of the primary production is an ocean, not land. Priroda 4, 56-62 [in Russian].

Krylov, I.N., 1975. Stromatolites of the Riphean and Phanerozoic of the USSR. pp. 1-243; Nauka, Moscow [in Russian].

Krylov, I.N., 1985. Stromatolites in the Upper Precambrian stratigraphy: problem - 85. Izvestiya Akad. Nauk SSSR, ser. geol. 11, 44-55 [in Russian].

Krylov, I.N. and Zavarzin, G.A., 1988. Sedimentation conditions for the carbonate rocks of the Late Riphean, South Urals. Doklady Akad. Nauk SSSR 300 (5), 1223-1225 [in Russian].

Lambert, I.B. and Donnelly, T.H., 1991. Atmospheric oxigen level in the Precambrian: a review of isotopic and geological evidence. Palaeogeogr., Palaeoclimatol., Palaeoecol. (Global and Planetary Change Section) 97, 83-91.

Leont'ev, O.K., 1982. Physical geography of the world ocean. pp. 1-198; Moscow State University, Moscow [in Russian].

Lowe, D.L., 1992. Major events in the geological development of the Precambrian Earth. In: Schopf, J.W. and Klein, C. (eds.), The Proterozoic Biosphere. A Multidisciplinary Study. pp. 67-75; Cambridge University Press, Cambridge.

Lowenstam, H.A. and Weiner, S., 1989. On Biomineralization. pp. 1-324; Oxford University Press, New York.

McLaughlin, P.J. and Dayhoff, M.O., 1973. Eucaryote evolution: a view based on cytochrome c sequence data. J. molecular evol. 2, 99-116.

Mendelson, K.V. and Schopf, J.W., 1992. Proterozoic and Early Cambrian acritarchs. In: Schopf, J.W. and Klein, C. (eds.), The Proterozoic Biosphere. A Multidisciplinary Study. pp. 219-232; Cambridge University Press, Cambridge.

Monster, J., Appel, P.W.U., Thode, H.G., Schidlowski, M., Carmichael, C.M. and Bridgewater, D., 1979. Sulfur isotope studies on early Archean sediments from Isua, West Greenland. Implications for the antiquity of bacterial sulfate reduction. Geochim. Acta 43, 405-413.

Monty, C.L.V., 1974. Precambrian background and Phanerozoic history of stromatolitic communities, an overview. Ann. Soc. Géol. Belg. 96, 585-624.

Muir, M.D., 1978. Microenvironments of some modern and fossil iron- and manganese-oxidizing bacteria. In: Krumbein, W.E. (ed.), Environmental Biogeochemistry and Geomicrobiology. Ann Arbor Sci. Pub., Ann Arbor M1, 937-944.

Ourisson, G., 1987. A Hypothetical phylogeny for membrain reinforces. Chimia 6, 12-14.

Ourisson, G., 1990. The general role of terpenes and their global significance. Pure and Applied Chemistry 60, 1401-1404.

Riding, R. and Voronova, L., 1982. Calcified cyanophytes and Precambrian-Cambrian transition. Naturwissenschaften 69, 498-499.

Riding, R. and Voronova, L., 1984. Assemblages of calcareous algae near the Precambrian-Cambrian boundary in Siberia and Mongolia. Geol. Mag. 121, 205-210.

Ronov, A.B., 1964. Common tendencies in the chemical evolution of the earth's crust, ocean and atmosphere. Geocemistry International 1, 713-737.

Rozanov, A.Yu., 1986. What had happened 600 million years ago? pp. 1-95; Nauka, Moscow [In Russian].

Rozanov, A.Yu. and Zhuravlev, A.Yu., 1992. The Lower Cambrian fossil record of the Soviet Union. In: Lipps, J.H. and Signor, P.W. (eds.), Origin and Early Evolution of the Metazoa. pp. 205-282; Plenum Press, New York.

Runnegar, B., 1982a. A molecular-clock data for the origin of the animal phyla. Lethaia 15, 199-205.

Runnegar, B., 1982b. Oxygen requirements, a biology and phylogenetic significance of the late Precambrian worm Dickinsonia and the evolution of the burrowing habit. Alcheringa 6, 223-239.

Schopf, J.W., 1992a. The oldest fossils and what they mean. In: Schopf, J.W. (ed.), Major Events in the History of Life. pp. 29-61; Jones and Bartlett Publishers Inc., Boston.

Schopf, J.W., 1992b. Patterns of Proterozoic microfossil diversity: an initial tentative analysis. In: Schopf, J.W. and Klein, C. (eds.), The Proterozoic Biosphere. A Multidisciplinary Study. pp. 529-552; Cambridge University Press, Cambridge.

Schopf, J.W., 1992c. Proterozoic procaryotes: affinities, geologic distribution, and evolutionary trends. In: Schopf, J.W. and Klein, C. (eds.), The Proterozoic Biosphere. A Multidisciplinary Study. pp. 195-218; Cambridge University Press, Cambridge.

Schopf, J.W. and Klein, C. (eds.), 1992. The Proterozoic Biosphere. A Multidisciplinary Study. pp. 1-1348; Cambridge University Press, Cambridge - New York.

Seilacher, A., 1984. Late Precambrian and early Cambrian Metazoa: preservational or real extinctions? In: Holland, H.D. and Trendall, A.F. (eds.), Patterns of change in earth evolution. pp. 159-168; Dahlem Konferenzen, Springer, Berlin Heidelberg New York.

Seilacher, A., 1989. Vendozoa: organismic construction in the Proterozoic biosphere. Lethaia 22, 229-239.

Seilacher, A., 1992. Vendobionta and Psammocorallia: lost constructions of Precambrian evolution. Journal of the Geological Society of London 149, 607-613.

Semikhatov, M.A., 1976. Experience in stromatolite study in the USSR. In: Walter, M.R. (ed.),

Stromatolites. Developments in Sedimentology 20, 337-358; Elsevier, Amsterdam.

Semikhatov, M.A., Fedonkin, M.A., Weiss, A.F., Volkova, N.A., Gnilovskaya, M.B., Golovenok, V.K., Sergeev, V.N., Sochava, A.V., Shenfil, V.Yu. and Yakshin, M.S., 1990. Paleontological method in the stratigraphy of the Precambrian. 2nd All-Union Conf. "General Questions of the Division of the Precambrian in the USSR", Ufa, Abstr., 35-45 [in Russian].

Semikhatov, M.A. and Raaben, M.E., 1993. Dynamics of systematic diversity of Riphean and Vendian stromatolites of northern Eurasia. Stratigraphy. Geological Correlation vol. 1 N2, 9-12 [in Russian].

Semikhatov, M.A. and Raaben, M.E., 1995, in press. Dynamics of global diversity in Proterozoic stromatolites. Stratigraphy. Geological Correlation.

Sochava, A.V., 1993. Isotopic geochemistry of sulfur, carbon and strontium, composition of atmosphere and evolutionary events of the Vendian and Early Cambrian. Stratigraphy. Geological Correlation [in Russian].

Sochava, A.V. and Podkovyrov, V.N., 1992. The evolution of the carbonate rocks composition during Meso- and Neoproterozoic. 29th Int. Geol. Congr., Kyoto, Abstr., p. 302.

Sochava, A.V. and Podkovyrov, V.N., 1993. Evolution of the composition of carbonate rocks in the Late Precambrian, Stratigraphy. Geological Correlation 1(4), 11-26 [in Russian].

Sokolov, B.S. and Iwanowski, A.B. (eds.), 1990. The Vendian System. Vol. 1. Paleontology, 1-383; Springer, Berlin Heidelberg New York.

Stebbins, G.L., Jr., 1974. Adaptive radiation and the origin of form in the earliest multicellular organisms. Systematic Zoology 22, 478-485.

Summons, R., 1988. Biomarkers: molecular fossils. Short Courses in Paleontology 1, 98-113.

Summons, R.E. and Walter, M.R., 1990. Molecular fossils and microfossils of procaryotes and protists from Proterozoic sediments. Amer. J. Sci. 290-A, 212-244.

Timofeyev, P.P., Kholodov, V.N. and Zverev, V.P., 1986a. Water balance of recent sedimentary process. Doklady AN SSSR 287(6), 1435-1439 [in Russian].

Timofeyev, P.P., Kholodov, V.N. and Zverev, V.P., 1986b. Sedimentary cover of Earth as possible source of the hydrosphere. Doklady AN SSSR 288(1), 197-200 [in Russian].

Timofeyev, P.P., Kholodov, V.N. and Zverev, V.P., 1988. Hydrosphere and evolution of Earth. Izvestiya AN SSSR 6, 3-19 [in Russian].

Towe, K.M., 1988. Early biochemical innovations, oxigen and earth history. In: Broadhead, T.W. (ed.), Molecular Evolution and the Fossil Record. Paleontol. Soc., Short Course 1, 114-129.

Towe, K.M., 1994. Earth's early atmosphere: constraints and opportunities for early evolution. In: Bengtson, S. (ed.), Early Life on Earth. Nobel Symp. N 84, Björkborn, Karskoga, 13-23; Columbia University Press.

Tucker, M.E., 1983. Diagenesis, geochemistry and origin of a Precambrian dolomite: the Beck Spring Dolomite of eastern California. J. sediment. Petrology 53, 1097-1119.

Valentine, J.W., 1991. Major factors in the rapidity and extent of the metazoan radiation during the Proterozoic-Phanerozoic transition. In: Simonetta, A.M. and Conway Morris, S. (eds.), The Early Evolution of Metazoa and the Significance of Problematic Taxa. pp. 11-13; Cambridge University Press, Cambridge.

Veizer, J. and Jansen, S.L., 1979. Basement and sedimentary recycling and continental evolution. J. Geol. 87, 341-370.

Veizer, J. and Jansen, S.L., 1985. Basement and sedimentary recycling, 2: time dimension to global tectonics. J. Geol. 93, 625-643.

Vidal, G., 1994. Early ecosystems - limitations imposed by the fossil record. In: Bengtson, S. (ed.), Early Life on Earth. Nobel Symp. N 84, Björkborn, Karskoga, 247-259; Columbia University Press.

Vidal, G. and Knoll, A.H., 1982. Radiations and extinctions of plankton in the Late Precambrian and Early Cambrian. Nature 297, 57-60.

Vidal, G. and Knoll, A.H., 1983. Proterozoic plankton. Geol. Soc. Amer. Mem. 161, 265-277.

Walker, J.C.G., 1985. Carbon dioxide on the early earth. Origin of Life 16, 117-127.

Walker, J.C.G., 1987. Was the Archaean biosphere upside down? Nature 329, 710-712.

Walter, M.R., 1992. Stratigraphic distribution of stromatolites and allied structures. In: Schopf, J.W. and Klein, C. (eds.), The Proterozoic Biosphere. A Multidisciplinary Study. pp. 507-509; Cambridge University Press, Cambridge.

Walter, M.R., Grotzinger, J.P. and Schopf, J.W., 1992. Proterozoic stromatolites. In: Schopf, J.W. and Klein, C. (eds.), The Proterozoic Biosphere. A Multidisciplinary Study. pp. 253-260; Cambridge University Press, Cambridge.

Walter, M.R. and Heys, G.R., 1985. Links between the rise of the Metazoa and the decline of the stromatolites. Precambrian Research 29, 149-174.

Wilde, P. and Berry, W.B.N., 1982. Progressive ventilation of the oceans - Potential for return to anoxic conditions in the post-Paleozoic. In: Schlanger, S.O. and Cita, M.B. (eds.), Nature and Origin of Cretaceous Carbon-rich Facies. pp. 209-224; Academic Press, London.

Yankauskas, T.V., ed., 1989. Microfossils from the Precambrian of the USSR. pp. 1-190; Nauka, Leningrad [in Russian].

Zakrutkin, V.E., 1993. On the rates of the organic matter accumulation in the Precambrian and Phanerozoic. In: Sokolov, B.S. and Rozanov, A.Yu. (eds.), The Problems of the pre-Antropogene evolution of biosphere; Nauka, Moscow.

Zavarzin, G.A., 1984. Bacteria and composition of atmosphere. pp. 1-199; Nauka, Moscow [in Russian].

Manuscript received May 1993
Revision received March 1994

Author's address:

Mikhail A. Fedonkin, Paleontological Institute, Russian Academy of Sciences, Profsoyuznaya ul. 123, Moscow 117647, Russia

The Basal Cambrian Transition and Cambrian Bio-Events (From Terminal Proterozoic Extinctions to Cambrian Biomeres)

Martin D. BRASIER

Abstract. Within the Cambrian about six globally traceable extinction events are recognised, of which those across the Mid Botomian through to Toyonian/Amgan are of highest order. All events are associated with facies changes, mostly combined with stepwise extinctions that preferentially affected nearshore and endemic taxa. The events appear to have coincided with changes in climatic and/or oceanographic parameters, such as sea level or fluctuation in oxygen-depleted water masses.

Contents

Introduction	113
Continental Reconstructions	114
Climate	117
Chrono- and Biostratigraphy	118
Geochronology	118
Sea Level	120
Biosphere and Chemostratigraphy	121
Data Bases and Diversity Change	124
Biodiversity - A Summary	124
Bioevents	127
1. The Terminal Proterozoic Extinction (PC/C Event)	127
2. Early Cambrian Radiation Events	128
3. The Botomian-Toyonian Crisis (M'Bo and Ty/Am Events)	131
4. Upper Cambrian Events	132

Introduction

The Cambrian System marks the traditional base of the Phanerozoic Erathem, distinguished from the preceding terminal Proterozoic (e.g. the 'Vendian System') by the appearance of distinctive skeletal fossils and trace fossil assemblages (Cowie and Brasier, 1989). It is now thought to have been a relatively short and dynamic period (ca 45 Ma) during which most modern invertebrate groups experienced their first evolutionary successes (the Cambrian Explosion) and setbacks (the Cambrian Biomere Events). It was a period of profound environmental change, with a major rise in sea level, continental rifting and postulated changes in ocean-atmosphere chemistry. Carbonate sediments are widespread on many cratons, while intervals of phosphorites, black alum shales, and metaliferous sedimentary ores indicate the influence of oxygen-depleted water masses. Major oil and gas source rocks and reservoirs are found across the Precambrian-Cambrian boundary interval in the Arabian Gulf, Siberian and Yangtze Platforms.

The skeletal fossil record of the Cambrian is notoriously patchy; fossils are often scarce, fragmentary or small, and confined to particular facies and beds. Endemism was also pronounced, so that precise global stratigraphy is often elusive. The system is rewarded, however, with an exceptional number of fossil lagerstätten deposits (e.g. the Burgess Shale, Chenjiang and Swedish Orsten faunas).

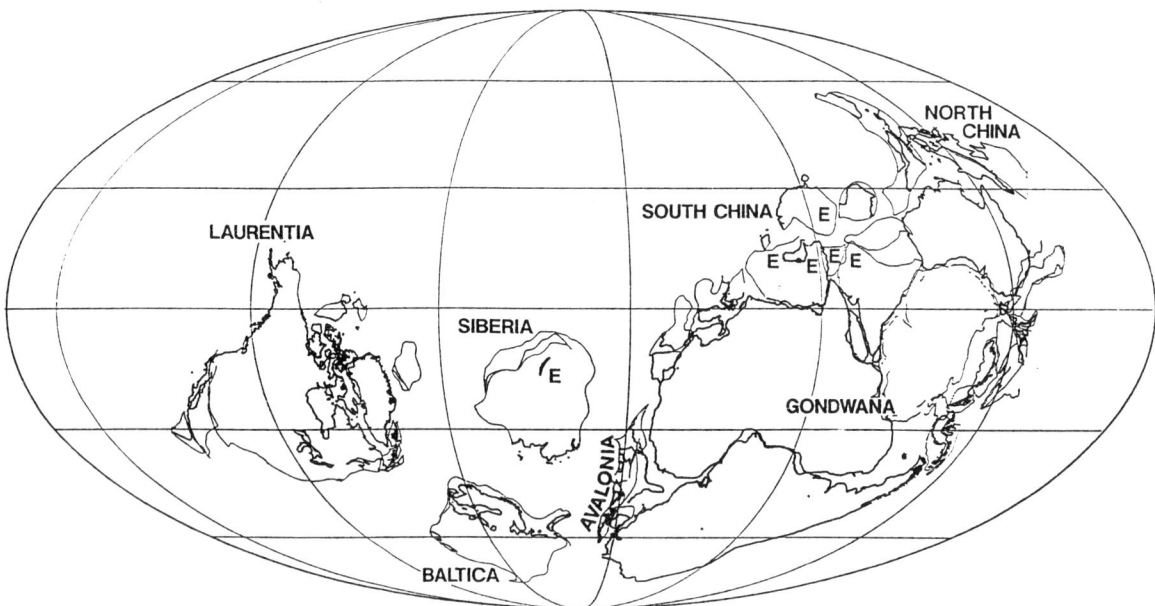

Fig. 1. Palaeogeographic reconstruction of continents in early Cambrian time: Atdabanian/Botomian stages (after McKerrow et al., 1992). E = evaporites

Continental Reconstructions

Recent attempts to reconstruct Cambrian configurations of the continents include those of Burrett et al. (1990), Moores (1991), Dalziel (1991, 1992), Kirschvink (1992), McKerrow et al. (1992) and Courjault-Rade et al. (1992). Most authors now infer that the main continental areas lay at relatively low latitudes during this time. Fig. 1, for example, shows the configuration in Late Atdabanian to Toyonian times.

Many recent papers have referred to a single late Precambrian Supercontinent (variously termed Pangea, Palaeopangea or Rodinia; e.g. Morel and Irving, 1978; Bond et al., 1984; Piper, 1983, 1987; McMenamin and McMenamin, 1989) and its supposed break up in the latest Precambrian to Cambrian. For example, a possible rifting of Laurentia, Australia and Antarctica is postulated (Moores, 1991; Dalziel, 1991). There is little palaeontological support for such an assembly in the Cambrian, however, and initial separation must have taken place well before this (Dalziel, 1992) Trilobite and palaeomagnetic data have also suggested a late Proterozoic supercontinental group formed by Laurentia, Baltica and Siberia: but this must also have rifted apart from each other prior to c. 600 Ma (McKerrow et al., 1992).

An Asiatic Gondwanaland region comprised the adjacent cratons of North and South China, Tarim, Qaidam, India-Pakistan, Iran-Oman (Arabia) and Kazakhstan, and probably Central Mongolia and Tuva-Mongolia. Palaeontological and lithological similarities have suggested that numerous rift basins in the Riphean-Vendian accompanied the break up of a supercontinent into blocks that may also have included Siberia (Ilyin, 1990). These were either separated by deep water volcanogenic intracontinental basins (Courjault-Radé et al., 1992) or by oceanic crust. Continued rifting in the Cambrian was accompanied by greater subsidence and evaporite-carbonate-phosphate facies (Brasier, 1989a; Ilyin, 1990).

Faunal similarities are shared between this region and Australia, which was almost certainly part of eastern Gondwanaland at this time. Courjault-Radé et al. (1992) have argued for an active margin along eastern Australia, Antarctica and western South America, characterised by a narrow steeply sloping shelf passing seaward into oceanic volcanics.

An Afro-European Gondwanaland region comprised the arcs of Avalonia and southern Europe, the African cratons and South American cratons. The latter (except the exotic Cambrian of the Precordilleran terrane of Argentina) is here placed adjacent to Africa. There is good evidence across much of this large region for latest Precambrian island arcs and deformation zones (the Avalonian, Cadomian, Pan African

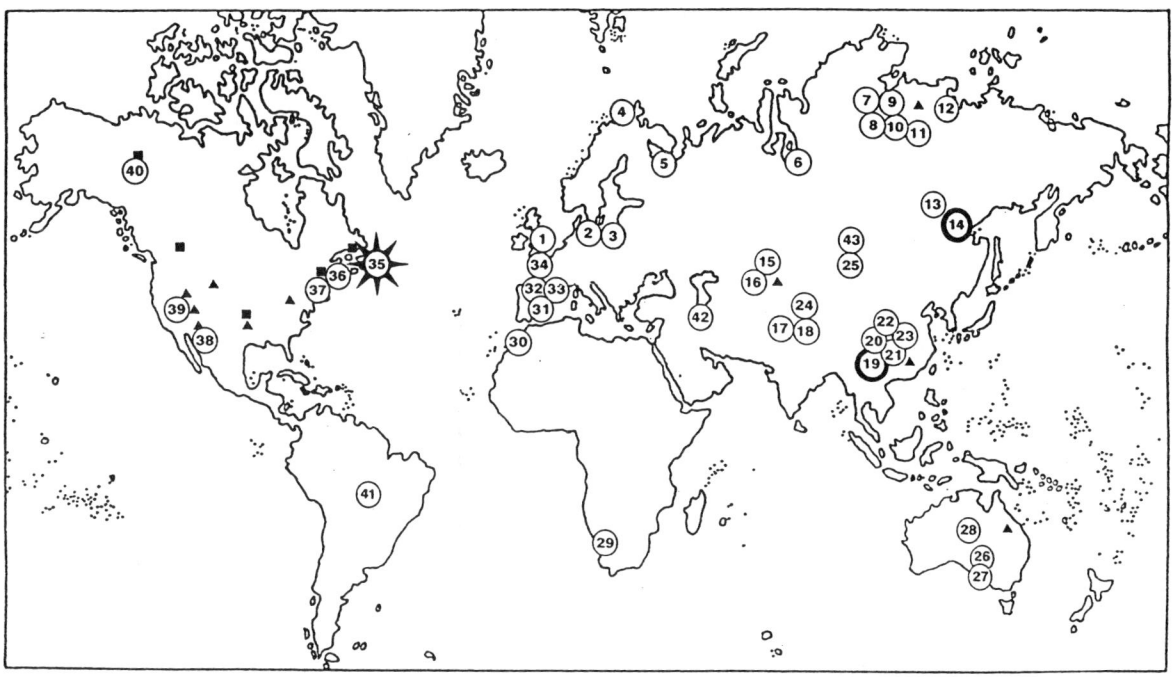

Fig. 2. Present day location of main sites spanning Cambrian events shown on Fig. 4. PC/C event: 7-10, Preanabar and Anabar regions of Siberia; 11, Olenek uplift, Siberia; 13, lena River (Siberia) 14, Dvortsy and Ulakhan Sulugur, Aldan River (Siberia), 19, Meishucun, Yangtze Platform (China); 25, Salanygol sections of SW Mongolia; 35, SE Newfoundland stratotype for PC-C marker and events. For further details of numbered localities see Cowie and Brasier (1989). M'Bo and Ty/Am Events: 13, middle reaches of Lena River, Siberia; 19-23, Yangtze Platform, south China; 26, South Australia; 32, N. Spain; 35, eastern Newfoundland and Labrador; 39, eastern California and Nevada; 40, Mackenzies, NW Canada. Tp/Ib event = black squares. Mi/Id and Dr/Fc Events = black triangles

and Brasiliano orogenies) which culminated in the aggregation of western Gondwanaland by about 540 Ma, followed by subsidence of passive (transtensional?) margins in the Cambrian. Both Avalonia and the south European regions were extended into intracontinental basins and blocks with distinct Cambrian successions bearing largely siliciclastic sediments.

The presence of archaeocyathan faunas in southern Europe, Morocco and the Siberian Platform, and their shared bigotinid trilobite faunas, suggest that these areas were adjacent to each other in the early Cambrian, and at low latitudes. Avalonia had close faunal links with western Gondwanaland, but lacked bigotinids and archaeocyathans and may have been situated off west Africa and Florida (McKerrow et al., 1992) or even off Columbia (D. Keppie in Moore, 1992).

Eastern and western Gondwanaland (in the modern sense) therefore have quite distinct geological histories, and they are faunally rather different as well (Fig. 1). Kirschvink (1992) has argued, on palaeomagnetic evidence, that these two parts of Gondwanaland did not come together until later in the Cambrian but there are still problems with this view. Although the shared faunal and lithological character of the Iran-Arabian, Indian and Chinese cratons may perhaps be discounted, (as may the similarity between Sardinian and Chinese trilobites; Pillola, 1990) there is evidence that the collision zone between India and the Mozambique belt is dated rather earlier, at c. 550 to 600 Ma (Prof. B. Windley, oral comm.).

During the early Cambrian, there was a postulated northward migration of Laurentia, towards the Equator as its separation from Baltica and Siberia increased (McKerrow et al., 1992). A southward movement of western Gondwana took it into high latitudes by the end of the Cambrian, so that warm water carbonates and biofacies are replaced by cooler water clastics and trilobites.

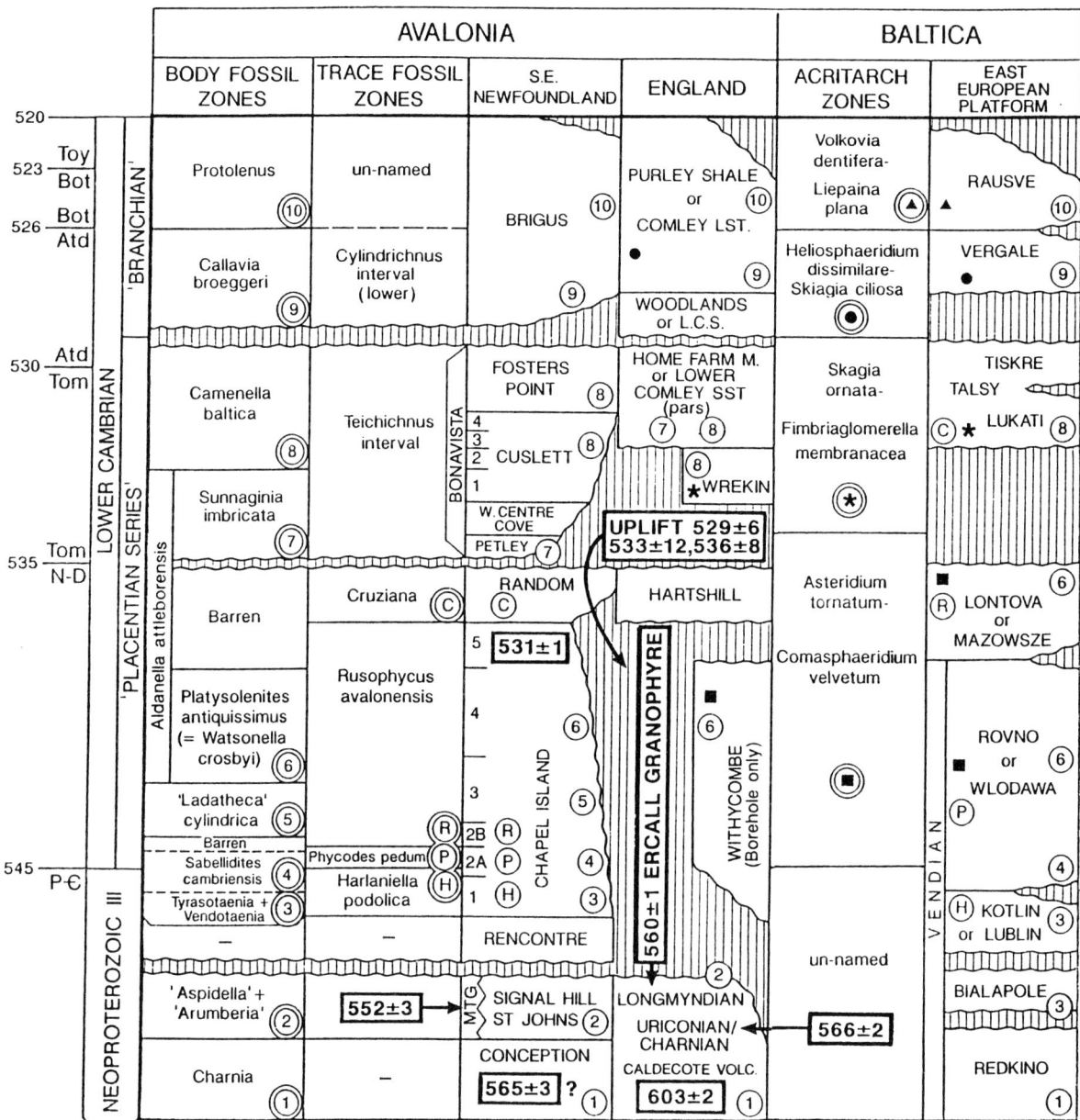

Fig. 3 (this and opposite page). Preliminary correlation between selected stages, biozones and lithostratigraphic units across the Precambrian-Cambrian transition. First appearance of selected fossil markers are shown. 1-10 (circled) = Avalonian markers; symbols = Baltic acritarch markers; H, R, P, C (circled) = trace fossil markers from Avalonia and elsewhere; N-D, T, A, B, E (circled) = archaeocyathan assemblages from Siberian stages and other localities. Numbers in boxes summarise the latest information on U/Pb and Pb/Pb zircon dates for the volcanic ashes in the latest Proterozic-early Cambrian (see text). Modified from Brasier (1992b)

The evidence from strontium isotopes is important to consider here. There is no evidence for a long-term increase in the 'hydrothermal' signal during the Cambrian, such as might seem consistent with major rifting. Rather, there is a steady increase in the signature from cratonic weathering between terminal Proterozoic and late Cambrian (Holser, 1984; Derry and Brasier, unpublished data).

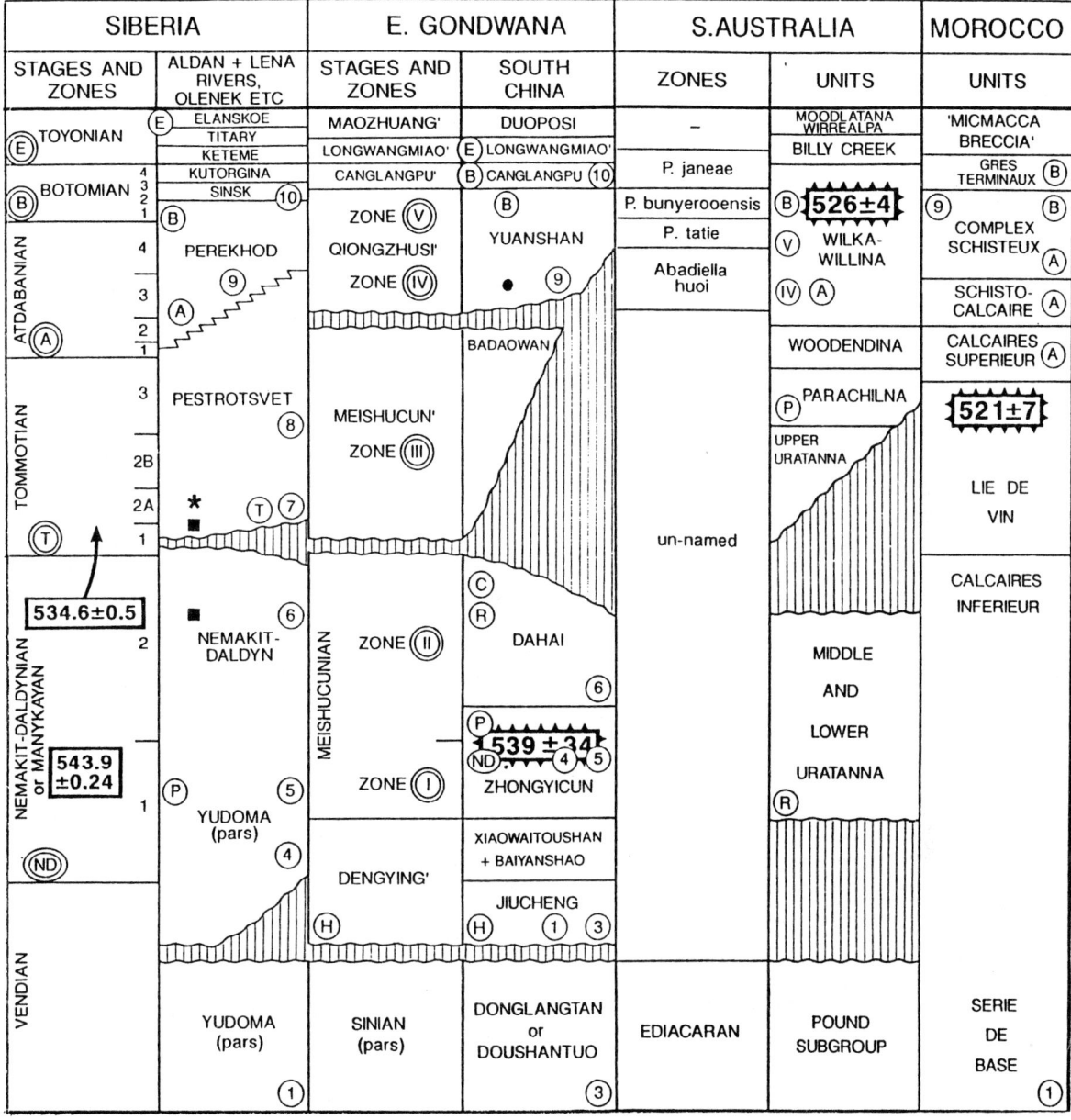

Climate

No certain glacial deposits are recorded in the Cambrian, in contrast to the extreme climates of the preceding terminal Proterozoic. The widespread occurrence of oolitic and dolomitic carbonates, evaporites and archaeocyathan reefs in the early Cambrian indeed suggests relatively warm climates over large areas (Fig. 1). The general impression, therefore, is one of ameliorating climate, from coldhouse to greenhouse, as summarised by Brasier (1992a), but with frequent climatic oscillations in the early Cambrian, as suggested by carbon isotopic fluctuations, and by sedimentary rhythms.

Most landmasses are thought to have lain at low latitudes through this time interval but Baltica and Avalonia probably lay at temperate latitudes, as perhaps did South America excepting the Precordillera terrain (McKerrow et al., 1992). Sedimentological and palaeomagnetic data also suggest a northward migration of Laurentia towards the equator and a southward movement of western Gondwana towards the poles during the Cambrian Period (ibid.).

Chrono- and Biostratigraphy

The present day location of major Precambrian-Cambrian boundary and Cambrian successions is shown in Fig. 2. Although the base of the Cambrian System has now been fixed at the base of the *Phycodes pedum* ichnofossil Zone in SE Newfoundland (Figs. 3, 4), there is as yet no agreed chronostratigraphic framework for global stratigraphy within the Cambrian. This arises partly from problems related to the endemism of polymeroid trilobites on which so much of the stratigraphic correlation depends. Hence the boundaries between lower/middle and middle/upper Cambrian differ between countries and are not strictly formal, nor are the 'stages' widely employed in different regions yet ratified (e.g. Fig. 4).

There is no formally agreed name for the basal stage of the Cambrian. Although Siberian stages have received widespread usage for the early Cambrian, it must be emphasized that their correlation beyond Siberia often rests on slender ground. Correlation hitherto has depended on assemblages of archaeocyathans, small shelly fossils, trilobites and acritarchs (Brasier, 1989b). One of the major aims of IGCP Project 303 has been to explore the potential of calibration against geochemical and sequence stratigraphic criteria.

Reference to early Cambrian events herein will be informally related to the Siberian stratigraphic scale, assuming that the base of the system in SE Newfoundland correlates with the base of the Nemakit-Daldynian (or Manykayan) Stage in Siberia, as inferred by Landing (1992, 1994) and Brasier (1992b).

Agnostoid trilobites have great potential for global middle and upper Cambrian time scale (Robison et al., 1977) and should be used where possible (e.g. Fig. 4) but regional polymeroid zones still predominate in inner shelf successions, where many of the bioevents take place.

Geochronology

Figure 3 brings together the latest information on U/Pb and Pb/Pb zircon dates for the volcanic ashes in the latest Proterozoic-early Cambrian. A date of 563 ± 3 Ma BP obtained by G. Dunning of the Royal Ontario Museum (Benus, 1988) from levels with Ediacara fauna in the Conception Group of SE Newfoundland is widely cited but awaits full documentation. Volcanics from the Marystown Group, below the unconformity beneath the Rencontre Formation (Fig. 4) have yielded a U-Pb age of 552 ± 3 Ma (Myrow and Hiscott, 1993). The date of 545 ± 5 from the putative pre-Tommotian *Rusophycus avalonensis* Zone of New Brunswick has been revised upward to near 530 myrs BP (Landing in Odin, 1993) and to 531 ± 1 (Isachsen and others cited in Bowring et al., 1993) but without documentation at the time of writing. Tommotian-type small shelly fossils (*Aldanella*, *Watsonella*, *Lapworthella*) occur below a similar level in southeastern Newoundland (Landing, 1992), and this date may therefore be of Tommotian age in this writer's opinion.

A similar suite of rocks in southern Britain yields dates that suggest two phases of events: metamorphism and intrusion of the (?Riphean) Mona Complex prior to 600 Ma BP; deformation and intrusion of terminal Proterozoic strata in England at about 560, calibrated by the Ercall granophyre at 560 ± 1 (Tucker and Pharaoh, 1991). In SE Newfoundland this 'Avalonian orogen' occurred before the Precambrian-Cambrian boundary. In England, the Ercall granophyre is overlain by the Wrekin Quartzite, which contains possible brachiopod remains (G. Odin, pers. comm.) and acritarchs of the *Skiagia ornata-Fimbriaglomerella mebranacea* Zone (Wright et al., 1993), with *Camenella baltica* and *Mobergella* in overlying Lower Comley Sandstone (Brasier, 1984). These data imply that the Wrekin Quartzite is no older than mid Tommotian and that part of the earlier story is missing here.

Several periods of uplift and basin reorganisation followed the 'Avalonian orogeny' (Landing, 1992), notably at the top of the Random Formation, and Fosters Point Formation and equivalents across much of Avalonia (Fig. 3). Some English Rb-Sr isochron dates may therefore relate to earliest Cambrian uplift, contemporaneous with deposition of the Chapel Island Formation in the Burin area of SE Newfoundland, and subsequent block movements prior to deposition of the Bonavista cycle: dewatering and uplift of the Longmyndian at 529 ± 6 (Bath, 1974); low temperature alteration of Ercall granophyre at 533 ± 12 (Patchett et al., 1980); uplift of Rushton Schist at 536 ± 8 (Patchett et al., 1980; Wright et al., 1993). These isochrons may be taken to postdate the Ercall granophyre and antedate deposition of the Wrekin Quartzite (Wright et al., 1993).

Isochrons obtained by Bowring et al. (1993) from lower Nemakit-Daldynian (or 'Manykayan', 543.9 ± 0.24) and from below supposed lower Tommotian (534.6 ± 0.5) in northern Siberia are now receiving wide attention, but it is emphasized here that their stratigraphic position is not very well proscribed. First, it is by no means certain that the lower isochron is

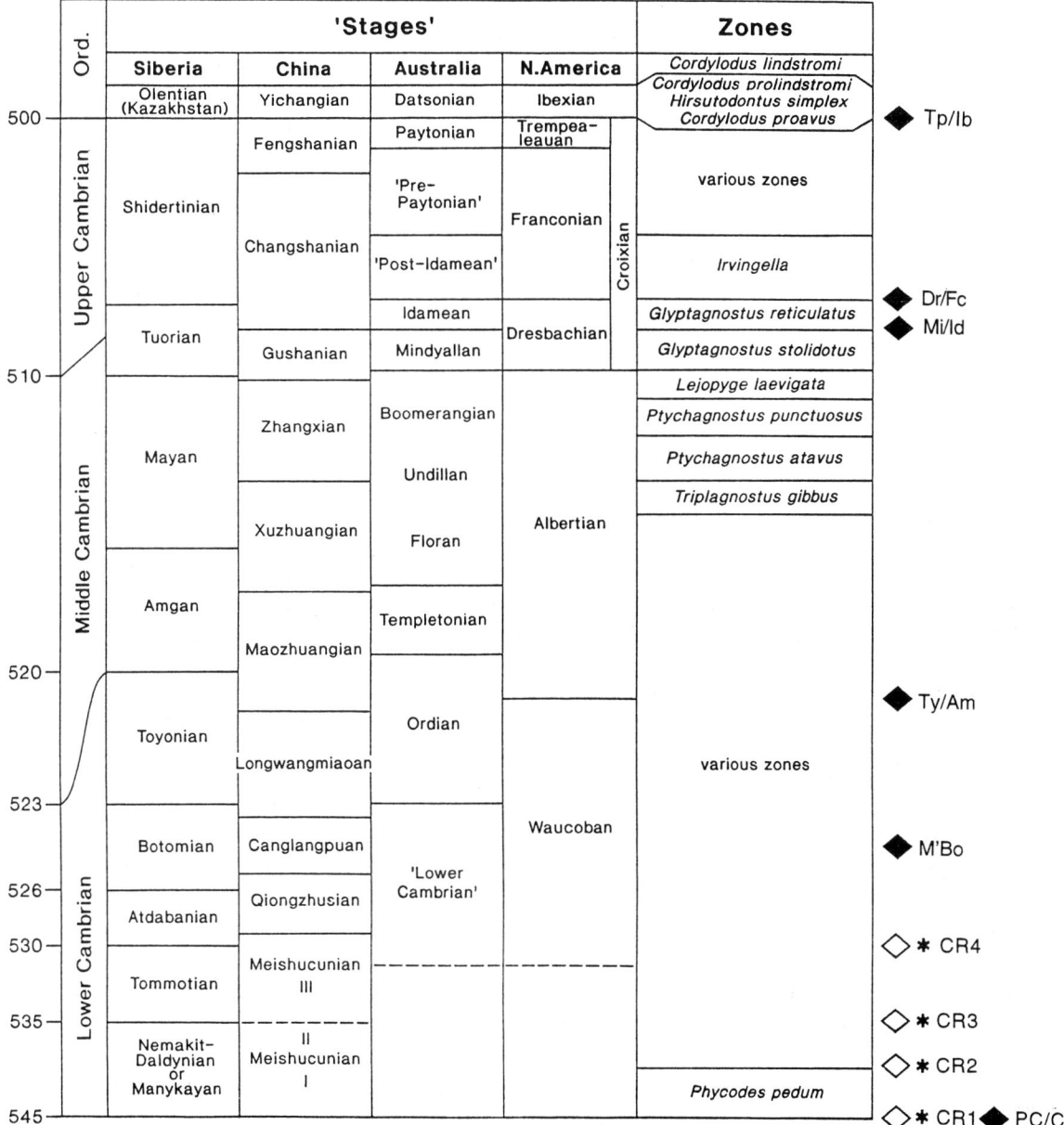

Fig. 4. Position of six major bio-events during the Cambrian Period (at right). Provisional and informal correlation between the schemes for Siberia, China, Australia and North America. At left, calibrated against a suggested new radiogenic isotope chronology (see text). At right, calibrated against pandemic zones, notably the *Phycodes pedum* ichnozone at the base, and agnostoid/polymeroid trilobite zones and conodont zones in the middle to upper part. PC/C: Precambrian/Cambrian boundary, as formally defined in Newfoundland. CR1 to CR4: Phase 1 to Phase 4 of the Cambrian Radiation Event. The following are provisionally named after Siberian stages: M'Bo: mid Botomian; Ty/Am: Toyonian/Amgan. The following is provisionally named after two Australian stages: Mi/Id: Mindyallan/Idamean. The following are provisionally named after North American stages: Dr/Fc: Dresbachian/Franconian; Tp/Ib: Trempealeauan/Ibexian. None of these stages have been formally ratified by the International Commission on Stratigraphy

coeval with the stratotype level in Siberia; our preliminary data from Mongolia indicate it could be older than the Precambrian-Cambrian boundary. Second, the younger isochron was obtained from conglomerates that lie within an extensive hiatus between Nemakit-Daldynian and *medial* Tommotian (Astashkin et al., 1991). Bowring et al. (1993) have suggested a minimum date for the base of the Cambrian of about 544 Ma and for the base of the Tommotian of 533 Ma. But the oldest date for the basal Tommotian could actually be several Ma older than 534.6 ± 0.5. Either way, it seems that the isochron of 531 from New Brunswick, referred to above, may be interpreted as a Tommotian date.

Data obtained by Compston et al. (1992) from ashes in the Zhongyicun phosphorite of Meishucun, Yunnan, south China, regrettably do not give a sharply defined isochron. Palaeontological and carbon isotopic data are consistent with a late Nemakit-Daldynian age (Brasier et al., 1990) and their isochron of 539 ± 39 broadly fits this interpretation.

The date of 526 ± 4 from south Australia (Cooper et al., 1992) can be closely tied to a level close to the Atdabanian-Botomian boundary. This means that the date of 521 ± 7 from the upper Serie Lie de Vin of Morocco, of Tommotian/Atdabanian boundary age (Compston et al., 1992) may lie close to the older part of the error bar.

The base of the Ordovician (earliest Tremadocian *Cordylodus proavus* Zone) is dated at Dayangcha in China at 501 ± 7 (Norford, 1991). A plausible framework for the latest Precambrian to Cambrian interval is therefore as follows: late Avalonian volcanism (560 Ma); base of Kotlinian interval (c. 555 Ma); Precambrian/Cambrian boundary (c. 545 Ma); Nemakit-Daldynian/Tommotian boundary (c. 535 Ma); Tommotian/ Atdabanian boundary (530 Ma); Atdabanian/Botomian boundary (526 Ma); Botomian/Toyonian boundary (523 Ma); top of Toyonian (c. top of lower Cambrian, 520 Ma); top of *Lejopyge laevigata* Zone (about top of middle Cambrian, 510 Ma); top of upper Cambrian (c. 500 Ma). This implies a much shorter time scale than employed in some recent compilations (e.g. Schopf and Klein, 1992).

Sea Level

A global sea level curve for the Cambrian has not yet been compiled by Cambrian specialists. Fig. 5 brings together data from facies changes on the relatively stable platforms of Siberia, south China and North America. Precise correlation between these changes is not yet possible, and only the North American data has been geophysically modelled (Bond et al., 1988).

Evidence from the onlap of Cambrian strata onto older Proterozoic or even Archaean rocks in cratonic areas (e.g. authors in Holland, 1971, 1974) certainly suggests that the Cambrian was a period of major sea level rise (Matthews and Cowie, 1979). Progressively deeper water facies on cratons of the Atlantic borderlands (Brasier, 1980; Notholt and Brasier, 1986) reflect this broad trend, but major episodes of subsidence on the cratonic margins as continental rifting proceeded (e.g. Cocozza and Gandin, 1990; Landing, 1992) make the eustatic component harder to recognise.

Varying rates of subsidence and sedimentation are shown in Fig. 6. Here, the cumulative thickness of early Cambrian sections from the Siberian, Indo-Pak and Yangtze Platform are plotted against the time scale outlined above. Assuming the latter to be correct, each shows a long-term trend towards an increased rate of basin subsidence and sedimentation from Vendian to Cambrian. The Nemakit-Daldyn and Tommotian, however, are characterised by sharply reduced rates. This interval might separate two phases of rifting and basin formation although coincidence with phosphatic intervals (Fig. 6) might appear to suggest another possibility: the suppression of carbonate precipitation by a stagnant, nutrient-enriched water mass during a major transgression (cf. Hallock and Schlager, 1986; Brasier, 1992d). However, a similar pause in basin subsidence is seen in data from clastic successions in Australia (e.g. Lindsay, 1993) so that the latter explanation seems less likely.

Within the context of a general rise in sea level, attention has begun to focus on sedimentological breaks and sequence boundaries. Some of those worthy of further attention are shown in Fig. 5. Studies of the Laurentian craton are important in this respect. Palmer (1981) noted widespread hiati across Laurentia at two levels: just above the regional lower/middle Cambrian boundary (the so-called 'Hawke Bay Event') and across the Dresbachian-Franconian boundary, dividing the Cambrian into three major sequences (Sauk I, II and III). Geophysical modelling by Bond et al. (1988) tended to confirm this pattern, superimposed on a long term rise in sea level of likely eustatic origin. On these sequences are superimposed short term cycles (e.g. the Grand Cycles of Aitken, 1981) that are more difficult to correlate (Bond et al., 1988).

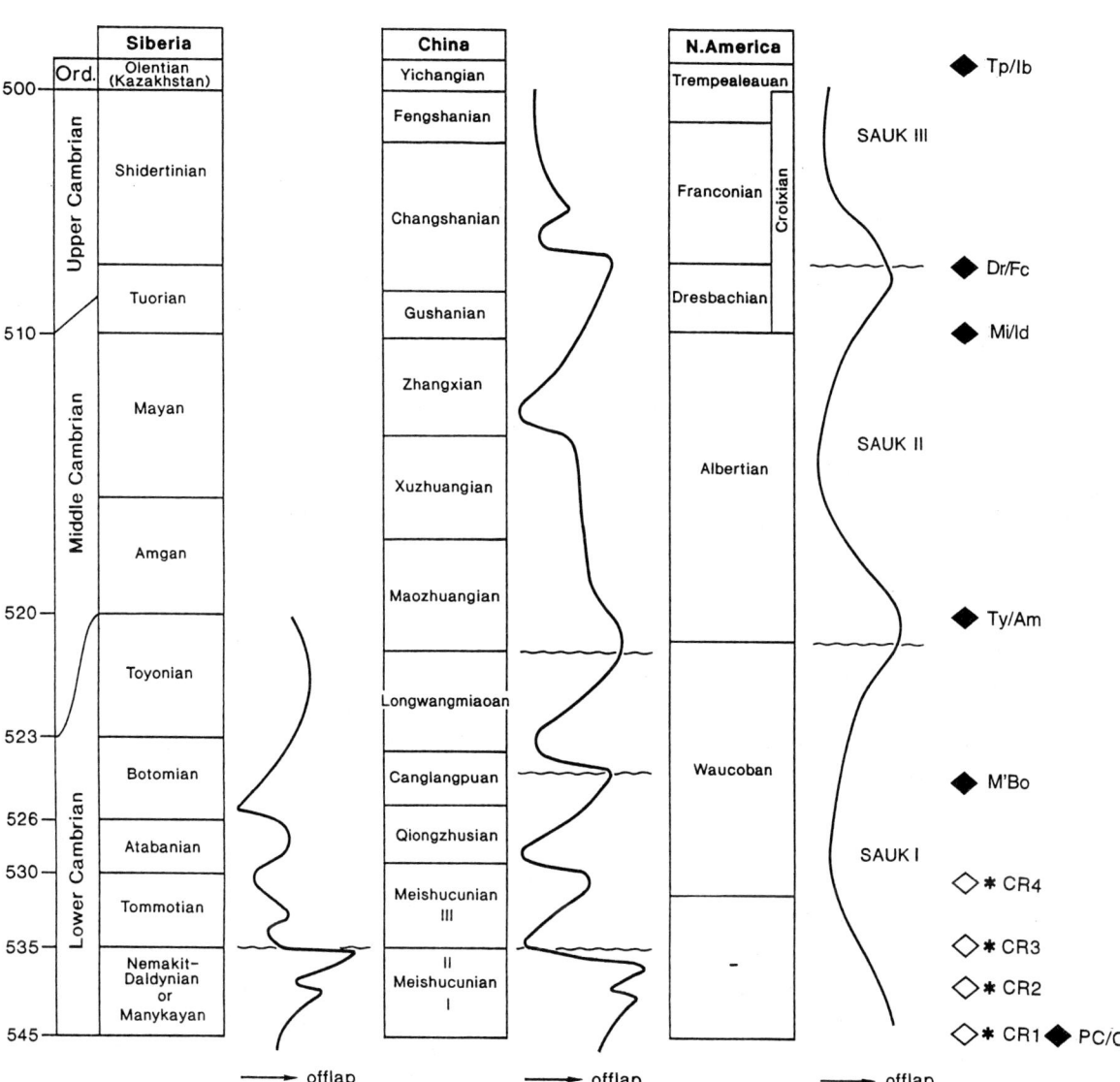

Fig. 5. Provisional sea-level curves for three stable cratonic areas: small-scale trends for the early Cambrian on the Siberian Platform (based on Zhuravleva et al., 1990; Brasier, 1992a): small to medium scale trends for the Cambrian of the Yangtze Platform, China (from data in Xiang and others, 1981; Chang, 1988; Lindsay, 1993); and medium-scale trends for the Cambrian of the Laurentian craton (based on Palmer, 1981; Bond et al., 1988). The large scale trend on each craton is for overall sea-level rise during the Cambrian. Wavy lines show the positions of reported, region-wide depositional breaks. "Offlap" refers to regressive facies; geophysical evidence has yet to be analysed

Biosphere and Chemostratigraphy

The geochemical background to the Cambrian explosion is shown in Fig. 7. The data of Strauss and others (in Schidlowski et al., 1992) on carbon isotopes from minimally altered kerogens (Fig. 7E) highlights the anomalous nature of the Upper Proterozoic biosphere. There was a gradual climb to a

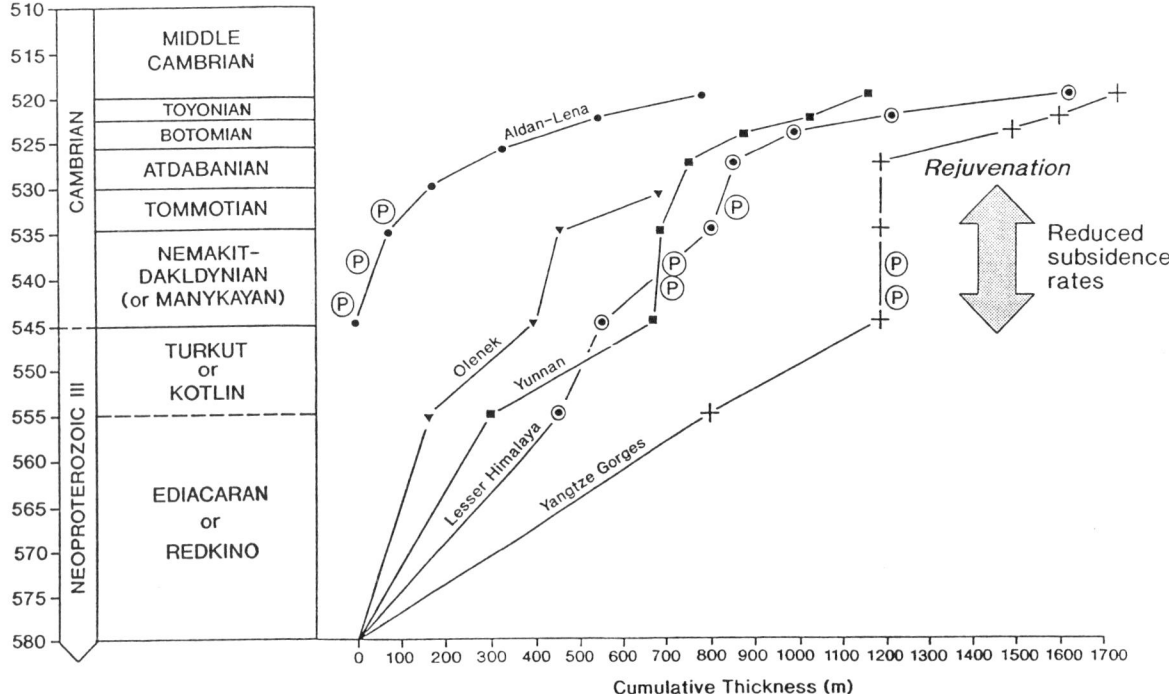

Fig. 6. Cumulative thickness (in metres) of several Asiatic successions, plotted against the revised time scale (not corrected for compaction). Siberian Platform in the Aldan-Lena, and Olenek regions (from data in Rozanov and Sokolov, 1984; Sokolov and Fedonkin, 1986; Khomentovsky, 1986; Brasier et al., 1993). Yangtze Platform in Yunnan, and the Yangtze Gorges (from data in Xiang and others, 1981; Chang, 1988). Indo-Pak Platform, in the Lesser Himalaya (from authors cited in Brasier, 1989a). In each area, the accumulation rate was lower during the deposition of phosphatic sediments (P) and increased greatly during the Botomian-Toyonian

major interval of widely dispersed and heavy $\partial^{13}C$ at about 800 Ma BP, declining to very light values before the Precambrian-Cambrian boundary. The $\partial^{13}C$ of carbonates studied by Strauss and others (Fig. 7D, cross hatched ornament) does not pick this trend out, however. These authors interpret the trend as representing a progressive reduction in organic discrimination against heavier ^{13}C through the Proterozoic, related to metabolic evolution and/or a reduction in the partial pressure of atmospheric carbon dioxide. The sulphur isotope data reviewed by Lambert and Donnelly (in Schidlowski et al., 1992; see Fig. 7C) even shows a peak in heavy $\partial^{34}S$ values, consistent with widespread anoxia and carbon burial in the late Proterozoic. This anomalous interval also coincides with the interval of Proterozoic glaciations at low latitudes (Fig. 7G), the last of which is associated with a major negative $\partial^{13}C$ anomaly (Knoll and Walter, 1992).

The latter authors have provided a sketch of $\partial^{13}C$ through the Ediacaran interval, during which values seem to have been relatively heavy, falling to lighter as the Precambrian-Cambrian boundary is approached. Major hydrocarbon source rocks were deposited on the cratons of Arabia, Asia and Siberia at this time (McKirdy and Imbus in Schidlowski et al., 1992). In Siberia, these remarkably light hydrocarbons (Fig. 7E) were generated in evaporitic basins with the involvement of green sulfur bacteria (Summons and Powell in Schidlowski et al., 1992). Indeed, there is now a great deal of evidence for the development of oxygen-depleted and nutrient-enriched waters prior to, and across, the Precambrian-Cambrian boundary interval (Brasier, 1992c, 1992d).

Carbon isotopic trends in the early Cambrian have been studied by numerous authors (Table 1). Siberian platform data are of good quality and well-calibrated by fossils, hence can provide a curve for comparison and

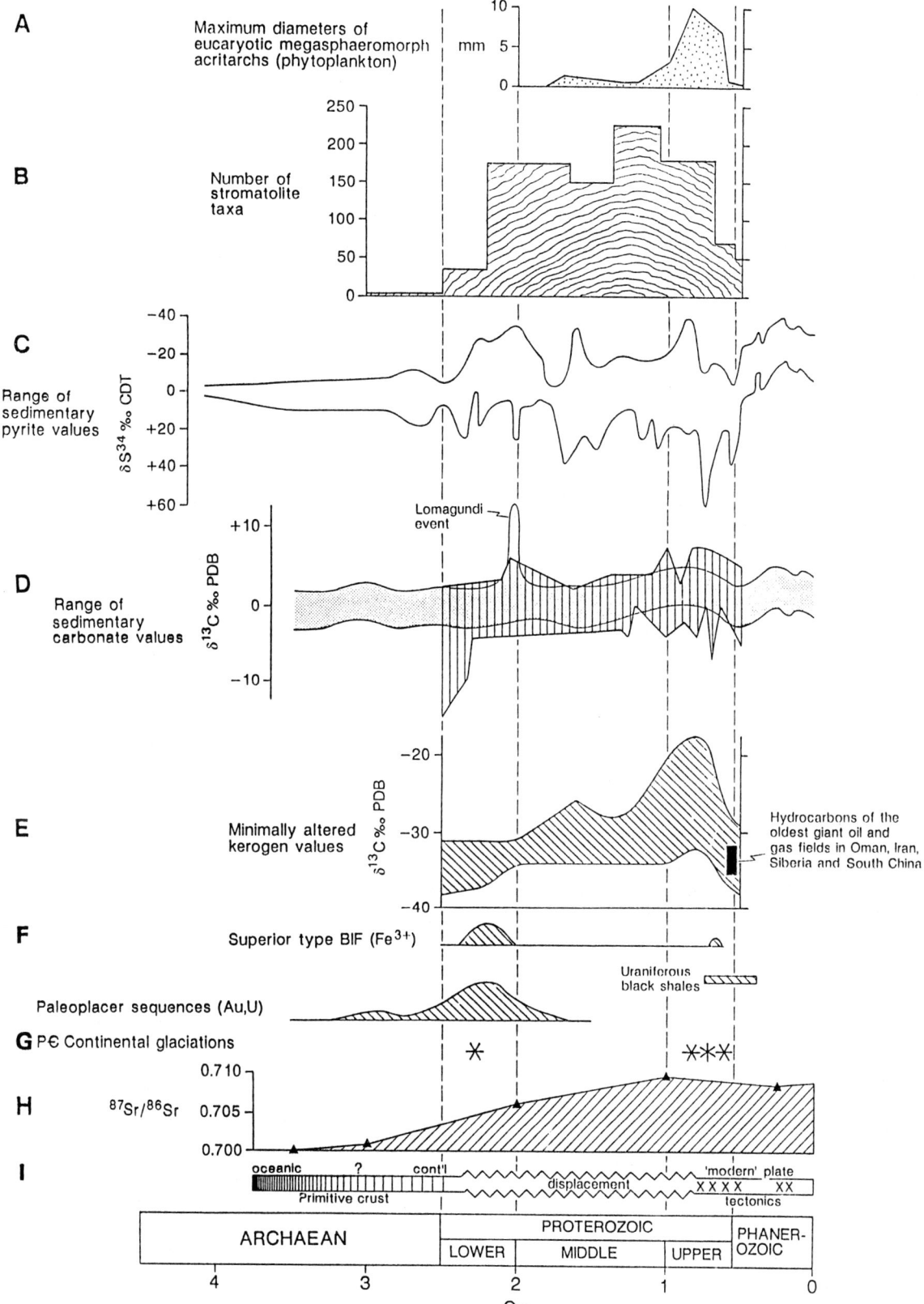

Fig. 7. Geochemical background to the Cambrian explosion at the start of the Phanerozoic. Based on Brasier (1993) from data in Schidlowski et al. (1992), Schopf and Klein (1992), Brasier (1992c). A to I refer to data bases cited in the text

correlation elsewhere. These data are summarised by Brasier et al. (1993), Brasier, Corfield et al. (1994) and Brasier, Rozanov et al. (1994) and full discussion is reserved for the final report of IGCP Project 303. Here, we emphasize the presence of not less than ten carbon isotopic cycles through the early Cambrian, some of which can be traced elsewhere (Table 1). Before this was known, there was a tendency to correlate the first-reported carbon cycle I of Siberia with excursions elsewhere. Some such correlations have little palaeontological support (e.g. those between Siberia and China made by Kirschvink, et al., 1991; and those between Siberia and Namibia made by Kaufman et al., 1991). These multiple carbon isotope excursions through the early Cambrian (Table 1) largely coincide with phases in the radiation of invertebrate groups.

Within this new framework, four episodes of distinctly negative carbon isotopes have also emerged (Table 1): Kotlinian (prior to PC/C Event), basal Tommotian-upper Tommotian, and upper Botomian-lower Toyonian (during M'Bo Event). These are discussed further below.

The carbon isotopic profile of higher levels in the Cambrian is only known in outline. Evidence from the Great Basin of western North America suggests the presence of several minor positive excursions in the middle Cambrian, plus a major excursion in the upper Cambrian, broadly coincident with the Sauk II/Sauk III regression and the Dr/Fc Bio-Event (Table 1, Brasier, 1993). A negative excursion also occurs higher, in the Payntonian of Australia (Ripperdan et al., 1992) but at a level below the Tp/Ib Event.

Too little is yet known about other geochemical trends (such as oxygen, strontium and sulfur) to be of much use for event stratigraphy in the Cambrian (e.g. Brasier, 1992c).

Data Bases and Diversity Change

The generic data base for Proterozoic and early Cambrian microfossils and skeletal fossils compiled by authors in Schopf and Klein (1992) provides a starting point for analysis of diversity changes through the interval. Although the taxonomic data are explicit, this compilation could not include unpublished occurrences of earliest skeletal fossils, and makes concealed assumptions about correlations between sections that are difficult to evaluate.

Figure 8A shows the patterns of diversity for genera from the upper Vendian through the Cambrian, modified from Sepkoski (1992) according to the time scale given here, with the Nemakit-Daldynian as basal Cambrian. There is a basic pattern of sigmoidal diversification, beginning with low diversity in the Vendian, rising rapidly through the Tommotian and Atdabanian, then a slow increase to nearly constant levels in the Botomian to upper Cambrian. This pattern is also seen in the data from families, orders and classes (Sepkoski, 1992).

Brasier (1979, 1982) first suggested that the Cambrian radiation is a composite event, involving successive faunas diversifying at different rates. A similar hypothesis gains support from Q-mode factor analysis (Sepkoski, 1992) which reveals the presence of the five main reaction groups shown in Fig. 8A: the Ediacaran fauna (Petalonamae, Cyclozoa, Scyphozoa); the Tommotian fauna (orthothecimorph hyoliths, monoplacophorans, sabelliditids, tommotiids and other problematica); the Archaeocyathan fauna (regular and irregular archaeocyathan sponges); the later, Trilobitic fauna (trilobites, ostracods, other arthropods, inarticulate brachiopods, all echinoderm classes); and Palaeozoic-Modern faunas.

Figure 8B shows the diversity of calcareous metaphytes (e.g. *Renalcis*, *Epiphyton*) through the same interval, based on the data of Mankiewicz (1992). This shows a pattern like that of archaeocyathans (Fig. 8A), declining after the M'B Event.

Figure 8C plots the number of ichnogenera through this interval, while Fig. 8D plots the diversity of acritarch phytoplankton. Neither shows a pattern like that of the Tommotian or Archaeocyathan Fauna, although there are indications of phytoplankton decline on the East European Platform (Fig. 8D). Work is required on factor analysis of trace fossil and acritarch assemblages.

Biodiversity - A Summary

The following outline can be derived from these data bases.
1. Diversification of the Ediacaran fauna took place during the Redkinian interval of the Vendian (Fig. 8A).
2. Possible extinction of much Ediacara fauna occurred through the Kotlinian interval of the Vendian (e.g. Seilacher, 1984; Sokolov and Fedonkin, 1986; McMenamin and McMenamin, 1989; Brasier, 1989c). The decline may have been prolonged, however, if the triradiate skeletal anabaritids are placed in the Trilobozoa (Sokolov and Fedonkin, 1984) and if there were survivors in the Burgess Shale (Conway Morris, 1989). A decline also took place in ichnogenera, with some turnover at the Precambrian/Cambrian boundary (PC/C Event, Fig. 8C).

Table 1. Positive carbon isotope cycles (Z and I-XIII) and negative excursions through the Cambrian System. Chronostratigraphy for the early Cambrian uses Siberian stage names; for middle and upper Cambrian uses agnostoid-polymeroid trilobite biozones (cf. Fig. 4). Carbon isotopic values are given to the nearest 0.5‰ (from Siberia where available)

Level and/or Stage	$\partial^{13}C$	Location	Author
UPPER CAMBRIAN:			
XIV-Franconian/Pre-Payntonian	+1.0	Australia:	Ripperdan et al. (1992)
		W. USA	Brasier (1992e)
XIII-*G. reticulatus* Dresbachian	+5.0	W. USA	Brasier (1992e)
MIDDLE CAMBRIAN:			
XII-*L. laevigata* Albertian	+1.5	W. USA	Brasier (1992e)
XI-*T. gibbus* Albertian	+0.5	W. USA	Brasier (1992e)
LOWER CAMBRIAN:			
X- upper Toyonian to Amgan	+1.0	Siberia	Brasier et al. (1994)
		Canada	Grant (1992)
		Australia	Donnelly et al. (1988)
IX-mid Toyonian	+1.0	Siberia	Brasier et al. (1994)
Upper Botomian-Toyonian	-2.0	Siberia	Brasier et al. (1994)
		?Canada	Grant (1992)
		W. USA	Brasier (1992e)
		Scotland	Brasier (1992e)
VIII-mid Botomian	+2.5	Siberia	Brasier et al. (1994)
VII-lower Botomian	+3.0	Siberia	Brasier et al. (1994)
VI-upper Atdabanian	+1.0	Siberia	Brasier et al. (1994)
V-mid Atdabanian	+3.0	Siberia	Brasier et al. (1994)
IV-lower Atdabanian	+1.5	Siberia	Kirschvink et al. (1991)
		Australia	Tucker (1989)
		Morocco	Latham & Riding (1990), Kirschvink et al. (1991)
		Canada	Brasier et al. (1992)
Upper Tommotian	-2.0	Siberia	Magaritz et al. (1986)
		Canada	Brasier et al. (1992)
		Australia	Tucker (1989)
		Morocco	Latham & Riding (1990), Kirschvink et al. (1991)
II/III-mid Tommotian	+1.5	Siberia	Magaritz et al. (1986)
Lower Tommotian	-2.0	Siberia	Magaritz et al. (1986)
		Australia	Tucker (1989)
		?China	Brasier et al. (1990)
		Canada	Brasier et al. (1992)
		?India	Aharon & Liew (1992)
		Morocco	Latham & Riding (1990), Kirschvink et al. (1991)
I-upper Nemakit-Daldynian	+3.5	Siberia	Magaritz et al. (1986), Brasier et al. (1993), Pokrovsky & Missarzhevsky (1993)
		India	Aharon & Liew (1992)
		China	Brasier et al. (1990)
		Iran	Brasier et al. (1990)
		Morocco	Latham & Riding (1990), Kirschvink et al. (1991)
		?Namibia	Kaufman et al. (1991)
		Oman	Burns & Matter (1993)
Z-lower Nemakit-Daldynian	+1.5	Siberia	Magaritz et al. (1986)
		India	Aharon & Liew (1992)
		Iran	Brasier et al. (1990)
Kotlinian	-4.5	Siberia	Magaritz et al. (1986), Pokrovsky & Missarzhevsky (1993)

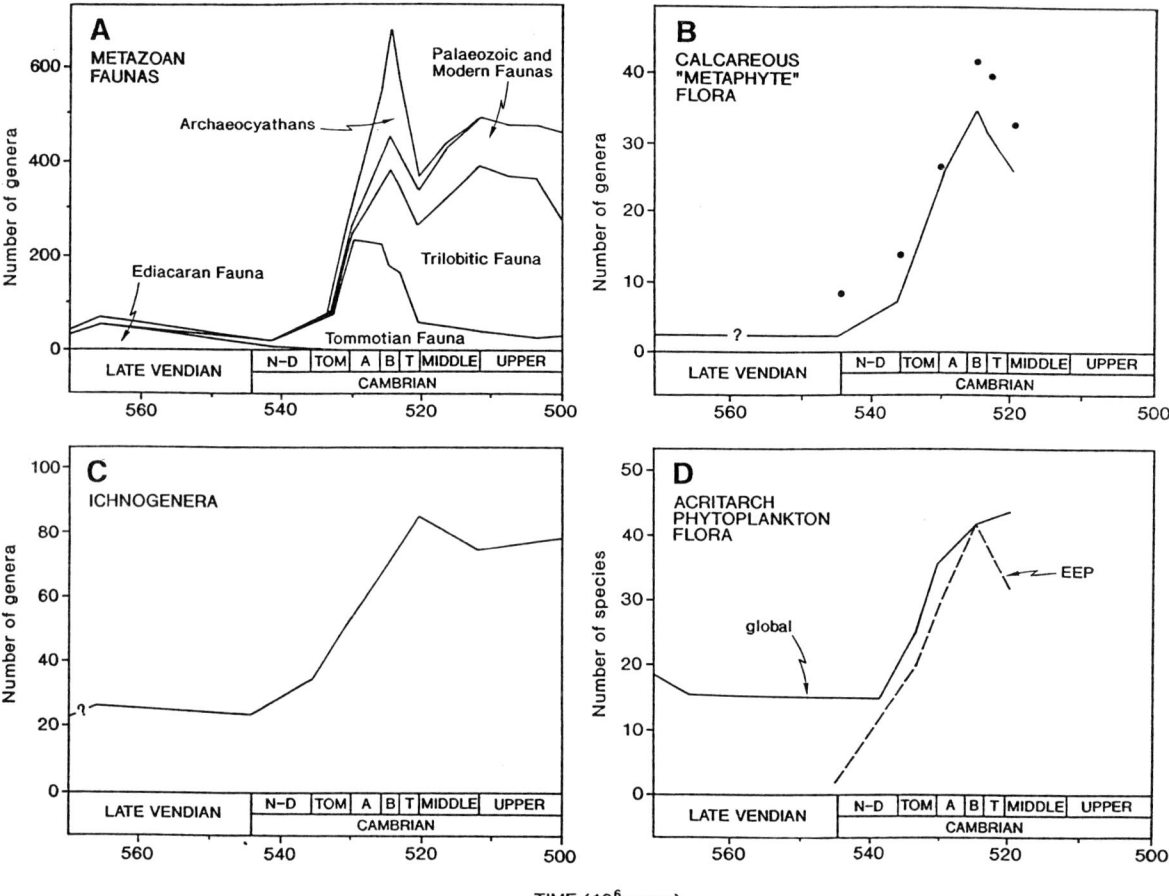

Fig. 8. Diversity calibrated against the revised time scale. A, metazoan faunas (based on data in Sepkoski, 1992). B, calcareous metaphyte flora (based on data in Mankiewicz, 1992; Sepkoski, 1992); dots represent upper limit, incorporating uncertain data. C, ichnogenera (based on data in Crimes, 1992a,b). D, acritarch phytoplankton (global, based on data in Schopf and Klein, 1992; EEP = East European Platform, based on data in Moczydlowska, 1991)

3. Rapid diversification of the 'Tommotian' Fauna took place during the Nemakit-Daldyn to Atdabanian, over an interval of some 20 Ma, and was followed by prolonged decline thereafter (Fig. 8A).

4. A peak in diversity of biohermal organisms belonging to the Archaeocyathan Fauna and Calcareous Metaphyte Flora is seen during the Botomian (Fig. 8A), c. 526 to 524 Ma BP. Recent data show an early Botomian acme for archaeocyathans (e.g. Debrenne, 1991). These sponge-grade organisms had geometrical skeletons that allow taxonomic rigour and are likely to provide a clear picture of evolutionary diversification within their 'reefal' habitat.

5. A substantial decline in diversity is recorded over the mid Botomian (M'Bo Event, c. 523-522 Ma BP) through to the Toyonian-Amgan boundary (T/A Event, c. 520 Ma BP). This decline, lasting only an estimated 2 to 3 Ma, was catastrophic for the Archaeocyathan Fauna and severe for Calcareous Metaphyte Flora (Figs. 8A, 8B) and the 'Tommotian' Fauna (Figs. 8A, 8B). Hyoliths were also greatly affected (Sepkoski, 1992). Such a pattern of diversification and decline appears to be different from the Trilobitic, Palaeozoic and Modern faunal elements (Sepkoski, 1992). Destruction of shallow shelf and biohermal ecosystems is implied.

Although some early forms of olenellid and redlichiid trilobites died out over this interval, this was a time of major diversification among the Trilobitic Fauna (Fig. 8A). Despite imprecision in correlation, this bioevent appears to be real and global. Signor (1992) has compiled an unpublished data base from which he has argued that this was one of the most severe extinctions in the Phanerozoic, echoing the view also put forward by Boucot (1990). Signor has argued that the high level of provincialism in the Botomian made faunas vulnerable to extinction.

6. A middle Cambrian peak in the diversity of families, orders and classes is related to special preservation in the Burgess Shale lagerstätten (e.g. Gould, 1991). Many of these taxa range down into the lower Cambrian, however (Conway Morris, 1989) and in China can occur in the Qiongzhusian (c. late Atdabanian-early Botomian).

7. The biomere extinctions of Palmer (1965, 1979, 1982) which affected cratonic trilobites in North America do not appear in Fig. 8 because each was of very short duration and followed by rapid radiations (Sepkoski, 1992). Three upper Cambrian biomere boundaries will be discussed further below: at the top of the Marjumid Biomere (Mi/Id Event), Pterocephalid Biomere (Dr/Fc Event) and Ptychaspid Biomere (Tp/Ib event; Fig. 4). The turnover at the top of the Olenellid Biomere mentioned by Palmer (1982) awaits full documentation but corresponds broadly to the Ty/Am Event (Fig. 4). The Mindyallan/Idamean (Mi/Id) and Dresbachian/Franconian (Dr/Fc) Events discussed below do appear to be global.

Each of these biomeres is characterised by evolutionary development and diversification of polymeroid trilobites, followed by abrupt extinction at the end of the biomere. Rapid migration then brought in forms from the outer shelf and slope, which progressively diversified in shelf habitats.

8. We can therefore conclude that the metazoan explosion was relatively rapid between about 535 and 526 Ma BP, i.e. over an interval of about 10 Ma. This is a short time relative to earlier and later radiations and, according to Sepkoski (1992), truly analagous rates may only be found in situations where taxa entered newly opened habitats. The basic sigmoidal patterns in Fig. 8 are consistent with the equilibrium or 'kinetic' model of taxonomic diversification (e.g. Sepkoski, 1984). This predicts that the early phases of radiations into ecologically vacant environments should be exponential, followed by declining diversification as the environment fills with new species.

Bioevents

1. The Terminal Proterozoic Extinction (PC/C) Event

The question of mass extinction and faunal turnover at the end of the Precambrian has been outlined elsewhere (Brasier, 1989c; Jenkins, 1989; McMenamin and McMenamin, 1989; Runnegar and Fedonkin, 1992; Sepkoski, 1992; Conway Morris, 1993). Several lines of evidence have been put forward that imply extinction events towards the very end of the Precambrian, and point towards the concept of a 'Kotlin Crisis' (Brasier, 1992b) or perhaps a 'Kotlinian Dead Zone'. Most of these changes took place over an interval herein suggested to span c. 555-545 Ma BP (Fig. 3), during which there may have been a succession of events but it should be emphasized that stratigraphic resolution during this time interval is currently rather poor, and that the fossil record is also highly enigmatic. These supposed events are as follows.

(i) Disappearance of diverse, large-sized acanthomorph acritarch assemblages, and their replacement by impoverished simple sphaeromorphs at about the time of the Ediacaran faunal radiation (Knoll and Walter, 1992). It is equally probable, given their poor fossil record, that this decline may have taken place during the later Kotlin interval (N. Butterfield, oral comm. 1993).

(ii) Decline in the abundance, size, and diversity of Ediacara fauna during the Kotlin interval (Sokolov and Fedonkin, 1986). Their presumed absence from Cambrian seas, and the possible uniqueness of body plan, have been advanced as reasons for mass extinction of the Ediacara fauna (Seilacher, 1984). The presence of rare, supposed Ediacaran survivors in the Burgess Shale fauna (Conway Morris, 1993), however indicates that this extinction may not have been total, and that body plans were not unique.

(iii) A decline also occurred in the size of trace fossils during the Kotlin interval (Brasier, 1992b) and perhaps in their diversity (Crimes, 1992a).

(iv) Perturbations in stable carbon isotopes and trace elements have been speculatively linked with extinction of the Ediacaran fauna (for example, Hsü et al., 1985; Xu et al., 1989; Magaritz, 1989) but the cited anomalies are now seen to lie within the Cambrian System. Kotlinian strata are, however, characterised by very light $\partial^{13}C$ isotopic values but heavy $\partial^{34}S$ values in Siberia, Iran and elsewhere (e.g. Holser, 1979; Magaritz et al., 1986; Brasier et al., 1990).

(v) Disappearance of ichnofossils such as *Harlaniella*, *Palaeopascichnus*, and *Nenoxites* at the end of the Kotlin stage, coincident with the appearance of the Cambrian ichnofauna (Narbonne et al., 1987). This corresponds to the boundary between Ichnofossil Zones I and II of Crimes (1989) and is labelled as the PC/C Event on Fig. 4.

(vi) Abrupt fall in diversity (for genera, families) or stasis (orders and classes) between the upper Vendian and Nemakit-Daldynian (Sepkoski, 1992).

C a u s e s : With the soft-bodied Ediacara fauna, its widespread preservation is perhaps as much as a puzzle as its disappearance, so that even the existence of its 'mass extinction' might be doubted by some (e.g. Runnegar and Fedonkin, 1992). Its disappearance invariably precedes the evidence for increased scavenging, predation and bioturbation in the fossil record at the base of the Cambrian. Taphonomic factors might explain the disappearance of the Ediacara fauna but not the disappearance of trace fossil taxa, and an extinction scenario must now be seriously considered.

It is suggested here, on admittedly slender evidence, that a pattern of extinctions and survivals may have taken place much like that during the Cambrian and later times; i.e. that nearshore, specialised forms (K-strategists such as the petalonamean fauna?) suffered mass extinction, and that the survivors were better adapted to cooler, oxygen-depleted outer shelf conditions. In favour of this possibility are the following observations.

(i) Elements of the Ediacaran fauna have a remarkably high surface area: volume ratio, consistent with update of oxygen by passive diffusion (e.g. Runnegar, 1982), or symbiotic adaptation to oligotrophic conditions (e.g. McMenamin and McMenamin, 1989). Given forms could have been vulnerable to a fall in oxygen or a rise in nutrient levels.

(ii) Skeletal elements of the earliest Cambrian fauna are very small, especially in comparison with the Ediacaran biota (Brasier, 1979). This may indicate their more opportunistic character.

(iii) These earliest skeletal elements are widely associated with phosphate, silica and the influence of more oceanic water masses (Brasier, 1992d). This suggests tolerance of cooler and more nutrient-enriched waters.

(iv) The earliest skeletal assemblages tend to be markedly pandemic. Provincialism and more mesotrophic to oligotrophic characteristics tend to increase to a maximum in the Botomian (Signor, 1992; Brasier, 1992d).

If these are so, then a palaeoceanographic-tectonic model should be considered. It certainly seems that widespread (?global) regression and emergence of the continental shelves took place at the end of Ediacaran (i.e. Kotlinian) times, probably in relation to crustal processes that encompassed the 'Avalonian orogeny'. Climatic changes are also likely to have taken place: there is putative evidence for a glaciation in Namibia between the Ediacara fauna and first Cambrian traces (e.g. Kaufman et al., 1991).

Speculations abound on changes in atmospheric oxygen over this interval. For example, massive rifting and extensional basin formation appear to have encouraged high rates of organic carbon burial (as signalled by $\partial^{13}C$) - during earlier Ediacaran times, perhaps generating significant amounts of oxygen (e.g. Derry et al., 1992; Knoll, 1992). Geochemical data for the Kotlinian interval, just prior to the Nemakit-Daldynian, however, indicate extremely light $\partial^{13}C$ values (Table 1) but heavy $\partial^{34}S$ values (Holser, 1984; Brasier, 1992c). This might actually indicate a *fall* in atmospheric oxygen levels during the Kotlinian, as organic rich sediments were exposed and oxidised during a major regression. (cf. Wignall and Hallam, 1992).

To summarise, it seems possible that the Ediacaran biota was largely extinguished by a sudden change in temperature, perhaps associated with stillstand to falling sea level and increased climatic gradients. The Cambrian biota may have contributed to their demise and, by implication, had its origins in deeper, cooler and more nutrient-enriched and oxygen-depleted water masses. This could explain the curiously 'transitional' nature of deeper water biotas below the boundary level (e.g. Brasier et al., 1979).

2. Early Cambrian Radiation Events

The difficulties of establishing global markers within the early Cambrian radiations are discussed in Cowie and Brasier (1989). About five successive and widespread assemblages were, however, recognised by Brasier (1989b), characteristic of intervals up to, and including the early Botomian. Each of these contains elements of lineages that can provide a backbone stratigraphy. But most groups appear abruptly, without known ancestors in the fossil record, in a manner suggestive of major migratory radiations, perhaps connected with transgressive episodes. These are summarised below as Phases 1 to 5.

Phase 1. The CR1 Event and the *Phycodes pedum* Assemblage

Trace fossils that show more complex behaviour, spreite, deeper burrowing, and larger body diameter, appear rapidly at the base of the Cambrian stratotype in SE Newfoundland, where *Phycodes pedum* first appears, and at correlated horizons in many other parts of the world (e.g. Crimes and Anderson, 1985;

Crimes, 1989; Narbonne et al., 1987). This marks an early phase in the steady increase of ichnofossil diversity, and behavioural complexity, that continued through to the end of the Atdabanian, followed by little change through the rest of the Lower Palaeozoic (Fig. 8c; Crimes, 1992b).

The following traces disappeared at or below this event: bilobed trails (*Bilinichnus*), spirally stuffed burrows (*Harlaniella*), unilobed pellet trails (*Nenoxites*, *Vimenites*, *Palaeopascichnus*), simple cnidarian resting traces (*Intrites*). Other early forms survived into this zone and beyond: simple sediment-filled burrows (*Planolites*, *Gordia*), stuffed burrows (*Neonereites*, *Scolicia*), and dwelling burrows (*Skolithos*). Others, such as *Arenicolites* (U-shaped dwelling burrow), *Cochlichnus* (meandering burrow) and *Didymaulichnus* (simple bilobed trail) are suspected to range below the *P. pedum* Zone elsewhere (Crimes, 1989).

New forms that appeared in the *Phycodes pedum* ichnofossil zone include: branched feeding burrows (*Phycodes*, *Palaeophycus*), anemone resting traces (*Conichnus*), trilobed trails (*Curvolithus*) and arthropod scratch marks such as *Monomorphichnus* (Narbonne et al., 1987).

In SE Newfoundland, a higher *Rusophycus avalonensis* assemblage contains: arthropod resting traces and scratch marks (*Rusophycus*, *Dimorphichnus*), grazing trails (*Taphrhelminthopsis*, *Cochlichnus*, *Helminthopsis*), branched feeding burrows (?*Hormosiroidea*), U-burrows with spreite (*Diplocraterion*) and others (Narbonne et al., 1987; Crimes, 1989). This is also claimed to be a global assemblage, equivalent to Ichnofossil Zone III of Crimes (1989).

Unfortunately, some recent papers by Crimes summarising ichnofossils changes (e.g. 1989, 1992b) may confuse the chronology by making premature use of archaeocyathan stages, referring the *P. pedum* assemblage to the Tommotian, and the *Rusophycus* assemblage to the upper Tommotian or Atdabanian. He also argues (1992b) that faunal changes in Newfoundland preceded those elsewhere, and that Avalonia lay at low latitudes at this time. These views are not followed here. The *P. pedum* event broadly coincided with the appearance of *Sabellidites* in Newfoundland, Baltica and Siberia (Narbonne et al., 1987), well before the first appearance of shelly faunas of Tommotian type. Hence, the *P. pedum* bioevent should rather be correlated with the base of the Nemakit-Daldynian stage in Siberia, the upper part of the Rovno interval in the Baltic region, and the base of the Meishucunian in China (Fig. 3; e.g. Fedonkin, 1987; Brasier, 1992b; Landing, 1994).

This revised correlation shows that the following kinds of skeletal fossil appeared more or less contemporaneously with the *P. pedum* ichnofauna, in the early Nemakit-Daldynian of Siberia, earliest Meishucunian of China and correlative strata in Iran: organic walled tubes (*Sabellidites*); phosphatic tubes (*Hyolithellus*); phosphatic protoconodont 'jaws' (*Protohertina*); calcareous tubes that were trilobate (*Anabarites*) or circular (*Cambrotubulus*, *Circotheca*, *Ladatheca*) in cross section; coeloscleritophorans (*Chancelloria*) and siliceous sponge spicules (*Protospongia*). This corresponds to the *Anabarites trisulcatus-Protohertzina anabarica* assemblage of Brasier (1989b) and *Anabarites trisulcatus* Zone of Khomentovsky and Karlova (1993) but its global synchroneity must be questioned. The latter authors (and references therein) show that this assemblage can be diachronous in appearance and strong ecological controls are implicated. Species dominance tends to be high in such assemblages, suggesting strong physical controls. Brasier (1990) also drew attention to the role of phosphogenesis in taphonomic enhancement of such early skeletal assemblages.

Phase 2. The CR2 Event and *Purella antiqua* and *Aldanella attleborensis* Assemblages

In Avalonian sections, the first skeletal assemblage includes the paragastropod *Aldanella attleborensis*, plus the primitive agglutinated ?foraminiferid *Platysolenites antiquissimus*, monoplacophoran molluscs (*Latouchella*, *Igorella*, *Archaeospira*), rostroconch molluscs (*Watsonella crosbyii*), halkieriid sclerites (*Halkieria* sp.), simple hyoliths (*Ladatheca*, allathecids) and sculptured calcareous tubes (*Coleoloides*; Narbonne et al., 1987). Phosphatic sclerites (*Eccentrotheca*, *Lapworthella*) also appear but there are no tommotiids with differentiation into right- and left-handed elements.

It is now emerging that such assemblages can be pre-Tommotian. In Eastern Siberia, the upper part of the Nemakit-Daldynian contains *Latouchella*, *Igorella*, *Archaeospira*, *Anabarella*, *Halkieria* sp., *Chancelloria* sp., cup-shaped *Purella* spp. and the paragastropod *Barskovia* in strata that lie demonstrably below the base of the type Tommotian base in the *Purella antiqua* Zone of the Nemakit-Daldynian (Brasier et al., 1993; Khomentovsky and Karlova, 1993). Simple orthothecimorph hyoliths (*Lophotheca*, *Ladatheca*, *Loculitheca*) and phosphatic tube *Torellella* also occur (op. cit.).

These Siberian assemblages also contain forms typical of Meishucunian Zone II in China: e.g. *Securiconus, Yangtzeconus, Archaeospira, Stenothecoides, Granoconus, Siphogonuchites, Palaeosulcachites, Olivooides* (Khomentovsky and Karlova, 1993). For this reason, and because of carbon isotopic similarities, this Chinese zone is placed here (Brasier et al., 1990; Brasier, 1992b).

Like the *Anabarites trisulcatus* fauna, there is evidence that taphonomic and ecologic controls brought in the *Aldanella* and *Purella* assemblages (e.g. Brasier, 1990; Landing, 1992) with the latter being diachronous across eastern Siberia (Khomentovsky and Karlova, 1993).

Phase 3. The CR3 Event and the *Sunnaginia imbricata* Assemblage

The Tommotian stage of Siberia is marked, in its type area, by a major, almost synchronous change in facies from extensive peritidal dolomites to deeper water argillaceous dolomites and limestones, often enriched in glauconite and bearing the first archaeocyathan-algal bioherms (Rozanov and Sokolov, 1984; Khomentovsky and Karlova, 1993). This is associated with a distinctive negative $\partial^{13}C$ shift (Magaritz et al., 1986), now widely traced (Table 1). No major hiatus is present, perhaps, since the isotopic shift actually lags behind the lithological change (Brasier et al., 1993) and evidence for deep downcutting is lacking (Khomentovsky and Karlova, 1993).

In each area, the base of the Tommotian occurs close to a major flooding surface and is associated with new faunal elements that usher in the distinctive and diverse Tommotian fauna (Rozanov, 1992). If we accept the correlation shown in Fig. 3, these new elements include: phosphatic sclerites with R- and L-handed symmetry (e.g. *Sunnaginia, Camenella*); inarticulate paterinid brachiopods, and lipped and triangulate hyoliths (e.g. *Burithes*). The first archaeocyathans appear in Siberia at this time (ibid.) and a further diversification occurred in the acritarch microflora at least by mid Tommotian times (Moczydlowska, 1991).

Such facies changes suggest that the higher diversity *Sunnaginia* assemblage was associated with more open-aspect facies brought in by a major transgression. This makes it possible, of course, that this fauna may have occurred a little earlier in as yet undiscovered outer cratonic settings.

Phase 4. The CR4 Event and the first trilobite Assemblages

The Atdabanian stage is characterised by changes in archaeocyathan architecture and by the appearance of biomineralised trilobites. Three carbon isotopic excursions have been found in Siberia, of which the first has now been traced into Morocco and SE Newfoundland (Table 1).

It has for long been clear that the appearance of trilobites is a diachronous event, for the following reasons.

(i) skeletons are preceded in most areas by arthropod traces that extend down into the sub-Tommotian and possibly into the Vendian.

(ii) The first trilobite groups differ around the globe: olenellids appear first in Siberia, Laurentia, Baltica and Avalonia; bigotinids appear first in parts of southern Europe and Morocco; redlichiids appear first in Asia and Australia (Brasier, 1989b).

(iii) Even the first olenellid genera differ in relation to geography: *Profallotaspis* appears first near the base of the Atdabanian in Siberia. *Schmidtiellus* in Baltica may be of similar age, or earlier (Moczydlowska and Vidal, 1988). *Fallotaspis* in Laurentia may be a little younger. *Callavia* in Avalonia is likely to be no older than mid Atdabanian on the basis of associated elements (Brasier, 1989b). Associated small shelly fossils and trilobites suggest that the first redlichiid in China (*Parabadiella*) may be mid Atdabanian and *Eoredlichia* may even be upper Atdabanian-lower Botomian (Brasier, 1989b).

An absence of trilobites from the earliest Atdabanian in E. Gondwanaland and Avalonia may be due in part to major hiati at this level. In the latter area, for example, the upper Smith (or Fosters) Point limestone shows evidence for emergence and time lapse traceable from SE Newfoundland to England (Landing, 1992; Brasier et al., 1992). Shallowing to dolomites and evaporites is also recorded in the lower Atdabanian of eastern Siberia (Zhuravleva et al., 1990).

A biotic turnover may have taken place in the mid Atdabanian: early forms of olenellids disappeared here in Siberia, Morocco and Laurentia (e.g. Fallotaspidae; Palmer and Repina, 1993) while various 'Tommotian' invertebrates died out in Avalonia (e.g. *Aldanella, Camenella, Watsonella;* Landing, 1992). This 'event' has yet to be confirmed and studied systematically.

Phase 5. The early Botomian Climax of Biohermal Biota

A peak in diversity of biohermal organisms such as archaeocyathans and calcareous metaphytes is seen during the Botomian (Fig. 8A). Recent data clearly show an early Botomian acme for archaeocyathan sponges (e.g. Debrenne, 1991). It appears that suitable carbonate habitats were at their most widespread in this interval (Brasier, 1982). This is not distinguished as an event marker in this study.

3. The Botomian-Toyonian Crisis (M'Bo and Ty/Am Events)

A substantial decline in diversity is recorded over the mid Botomian (M'Bo Event) through to the Toyonian-Amgan boundary (Ty/Am Event). This decline (the 'Toyonian Crisis' of Brasier, 1992b) was catastrophic for archaeocyathan sponges and severe for calcareous metaphytes (Figs. 8A, B) and the 'Tommotian' Fauna (Fig. 8A). Hyoliths were also greatly affected (Sepkoski, 1992). Such a pattern of diversification and decline appears to be different from the Trilobitic, Palaeozoic and Modern faunal elements (Fig. 8A; Sepkoski, 1992). Destruction of shallow shelf and biohermal ecosystems is implied. According to the time scale used here, this Botomian-Toyonian crisis event lasted from mid Botomian (c. 523) at least into late Toyonian (c. 520 Ma BP).

Current information indicates a stepwise extinction, spanning points from M'Bo to Ty/Am (Fig. 4). Archaeocyathans from the Siberian Platform clearly show a drop in speciation and a rise in the percent extinction within the Botomian (i.e. between the first, *B. micmacciformis* Zone and the third, *B. asiaticus* Zone; Debrenne, 1991). Elements of the Tommotian assemblage were already in decline at this time (Fig. 8A; Brasier, 1992d). Olenellid trilobites also suffered extinctons in all regions except Laurentia, leading to the demise of Nevadiidae, Judomiidae and Holimiidae (Palmer and Repina, 1993). In Siberia and Avalonia, other trilobites also experienced rather rapid turnover during this interval, with several pandemic pagetiids (*Triangulaspis* spp., *Dipharus* sp., *Calodiscus* sp.) making their last appearance (Repina, 1981; Landing, 1992). This faunal turnover broadly coincides with a rapid deepening which brought in dark shales of the Sinyaya Formation bearing diverse protolenids such as *Bergeroniellus* sp.

Precise correlation is difficult to achieve with Siberia, though a post Botomian drop in archaeocyathan diversity is clear on a global scale (Debrenne, 1991). On the Laurentian craton, the M'Bo Event is believed to lie near the base of the lower *Bonnia-Olenellus* Zone, below the appearance of protolenids and *Salterella* which arguably mark the arrival of pandemic and/or cooler water faunas. Olenellids managed to survive here, in the form of *Olenellus* spp., until the Ty/Am Event (see below, Palmer and Repina, 1993), when there was shoaling or emergence across the Sauk I/Sauk II sequence boundary.

Documentation of the M'Bo and Ty/Am Events in Avalonia is still hampered by a lack of detailed published data on trilobites. A comparable faunal turnover takes place just below the top of the *Callavia broeggeri* Zone (Landing, 1992) where this olenellid disappears (as apparently do *Triangulapsis vigilans* and *Acanthomicmacca walcotti*, Brasier, 1989b) and red mudstones give way to less diverse green mudstones. These events are followed at the top of the *Callavia broeggeri* Zone by extinction of the pagetiids *Serrodiscus bellimarginatus*, *Ladadiscus llarenai*, *Dipharus planus* and the polymeroid *Strenuella strenua* (Landing, 1992). The latter author suggests that the disappearance of the *Callavia broeggeri* Zone fauna, and associated appearance of *Protolenus howleyi* Zone faunas, may be linked to the disappearance of carbonate substrates and the onset of red, purple and pyritiferous green siliciclastic mudstones.

Palmer (1982) has speculated that the first major trilobite extinction event in North America may occur at the top of the *Bonnia-Olenellus* Zone (i.e. close to the Ty/Am boundary, Fig. 4). Such evolutionary turnovers have been used by him to define biostratigraphic intervals called 'biomeres'. In this case, the turnover lies between the 'Olenellid biomere' of the early Cambrian and the Corynexochid biomere of the middle Cambrian, though there may have been some overlap between stocks (A.R. Palmer written comm.). Although the last of the 'primitive' olenellid and redlichiid trilobite stocks died out during the Ty/Am Event, this was a time of major diversification among other trilobitic elements (Fig. 8A).

C a u s e s . Despite imprecision in correlation, the M'Bo to Ty/Am bio-events appear to be real and global. Signor (1992) has compiled an unpublished data base from which he argued that this was one of the most severe extinctions in the Phanerozoic, echoing a view also put forward by Boucot (1990).

The causes for this extinction event are not yet clear. Brasier (1982) drew attention to the reduction in suitable (i.e. mid shelf carbonate) habitats in the Toyonian, towards the end of the crisis. Signor (1992) has argued that the high level of provincialism in the Botomian made faunas more vulnerable to extinction.

Carbon isotopic data from the Siberian Platform show a change in pattern, from oscillating and heavy, to more stable and light $\partial^{13}C$ through the Botomian-Toyonian crisis (Table 1). A similar shift is recorded at this level in the western USA (Brasier, 1993) and a drop in palaeoproductivity and/or fall in carbon burial rate, may be suspected. As mentioned above, the turnover may have begun not with shallowing but with a major deepening, bringing black shales onto the Siberian Platform and pyritiferous green shales into Avalonia. These, and the isotopic shifts, may indicate a major flooding event, which stressed shallow carbonate platform biotas, reducing their diversity and provincialism. The global spread of protolenid trilobites at this time arguably indicates the influence of cooler, deeper waters.

4. Upper Cambrian Events

Three mass extinctions have been identified in the upper Cambrian sequence of trilobite faunas in North America (Stitt, 1971, 1977; Palmer, 1965, 1979, 1982), at the boundaries between the Marjumiid-Pterocephalid, Pterocephalid/Ptychsaspid, and Ptychaspid-'post Ptychaspid' biomeres. Ludvigsen and Westrop (1985) proposed stage names for each of these intervals. These turnovers are here termed the Mi/Id, Dr/Fc and Tp/Ib Events (Fig. 4).

Palmer (1982) has given details of trilobite changes for several metres on either side of the Pterocephalid biomere boundaries in North America (i.e. the Mi/Id and Dr/Fc Events). Assemblages of shelf trilobites are abruptly replaced, within 0.1 to 0.2m, by an assemblage made up mostly of new trilobites of different families. These changes took place without any accompanying lithological markers. Within a metre of abrupt faunal change, nearly all representatives of older families are gone and for many metres thereafter, the trilobite assemblages consist of only a few species, one of which often constitutes more than 90 percent of the fauna.

The turnover at the base of the Pterocephaliid biomere (the Dr/Fc Event) occurs across the base of the *Aphelaspis* (polymeroid) Zone in Laurentia, i.e. at about the *Glyptagnostus stolidotus*/*G. reticulatus* (agnostoid) zonal boundary (Fig. 4). This allows correlation beyond Laurentia into areas with Chinese-type faunas. In Kazakhstan, the basal *G. reticulatus* Zone is taken as the middle-upper Cambrian boundary; polymeroids (e.g. *Prodamesella*, *Blackwelderia*, *Blountia*, *Liostracina*, *Cyclolorenzellina*) and agnostoids (e.g. *Grandagnostus*, *Ammagnostus*) become extinct across this interval, replaced by new forms such as *Pterocephalia* and *Prochuangia* and a typical late Cambrian agnostoid fauna (Ergaliev, 1981). In China, this turnover can be traced to the base of the Changshanian stage (e.g. Xiang et al., 1981; Lu and Lin, 1981; Chang, 1988). In Australia, a contemporaneous mass extinction takes place across the Mindyallan-Idamean stage boundaries (Opik, 1961), including a drastic reduction in agnostoid genera (Shergold, 1981). The latter has suggested that the crisis followed a change from stable lithosphere to active plate margins, reducing provincialism and bringing about deepening and cooling over former carbonate platforms.

The turnover at the base of the Ptychaspid biomere occurs in the upper part of the *Elvinia* Zone in Laurentia (Palmer, 1982), for which the first appearance of *Irvingella major* provides a global marker near the top. According to Taylor (1985) trilobite extinction followed a two-step pattern: 1, shallow shelf genera and families began to drop out at the end of the early *Elvinia* Zone, when phosphatic limestones and exotic olenids and other trilobites of the late *Elvinia* Zone first appear. 2, remaining shallow shelf genera and families became extinct at the end of the late *Elvinia* Zone, after which a diverse olenid fauna appears associated with the North American eurytopic trilobite *Parabolinoides* at the beginning of the *Taenicephalus* Zone. Taylor (1985) suggests that the coincidence between olenid trilobites and phosphatic limestones suggests migration in association with oxygen-depleted water masses, perhaps associated with lowered temperatures that led to extinctions. A shift from heavy to light carbon isotopes in the *Elvinia* Zone (Brasier, 1993) is consistent with this hypothesis. Inarticulate brachiopods, which flourished in outer shelf sites, were notably less affected by the Pterocephaliid extinction event (Rowell and Brady, 1975).

Westrop (1989) has provided a detailed analysis of trilobite changes across the latest, Ptychaspid-'post Ptychaspid' biomere, which lies just below the Cambro-Ordovician boundary interval between the *Saukia* and *Missisquoia* trilobite zones. He makes the following points: 1, about half of North American shelf trilobite families disappeared through an interval of up to 26m of strata. 2, Concurrent migration of off-shelf and shelf margin taxa took place towards the inner craton, leading to a reduction in biofacies differentiation and substrate control. 3, survival of families with wide habitat or biogeographic ranges was favoured. 4, group selection, rather than natural selection of individuals, is inferred. 5, these changes were initiated by onlap in the outer part of the shelf, probably in response to sea level rise. 6, there is no evidence for a shelf-wide change in physical conditions.

C a u s e s . As mentioned above, each of these 'biomere events' shows clear evidence for the inshore migration of more pandemic, cool, outer shelf to slope faunas and the extermination of more endemic, warmer, inner shelf faunas. Rapid sea level rise, or upwelling episodes are thought to have brought about a drop in temperature, a fall in dissolved oxygen concentrations, and a rise in nutrient level, leading to extinctions. They confirm the picture, already familiar from the Precambrian/Cambrian boundary interval, of the episodic influence of stagnant, outer shelf water masses on inner shelf biota. The ultimate cause should no doubt be sought in crustal processes related to the opening of ocean basins, following break-up of the Proterozoic supercontinents.

References

Aharon, P. and Liew, T.C., 1992. An assessment of the Precambrian/Cambrian transition events on the basis of carbon isotope records. In: Schidlowski et al. (1992). pp. 212-223.

Aitken, J.D., 1981. Generalizations about Grand Cycles. United States Department of the Interior, Geological Survey, Open-File Report 81-743, 8-14.

Astashkin, V.A. et al., 1991. The Cambrian System on the Siberian Platform; correlation chart and explanating notes. Publication Intern. Union of Geological Sciences 27, 133 pp.

Bath, A.H., 1974. New isotopic data on rocks from the Long Mynd, Shropshire. Journal of the Geological Society 130, 567-574.

Benus, A.P., 1988. Sedimentological context of a deep-water Ediacaran fauna (Mistaken Point Formation, Avalon Zone, Eastern Newfoundland). Bulletin of the New York State Museum 463, 8-9.

Bond, G.C., Nickeson, P.A. and Kominz, M.A., 1984. Breakup of a supercontinent between 625 Ma and 555 Ma: new evidence and implications for continental histories. Earth and Planetary Science Letters 70, 325-345.

Bond, G., Kominz, M.A. and Grotzinger, J.P., 1988. Cambro-Ordovician eustasy: evidence from geo-physical modelling of subsidence in Cordilleran and Appalachian passive margins. In: Kleinspehn, K.L. and Paola, C. (eds.), New Perspectives in Basin Analysis. pp. 129-160. Springer, Berlin Heidelberg New York.

Boucot, A., 1990. Phanerozoic extinctions: how similar are they to each other? In: Kauffman, E.G. and Walliser, O.H. (eds.), Extinction Events in Earth History. pp. 5-30. Springer, Berlin Heidelberg Berlin.

Bowring, S.A., Grotzinger, J.P., Isachsen, C.E., Knoll, A.H., Pelechaty, S.M. and Kolosov, P., 1993. Calibrating rates of early Cambrian evolution. Science 261, 1293-1298.

Brasier, M.D., 1979. The Cambrian radiation event. In: House, M.R. (ed.), Origin of Major Invertebrate Groups. pp. 103-159. Academic Press, London.

Brasier, M.D., 1980. The Lower Cambrian transgression and glauconite-phosphate facies in western Europe. Journal of the Geological Society 137, 695-703.

Brasier, M.D., 1982. Sea level changes, facies changes and the late Precambrian-early Cambrian evolutionary explosion. Precambrian Research 17, 105-123.

Brasier, M.D., 1984. Microfossils and small shelly fossils from the Lower Cambrian Hyolithes Limestone at Nuneaton, English Midlands. Geological Magazine 121, 229-253.

Brasier, M.D., 1989a. China and the Palaeotethyan Belt (India, Pakistan, Iran, Kazakhstan, and Mongolia). In: Cowie, J.W. and Brasier, M.D. (eds.), The Precambrian-Cambrian boundary. pp. 40-74. Clarendon Press, Oxford.

Brasier, M.D., 1989b. Towards a biostratigraphy of the earliest skeletal biotas. In: Cowie, J.W. and Brasier, M.D. (eds.), The Precambrian-Cambrian boundary. pp. 117-165. Clarendon Press, Oxford.

Brasier, M.D., 1989c. On mass extinction and faunal turnover near the end of the Precambrian. In: Donovan, S.K. (ed.), Mass Extinctions. Processes and Evidence. pp. 73-88. Belhaven Press, London.

Brasier, M.D., 1990. Phosphogenic events and skeletal preservation across the Precambrian-Cambrian boundary interval. In: Notholt, A.J.G. and Jarvis, I. (eds.), Phosphorite Research and Development. Geological Society Special Publication 52, 289-303.

Brasier, M.D., 1992a. Global ocean-atmosphere change across the Precambrian-Cambrian transition. Geological Magazine 129, 161-168.

Brasier, M.D., 1992b. Introduction. Background to the Cambrian Explosion. Journal of the Geological Society 149, 585-587.

Brasier, M.D., 1992c. Palaeoceanography and changes in the biological cycling of phosphorus across the Precambrian-Cambrian boundary. In: Lipps, J.H. and Signor, P.W. (eds.), Origin and Early Evolution of the Metazoa. pp. 483-523. Plenum Press, New York.

Brasier, M.D., 1992d. Nutrient-enriched waters and the early skeletal fossil record. Journal of the Geological Society London 149, 621-629.

Brasier, M.D., 1992e. Towards a carbon isotope stratigraphy of the Cambrian System: potential of the Great Basin succession. In: Hailwood, E.A. and Kidd, R.B. (eds.), High Resolution Stratigraphy. Geological Society Special Publication 70, 341-350.

Brasier, M.D., 1993. Early organic evolution. Terra Nova 5, 310-311.

Brasier, M.D., Perejon, A. and de San José, M.A., 1979. Discovery of an important fossiliferous Precambrian-Cambrian sequence in Spain. Estudios Geologicos 35, 379-383.

Brasier, M.D., Magaritz, M., Corfield, R., Luo Huilin, Wu Xiche, Ouyang Lin, Jiang Zhiwen, B. Hamdi, He Tinggui and Fraser, A.G., 1990. The carbon- and oxygen-isotope record of the Precambrian-Cambrian boundary interval in China and Iran and their correlation. Geological Magazine 127, 319-332.

Brasier, M.D., Anderson, M.M. and Corfield, R.M., 1992. Oxygen and carbon isotope stratigraphy of early Cambrian carbonates in southeastern Newfoundland and England. Geological Magazine 129, 265-279.

Brasier, M.D., Khomentovsky, V.V. and Corfield, R.M., 1993. Stable isotopic calibration of the earliest skeletal fossil assemblages in eastern Siberia (Precambrian-Cambrian boundary). Terra Nova 5, 225-232.

Brasier, M.D., Corfield, R.M., Derry, L.A., Rozanov, A.Yu. and Zhuravlev, A.Yu., 1994. Multiple $\partial^{13}C$ excursions spanning the Cambrian explosion to the Botomian crisis in Siberia. Geology 22, 455-458.

Brasier, M.D., Rozanov, A.Yu., Zhuravkev, A.Yu., Corfield, R.M. and Derry, L.A., 1994. A carbon isotope reference scale for the Lower Cambrian succession in Siberia: report of IGCP Project 303. Geological Magazine 131, 767-783.

Burns, S.J. and Matter, A., 1993. Carbon isotopic record of the latest Proterozoic from Oman. Eclogae Geologicae Helvetiae 86/2, 595-607.

Burrett, C., Long, J. and Stait, B., 1990. Early-Middle Palaeozoic biogeography of Asian terranes derived from Gondwana. In: McKerrow, W.S. and Scotese, C.F. (eds.), Palaeozoic Palaeogeography and Biogeography. Geological Society Memoir 12, 163-174.

Chang, W.T., 1988. The Cambrian System in Eastern Asia. (eds. Shergold, J.H. and Palmer, A.R.). International Union of Geological Sciences, Publication 24.

Cocozza, T. and Gandin, A., 1990. Carbonate deposition during early rifting: the Cambrian of Sardinia and the Triassic-Jurassic of Tuscany, Italy. Special Publications of the International Association of Sedimentologists 9, 9-37.

Compston, W., Williams, I.S., Kirschvink, J.L., Zhang Zichao and Ma Guogan, 1992. Zircon U-Pb ages from the Early Cambrian time-scale. Journal of the Geological Society London 149, 171-184.

Conway Morris, S., 1989. Burgess Shale faunas and the Cambrian explosion. Science 246, 339-346.

Conway Morris, S., 1993. Ediacaran-like fossils in Cambrian Burgess Shale-type faunas of North America. Palaeontology 36, 593-635.

Cooper, J.A., Jenkins, R.J.F., Compston, W. and Williams, I.S., 1992. Ion-probe dating of a mid-Early Cambrian tuff in South Australia. Journal of the Geological Society London 149, 185-192.

Courjault-Radé, P., Debrenne, F. and Gandin, A., 1992. Palaeogeographic and geodynamic evolution of the Gondwana cotinental margins during the Cambrian. Terra Nova 4, 657-667.

Cowie, J.W. and Brasier, M.D., 1989. The Precambrian-Cambrian boundary. Clarendon Press, Oxford.

Crimes, T.P., 1989. Trace Fossils. In: Cowie and Brasier (1989). pp. 166-185.

Crimes, T.P., 1992a. Changes in the trace fossil biota across the Proterozoic-Cambrian boundary. Journal of the Geological Society London 149, 637-646.

Crimes, T.P., 1992b. The record of trace fossils across the Precambrian-Cambrian boundary. In: Lipps, J.H. and Signor, P.W. (eds.), Origin and Early Evolution of Metazoa. pp. 177-202. Plenum, New York.

Crimes, T.P. and Anderson, M.M., 1985. Trace fossils from Late Precambrian-Early Cambrian strata of southeastern Newfoundland (Canada): temporal and environmental implications. Journal of Palaeontology 50, 310-343.

Dalziel, I.W.D., 1991. Pacific margins of Laurentia and East Antarctica-Australia as a conjugate rift pair: evidence and implications for an Eocambrian supercontinent. Geology 19, 598-601.

Dalziel, I.W.D., 1992. Antarctica: a tale of two supercontinents? Annual Reviews of Earth and Planetary Science 20, 501-526.

Debrenne, F., 1991. Extinction of the Archaeocyatha. Historical Biology 5, 95-106.

Derry, L.A., Kaufman, A.J. and Jacobsen, S.B., 1992. Sedimentary cycling and environmental change on the late Proterozoic: evidence from stable and radiogenic isotopes. Geochimica et Cosmochimica Acta 56, 1317-1329.

Donnelly, T.H., Shergold, J.H. and Southgate, P.N., 1988. Anomalous geochemical signals from phosphatic middle Cambrian rocks in the southern Georgina Basin, Australia. Sedimentology 35, 549-570.

Ergaliev, G.K., 1981. Upper Cambrian biostratigraphy of the Kyrshababakty section, Maly Karatau, southern Kazakhstan. USGS Open File Report 81-743, 82-88.

Fedonkin, M.A., 1987. Paleoichnology of the Precambrian-Cambrian transition in the Russian Platform and Siberia. In: Landing, E., Narbonne, G.M. and Myrow, P. (eds.), Trace Fossils, Small Shelly Fossils and the Precambrian-Cambrian Boundary. Bulletin of the New York State Museum 463, 12.

Gould, S.J., 1991. Wonderful Life. Penguin books.

Grant, S.W.F., 1992. Carbon isotopic vital effect and organic diagenesis, Lower Cambrian Forteau Formation, northwest Newfoundland: implications for $\partial^{13}C$ chemostratigraphy. Geology 20, 243-246.

Hallock, P. and Schlager, W., 1986. Nutrient excess and the demise of coral reefs and carbonate platforms. Palaios 1, 389-398.

Holland, C.H. (ed.), 1971. Cambrian of the New World. Wiley, London.

Holland, C.H. (ed.), 1974. Cambrian of the British Isles, Norden and Spitsbergen. Wiley, London.

Holser, W.T., 1979. Catastrophic chemical events in the history of the ocean. Nature 267, 403-407.

Holser, W.T., 1984. Gradual and abrupt shifts in ocean chemistry during Phanerozoic time. In: Holland, H.D. and Trendall, A.F. (eds.), Patterns of Change in Earth Evolution. pp. 123-143.

Springer, Berlin Heidelberg New York.

Hsü, K.J., Oberhansli, H., Gao, J.Y., Sun Shu, Chen Haihong and Krahenbuhl, U., 1985. 'Strangelove ocean' before the Cambrian explosion. Nature 316, 809-811.

Ilyin, A.V., 1990. Proterozoic supercontinent, its latest Precambrian rifting, breakup, dispersal into smaller continents, and subsidence of their margins: evidence from Asia. Geology 18, 1231-1234.

Jenkins, R.J.F., 1989. The 'supposed terminal Precambrian extinction event' in relation to the Cnidaria. Memoirs of the Association of Australasian Palaeontologists 8, 307-317.

Kaufman, A., Hayes, J.M., Knoll, A.H. and Germs, G.J.B., 1991. Isotopic composition of carbonates and organic carbon from upper Proterozoic successions in Namibia: stratigraphic variation and the effects of diagenesis and metamorphism. Precambrian Research 49, 301-327.

Khomentovsky, V.V., 1986. The Vendian System of Siberia and a standard stratigraphic scale. Geological Magazine 123, 333-348.

Khomentovsky, V.V. and Karlova, G.A., 1993. Biostratigraphy of the Vendian-Cambrian beds and lower Cambrian boundary in Siberia. Geological Magazine 130, 29-45.

Kirschvink, J.L., 1992. Chapter 12.1. A Palaeogeographic Model for Vendian and Cambrian Time. In: Schopf and Klein (1992). pp. 569-581.

Kirschvink, J., Magaritz, M., Ripperdan, R.L., Zhuravlev, A.Yu. and Rozanov, A.Yu., 1991. The Precambrian/Cambrian boundary: Magnetostratigraphy and carbon isotopes resolve correlation problems between Siberia, Morocco and south China. GSA Today 1, 69-71, 87, 91.

Knoll, A.H., 1992. Biological and biogeochemical preludes to the Ediacaran Radiation. In: Lipps, J.H. and Signor, P.W. (eds.), Origin and Early Evolution of the Metazoa. pp. 53-86. Plenum, New York.

Knoll, A.H. and Walter, M.R., 1992. Latest Proterozoic Stratigraphy and Earth History. Nature 356, 673-678.

Landing, E., 1992. Lower Cambrian of southeastern Newfoundland: epeirogeny and Lazarus faunas, lithofacies - biofacies linkages, and the myth of a global chronostratigraphy. In: Lipps, J.H. and Signor, P.W. (eds.), Origin and Early Evolution of the Metazoa. pp. 283-310. Plenum, New York.

Landing, E., 1994. Precambrian-Cambrian global

stratotype ratified and a new perspective of Cambrian time. Geology 22, 179-182.

Latham, A. and Riding, R., 1990. Fossil evidence for the location of the Precambrian/Cambrian boundary in Morocco. Nature 344, 752-754.

Lindsay, J.F., 1993. Sequence stratigraphic comparisons of the Neoproterozoic and Cambrian sections of the Yangtze Platform, China and Amadeus Basin, Australia. AGSO Record, Canberra, 1993/02.

Lu Yanhao and Lin Huanling, 1981. Zonation of Cambrian faunas in western Zhejiang and their correlation with those in North China, Australia and Sweden. USGS Open-File Report 81-743, 118-120.

Ludvigsen, R. and Westrop, S.R., 1985. Three new Upper Cambrian stages for North America. Geology 13, 139-143.

Magaritz, M., 1989. ^{13}C minima follow extinction events: a clue to faunal radiation. Geology 17, 337-340.

Magaritz, M., Holser, W.T. and Kirschvink, J.L., 1986. Carbon-isotope events across the Precambrian-Cambrian boundary on the Siberian Platform. Nature 320, 258-259.

Mankiewicz, C., 1992. Proterozoic and Early Cambrian Calcareous algae. In: Schopf and Klein (1992). pp. 359-367.

Matthews, S.C. and Cowie, J.W., 1979. Early Cambrian transgression. Journal of the Geological Society London 136, 133-136.

McKerrow, W.S., Scotese, C.R. and Brasier, M.D., 1992. Early Cambrian geological reconstructions. Journal of the Geological Society London 149, 599-606.

McMenamin, M.A.S. and McMenamin, D.L.S., 1989. The Emergence of Animals. The Cambrian Breakthrough. Columbia University Press, New York.

Moczydlowska, M., 1991. Acritarch biostratigraphy of the Lower Cambrian and the Precambrian-Cambrian boundary in southeastern Poland. Fossils and Strata 29.

Moczydlowska, M. and Vidal, G., 1988. How old is the Tommotian? Geology 16, 166-168.

Moore, G.W., 1992. Tectonic assembly of South America. Episodes 15, 204-206.

Moores, E.M., 1991. Southwest U.S.-east Antarctic (SWEAT) connection: a hypothesis. Geology 19, 425-428.

Morel, P. and Irving, E.C., 1978. Tentative paleocontinental maps for the early Phanerozoic and Proterozoic. Journal of Geology 86, 535-561.

Myrow, P.M. and Hiscott, R.N., 1993. Depositional history and sequence stratigraphy of the Precambrian-Cambrian boundary stratotype section, Chapel Island Formation, Southeast Newfoundland. In: Geldsetzer et al. (eds.), Event markers in Earth history. Palaeogeography, Palaeoclimatology, Palaeoecology 104, 13-35.

Narbonne, G.M., Myrow, P.M., Landing, E. and Anderson, M.M., 1987. A candidate stratotype for the Precambrian-Cambrian boundary, Fortune Head, Burin Peninsula, southeastern Newfoundland. Canadian Journal of Earth Sciences 24, 1277-1293.

Norford, B.S., 1991. The international working group on the Cambrian-Ordovician boundary: report of progress. Geological Survey of Canada Paper 90-9, 31.

Notholt, A.J.G. and Brasier, M.D., 1986. Proterozoic and Cambrian phosphorites - regional review: Europe. In: Cook, P.J. and Shergold, J.H. (eds.), Phosphate deposits of the world. Volume 1. Proterozoic and Cambrian phosphorites. pp. 91-100. Cambridge University Press.

Odin, G.S. (ed.), 1993. Phanerozoic time scale. Bulletin of Liason and Information, IUGS Subcommission on Geochronology, offset Paris, volume 11.

Opik, A.A., 1961. The geology and palaeontology of the headwaters of the Burke River, Queensland. Commonwealth of Australia Bureau of Mineral Resources, Geology and Geophysics Bulletin 64, 133 pp.

Palmer, A.R., 1965. Biomere - new kind of biostratigraphic unit. Journal of Palaeontology 39, 149-153.

Palmer, A.R., 1979. Biomere boundaries re-examined. Alcheringa 3, 33-41.

Palmer, A.R., 1981. Subdivision of the Sauk sequence. USGS Open-File Report 81-743, 160-162.

Palmer, A.R., 1982. Biomere boundaries: a possible test case for extraterrestrial perturbation of the biosphere. Geological Society of America Special Paper 190, 469-475.

Palmer, A.R. and Repina, L.N., 1993. Through a glass darkly: taxonomy, phylogeny, and biostratigraphy of the Olenellina. The University of Kansas Paleontological contributions. New Series. Number 3.

Patchett, P.J., Gale, N.H., Goodwin, R. and Humm, M.J., 1980. Rb-Sr whole-rock isochron ages of late Precambrian to Cambrian igneous rocks from southern Britian. Journal of the Geological Society London 137, 649-656.

Pillola, G.L., 1990. Lithologie et trilobites du Cambrien inférieur du SW de la Sardaigne (Italie): implications paléobiogéographiques. Comptes Rendus de l'Academie des Sciences, Paris, 310 Series II, 321-328.

Piper, J.D.A., 1983. Proterozoic palaeomagnetism and single continent plate tectonics. Geophysical Journal of the Royal Astronomical Society 74, 163-197.

Piper, J.D.A., 1987. Palaeomagnetism and the continental crust. Wiley, New York.

Pokrovsky, B.G. and Missarzhevsky, V.V., 1993. Isotopic correlation of Precambrian and Cambrian of the Siberian Platform. Doklady Akademy Nauk 329, 768-771 [In Russian].

Repina, L.N., 1981. Trilobite biostratigraphy of the Lower Cambrian Stages in Siberia. USGS Open-File Report 81-743, 173-180.

Ripperdan, R.L., Magaritz, M., Nicoll, R.S. and Shergold, J.H., 1992. Simultaneous changes in carbon isotopes, sea level, and conodont biozones within the Cambrian-Ordovician boundary interval at Black mountain, Australia. Geology 20, 1039-1042.

Robison, R.A., Rosova, A.V., Rowell, A.J. and Fletcher, T.P., 1977. Cambrian boundaries and divisions. Lethaia 10, 257-262.

Rowell, A.J. and Brady, M.J., 1975. Brachiopods and biomeres. Brigham Young University Geology Studies 23, 165-180.

Rozanov, A.Yu., 1992. Some problems concerning the Precambrian-Cambrian transition and the Cambrian faunal radiation. Journal of the Geological Society London 149, 593-598.

Rozanov, A.Yu. and Sokolov, B.S. (eds.), 1984. Lower Cambrian stage subdivision. Stratigraphy. Akademii Nauk SSSR, Nauka, Moscow [In Russian].

Runnegar, B., 1982. Oxygen requirements, biology and phylogenetic significance of the late Precambrian worm *Dickinsonia*, and the evolution of the burrowing habit. Alcheringa 6, 223-239.

Runnegar, B. and Fedonkin, M.A., 1992. Proterozoic metazoan body fossils. In: Schopf & Klein (1992). pp. 369-388.

Schidlowski, M., Golubic, S., Kimberley, M.M., McKirdy, D.M. and Trudinger, P.A., (eds.) 1992. Early Organic Evolution. Springer-Verlag, Berlin.

Schopf, J.W. and Klein, C., 1992. The Proterozoic Biosphere. Cambridge University Press.

Seilacher, A., 1984. Late Precambrian and early Cambrian metazoa: preservation or real extinction? In: Holland, H.D. and Trendall, A.F. (eds.), Patterns of Change in Earth Evolution. pp. 159-168. Springer, Berlin Heidelberg New York.

Sepkoski, J.J., jr., 1984. A kinetic model of Phanerozoic taxonomic diversity. III. Post-Palaeozoic families and mass extinction. Palaeobiology 10, 246-267.

Sepkoski, J.J., jr., 1992. Proterozoic-Early Cambrian diversification of metazoans and metaphytes. In: Schopf and Klein (1992). pp. 553-561.

Shergold, J.H., 1981. Towards a global late Cambrian agnostid biochronology. USGS Open-File Report 81-743, 208-214.

Signor, P., 1992. Taxonomic diversity and faunal turnover in the early Cambrian: did the most severe mass extinction of the Phanerozoic occur in the Botomian stage? Fifth North American Palaeontological Convention, Abstracts with Programs, p. 272.

Sokolov, B.S. and Fedonkin, M.A., 1984. The Vendian as the terminal system of the Precambrian. Episodes 7, 12-19.

Sokolov, B.S. and Fedonkin, M.A., 1986. Global biological events in the late Precambrian. In: Walliser, O.H. (ed.), Global Bio-Events. Lecture Notes in Earth Sciences 8, 105-108. Springer, Berlin Heidelberg New York.

Stitt, J.H., 1971. Repeating evolutionary patterns in Late Cambrian trilobite biomeres. Journal of Palaeontology 45, 178-181.

Stitt, J.H., 1977. Late Cambrian and earliest Ordovician trilobites, Wichita Mountains area, Oklahoma. Oklahoma Geological Survey Bulletin 124.

Taylor, M.E., 1985. Late Cambrian trilobite mass extinction coincident with oxygen depletion of outer shelf benthic habitats in central Nevada. SEPM Midyear Meeting (Golden, CO), Abstracts v. 2, 88-89.

Tucker, M.E., 1989. Carbon isotopes and Precambrian-Cambrian boundary geology, South Australia: ocean-basin formation, seawater chemistry and organic evolution. Terra Research 1, 573-582.

Tucker, R.D. and Pharaoh, T.C., 1991. U-Pb ages for Late Precambrian igneous rocks in southern Britain. Journal of the Geological Society London 148, 435-443.

Westrop, S.R., 1989. Trilobite mass extinction near

the Cambrian-Ordovician boundary in North America. In Donovan, S.K. (ed.), Mass Extinctions. Processes and Evidence. pp. 89-103. Belhaven Press, London.

Wignall, P.B. and Hallam, A., 1992. Anoxia as a cause of the Permian/Triassic mass extinction: facies evidence from northern Italy and the western United States. Palaeogeography, Palaeoclimatology, Palaeoecology 93, 21-46.

Wright, A.E., Fairchild, I.J., Moseley, F. and Downie, C., 1993. The Lower Cambrian Wrekin Quartzite and the age of its unconformity on the Ercall Granophyre. Geological Magazine 130, 257-264.

Xiang Liwen and others, 1981. The Stratigraphy of China. Volume 4. The Cambrian System. Geological Publishing House, Beijing [In Chinese].

Xu Daoyi, Yan Zheng, Sun Yiyin, He Jinwen, Zhang Qinwen and Chai Zhifang, 1989. Astrogeological Events in China. Scottish Academic Press, Edinburgh, and Geological Publishing House, Beijing.

Zhuravleva, I.T., Repina, L.N. and Rozanov, A.Yu., 1990. Stage subdivision of the lower Cambrian. In: Repina, L.N. and Zhuravlev, A.J. (eds.), Third International Symposium on the Cambrian System. Abstracts, Novosibirsk, pp. 178-179.

Manuscript received October 1992
Revision received March 1994

Author's address:

Martin D. Brasier, Department of Earth Sciences, University of Oxford, Parks Road, Oxford OX1 3PR, U.K.

The Pattern of Global Bio-Events During the Ordovician Period

Christopher R. BARNES, Richard A. FORTEY, and S. Henry WILLIAMS

Abstract: The 70 Ma Ordovician Period is characterized by extensive epeiric seas, paleocontinent dispersal, intervals of intense volcanism and black shale deposition, a greenhouse climate state deteriorating to a brief icehouse state, strong faunal provincialism, and profound changes to the biota including the changeover from the Cambrian Fauna to the Paleozoic Fauna. Although many invertebrate phyla diversify during the Ordovician, precise biostratigraphic and global biogeographic data are provided best by conodonts, trilobites and graptolites. These three groups are used in this chapter to recognize five major bio-events four of which correspond closely to Series boundaries: Basal Tremadoc (B'Tc), Basal Arenig (B'Ag), Basal Llanvirn (B'Ln), Basal Caradoc (B'Cc) and Upper Ashgill (U'Al). Most of these correspond to significant eustatic events and the latter to the terminal Ordovician glaciation. The first four are each characterized by extinctions but these are overshadowed by a rapid innovation event with a radiation of a more diversified fauna; the U'Al is a severe extinction event, second only to the terminal Permian event in the entire Phanerozoic. Compared to many other Phanerozoic systems, the Ordovician is a period of considerable biologic, climatic and oceanographic complexity within which the balance between the forcing processes that produced the major and minor events is still not well understood.

Contents

Introduction	139
Basal Tremadoc Bio-Event (B'Tc)	141
Basal Arenig Bio-Event (B'Ag)	145
Basal Llanvirn Bio-Event (B'Ln)	150
Basal Caradoc Bio-Event (B'Cc)	154
Upper Ashgill Bio-Event (U'Al)	157
Ordovician Event Stratigraphy	161
Conclusions	163

Introduction

The Ordovician Period was a time of rather special conditions and events which undoubtedly influenced the global biota resulting in five principal global bio-events discussed in detail in this chapter. The Ordovician is characterized by a) widespread epeiric seas in which extensive carbonates accumulated on the low latitude paleocontinents, b) wide dispersal of paleocontinents following the breakup of the Rodinia Supercontinent in the late Proterozoic – early Cambrian, c) several intervals of intense volcanism and periods of deep ocean anoxia with extensive black shale deposition, d) a greenhouse state climate with rapid deterioration toward an icehouse state with the development of the North African continental glaciation in the Ashgill, e) major changes in the composition of the biota with the rapid replacement of the Cambrian Fauna by the Paleozoic Fauna (of Sepkoski, 1981). Several of these developments were noted earlier by Jaanusson (1984). The Period marks a

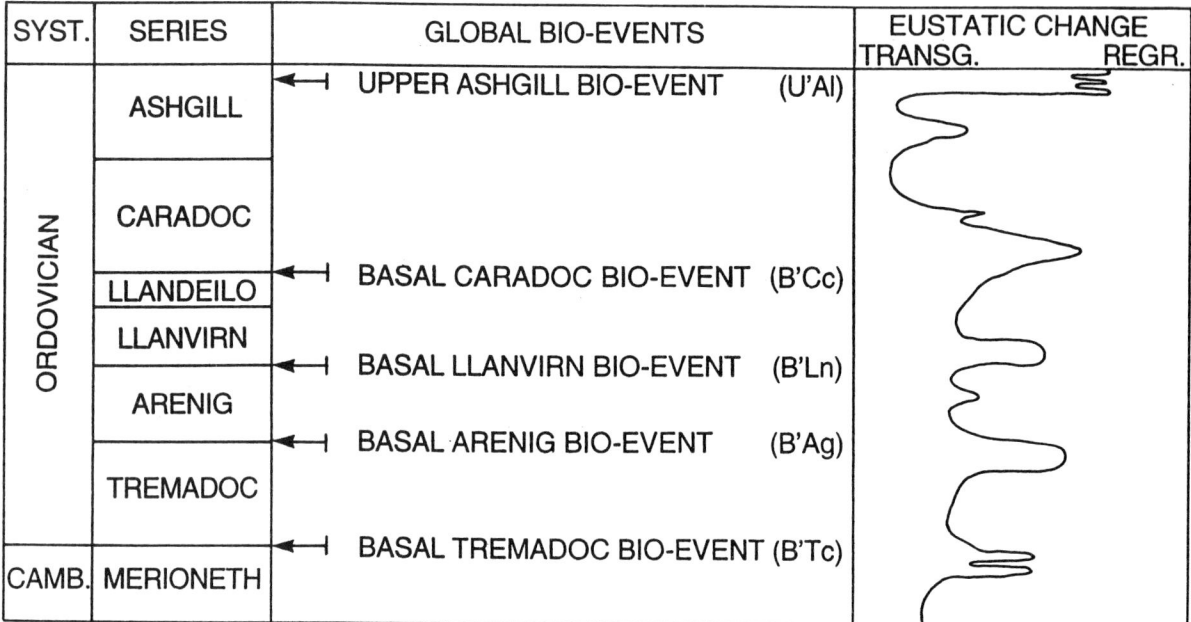

Fig. 1. Series chronostratigraphy, level of Ordovician global bio-events and first-order eustatic sea-level curve for the Ordovician

critical phase in the Early Paleozoic when the rapidly diversifying global biota was being modified to adapt to the evolving physical, chemical and climatic influences of the world's oceans. Major steps in the development of Lower Paleozoic stratigraphy were considered by McKerrow (1993). Attempts to model Ordovician oceanographic circulation (Ross, 1975; Wilde et al., 1986; Wilde, 1991) and initial work on determining the isotopic proxy record (see below) are all in their infancy.

The record of Ordovician global bio-events is complex and, given the space restraints imposed on this chapter, must of necessity focus only on the higher order bio-events herein. Five such bio-events are recognized: the Basal Tremadoc Bio-Event (B'Tc), Basal Arenig Bio-Event (B'Ag), Basal Llanvirn Bio-Event (B'Ln), Basal Caradoc Bio-Event (B'Cc), and Upper Ashgill Bio-Event (U'Al) (Fig. 1). Not surprisingly the early researchers investigating Ordovician stratigraphy and paleontology identified major faunal changes which were, in turn, used to define the five Standard British Series for the Ordovician System.

For each of the five bio-events we define the stratigraphic level of the bio-event and discuss the faunal changes occurring at or close to this level for conodonts, trilobites and graptolites. These three fossil groups are the best studied of the preserved Ordovician biota, each provides biostratigraphic data together with other studies that have documented evolutionary and paleoecological changes. We note some important changes in some other invertebrate groups at these bio-events but in general the database available is uneven and they appear to mirror the changes recognized for the three faunal groups considered herein.

For each major bio-event, we also attempt to interpret the possible causes of faunal change and to consider the other significant physical and chemical events occurring at or close to these five levels. The Ordovician System displays a dynamic character such that the effects of eustatic, anoxic, volcanic, climatic, and tectonic events are imprinted into the stratigraphic record. While stratigraphic, sedimentologic and paleobiologic studies are well advanced, it is unfortunately the case that paleomagnetic, geochemical and geochronological studies are, relatively, in their infancy. Thus, the interpretations presented herein and the issues raised in the summary sections for each bio-event must be in part speculative. In this review we raise these speculations in order to stimulate and focus further research in order to provide more quantitative and better constrained data and interpretations.

We emphasize that the Ordovician System is not only characterized by profound bio-events of first, second, third and lower orders of magnitude. The System is not only complex in this regard compared to, say, the Silurian System, but offers remarkable if challenging research opportunities to analyse the variable hierarchies of physical and biotic change in the Lower Paleozoic.

We note that the five bio-events must also be viewed in the light of the evolution of the Early Paleozoic skeletonized invertebrate fauna and of possible environmental threshold limits. The emerging isotopic and geochemical data for the Ordovician reveal some profound changes that suggest a continuation of major excursions experienced in the late Proterozoic and the Cambrian, perhaps with an attenuation in the Silurian.

Basal Tremadoc Bio-Event (B'Tc)

The first major bio-event recognized in the Ordovician coincides with the base of the System and therefore also at the base of the lowest series, the Tremadoc Series (Fig. 1). The Cambrian-Ordovician Boundary Working Group was established in 1974 to recommend a definition of the base of the System (horizon; boundary stratotype) to the IUGS Commission on Stratigraphy. After a large number of field meetings, workshops, newsletter reports and symposia publications (reviewed by Norford, 1991), the Working Group failed to achieve a satisfactory majority consensus recommendation. The Working Group had agreed to select a level defined by conodonts at or near the first appearance of nematophorous graptolites. The two final candidate sections were located at Green Point, western Newfoundland (Barnes, 1988) and at Dayangcha, Jilin Province, north-west China (Chen and Gong, 1986; Chen et al., 1988) with the latter gaining preference. The proposed horizon was the base of the *Cordylodus lindstromi* Zone which in both sections occurred a few metres below the first appearance of nematophorous graptolites. The Subcommission on Ordovician Stratigraphy has established a new Working Group to reexamine the issues. Despite this problem, a great deal of focussed research was undertaken across the boundary interval in many key sections around the world. For the purpose of this chapter the base of the Tremadoc is taken as the base of the *Cordylodus lindstromi* Zone in order to discuss the relative timing of bio-events.

Conodonts undergo a major evolutionary radiation across the Cambrian-Ordovician boundary. In the Cambrian, protoconodonts and paraconodonts which possess a different skeletal ultrastructure (Sweet, 1988) predominate. In the late Cambrian, the true conodonts, euconodonts, evolved. Faunas are commonly sparse and probably are not as yet fully documented. In a series of pioneering studies, both Müller (1973) and Miller (1980, 1984) established the general evolutionary framework. *Proconodontus* is an important parent stock from which other key taxa arose, such as *Eoconodontus* which in turn appears to have spawned the widespread and important genus *Cordylodus*. Some of the early speciation of *Cordylodus* was documented by Bagnoli et al. (1987), Barnes (1988) and most recently by Nicoll (1992). In the deep water slope facies (e.g. Cow Head Group, western Newfoundland), it is one of the most abundant taxa, represented by several species; other taxa are mainly coniform (e.g. *Teridontus, Semiacontiodus*).

In platform carbonate facies a dramatic radiation also occurs, perhaps further accentuated by a shallowing or regressive event in latest Cambrian time (late Trempealeauan) where more dolomitic facies with paraconformities rarely provide a complete faunal record (Fig. 1). The faunal diversification is best known from Australia (Druce and Jones, 1971), western United States (Miller, 1969, 1984), eastern Canada (Ji and Barnes, 1990, 1994) and China (Chen and Gong, 1986). Ji and Barnes (1990, 1994) documented the initial radiation and development of nine major conodont lineages in the earliest Ordovician based on *Clavohamulus, Semiacontiodus, Teridontus, Loxodus, "Acodus", Cordylodus, Drepanodoistodus, Utahconus* and *Variabiloconus*. These radiate primarily from the *Teridontus* and *Proconodontus* lineages that have their ancestry in the late Cambrian (Ji and Barnes, 1994).

Three aspects of this conodont radiation at the B'Tc are noteworthy; mineralization pattern, apparatus modification, and habitat differentiation. Although conodonts have been considered as fossil representatives of chaetognaths (e.g. Szaniawski, 1987), there is increasing evidence that they belong to early chordates, possibly close to myxinoids (hagfish) (Briggs et al., 1983; Aldridge and Briggs, 1986;

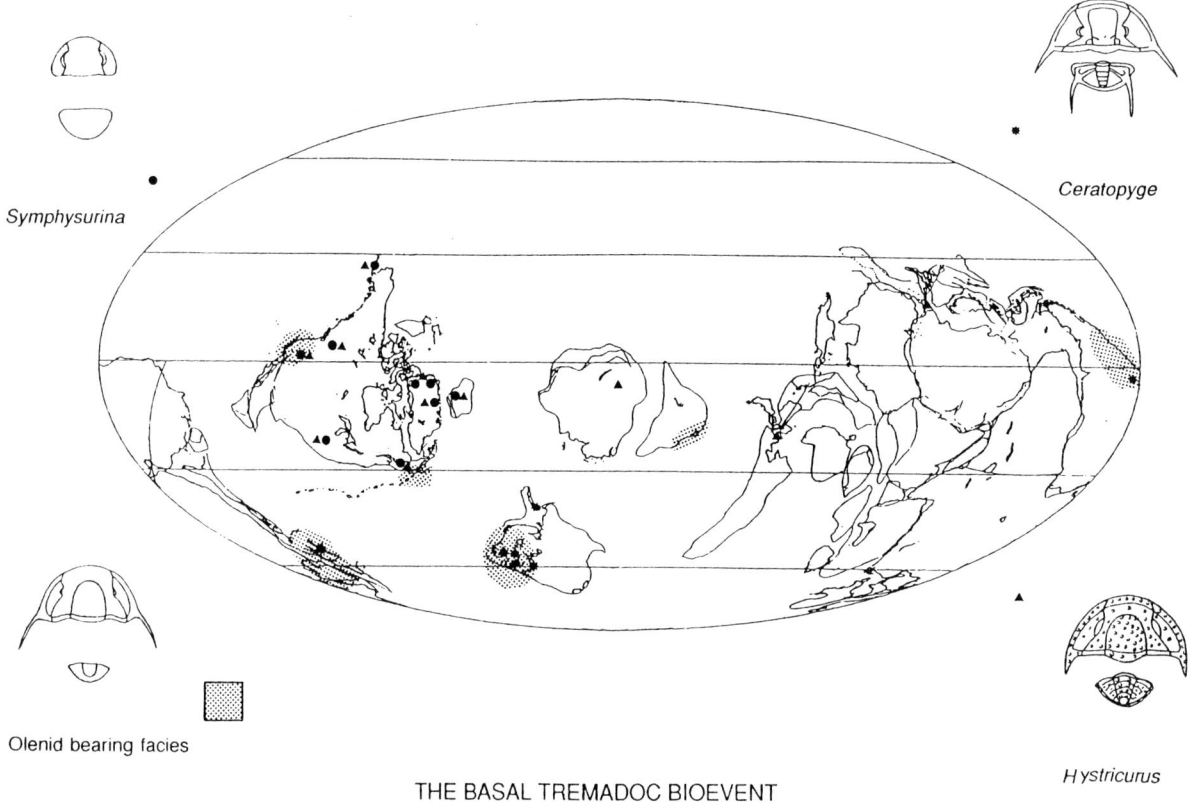

Fig. 2. Paleogeographic reconstruction (from McKerrow and Scotese, 1990) for the Basal Tremadoc Bio-Event, with the general distribution of key trilobite genera

Aldridge, Briggs, Smith et al., 1993). Sansom et al. (1992) have recently argued for the presence of cellular bone in *Cordylodus*, supposedly predating the earliest previous reported occurrence of vertebrate hard tissues by about 40 Ma. There is clearly much more skeletal ultrastructure work necessary on early conodonts, but the B'Tc Bio-Event may be important in the evolution of vertebrate tissue.

In apparatus evolution, the B'Tc is likewise a horizon of profound change, but with important modifications occurring over several short-duration conodont zones. The late Cambrian apparatuses (e.g. of *Protoconodontus*; Landing, 1977) are largely of a battery of simple coniform elements exhibiting subtle symmetry modification. The *Eoconodontus* apparatus displays at least two different morphotypes with additional variation within each. Just below the B'Tc level (base of the *C. lindstromi* Zone), a variety of new apparatus types evolved in which septimembrate apparatuses become established for the first time (Ji and Barnes, 1990, 1994; Nicoll, 1992).

This development of different apparatuses presumably reflects a diversified range of habit and diet, assuming conodonts to be a dental or pharangial apparatus (Briggs et al., 1983; Aldridge, Briggs, Smith et al., 1993). There is clearly habitat differentiation of a number of taxa with specialized apparatus. Ji and Barnes (in press) demonstrated, for example, the preference of *Clavohamulus* species to small algal-thrombolite biohermal complexes and of *Loxodus* species to certain restricted peritidal facies. By the B'Tc level, the coarse pattern of faunal realm differentiation of conodonts into the Midcontinent Realm and the North Atlantic Realm had developed (Barnes et al., 1973; Miller, 1984; Pohler and Barnes, 1990; Bergström, 1990), the former occurring in the shallow, hypersaline, epeiric seas of the carbonate platforms and the latter in the open circulation, outer shelf to slope and oceanic environments. The protoconodont-paraconodont fauna that characterized the late Cambrian suffered in abundance and taxonomic diversity. It extended into the Tremadoc but is more characteristic of deep water environments. Even there it tends to occur separately from euconodont faunas (e.g. Barnes, 1988), suggesting a preference for lower oxygenated environments.

Trilobite faunas are known through numerous sections spanning the Cambrian-Ordovician boundary, and have recently been studied in detail in connection with the search for a stratotype suitable for the definition of the lower boundary of the Ordovician System. There is a clear distinction between trilobite faunas that lived on platform areas and those that lived in deeper water sites fringing the paleocontinents. The latter include genera, and even species, which are widespread (Shergold, 1988), while the former include more endemic taxa, which may be extremely useful in the correlation of strata within the bounds of a single paleocontinent, but do not extend far outside it. The Basal Tremadoc Bio-Event records an abundant appearance of new platform trilobites, which were very rapidly propagated over wide areas.

By contrast, there is phylogenetic continuity between late Cambrian and early Tremadoc off-shelf faunas. Among trilobites belonging to the Agnostina, Asaphidae, Olenidae, Remopleurididae and Shumardiidae, for example, there are closely related taxa passing through from the Cambrian to the Tremadoc. The olenid biofacies, in particular, is also geographically widespread (Fig. 2). The maps adopted (Figs. 2, 3, 5-7) are from McKerrow and Scotese (1990), and are not critically reappraised here, even though they could be adjusted in some respects to better accord with faunal data. There is some evidence of a shoreward onstep of such off shelf trilobite biofacies at the B'Tc, manifest by the appearance of such taxa in shelf successions, which has also been claimed as the cause of 'mass extinction' of Upper Cambrian platform faunas (Westrop and Ludvigsen, 1987). As one example, the olenid trilobite *Jujuyaspis* appears in many sections, including platformal ones, at or very close to the eustatic transgression at the base of the Tremadoc. So widespread is its occurrence that it has been suggested as a possible index fossil for the base of the Ordovician (Anceñolaza and Anceñolaza, 1992). Olenid trilobites are often associated with dysaerobic facies (Henningsmoen, 1957; Fortey, 1975), and if this were true of *Jujuyaspis* then it may imply shelfward spread of oxygen-poor water for a short interval at the B'Tc.

In shelf successions - particularly those in former paleoequatorial limestone sequences - there is a sharp definition of the B'Tc associated with the first appearance of new genera, as in Laurentia (Stitt, 1971, 1977), the North China Platform (Zhou and Zhang, 1984), or central Queensland, Australia (Druce et al., 1982; Shergold and Nicoll, 1992). Several of these genera were confined to a single paleo-platform, but are abundant over it. An example is *Symphysurina* (Fig. 2), which turns up in virtually every earliest Ordovician section in Laurentia, but is unknown outside it. *Ceratopyge* is almost as common in Baltica, but is extremely rare outside mid to high paleolatitudes. Dikelokephalinids are especially typical of inshore sediments in eastern Gondwana. This repopulation of the craton seems to conform to a continuation of the Cambrian 'biomere' style of radiation, and was explicitly described as such by Stitt (1975). Other genera are found more widely distributed through the paleoequatorial zone - *Hystricurus* is a particularly widespread example (Fig. 2) (Fortey and Peel, 1989) - in a fashion comparable with that of many Midcontinent Province conodonts. The diminutive earliest Ordovician guide fossil *Missisquoia* is similarly distributed. The implication is that dispersion after the B'Tc was rapid, and presumably of forms that were to some extent pre-adapted to conditions pertaining in the early Ordovician.

The end of the Cambrian is often described as a mass extinction (Newell, 1967), and it is indeed true that Cambrian platform trilobite genera do not, with very few exceptions, pass through into the Ordovician. Fortey (1989) noticed that it would be more correct to describe the basal Ordovician as a time of o r i g i - n a t i o n of clades, but that the Cambrian ancestors of these clades, many of which would come to dominate the rest of the Ordovician, were unknown in many cases. Calymenids, hystricurids and nileids all appear in early Tremadoc rocks, from unknown origins. Yet it is evidently true that they will have had Cambrian sister taxa, and identification of such phylogenetic connections will serve to reduce the magnitude of the end-Cambrian extinction. The most major trilobite extinction was of the Superfamily Dikelocephaloidea, comprising typical inshore genera. This point is worth emphasizing, because Sepkoski (1981) showed a diversity "spurt" commencing at the Ordovician, which is arguably one of the formative events in the marine biosphere. Flower (1964) identified the radiation of the Ellesmeroceratida after the B'Tc as the origin of nautiloid (and hence cephalopod) diversification, and the similar pattern in the graptolites is commented on below. The B'Tc signals a punctuation in the record, followed by a succession of new appearances. The transgression spreads these novelties widely, but it is an interesting question why the B'Tc in particular should be associated with major innovation (expressed in both taxonomy and morphologic design) in several groups when other, apparently similar eustatic events were not so characterised.

Benthic, dendroid graptolites first appeared during the Middle Cambrian, and became widespread with a relatively high diversity during the late Cambrian. These sessile forms were rooted to the substrate with a basal holdfast structure which enveloped the sicula; because of their benthic mode of life, they are most commonly found in clastic sediments deposited under relatively shallow marine conditions, or as reworked fragments in transported, deeper water clastics. It is thus unusual to find them preserved together with planktonic graptoloids in "typical" graptolitic black shales, these having been deposited under quiet, oxygen and sediment-starved bottom conditions.

At or near the base of the Ordovician, the first planktonic graptolites evolved. Until recently these multiramous, bithecate rhabdosomes with a well-developed stolon system were assigned to the Dendroidea (e.g. Bulman, 1970). Cooper and Fortey (1983) among others demonstrated, however, that the proximal development of the planktonic Anisograptidae was essentially similar to that found in the Graptoloidea, and following Fortey and Cooper (1986) the term "dendroid graptolite" is now restricted to originally benthic taxa lacking a nematophorous sicula in the mature rhabdosome. Thus defined, it may be stated that major faunal changes at the B'Tc include the first occurrence of the Graptoloidea. The exact order of appearance of the major anisograptid taxa is still somewhat contentious; quadriradiate members of *Radiograptus* are considered to be the primitive ancestral stock by some authors (e.g. Fortey et al., 1982; Erdtmann, 1982; Fortey and Cooper, 1986), while others (e.g. Erdtmann, 1988) conclude that quadriradiate "staurorhabdinoporids", often possessing multi-fibrous nematophorous structures reminiscent of dendroids, are the earliest graptoloids. There does, however, appear to be general agreement that a quadriradiate proximal development, with two consecutive dichotomies, represents the primitive state, and that the loss of one and subsequently two dichotomies gave rise rapidly and successively to triradiate and biradiate anisograptids. Early graptoloid taxa include members of the genera *Rhabdinopora, Staurograptus, Radiograptus* and *Aletograptus* (Yu et al., 1982; Wang and Erdtmann, 1986, 1987; Erdtmann, 1988; Cooper and Lindholm, 1990; Cooper, 1992).

The first occurrence of graptoloids at or near the base of the Tremadoc undoubtedly marked a major bio-event; the number of continuous sections which include both late Cambrian dendroids and earliest Ordovician graptoloids is, however, small owing to the contrasting original paleoenvironments and lithologies required for the habitation and preservation of each group. Most of these sections are in slope deposits, where silts and grainstones containing dendroids have been transported into relatively deep water, anoxic settings suitable for the preservation of graptoloids through turbidite activity. The best known sections yielding both groups of graptolite are in the Cow Head Group of western Newfoundland, Canada and at Dayangcha, Jilin Province, China (Erdtmann, 1988). Other sections either lack graptolites below the first occurrence of graptoloids, such as at Naersnes, southern Norway (Bruton et al., 1982; Erdtmann, 1988), or are not continuously graptolitic across the Cambrian-Ordovician boundary, such as in Victoria, Australia (Cooper and Stewart, 1979; Cooper and Lindholm, 1990).

If the occurrence of the two groups of graptolites (i.e. benthic dendroids and planktonic graptoloids) was so facies dependent, can it be established whether the first occurrence of graptoloids was a near synchronous event or was it a widely diachronous bio-event controlled in large part by sedimentological change? The appearance of benthic trilobite species, especially those in deeper water facies, supports the notion that the appearance of graptolites was a synchronous event. Similar species of olenid trilobites, including *Parabolinella* and *Jujuyaspis*, occur in a variety of successions at the same time, or in consistent sequence, with the nematophorous graptolites (Shergold, 1988).

If the first occurrence was controlled by facies changes which were more-or-less globally synchronous, it implies the presence of a major physical or chemical event during that interval, probably related to a global eustatic transgression at or near the Cambrian-Ordovician boundary. Support for such an event is certainly provided by the almost ubiquitous "*Dictyonema*" (i.e. *Rhabdinopora*)-bearing black shale deposition around the world during the Tremadoc and its onstep onto "platform" areas, and it was possibly this event which enhanced the rapid evolution and diversification of the Anisograptidae as suitable new planktonic niches became available. Although the graptoloids soon became the dominant group of graptolites, both numerically and taxonomically, the dendroids continued to survive until the late Carboniferous, many remaining almost unchanged in morphology during their range except for a peak of thecal complexity during the Silurian (Bull, 1991).

Basal Tremadoc Bio-Event: Summary and Issues

From the above it is clear that trilobite faunas were significantly modified and that euconodonts and graptolites first appeared and diversified at or close to the Basal Tremadoc Bio-Event. There is no doubt that the early Tremadoc was a time of transgressive-regressive cycles in this interval (e.g. Lange Range and Black Mountain events; Miller, 1992). Miller (1984) has also argued that some of the eustatic changes at this level may have been generated by a minor glaciation in the South American segment of Gondwanaland. The sedimentological evidence for a definite glacial phase remains equivocal, but, if true, a minor glaciation may have been influential in the early ventilation of the deeper oceans. One may question whether such eustatic changes are a sufficient explanation to account for a number of paleobiological innovations near the B'Tc such as the origination of euconodonts, graptolites and the early radiation of inarticulate brachiopods and nautiloid cephalopods.

One possible explanation is that the bio-event is not only a reflection of a eustatic event but possibly of a critical threshold in ocean ventilation (oxygenation). It may be argued that the shallow platform waters attained a level of oxygenation to allow a more diversified invertebrate fauna and created a greater segregation from those elements evolving in the Cambrian that were adapted to relatively low levels of oxygenation. The alternation of moderately and poorly oxygenated upwelling waters may best explain the pattern of late Cambrian-early Ordovician biomere extinctions. The progressive ventilation of the deep oceans was likely a principal factor in the replacement of the Cambrian Fauna by the Paleozoic Fauna (of Sepkoski, 1981). The elements of the Cambrian Fauna, including the protoconodont-paraconodont fauna, were seemingly displaced into progressively deeper, poorly oxygenated habitats in the early Ordovician (cf. Fortey and Barnes, 1977) with a sharp initial radiation of the Paleozoic Fauna occurring at or close to the B'Tc. Thus new groups such as the euconodonts and planktonic graptolites radiated rapidly. The widespread distribution of these elements (and of various trilobite taxa noted above) was induced by the transgression of the early Tremadoc, at or close to the B'Tc. Thus the B'Tc may be considered a consequence of an evolutionary response to both chemical and physical events in the earliest Ordovician oceans.

Future work needs to clarify the full nature of the replacement of the Cambrian Fauna by the Paleozoic Fauna through analysis of different environmental transects. Attention could profitably be paid to interpreting the changes in trophic structures. Certainly, more isotope analyses are required to better constrain the possible chemical changes at or near the B'Tc.

Basal Arenig Bio-Event (B'Ag)

The Basal Arenig Bio-Event (B'Ag) (Fig. 1) occurs at or close to the base of the Arenig Series which, although not formally defined, is taken as the base of the *Tetragraptus approximatus* Zone. This horizon has long been used informally as the base of the Arenig even though the nominate species does not occur in the clastic sequence of the type Arenig Series in Wales. Berry (1992), as Chair of the Arenig Working Group, has recently documented the arguments and evidence in moving towards a formal recommendation for definition. One of the potential stratotypes occurs in the Cow Head Group, western Newfoundland, best exhibited at the Ledge section, Cow Head Peninsula (Williams et al., 1994) for which the conodonts (Stouge and Bagnoli, 1988) and graptolites (Williams and Stevens, 1991) are documented provisionally.

The faunal contrasts introduced at the base of the Arenig Series were one reason why Lapworth (1879) originally proposed this horizon as the base of the Ordovician System. In the type area, and elsewhere in England and Wales, the Series is marked at the base by a transgressive sandstone, which is probably the local manifestation of a global eustatic event at this level (Fig. 1). Such an event was originally proposed by Fortey (1979, p.67), and further evidence for it was summarised by Fortey (1984) and Barnes (1984). Because the Tremadoc in the type area is mostly developed in deeper shelf facies, including both olenid and agnostid trilobites among its fauna, it seemed in greater continuity with the Cambrian to the earlier workers, by comparison with the Arenig and younger rocks.

The conodont changes at or near the B'Ag can best be discussed within each of the two faunal realms. In the Midcontinent Realm, one of the most dramatic extinction bio-events occurs in the latest Tremadoc. This level is between conodont Faunas C and D of Ethington and Clark (1971) which equates approximately to the base of trilobite Fauna D of Utah-Nevada (Ross et al., 1982) and the base of the *Adelograptus antiquus* graptolite Zone (the zone below the *T. approximatus* Zone).

Ethington and Clark (1971, 1981) recognized this significant change in conodont faunas and the abrupt change was documented in more detail by Ethington et

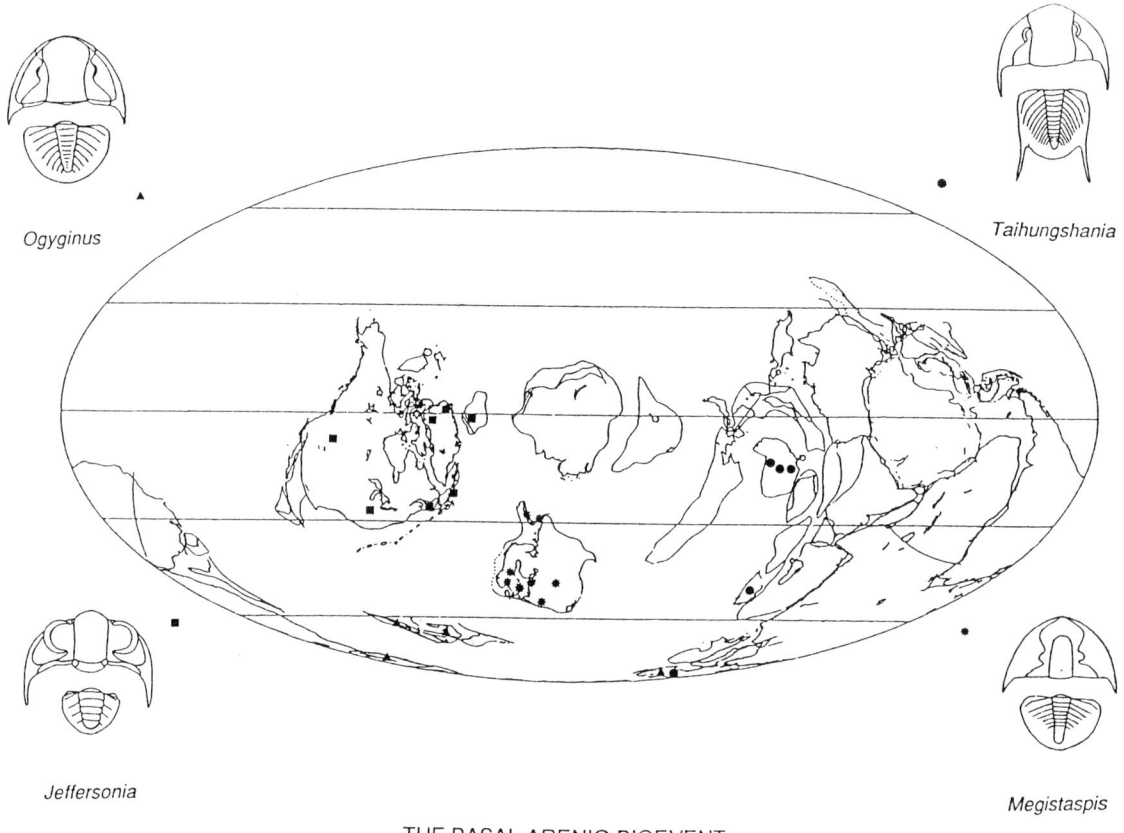

Fig. 3. Paleogeographic reconstruction (from McKerrow and Scotese, 1990) for the Basal Arenig Bio-Event, with the general distribution of key trilobite genera

al. (1987). They noted the dominance of taxa such as *Rossodus manitouensis, Variabiloconus bassleri, Polycostatus oneotensis* and *Acanthodus lineatus* in Fauna C (or *R. manitouensis* Zone of Landing et al., 1986). The replacement Fauna D has an initially low diversity fauna followed by a radiation; *Glyptoconus bolites, G. floweri,* and *Macerodus dianae* are characteristic species. The faunal change has been recognized within the Midcontinent Realm on several paleoplates (Laurentia: Landing and Barnes, 1981; Ji and Barnes, 1993; Siberia: Moskalenko, 1967; Abaimova, 1975; North China: An et al., 1983).

A detailed examination of this extinction bio-event has been undertaken by Ji and Barnes (1993) from the St. George Group, western Newfoundland. At a level 44 m above the base of the Boat Harbour Formation, seven major conodont lineages (*Clavohamulus, Loxodus, Polycostatus, Rossodus, Semiacontiodus, Teridontus* and *Variabiloconus*) involving over 30 multielement species become extinct. The replacement taxa include predominately hyaline coniforms (*Drepanoistodus, Macerodus, Glyptoconus, Striatodontus*) with the later radiation generating new lineages such as *Acodus, Oepikodus, Protopanderodus, Scolopodus* and *Stultodontus*. In the western Newfoundland sequence there appears to be a minor disconformity at the extinction level, possibly giving some accentuation. In terms of timing and event stratigraphy (Ji and Barnes, 1993), the extinction level appears to correspond to the peak of eustatic sea level fall in the late Tremadoc, the low diversity Fauna D corresponding to the low stillstand, and the radiation in upper Fauna D equating to the initial phase of a prominent eustatic sea level rise in the earliest Arenig. The interpretation of eustatic events for this area at this time have been given by Barnes (1984) and James et al. (1989).

The same conodont extinction event in eastern Canada has recently been recognized in faunas from the Survey Peak Formation, Wilcox Pass, Alberta in western Canada (Ji and Barnes, in prep.). Many species disappear 60 m above the base of the middle member.

In the North Atlantic Realm, significant faunal

changes also occur in conodont faunas at the B'Ag. The latest Tremadoc faunas still contain representatives of *Cordylodus* and associated coniforms. In the *Paroistodus proteus* Zone there is a low diversity fauna with species such as *Coelocerodontus bicostatus, Acodus deltatus, Parapanderodus gracilis* and the distinctive *Paracordylodus gracilis*. This occurs close to the base of the *T. approximatus* graptolite Zone, for example in the Cow Head Group, western Newfoundland. In these sections, the thin sequence of red shale in the earliest Arenig yields few conodonts. The first carbonates above these produce species characteristic of the early Arenig (*Prioniodus elegans* Zone) such as *P. elegans*, the earliest representative of *Periodon, Walliserodus australis* and, a little higher, *Protoprioniodus simplissimus, Oepikodus communis* and *Bergstroemognathus* cf. *B. extensus* (Johnston, 1986; Stouge and Bagnoli, 1988). A generally similar pattern of faunal replacement at the B'Ag is seen in other parts of the world, such as Argentina (Serpagli, 1974) and Baltoscandia (Löfgren, 1978).

These changes in conodont faunas in the North Atlantic Realm suggest three phases: the demise of the rich Tremadoc *Cordylodus* fauna, the entry of the low diversity *Paroistodus proteus* Zone fauna perhaps just below and across the base of the Arenig and the marked radiation in the *Prioniodus elegans* Zone in the early Arenig. These changes may reflect the tripartite pattern of the platform Midcontinent Realm faunas with the regression, the stillstand, and the transgression producing a similar tripartite sequence in the deeper oceanic faunas. It must be emphasized that the new Arenig conodont fauna contains dramatically different apparatus types – septimembrate apparatuses in which processes become prominent, extended and highly denticulated with the alternating hindeodellid dentition pattern appearing prominently for the first time. Species in some lineages commence with undenticulated processes or bearing only germ denticles, others develop prominent blade elements. Coniform taxa become much reduced in proportion. The radiation suggests a significant niche partitioning with the evolution of many new nektobenthic taxa with a reduction of the pelagic coniform taxa; the early Arenig transgressive onlap onto the cratons (Barnes, 1984; Fortey, 1984) produced a habitat expansion with taxa such as *Oepikodus* being widespread in deeper shelf facies (Sweet et al., 1971; Barnes and Fåhraeus, 1975; Ethington, 1972; Ethington and Repetski, 1984).

The transgressive pulse at the base of the Arenig (Fig. 1) stimulated endemic trilobite speciation on the separate paleocontinents (Fig. 3). Trilobite provinciality was at a maximum at this time (Whittington and Hughes, 1972), which may be attributed to a well developed climatic gradient, coupled with wide continental dispersal.

On Laurentia, the main diversification of the endemic family Bathyuridae was at this time, as shown in Utah and Nevada in such formations as the Ninemile and Wahwah (Hintze, 1953), and in the Catoche Formation, western Newfoundland (Fortey, 1979). This fauna extends into Greenland, north-west Scotland and the Canadian arctic; bathyurids are accompanied by a whole suite of other endemic trilobites, asaphids and pliomerids particularly. In the inner shelf limestone facies such trilobites are often associated with *Calathium* sponge 'bioherms', Nuia and other algae.

Elements of this bathyurid fauna are known from similar carbonate platform rocks in north-east Siberia, and probably also the North China platform (this is not true of the genus shown on Fig. 3, but probably applies to the early Arenig bathyurid *Peltabellia*). However, that part of China and Australia which was also adjacent to the Ordovician paleoequator has a largely endemic, different fauna from that of Laurentia, with distinctive dikelokephalinids, taihungshaniids and prosopiscids (Lu, 1975; Fortey and Shergold, 1984). This implies that even within the same climatic belt different plates responded to the B'Ag distinctively.

At high paleolatitudes, confined to that part of Gondwana adjacent to the Ordovician south pole, carbonates are almost completely lacking (this is therefore the most problematic area with regard to conodonts). The B'Ag is recorded in widespread clastics of the Grès Armoricain facies extending over Armorica, Iberia, eastern Newfoundland, and as far east as Saudi Arabia and Oman (?Yunnan). These shallow water clastics are associated with a consistent suite of trace fossils, especially *Cruziana furcifera* (Bergström, 1976). These are attributed to the work of grazing trilobites by most authorities. The inshore trilobites that occur (rather rarely) in beds adjacent to *Cruziana* include especially the earliest true calymenid *Neseuretus* and the asaphid *Ogyginus*, neither of which are found outside Gondwana. This, then, is a third, inshore endemic association produced in response to the B'Ag, and one which is particularly distinctive if taken in conjunction with the associated ichnofossils. In the type Ordovician area in Wales, Fortey and Owens (1978, 1987) have described a varied stratigraphy following the *Neseuretus* biofacies upwards in the Arenig section from the B'Ag, but such deeper water facies are rarely preserved in areas peripheral to Gondwana. However, they described the earliest occurrence of the deep water cyclopygid biofacies (typified by the combination of mesopelagic cyclopygid trilobites and associated benthic blind, or atheloptic forms) which was at that time not known in more equatorial regions. The earliest history of trinucleid trilobites was also confined to Gondwana, and particularly to this region. Note that there is some dispute about whether or not Avalonia was welded to the rest of Gondwana in the Arenig, but it was assuredly at high paleolatitudes. In Baltica, the condensed limestone sequences of Sweden, Estonia, Latvia and the Russian platform (eastward to northern Poland) contain another suite of trilobites which radiated at the B'Ag, especially megalaspid asaphids, raphiophorids and nileids. Neither bathyurids, nor calymenids, nor taihungshaniids are known from this paleocontinent. This presents a fourth response to the B'Ag, one which is thought to have taken place at temperate paleolatitudes.

The deep water, olenid biofacies included widespread taxa, with distribution independent of continental disposition (Fortey, 1980). This is a continuation of the Tremadoc pattern.

In general, there is a striking correspondence between shelf trilobite faunas and paleogeographic boundaries; Fortey and Mellish (1992) observed that trilobites probably reflect such boundaries most precisely among common fossils. For example, many conodont species extend from Baltica to North America in the North Atlantic Realm in a way which is rare in shelf trilobites. This means that the B'Ag was a particular source of vicariance in trilobite speciation; total diversity of taxa was far greater than if there had been climatic and geographic uniformity. This vicariance effect was to have important implications for the fate of trilobites later in the Ordovician.

During the Tremadoc, graptoloid diversity rapidly increased, with a gradual shift in composition from an assemblage dominated by multiramous, dissepimentous forms such a *Rhabdinopora* to one in which rhabdosomes with fewer stipes and less complex stolon structures were most diverse. This evolutionary shift culminated in a high diversity fauna during the late Tremadoc (Fig. 4), with many anisograptid taxa belonging to genera such as *Clonograptus, Kiaerograptus* and *Adelograptus* which are almost indistinguishable from Arenig dichograptid and sinograptid forms. This distinctive assemblage has been recognized in many parts of the world (see Cooper, 1979b; Mu et al., 1979; Erdtmann and VandenBerg, 1985; Wang and Erdtmann, 1987; Williams and Stevens, 1991), and has recently been redescribed in detail from western Newfoundland (Williams and Stevens, 1991), southern Scandinavia (Lindholm and Maletz, 1989) and Australia (Morris, 1988). Despite the apparently simple nature of thecal style and occurrence of rhabdosomes with relatively few stipes, Williams and Stevens (1991) have shown that all taxa from this interval possess at least one bitheca (associated with the sicula), and thus differ from known Arenig taxa whose rhabdosomes are composed solely of autothecae.

At the end of the Tremadoc, practically all of these distinctive anisograptid taxa disappear over a very short interval (Fig. 4), to be replaced in the basal Arenig *T. approximatus* Zone by a low diversity dichograptid and sinograptid fauna including *Tetragraptus approximatus, Pendeograptus* cf. *pendens* and a number of other *Tetragraptus* and *Didymograptus (Expansograptus)* species (e.g. Tzaj, 1974; Mu et al., 1979; Chen et al., 1983; Cooper and Lindholm, 1985; Cooper, 1979a; Williams and Stevens, 1988; Lindholm, 1991; Berry, 1992). A number of the "typically Tremadoc", multiramous anisograptids such as *Rhabdinopora* and *Clonograptus* do, however, range upward into the basal Arenig, reaching into the *P. fruticosus* Zone or higher (Williams and Stevens, 1988).

Critical study of isolated graptolite material from both the latest Tremadoc and early Arenig of the Cow Head Group, western Newfoundland, appears to suggest that several of the late Tremadoc taxa gave rise directly but independently to early Arenig forms through loss of the sicular bitheca, and in some cases other bithecae, suggesting a polyphyletic origin for the Dichograptidae and Sinograptidae (Williams and Stevens, 1991; SHW, pers. observ.). If this hypothesis, which still requires further critical study of isolated material, is correct, it suggests that loss of bithecae occurred simultaneously in several independent graptoloid lineages. What physical or biological pressure could have enhanced such evolution is unknown; before any models can be considered, the exact function of the zooids housed in the bithecae must be understood (e.g. sex-linked?).

Whatever the cause or mechanism, the almost total replacement of the graptoloid fauna in the basal Arenig marks a major bio-event which is readily identifiable in all complete graptolitic sections containing a "Pacific province" (i.e. low latitude) graptolite fauna (Cooper et al., 1991). The only clue to a possible physical cause evident in western Newfoundland is the presence of an interval composed either of non-grapto-

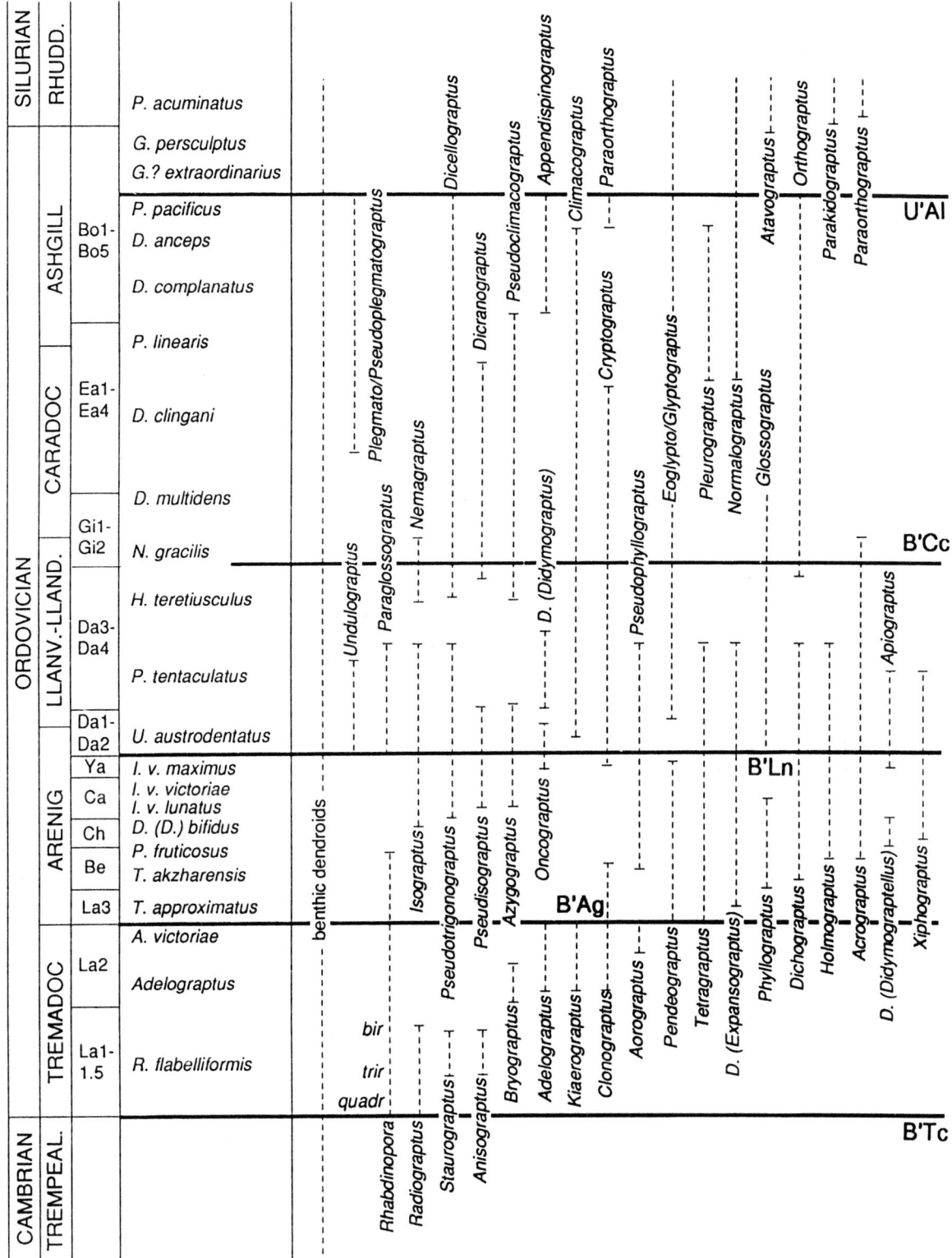

Fig. 4. Stratigraphic ranges of representative Ordovician graptolite genera, using data largely taken from Mitchell (1987), Williams and Stevens (1988), Cooper and Lindholm (1990) and VandenBerg and Cooper (1992)

litic, pale grey-green, dolomitic shale and siltstone (proximal sections) or red shale (distal sections) lying between the black shales and limestones of the late Tremadoc *A. victoriae* Zone and those of the following *T. approximatus* Zone (Williams and Stevens, 1991). This lithological change suggests increased oxygen content which might have resulted from an eustatic fall in sea level. It remains to be seen whether this was a final culmination of continued eustatic lowering throughout the Tremadoc, or whether it represented a relatively short interval with temporary low sea level stand. In some areas, such as the type area in Wales and the Welsh Borderlands, there is evidence of a rather prolonged interval of shallowing towards the end of the Tremadoc (Fortey and Owens, 1978), and the same phenomenon might account for the absence of younger Tremadoc graptolites in that area. Indeed, the later Lancefieldian faunas are rare on a global scale.

Basal Arenig Bio-Event (B'Ag): Summary and Issues

The Basal Arenig Bio-Event (B'Ag) reflects rather significant changes in the early development of the Paleozoic Fauna which occur over the span of about one graptolite zone just below and cross the base of the *T. approximatus* Zone. Although there are many second and third order eustatic changes during the Tremadoc and Arenig (Miller, 1992; Nielsen, 1992a,b; Ross and Ross, 1992), it would appear that the first-order regression during the close of the Tremadoc had major effects on the conodont, trilobite and graptolite faunas.

The continued phase of continental plate dispersal, following the break-up of the supercontinent Rodinia during the late Proterozoic-early Cambrian (e.g. Hoffman, 1991), suggests that the time of the B'Ag was one of peak separation of continental plates. This may well explain some of the contrasts of trilobite and conodont faunas between different plates in the low latitudes (Bergström, 1990). The first major regression experienced by the initial Paleozoic Fauna would most likely have a profound effect and seemingly produced the extinction B'Ag. It may explain the brief interval near the B'Ag with low diversity faunas.

The earliest Arenig may have been a time of increased oxygenation in the deep ocean as suggested at least by the Cow Head sequence in Newfoundland. The early Arenig transgression (Barnes, 1984; Fortey, 1984; Neilsen, 1992a; Ross and Ross, 1992) is certainly a period of adaptive radiation in all three groups considered, particularly demonstrated by the progressive diversification of trilobites and several other invertebrates (brachiopods, nautiloid cephalopods). However, the causes for the sudden evolution of non-bithecate graptolites and of the more complex denticulation patterns and septimembrate conodont apparatuses (e.g. *Prioniodus* lineage) remain an enigma.

Basal Llanvirn Bio-Event (B'Ln)

As with the two earlier series of the Ordovician, the base of the Llanvirn Series is yet to be formally defined. In the type area, the base of the *Didymograptus artus* Zone defines the lower boundary; elsewhere, particularly in areas with Pacific Province graptolite faunas, a slightly lower level at the base of the *Undulograptus austrodentatus* has been advocated. For the purposes of this chapter the B'Ln (Fig. 1) is taken as occurring at or close to this latter horizon. As will be noted below the changes correlate with a brief eustatic sea level change that largely spans the short duration between these two zonal levels (Fig. 1).

The top of the Arenig Series in the type area and elsewhere is marked by a short-lived regressive episode, which may be complex in detail (Neilsen, 1992). This is marked by a biofacies shift in the lower part of the Llanfallteg Formation in South Wales (Fortey and Owens, 1987); it is expressed by the Komstad Formation and correlative formations ('Orthoceras Limestone' partim) over Scandinavia; and by the Juab Limestone in the standard Ibex section in Utah. Throughout Ordovician Gondwana it is frequently marked by an ironstone horizon (Young, 1990). It seems that this regression corresponds closely with the base of the Whiterock Series as recently defined in North America by Ross et al. (1991), which Fortey earlier termed the Valhallan Stage. Biserial graptoloids apparently appear at the same interval, which was a time of great innovation in the Graptoloidea, as described below. The base of the Llanvirn marks a subsequent flooding of the Gondwana craton by graptolitic rocks bearing 'tuning fork' graptolites of the *D. artus* and *D. spinulosus* groups. Thus it appears to correspond with the maximum flooding surface (MFS) of the ensuing transgression. This opens the question whether the B'Ln should be considered as the couplet of regression plus transgression.

For conodonts, the Basal Llanvirn Bio-Event (B'Ln) is rather muted compared to the earlier B'Tc and B'Ag. It is marked less by profound extinction but rather by the originations of many new and morphologically different taxa. The change conveys a further significant modification to conodont faunas in both realms.

In the cratonic areas bearing Midcontinent Realm

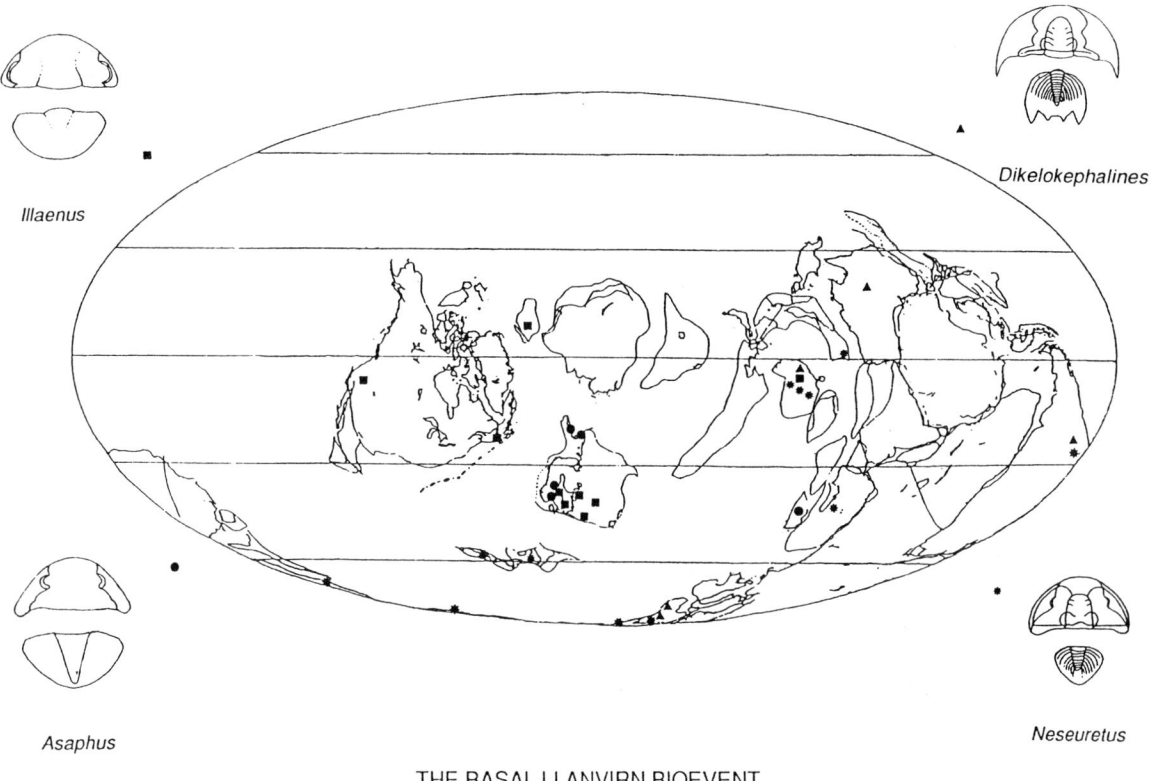

Fig. 5. Paleogeographic reconstruction (from McKerrow and Scotese, 1990) for the Basal Llanvirn Bio-Event, with the general distribution of key trilobite genera

faunas, the latest Arenig (early Whiterockian Series) is marked by a regression. In western Newfoundland where the conodont faunas are now well known for this interval the regressive phase is expressed in the Aguathuna Formation with the transgressive phase represented by the overlying Table Head Group (deepening upward sequence commencing with the Table Point Formation). The conodonts from the Aquathuna Formation (Stait and Barnes, 1991) belong to a peritidal community characterized by *Diaphorodus emanuelensis, Eucharodus parallelus, Oneotodus costatus, Oistodus multicorrugatus* as well as *Pteracontiodus cryptodens*. These are predominantly coniform taxa, variably costate and several of which are hyaline in structure. They are replaced within the upper 15 m of the formation by more complex apparatuses, many still hyaline, but in which lateral processes, commonly denticulate, occur (e.g. *Paraprioniodus costatus, Leptochirognathus quadrata, Multioistodus subdentatus* together with coniforms such as *Scandodus sinuosus, Drepanoistodus angulensis* and *Protopanderodus strigatus*). This upper fauna typically occurs above an omission surface and the two faunas are correlated by Stait and Barnes (1991) with Fauna 2 and Fauna 4 respectively of Sweet et al. (1971). Despite a possible small hiatus the two faunas demonstrate the significant taxonomic and morphological changes in conodont apparatuses within the peritidal communities of the Midcontinent Realm at the B'Ln. The upper fauna continues into the lower strata of the Table Point Formation. As the Llanvirn transgression continues the new subtidal faunas radiate (Stouge, 1984). In these environments are found new genera such as *Histiodella, Belodella, Eoneoprioniodus, Trigonodus* and possibly early species of *Plectodina* and *Erraticodon*. The similar conodonts of Fauna 4 in Utah occur in the Kanosh and Lehman formations (Ethington and Clark, 1981). The correlation of the underlying Whiterock faunas has recently been discussed by Ross et al. (1991) and Ross and Ethington (1992).

The North Atlantic Realm faunas are characterized

less by extinction near the B'Ln than by the development of new apparatuses which include distinctive platform elements. Genera such as *Pygodus, Cahabagnathus, Eoplacognathus,* and *Polonodus*, originated and others, such as *Amorphognathus* and *Prioniodus* which flourished in the Arenig, fully developed their Pa and/or Pb elements into platforms (Bergström, 1971; Löfgren, 1978; Dzik, 1983). This increased differentiation within conodont apparatuses is one of the most important stages in conodont evolution particularly since these elements have usually been the most useful for biozonation. There are other platform conodonts that seem to occur at or shortly after the B'Ln but which have yet to be fully described. Within the North Atlantic Realm, the changes at the B'Ln generally set the stage in creating a conodont fauna that is not drastically modified by innovations through the rest of the Ordovician.

The basal Llanvirn is usually marked by an abrupt change in trilobite biofacies. In the type area the deeper shelf cyclopygid biofacies spread shelfwards into Shropshire (Hope Formation), and a comparable change occurs above the ferruginous base of the Sarka Formation in Bohemia. In more inner shelf sites in Armorica, Iberia, Saudi Arabia, Libya, north Africa, a single species of the genus *Neseuretus, N. tristani*, is extremely widespread (Fig. 5), and in many localities the diagnostic pendent graptolites can be collected in adjacent beds (e.g. Saudi Arabia). The flooding of Armorica/Iberia and Morocco at comparatively high paleolatitudes stimulated a great burst of evolution among calymenoid, homalonotid and dalmanitacean trilobites (Cocks and Fortey, 1990) with many endemic genera, for example, zelizskellines and plaesiacomiids, unknown in Baltica, Laurentia or tropical Gondwana. Whittington and Hughes (1972) termed this the *Selenopeltis* Province, after a characteristic odontopleurid.

In Baltica, the genus *Asaphus* and its allies are widespread over the carbonate platform, together with endemic late megalaspids. The Chinese faunas described by Lu (1975) include some endemics (such as the trinucleid *Ninkiangolithus*), but also forms such as *Neseuretus* and *Taihungshania* which are known further west on Gondwana. The North China platform (tropical) includes a meagre, largely endemic fauna. In North America, endemic bathyurid history continued, but the Arenig-Llanvirn boundary interval also marks the commencement of an important restructuring of the marine communities to become more typical of the Paleozoic in general. Corals and bryozoans became appreciably more numerous at this level, for example,

and by the Chazy (late Llanvirn) linear reefal structures with low topography had appeared, together with a suite of trilobites which were to persist with little change until the Devonian. There is an increase in the pervasiveness of bioturbation. Continuous sections are best displayed in the western USA, Newfoundland and Spitsbergen; on the eastern seaboard there is a widespread tectonic unconformity which punctuates the record (James et al., 1989). The first appearance of calymenids (*Protocalymene*), lichids (*Apatolichas*), dalmanitaceans (*Calyptaulax*), proetides (*Decoroproetus, Oenonella*), encrinurids (*Cybelurus*), harpids (*Selenoharpes*), and diverse odontopleurids in Laurentia happens at this level (i.e. in the regression plus transgression interval). *Bathyurus* and *Illaenus* become widespread. All these taxa are typical of later Ordovician - indeed, Silurian and Devonian - trilobite communities. Many have Arenig or even Tremadoc sister taxa elsewhere, but this is the time when they expand rapidly.

There is also evidence of wider distribution of some of these inshore genera, by comparison with the Tremadoc or Arenig. For example, *Bathyurus* is widespread in Baltica but also occurs in North America and China, and may be a novelty in both. *Nileus* and dimeropygids show a similar distribution. It was at this time that trinucleids appeared in Scandinavia, after an earlier history that appears to have been exclusively Gondwanan. The B'Ln thus includes an admixture of taxa originally generated by vicariance, a process which was to continue further as geography and sea level continued to change in the later part of the period.

During the Arenig, the low diversity graptoloid fauna of the *T. approximatus* Zone (Fig. 4) radiated to produce a series of abundant, high diversity assemblages which enable precise biostratigraphic global correlation. A number of distinctive lineages (e.g. the *Isograptus victoriae* group, see Cooper, 1973) have also proved invaluable biostratigraphically. The next radical faunal event, however, does not occur until the late Arenig when biserial graptoloids first appear. Although these occur initially as a relatively minor constituent of the assemblage, they diversify rapidly and quickly become numerically dominant at many localities.

It is well known that the bio-event marked by the appearance of the Diplograptidae, Glossograptidae and Cryptograptidae is at a level slightly below the traditional base of the Llanvirn (e.g. Fortey et al., 1990). That series boundary was originally based on the incoming of pendent didymograptids; although such a level may be recognized in sequences originally deposited on the Gondwanan margin (i.e. "Atlantic province", high latitude), pendent didymograptids first occur in the mid Arenig (*D. bifidus* Zone and equivalents) elsewhere in lower paleolatitudes (e.g. Williams and Stevens, 1988). Following examination of isolated, three-dimensional material from Spitsbergen, Cooper and Fortey (1982) realized that the Arenig pendent forms had a different proximal development from those found in the Llanvirn, and assigned them to two subgenera, namely *Didymograptus* (*Didymograptellus*) and *Didymograptus* (*Didymograptus*) respectively. Since that time, greater emphasis has been placed on the understanding of early biserial evolution, which is considered by most workers to represent a more fundamental event in graptolite evolution.

In many sections the level marked by the appearance of *U. austrodentatus* shows evidence of being a regressive interval, while the incoming of *D. (Didymograptus) artus* appears to mark the Maximum Flooding Surface of the ensuing transgression. This is clearly shown in the Lake District of northern England, Wales and elsewhere in Gondwanan regions, and also at the base of the Upper *Didymograptus* Shale in Scandinavia where *U. austrodentatus* and other diplograptids occur within the underlying limestones. Similar evidence for a regressive-transgressive interval is shown by the trilobites which show a shoreward shift in the *U. austrodentatus* Zone followed by oceanward movement at the appearance of *D. (D.) artus*. The occurrence of "*D. bifidus*" recorded by Braithwaite (1976) in Utah also appears to mark this same transgressive event (R.A.F., pers. observ.). Thus both the regressive and transgressive events may be considered bio-events, the latter equating more closely with the traditional base of the Llanvirn as used in the U.K.

The majority of recent studies (Williams and Stevens, 1988; Maletz, 1992; Mitchell, 1992) do not support the idea of a somewhat diachronous appearance of *U. austrodentatus* as suggested by Fortey et al. (1990), although Finney and Ethington (1992) report some apparently anomalous occurrences from Nevada. Ongoing work by Mitchell (1992) suggests that *U. sinodentatus* is probably the first diplograptid species to appear, sometime during the Yapeenian, and is quickly followed by the more widespread *U. austrodentatus* at the base of the Darriwilian, which is also marked by the appearance of *Paraglossograptus tentaculatus* and *Cryptograptus* spp. (e.g. Williams and Stevens, 1988; Cooper and Lindholm, 1990; Maletz, 1992; VandenBerg and Cooper, 1992).

Unlike the faunal change occurring at the basal Arenig, few faunal elements disappear during the B'Ln, most genera continuing up into middle or late Llanvirn. The bio-event is thus marked by addition, rather than by replacement, in terms of the graptolite fauna. The origin of the diplograptids and other biserials is still uncertain; Jenkins (1980) proposed *Maeandrograptus schmalenseei* as a possible ancestor, and while this remains a possibility no further proof has come to light. Part of the problem now hinges on acceptance of Fortey and Cooper's (1986) bifold division of the graptoloids into the Virgellina and Dichograptina, based on presence or absence of a virgella. If their hypothesis, whereby the virgella is considered to be a monophyletic feature, is accepted, the diplograptids probably must have evolved from a scandent virgelline such as *Phyllograptus*, although the simple thecal style of that genus is totally unlike that of the diplograptids. Williams and Stevens (1988), however, argued that the virgella may have arisen polyphyletically, and that the ancestor of the diplograptids might, therefore, have lacked a virgella. Should this prove to be the case, the complex, somewhat bulbous thecae might suggest a sinograptid ancestor, perhaps with an appearance similar to *Perissograptus pygmaeus* (although this particular mid Arenig species is much too early to be considered as a direct ancestor - see Williams and Stevens, 1988).

Basal Llanvirn Bio-Event (B'Ln): Summary and Issues

In the review of events associated with the Basal Llanvirn Bio-Event (B'Ln), it is shown above that this horizon corresponds to a regressive-transgressive couplet, likely with a brief stillstand period. With all three faunal groups, the B'Ln is characterized by minor extinction but more especially by addition to or significant modification of existing faunas. The B'Ln marks a period of origination and rapid diversification.

To refer again to the evolution of the Paleozoic Fauna (of Sepkoski, 1981), the B'Ln represents a stage in the marked acceleration of development and radiation of this Fauna. As noted above, many groups of invertebrates proliferate taxonomically and in abundance; infaunal bioturbation is more prevalent; algae bioherms become more prominent prior to the development of small bryozoan-coral bioherms in the Chazyan.

Although the subsystem classification of the Ordovician System has yet to be formally defined if there is to be a Middle Ordovician the B'Ln will most likely mark its lower boundary. This bio-event reflects the time of diversification of the relatively primitive Paleozoic Fauna of the Early Ordovician (Tremadoc-Arenig) into the more abundant and complex faunas of the early maturing phase of the Paleozoic Fauna (Llanvirn to Ashgill).

It is interesting to note that this level is approximately the time of a major excursion in the $^{87}Sr/^{86}Sr$ isotope curve (e.g. Burke et al., 1982; Keto and Jacobsen, 1987) suggestive of a relative increase in strontium derived from oceanic ridges. This may have resulted from a combination of partial flooding of the cratons (Llanvirn transgression) combined with increased sea floor spreading and ridge activity. The early Middle Ordovician displays substantially more volcanic activity than much of the early Ordovician (Stillman, 1984) with deformation along subduction zones (e.g. early Taconic Orogeny in Iapetus Ocean).

Much work remains on refining and interpreting the Sr isotope record near the B'Ln. The faunal changes suggest that a new threshold restraint was surmounted allowing a surge of evolutionary development, both taxonomically and in community structure. This was also expressed functionally within fossil groups, such as in the development of biserial graptolites and elaborate platform conodonts.

This interval is therefore one of profound reorganization of the Ordovician biota and the causes are not at all understood. Much more attention is merited in future research to explain this fundamental restructuring.

Basal Caradoc Bio-Event (B'Cc)

Faunal changes for conodonts, graptolites and trilobites are more gradational such that the Basal Caradoc Bio-Event (B'Cc) (Fig. 1) spans one or two conodont zones from the Llandeilian into the earliest Caradocian. The previous bio-events were all classically eustatic, being sharply bounded and of comparatively short duration relative to biostratigraphic calibration. The Basal Caradoc Bio-Event, is, by contrast, more drawn out (Fig. 1). It corresponds with the *Nemagraptus gracilis* Biozone, which spans the later Llandeilo and early Caradoc Series in the type area of Wales and the Welsh Borderlands. We might have better referred to this bio-event as the Llandeilo-Caradoc Bio-Event but it is likely that the name Llandeilo will disappear as a series name with ongoing revisions to Ordovician chronostratigraphy.

In cratonic areas this interval is typically expressed as a regression, or minor hiatus. In Laurentia, for example, it occurs in the late Chazy – early Black River interval. This regression is followed by the profound mid-late Caradoc (Rocklandian to Edenian) transgression that virtually drowned the entire craton. This relatively slow (if oscillatory) transgression resulted in predominantly restricted peritidal sequences close to the B'Cc. Only in the mid-Caradoc do subtidal facies become widespread (e.g. Trenton Group of New York State, Ontario and Quebec and regional equivalents in Laurentia; type Caradoc onlap in the Welsh Borderlands, U.K.).

In the Midcontinent Realm the conodont faunas found close to the B'Cc represent the peak development of the neurodont (hyaline) conodonts with genera such as *Erismodus, Microcoelodus, Polycaulodus, Trucherognathus* and *Chirognathus*. These suffer only minor modification through the B'Cc but diminish numerically and in diversity once the later transgressive phase occurred. The dominant change is seen in the subtidal facies where new species of *Panderodus, Belodina, Plectodina, Oulodus, Scyphiodus, Bryantodina* and *Phragmodus* produce significantly different faunas above and below the B'Cc. Most of these genera have their origins within one or two stages below the B'Cc but are all more prominent and diversified above this level.

The same pattern is seen with the flooding of the Siberian Platform at the start of the Baksian Stage. Many of the genera noted for Laurentia are present but some are new (*Columbodina, Acanthocordylodus*) and others differ substantially at the species level (Moskalenko, 1976, 1983). In the Australian segment of Gondwanaland, shallow water sequences possessed rather different faunas characterized by *Tasmanognathus* in the Tasmanian sequence (Burrett, 1979; Burrett et al., 1984) and by *Aphelognathus, Belodina, Taoqupognathus* and *Yaoxianognathus* in the Cliefden Caves Group, New South Wales (Savage, 1990; CRB unpublished collections). These latter faunas bear close resemblance to equivalent age faunas from the North China Platform (An, 1985) which also lay in a low latitude position.

At, or close to, the B'Cc, the North Atlantic Realm faunas also contain some elements of Midcontinent affinity. This was noted in a comparison of faunas between Laurentia and Baltica across the Iapetus Ocean (Sweet and Bergström, 1974; Bergström, 1990). It is not known whether this was in response to climatic change, the progressive equatorial drift of Baltica, or the development of island arc complexes with the increasing volcanism and deformation in Iapetus.

Overall, the North Atlantic Realm faunas also change gradually and modestly across the B'Cc.

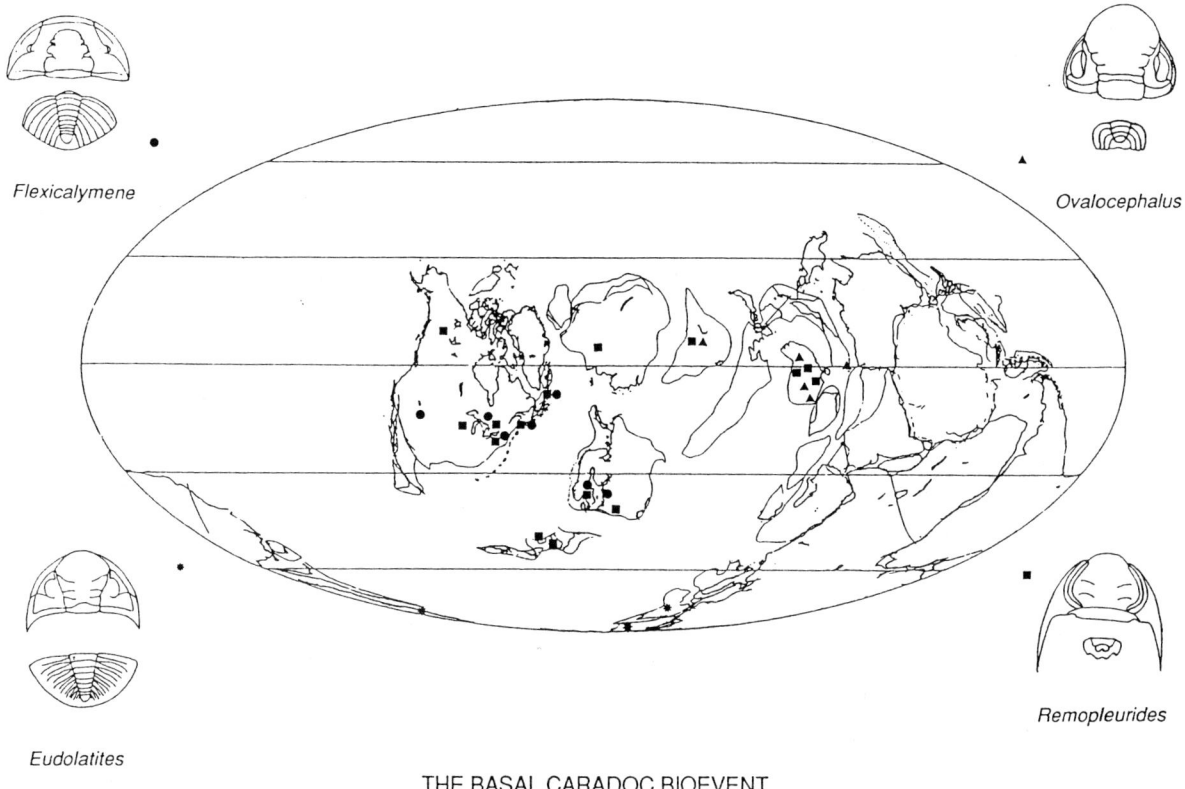

Fig. 6. Paleogeographic reconstruction (from McKerrow and Scotese, 1990) for the Basal Caradoc Bio-Event, with the general distribution of key trilobite genera

Important components of earlier faunas such as *Periodon, Eoplacognathus, Protopanderodus* and *Amorphognathus* continue higher but new genera such as *Hamarodus, Polyplacognathus, Icriodella* and *Scabbardella* occur and important earlier taxa such as *Pygodus, Cahabagnathus* and *Complexodus* become extinct. Dzik (1983) noted that this interval was a particular time of rebuilding of the North Atlantic faunas.

As noted for both graptolites and trilobites, the pattern of conodont realms and provinces is dramatically modified soon after the B'Cc as a result of the mid Caradoc transgression (Fig. 1), possibly the most extensive in Phanerozoic history. The main consequence is a blurring of ecologic boundaries and in particular a penetration into cratonic areas of North Atlantic Realm faunas (Sweet et al., 1971; Barnes and Fåhraeus, 1975; Bergström, 1990) such as species of *Periodon, Icriodella* and *Amorphognathus*. This pattern is seen on virtually all low latitude cratons.

The B'Cc Event marks an interval in Ordovician history which heralds several major changes. The impressive mid Caradoc transgressive cycle has been well known for over a century. On low latitude platforms it resulted in widespread precipitation of carbonates and some local evaporites. In deeper basins, notably with the closing Iapetus Ocean, the Caradoc is an interval also long well-known for black shale deposition as noted below under the discussion of graptolite faunas. Thus, in both carbonate platforms and in anoxic basins, carbon was locked into the stratigraphic record in major proportions. The early Caradoc is also a time of widespread volcanism especially in Laurentia which was half covered at least twice in bentonite ash falls sourced from major volcanic centres in the deformed Appalachian orogen, mainly in the Carolinas.

This pattern of vast transgression, deep ocean anoxia, and major volcanism are also key ingredients present in the mid Cretaceous where the occurrence of a giant mantle plume, or superplume, has been invoked (Larson, 1991). In the Cretaceous interval there was also an associated prolonged interval in which no magnetic reversals occurred. An accurate picture of the magnetostratigraphic record for the Ordovician has yet to be documented. It seems possible, and worthy of future testing, that the mid to late Ordovician was a period of superplume development creating excessive eustatic rise, excess CO_2 in the oceans and atmosphere, and severe regional volcanic activity. The plume activity may even have started on a smaller scale in the Llanvirn (B'Ln) where many of these events occur on a more modest scale. The enhanced phytoplankton productivity argued for the Cretaceous may have generated the increased nutrient supply to fuel the development of the radiating Paleozoic Fauna.

The B'Cc interval was one of intense volcanic and tectonic activity, especially in the Iapetus Ocean, and it is often difficult to disentangle regional from global events. However, global onstep of deeper water facies is well established, including from seismic stratigraphic studies. This was one of the greatest floodings of stable cratons in the Phanerozoic. As far as trilobites were concerned, this is accompanied by a much more widely spread and less geographically restricted fauna, a fact recognised by Whittington and Hughes (1972) when they proposed a unified Remopleuridid Province to encompass several provinces of the earlier Ordovician. The distribution of *Remopleurides* itself is shown on Fig. 6. This more uniform fauna may in part simply be the result of shelfward spread of an already more widespread deep shelf fauna. More widely distributed trilobites had always occupied deeper water biofacies fringing paleocontinents since the Tremadoc. There was also a continued convergence of the Avalonian and Baltic plates towards the paleoequator. However, common elements in the Remopleurid fauna include a mixture of those having a previous phylogenetic history in either Gondwana, or Laurentia. Thus *Remopleurides* itself is almost certainly of Laurentian (Bathyurid Province) ancestry (also *Lonchodomas*, dimeropygids), whereas *Flexicalymene* (Fig. 6) was derived, with other calymenids, from Gondwanan stocks. Trinucleids, ultimately of Gondwanan origin, became widely distributed around Laurentia; and the cyclopygid biofacies (traced to peri-Gondwana origins in the Arenig) appears there for the first time. The B'Cc Event mixed these various elements that may have originated as the vicariant products of earlier bio-events.

As the transgression proceeded, the inshore biofacies in Laurentia migrated to such redoubts as the upper Mississippi Valley, where endemic Laurentian and tropical trilobites continued to evolve (DeMott, 1987). Conversely, in the boreal regions of Gondwana (Spain, Morocco, Armorica) surrounding the South Pole the B'Cc stimulated further evolution of the endemic calymenoid-homalonotid-dalmanitacean faunas (e.g. Hammann, 1983), a restricted continuation of the *Selenopeltis* Province. In South China (extending to Kazakhstan), *Ovalocephalus* (=*Hammatocnemis*) is a characteristic genus endemic to that area, as it had been since the Arenig. Thus, although the Llandeilo-Caradoc "flooding" broke down provincial barriers, and in a statistical sense homogenised the trilobite faunas, provinciality survived, especially for inshore sites.

It might be noted briefly that the Chinese nautiloid-rich Pagoda Limestone, which embraces this interval, used to be regarded as of shallow water origin, but that it is now recognised as deeper, consistent with B'Cc; alleged mud cracks, for example, have been re-interpreted as synaeresis structures. In the Ashgill some of the Chinese endemics at last appeared in the European part of Gondwana (e.g. *Ovalocephalus* in Spain and Sardinia).

Following the incoming of the diplograptids, graptoloid evolution went through an interval characterized by only relatively minor changes at generic level until sometime during the "Llandeilo" (late Llanvirn) when the families Dicranograptidae and Nemagraptidae appeared. Graptolite biostratigraphy during this interval reflects the unusually slow period of evolution and extinction, with few reliably distinguishable biozones (e.g. Cooper and Lindholm, 1990).

The first occurrence of the dicranograptids and nemagraptids appears to be approximately synchronous, lying within the rather vaguely defined *H. teretiusculus* Zone. Both are represented initially by only one or two species; the families soon, however, diversify, giving rise to numerous species belonging to the genera *Dicellograptus, Dicranograptus, Leptograptus* and *Nemagraptus* by the beginning of the *N. gracilis* Zone (Fig. 4).

The scope and definition of the *N. gracilis* Zone was discussed fully by Finney and Bergström (1986), who summarized its occurrences and assemblages from around the globe. Correlation via conodonts (e.g. Bergström, 1986) suggests that the base of the *N. gracilis* Zone lies somewhere within the Llandeilo Series of the U.K. Following and during the appearance and diversification of the dicranograptids and nemagraptids, new diplograptid and retiolitid genera appear and the last dichograptids/sinograptids (e.g. *Acrograptus superstes*) vanish. By the end of the *N. gracilis* Zone or beginning of the overlying biostratigraphic intervals within the early Caradoc (e.g. *D. multidens* or *C. bicornis* zones) these changes have resulted in a totally different faunal composition from that of the Llanvirn or earlier.

There is, therefore, a profound change in graptolite evolution during this interval, but it appears to be a relatively long-drawn-out event in contrast to those bio-events occurring earlier in the Ordovician. Many Ordovician stratigraphers have long considered the base of the *N. gracilis* Zone to mark a major, relatively sudden faunal change, linked with the widespread onset of black shale deposition. From what is now known of the events summarized above, is this a valid observation?

Finney and Bergström (1986) proposed Berrybush Burn in southern Scotland as the general reference section for the *N. gracilis* Zone, due to Lapworth's (1878) original work in that area. They did, however, recognize that the region was unsuitable for defining the base of the zone, as no contact with the underlying *H. teretiusculus* Zone was present. Finney and Bergström (1986) also proposed secondary reference sections for the *N. gracilis* Zone in Scania, southern Sweden and in New York State. Ongoing research by SHW on the biostratigraphy of the lower part of the Moffat Shale Group of southern Scotland, and on the equivalent Lawrence Harbour Formation of central Newfoundland, suggests that the sudden onset of black shale deposition actually occurs fairly late in the *N. gracilis* Zone. This is because all levels within the zone, including the basal black shale interval, yield *Climacograptus bicornis* and *Orthograptus calcaratus* in association with *Nemagraptus gracilis* and a number of *Dicellograptus* and *Dicranograptus* species. Elsewhere (e.g. Scandinavia and North America), this assemblage either occurs late in the *N. gracilis* Zone, or is used to characterize an overlying biostratigraphic interval, such as the *C. bicornis* Zone of the U.S. (see Finney and Bergström, 1986, for summary). The well known transgression which led to widespread deposition of black shale following non-graptolitic sedimentary and volcanic sequences, at least in arc/forearc sequences formed towards the Laurentian margin of the Iapetus Ocean, was thus considerably later than the bio-event which gave rise to the *N. gracilis* Zone graptolite fauna, and might well coincide approximately with the Llandeilo-Caradoc boundary. Based on a recent summary of Ordovician graptolite sequences in Australia (VandenBerg and Cooper, 1992) it appears that the Darriwilian-Gisbornian boundary also occurs at this level, and was perhaps related to the same physical event.

Basal Caradoc Bio-Event: Summary and Issues

If the Basal Llanvirn Event (B'Ln) was the most significant event within the Ordovician System in restructuring the global biota, the Basal Caradoc Bio-Event (B'Cc) represented the peak enhancement of many of those changes. The B'Cc level marks another significant regressive phase. The mid Caradoc transgression, the increased volcanic activity, extensive anoxic deep basins with black shales, and a reduction in faunal provincialism all characterize the interval after the B'Cc. With the transgression being so vast, inundating most cratons of modest relief, extensive epeiric seas were established that were probably more extensive than at any time in the Phanerozoic. These created large tracts of new habitat for benthic communities. The transgression also brought deep or cool water faunas on to the margins and outer platforms of the cratons, blurring formerly sharp faunal provincial boundaries.

With the inundation of many of the cratons there must have been a significant reduction of inorganic nutrients into the marine realm derived from the continental coast. How was high marine productivity sustained? One possibility at this particular time is by the effects of a mantle superplume (cf. Larson, 1991, for the mid Cretaceous). The idea of a mid to late Ordovician superplume is, of course, speculative, but some close comparisons can be made with mid Cretaceous events. It would be helpful to determine more precisely the Ordovician magnetic reversal history and to gain an improved isotope history (e.g. Sr) for this interval. The cause of this mid Caradoc transgression, which influenced so much the faunal radiations and dispersal after the B'Cc, is one of the great unknowns in Ordovician earth history.

Upper Ashgill Bio-Event (U'Al)

The last major bioevent within the Ordovician occurs near the top of the System at or close to the base of the *G.? extraordinarius* graptolite Zone. Historically, it marked the base of the Silurian since it represented the interval of profound faunal turnover between the Ordovician and the Silurian. However, the base of the Silurian has been formally defined within the slightly higher and more continuous graptolitic sequence at the base of the *A. acuminatus* Zone at Dob's Linn, Scotland (Holland, 1989). The U'Al (Fig. 1) occurs at a level of peak impact on the global biota by the terminal Ordovician continental glaciation in North Africa (Brenchley, 1984; Barnes, 1986; Barnes

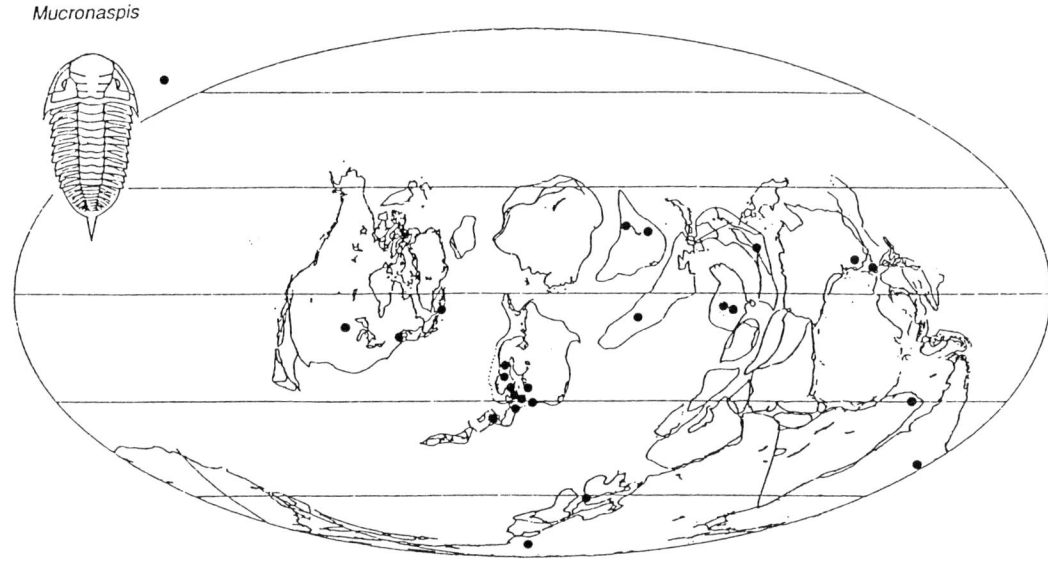

THE UPPER ASHGILL BIOEVENT

Fig. 7. Paleogeographic reconstruction (from McKerrow and Scotese, 1990) for the Upper Ashgill Bio-Event, with the general distribution of key trilobite genera

and Williams, 1990), whereas the base of the Silurian lies within the initial post-glaciation period marked by widespread deep basin anoxia, transgresssive onlap, and early faunal adaptive radiation. The U'Al therefore lies high within the Ashgill Series. The Ashgill records the progressive influence of the glaciation is recorded, climaxing in the Hirnantian Stage (Cocks and Rickards, 1988; Barnes, 1992).

The U'Al is a profound extinction bio-event. Sepkoski (1981) calculated that 28% of the families became extinct making this the second most severe extinction event in the Phanerozoic, after that at the end of the Permian.

A review of the extinction of conodonts at the end of the Ordovician was given by Barnes and Bergström (1988) with a more specific correlation with the graptolite record by Melchin et al. (1991). The nature of the Silurian recovery was discussed by Barnes (1989) and Aldridge and Schönlaub (1989). The precise documentation of change across the U'Al has not been easy because of the dramatic eustatic sea level drop in the latest Ordovician (Hirnantian; Gamachian) which effectively drained nearly all of the cratonic platforms (Fig. 1). Continuous sections, especially in carbonate sequences, are rare; Anticosti Island, Quebec being the best example (Barnes, 1988). Conodonts are rare and, when present, are not commonly diagnostic in graptolitic basinal sequences through the U'Al (e.g. Barnes and Williams, 1988). From present evidence the main extinction in conodonts occurs in the upper *G. persculptus* Zone – a little higher than the main graptolite extinction event.

The decline in faunal diversity for conodonts across the U'Al is severe, changing from about 75–100 species in the lower–middle Ashgill to about 20 species in the lower Llandovery (Sweet, 1988). Within the best documented boundary section for conodonts (Anticosti Island; McCracken and Barnes, 1981) the Gamachian fauna (Ellis Bay Formation) contains some 38 species. Above the conodont-based systemic boundary (or faunal turnover), the uppermost Ellis Bay strata of lowest Silurian age yields only 21 species of which most (16) are not known from older strata; those that do cross the extinction level are dominantly coniform taxa (representatives of *Dapsilodus, Decoriconus, Panderodus, Pseudooneotodus, Walliserodus*). Elsewhere, representatives of genera with more compound apparatuses which cross this boundary belong to *Icriodus, Icriodella, Oulodus* and *Ozarkodiana*. For other genera, such as the less well known, more recently established genera *Birksfieldia* and *Gamachignathus,* the situation is less clear as to their precise stratigraphic range and relationships.

For conodonts, the division into two faunal realms (Midcontinent and North Atlantic) persisted until this extinction event, thereafter faunal differentiation at realm and provincial level in the early Silurian was absent or greatly subdued. During the Ashgill, as Baltica moved toward Laurentia with the closure of the Iapetus Ocean and into a low latitude belt, some Midcontinent taxa penetrated into European Ashgill faunas. After the principal extinction event, in the upper *G. persculptus* Zone, the radiating stocks appear to be derived mainly from ancestors in the low latitude warm-water Midcontinent Realm; while the stocks of the cold-water North Atlantic Realm nearly all disappeared (Barnes and Bergström, 1988). The coniform taxa noted above, interpreted as being predominantly pelagic in habit (Barnes and Fåhraeus, 1975), appear to have been affected the least although some taxa (e.g. *Drepanoistodus, Protopanderodus)* suffered extinction with many of these existing predominantly within the North Atlantic Realm.

The trilobite extinction at the end of the Ashgill was probably the most considerable in the history of the group. Such long-lived clades as Agnostida, Olenina, Remopleurididae and Asaphidae were terminated, as well as many typical Ordovician families, such as Cyclopygidae, Telephinidae, Trinucleidae, Nileidae and Dimeropygidae. One raphiophorid genus survived into the Silurian. Literature pertaining to rock sections and fossil groups has been summarised in Cocks and Rickards (1988). It is clear in the brachiopods, for example, that there was a decline in diversity through the later Rawtheyan and into the Hirnantian (Brenchley, 1984). Owen (1986), however, has shown that rare examples of these Ordovician trilobite families did survive into the Hirnantian Stage. It seems likely, therefore, that the extinction event which affected graptolites was close in time to the extirpation of trilobite clades, although exact synchronicity is hard to prove. The decline or changes associated with climatic deterioration earlier in the Ashgill were not a single event, nor so drastic.

The Hirnantian shelly faunas are dominated by brachiopods (Rong and Harper, 1988), and are usually referred to as the *Hirnantia* fauna after one of the characteristic genera; this broad term embraces several different faunas of varying diversity. As Jaanusson (1969) pointed out, the *Hirnantia* fauna is accompanied by a sparse trilobite fauna, most widespread of which is the genus *Mucronaspis* (formerly *Dalmanitina*, partim) (Fig. 7). This genus has high latitude Gondwana ancestry (Morocco, Destombes, 1972; *Hirnantia* had a similar origin); its equatorwards spread is associated with the glacioeustatic regression(s) at the end of the Ordovician, and with the climatic crisis. This affords a clear example of how the earlier phases of vicariant evolution only achieved wider geographical propagation when the appropriate conditions became widespread. Note that the spread of *Mucronaspis* took place over a longer time period than the bio-event represented by the extinction of trilobite families.

Fortey and Chatterton (1988), Fortey (1989) and Chatterton and Speyer (1989) have noticed that those trilobites which became extinct at the U'Al were mostly either oceanic, pelagic forms (e.g. telephinids, cyclopygids and probably agnostids), or had asaphoid larvae with planktonic habits. The Order Asaphida was preferentially extinguished for this reason. One genus alone survived to the Silurian (*Raphiophorus*) of a group which often dominated Ordovician assemblages. The implication was that a major extinction ("the bioevent") preferentially affected species with obligate connexions to oceanic environments; i.e. in spite of glaciation drastically affecting continental shelf seas, the extinction operated especially in the oceans, affecting trilobites and graptolites alike. This may have been connected with changes in oceanic oxygenation and/or stratification accompanying the deglaciation.

The tropical belt doubtless shrunk to a narrow zone, but did not disappear entirely. It is interesting to note (e.g. Paul, 1982; Cocks in Cocks and Rickards, 1988; Fortey, 1989) that there are many 'Lazarus taxa' among echinoderms, brachiopods and trilobites. These disappear late in the Ordovician (prior to the first glacial beds) only to reappear at some stage in the Silurian, usually in the late Llandovery, but Wenlock examples are known. These forms must have existed in refugia which guaranteed the persistence of inshore faunas during the glacial maxima.

During the Ashgill, graptoloid evolution continued at a moderate pace, with retention of relatively high diversity biozone assemblages. Sometime during the mid Ashgill (probably Rawtheyan Stage of the U.K.)

a number of new genera appear, particularly *Paraorthograptus*, and the fauna becomes rather more varied. This interval, which has been recognized as the *P. pacificus* Zone or Subzone in many parts of the world (see Williams, 1982), immediately predates the sudden and almost complete extinction of the Graptoloidea in the late Ashgill, which probably occurred during the late Rawtheyan or early Hirnantian and is here taken as marking the Upper Ashgill Bio-Event. At this time, all graptoloids became extinct with the exception of three or four diplograptid species currently assigned to *Normalograptus* and *Glyptograptus*. It was these few taxa which subsequently gave rise to a totally new, rapidly evolving fauna in the Silurian.

It should be emphasised that graptolite evolution does not suggest gradual decline in diversity throughout the Ashgill as stated by several authors (e.g. Rickards, 1978) prior to detailed biostratigraphic and taxonomic studies on late Ashgill graptolites during the past decade (e.g. Williams, 1982, 1987; Koren and Sobolevskaya, 1983; Melchin, 1987; Chen and Rong, 1991). It is now clear that graptoloid diversity actually increased during mid to late Ashgill prior to the mass extinction of the U'Al.

The bio-event was almost certainly related to the late Ordovician glaciation, perhaps due to increased oceanic circulation, overturn and cooling of surface waters; the faunal diversification immediately prior to the event may have been in response to commencement of deteriorating conditions. Evidence for a late Ordovician-earliest Silurian regressive-transgressive event is present in almost all sections throughout the world, marked by either shallow water deposits or an interval of non-deposition in the *D. anceps* or *G.? extraordinarius* Zone (i.e., late Rawtheyan-Hirnantian) commonly followed by black shale deposition during the *G. persculptus* or *P. acuminatus* Zone (Williams, 1988). The beginning of the regressive phase is unclear, but might perhaps be as early as *D. complanatus* Zone (?early Cautleyan) if the onset of grey, barren mudstone deposition as seen at Dob's Linn, southern Scotland (Barnes, 1986; Williams, 1988) is considered significant.

Upper Ashgill Bio-Event (U'Al): Summary and Issues

As Sepkoski (1981) noted, and a host of authors documented in the summary volume edited by Cocks and Rickards (1988), the Upper Ashgill Bio-Event is characterized by profound extinction in virtually all well known fossil groups. The glacial period ranges through the Ashgill Series with perhaps three or four main oscillatory glacial phases within the Hirnantian Stage (Fig. 2). This final climax phase was characterized by low sea level stands, aggressive ocean circulation, and deep ocean erosion and oxygenation (Brenchley, 1984; Barnes, 1986). These factors result in few complete stratigraphic sections and hence a limited opportunity to precisely document the faunal changes of the U'Al.

Although the peak extinction for most groups lies within the Hirnantian, it appears to vary slightly for different groups (cf. conodonts and graptolites), and may have a step-wise extinction pattern for some groups. Although many authors attribute the terminal continental glaciation in North Africa as the cause of the extinction, McLaren and Goodfellow (1990) have suggested that the faunal extinction patterns favour a biolide impact to enhance the effects of glaciation. Despite several studies to check for high levels of iridium near the U'Al (Orth et al., 1986; Wang et al., 1993) no definitive evidence exists to substantiate a biolide impact hypothesis.

It is worth noting that if the possible minor glaciation near the base of the Ordovician (Miller, 1984) is disregarded, the evolution of the Cambrian Fauna and then the early diversification of the Paleozoic Fauna occurred within a greenhouse climate state. The skeletonized global biota had not experienced the influence of cold polar areas, aggressive deep ocean circulation, oxygenated deep ocean currents, and rigid climatic differentiation typical of an icehouse state. Added to this was the immense contrast between the vast transgressions of the mid-late Caradoc and the marked regressions of the Ashgill with the consequent severe habitat area effect. The combination of these forces produced drastic consequences on a biota that had evolved for some 200 Ma within a greenhouse state.

Whereas much attention has been given in recent years to the nature of this extinction, far less study has gone into the recovery and adaptive radiation of the early Silurian. As noted above, graptolites were virtually obliterated, conodonts diversified from Midcontinent Realm warm-water stocks, whereas Silurian trilobites appeared to have their ancestry in cool-water high latitude stocks. Although the Silurian is a classic example of an equable, greenhouse state interval, recent detailed studies of conodonts suggests an alternation of two climate states (Jeppsson, 1990). In this model, the primo episode has a colder humid climate in contrast to the warm dry climate of the secundo episode. The actual period of reversal of these two states creates an extinction event in the marine biota.

Compared to the bio-events discussed for the Ordovician, these Silurian bio-events are perhaps of second order rank but the concept of oscillating climate/ocean states is a useful model to consider. Jeppsson (1990) bases much of his arguments on the evidence of the patterns of conodont distribution and extinction; one of us (CRB) is collaborating with Jeppsson to test for similar second order events in the Ordovician.

The consequences of the terminal Ordovician glaciation and the U'Al extinction were also severe in terms of total biomass. With such drastic extinctions, much of the biomass must have been removed. Most may have been rapidly oxidized and recycled during peak glacial phases; some may have contributed carbon to the black shales. What is unclear is the rate and extent of recovery in the early Silurian; carbon isotope studies underway (e.g. Middleton et al., 1991) should help in this regard. Being such a marked climate shift, the terminal Ordovician glaciation and the U'Al still remain exceptional research problems for the elucidation of ocean-climate-biological processes in the Early Paleozoic.

Ordovician Event Stratigraphy

In this section, we reiterate and expand on those processes imprinted on Ordovician stratigraphy that have some significant effect, at a relatively high order, on the biota with the potential to create global bio-events. The physical changes or processes discussed are those of plate tectonics, eustasy, climate change, and volcanic and possible superplume activity.

Plate Tectonics

The illustrations (Figs. 2, 3, 5-7) included herein show the changes in paleogeography during the Ordovician as interpreted by McKerrow and Scotese (1991). Variations of the McKerrow and Scotese maps are widely adopted, including the major symposium volume on Paleozoic paleogeography and biogeography (McKerrow and Scotese, 1990). The sequence of maps shows the continued dispersal of Ordovician paleoplates following the Late Proterozoic rifting and breakout of Laurentia and the assembly of Gondwana (e.g. Hoffman, 1991). Most Ordovician paleogeographic reconstructions show the vast Panthalassic (Paleo-Pacific) Ocean. The two other oceans -- Iapetus and Paleotethys (McKerrow and Scotese, 1991, Fig. 1) -- lie between Laurentia-Siberia in the equatorial belt and the North African margin of Gondwana (close to the south pole). Separating these two oceans are Baltica and part of the Avalon Terrane (New England, Maritime Canada) and southern England. These two areas moved northward towards the equator as Iapetus closed. Part of the Iapetus closure resulted in the Taconic Orogeny of the Northern Appalachian Orogen. The effects of this orogenic phase are seen in the event stratigraphy of the platform margin which experiences local unconformities, the migration of a peripheral bulge, and the development of a foreland basin. This sequence of events is seen slightly earlier in Newfoundland (late Arenig through early Caradoc; James et al., 1989; Knight et al., 1991) compared to New York State (Caradoc; Read, 1989; Goldman et al., 1994). This variation in timing partly reflects the deformation along an irregular conjugate margin (Stockmal et al., 1987).

More recently, a different paleogeographic interpretation by Dalziel et al. (1994) has aligned eastern Laurentia close to the western margin of South America. Through the Ordovician, Laurentia slid along the margin of South America, leaving behind part of the Onachita region as the Occidentalia Terrane, now preserved as an anomalous belt of carbonates in the Argentina Precordillera (Dalla Salda et al., 1992). The implications of these two significantly different paleogeographic reconstructions (i.e. Scotese and McKerrow, 1991; Dalziel et al., 1994) are to caution against simplistic or sweeping interpretations of regional event stratigraphy based on a particular plate tectonic and/or paleogeographic reconstruction. Patzkowsky and Holland (1993) have argued that regional bio-events account for a greater proportion of extinction in the fossil record, using four examples from the Middle Ordovician where tectonic processes created subsidence and paleoceanographic disruption.

Eustasy

A wide array of papers have attempted to document global or regional patterns of eustasy during the Ordovician. There is surely no doubt that the first order Caradoc transgression can be recognized on most paleoplates and represents for Laurentia the greatest flooding during the entire Phanerozoic. Other important transgressive phases, perhaps of second order rank are seen near the base of the Tremadoc, Arenig, and Llanvirn (Fig. 1) (Barnes, 1984; Fortey, 1984). Third order and lower cycles or eustatic events are seen in many stratigraphic sequences (e.g. Read, 1989; James et al., 1989; Ross and Ross, 1992). However, at lower orders it becomes difficult to establish potential global extent because of the temporal and spatial limits of biostratigraphic correlation for time intervals of less than 1 Ma in the Ordovician (e.g. Holland, 1993). Seismic stratigraphy has been employed (e.g. for the Exxon curves) but the limits of resolution in correlation are again a significant problem (Miall, 1992). In the Ordovician one of the more courageous attempts at inter-paleoplate correlation of eustatic events has been by Neilsen (1992a,b) for the Arenig sequences of Australia and the Baltic. This work is based on both event stratigraphy and ecostratigraphy and perhaps needs further verification.

Several studies of eustatic events have been made close to both systemic boundaries following the detailed studies of many sections globally in the work to define the base of both the Ordovician and the Silurian systems. In the late Cambrian through early Arenig, Nicoll et al. (1992) recognized 11 sea-level events, most of which have been named. Miller (1988, 1992) likewise recognized several events (e.g. Lange Ranch and Black Mountain eustatic events among others) close to the Cambrian-Ordovician boundary, seemingly on several paleoplates. Such correlations need to be supported by high resolution chemostratigraphy (e.g. carbon, strontium isotopes). Initial work (e.g. Wright et al., 1987) involved too few samples for accurate correlation compared to more recent regional studies (Magaritz, 1991; Ripperdan et al., 1992). More widespread and intensive sampling is required to establish acceptable isotope curves before they can be used routinely for event stratigraphic correlation.

For the Ordovician-Silurian boundary interval, event stratigraphy has been used to establish the number and severity of eustatic change induced by the terminal Ordovician glaciation (e.g. Brenchley, 1988; Sheehan, 1988; Brenchley et al., 1991; Johnson et al., 1991). In the clastic sequence in the Oslo region, Norway, two main regressive phases are evident whereas in the tropical carbonate belt of Anticosti Island, Quebec, four regressive phases appear to be present (Barnes, 1986), a number similar to the main glacial deposits preserved in the Morrocan Sahara. With collapsing ecosystems at this time, precise biostratigraphic correlation is again difficult and chemostratigraphy is still in its early stages (Orth et al., 1986; Middleton et al., 1991; Long, 1993; Wang et al., 1993).

Climate Change

This field probably holds the most promise for future research as it relates to event stratigraphy. To date, the Ordovician has received relatively few studies documenting climate-induced event stratigraphy. The best examples, as cited above, concern the terminal Ordovician glaciation with the change from a greenhouse to an icehouse state. The events are recognized particularly well in the deep ocean record by pale oxidized sediments, erosion intervals, and evidence of strong bottom currents, particularly within the peak glacial phase corresponding to the *C. extraordinarius* graptolite Zone.

Major changes in climate will be best recognized in the imprint within the marine sedimentary record, especially those reflecting basic changes in the ocean state (e.g. Wilde and Berry, 1984). The pioneering Silurian work of Jeppsson (1990) and Aldridge, Jeppson and Dorning (1993) have yet to be applied in similar detail in the Ordovician. These studies attributed biotic diversity cycles (especially those for conodonts) to alternations of primo and secundo events; the former being characterized by cool high-latitude climates, cold oceanic bottom waters and high nutrient supply, while the latter were characterized by warmer high-latitude climates, salinity-dense oceanic bottom waters, and carbonate formation in shallow waters.

Volcanism and Possible Superplume Activity

The Ordovician displays a record of moderate to high levels of volcanism. As in the Cretaceous, widespread bentonite ash beds in sedimentary sequences have been employed to locate volcanic centres, to provide chronostratigraphic marker horizons, and to provide isotope age control.

Huff et al. (1992) have estimated that the Middle Ordovician K-bentonites dated at 454 Ma (e.g. Big Bentonite and Millbrig bentonite beds) that are found in both the Baltic area and eastern North America represented an eruption of 1000 km^3 dense rock equivalent of volcanic ash. This is most likely the largest eruption known in the Phanerozoic. Interestingly, the widespread deposition of ash had little effect on the level bottom benthic communities. Stratigraphically higher studies in New York State by Goldman et al. (1994) have likewise used bentonites as time places to precisely clarify sequence stratigraphy following the earlier classic studies by Cisne and Chandlee (1982) and Cisne et al. (1982).

As speculated earlier in this paper, the high level of volcanism in the Middle Ordovician may be attributed in part to a possible mantle superplume. Larson

(1991) argued that in the Cretaceous a superplume generated a major pulse of ocean crust formulation, mid-plate volcanism, ocean plateau production, and higher sea levels. Outgassing increased primary productivity levels which in turn led to widespread black shale deposition. The 20 Ma interval in the mid-Cretaceous is characterized by a lack of magnetic reversals. Caldeira and Rampino (1991) further attribute high CO_2 levels in the Cretaceous to superplume activity and a resulting global warming of 2.8 to 7.7 °C over today's global mean temperature.

Ordovician event stratigraphy holds some interesting parallels to the Cretaceous record, if less complete in lacking the ocean crust record. Caradoc sea-levels were among the highest in the Phanerozoic; carbonate sedimentation was widespread; global CO_2 levels were especially high (Berner, 1990); volcanicity was at a high level as just discussed; the magnetic reversal record is sparse but shows a dominantly normal polarity for most of the Llandeilo late Caradoc (Trench et al., 1992); black shales are especially widespread during the Llanvirn and Caradoc. This association of indicators suggests that superplume activity may well have existed during the Middle Ordovician, possibly initiated in the Llarvirn and peaking in the Caradoc. Although Ordovician ocean crust is not widely preserved, compared to the Cretaceous, a proxy indicator for increased ocean ridge activity is the strontium isotope record. This shows a marked excursion in the Llanvirn through Caradoc from 0.709 to 0.708 (Burke et al., 1982; CRB unpublished; Keto and Jacobsen, 1987).

In summary, many aspects of event stratigraphy are potentially of great interest interpreting relationships with Ordovician global bio-events, but most lack a sufficiently systematic database at this stage, compared with the rich paleontological database. However, the rapidly accumulating isotope data will complement the paleontologic data to better constrain the interpretations of the causes and interrelationships between these events.

Conclusions

The Ordovician Period is a complex time interval characterized by widespread epeiric seas, wide dispersal of paleocontinents, intervals of intense volcanism and oceanic anoxia, a general greenhouse climate state deteriorating to an icehouse state, and major changes to the global biota. Five high order global bio-events are recognized: Basal Tremadoc Bio-Event (B'Tc), Basal Arenig Bio-Event (B'Ag), Basal Llanvirn Bio-Event (B'Ln), Basal Caradoc Bio-Event (B'Cc), and Upper Ashgill Bio-Event (U'Al). These are based mainly on the detailed studies available for conodonts, trilobites and graptolites, but other faunal changes are also considered.

Most of these five global bio-events are correlated to eustatic changes; some are characterized by extinctions, others by faunal radiations. The latest bioevent is related to the effects of the terminal Ordovician glaciation. The Ordovician faunas also show lower order bio-events but space restrictions have not allowed their consideration herein. The Ordovician was an interval of considerable biologic, climatic and oceanographic complexity and more detailed paleontologic, stratigraphic and isotopic studies are required to further understand the forcing processes that have produced these major and minor global bio-events.

References

Abaimova, G.P., 1975. Early Ordovician conodonts of the middle fork of the Lena River. Trudy Sibirskogo Nauchno-Issledovatel skogo Instituta, Geologii, Geofiziki i Mineralnogo Sirya (SNIGGIMS) 207, 129 [In Russian].

Aldridge, R.J., Briggs, D.E.G., Clarkson, E.N.K. and Smith, M.P., 1986. The affinities of conodonts-new evidence from the Carboniferous of Edinburgh, Scotland. Lethaia 19, 279-291.

Aldridge, R.J., Briggs, D.E.G., Smith, M.P., Clarkson, E.N.K. and Clark, N.D.L., 1993. The anatomy of conodonts. Philosophical Transactions of the Royal Society of London Series B340, 405-421.

Aldridge, R.J., Jeppsson, L. and Dorning, K.J., 1993. Early Silurian oceanic episodes and events. Journal of the Geological Society London 150, 501-513.

Aldridge, R.J. and Schönlaub, H.P., 1989. Conodonts. In: Holland, C.H. and Bassett, M.G. (eds.), A global standard for the Silurian System. National Museum of Wales, Geological Series 9, 274-279.

An Taixiang, 1985. Ordovician conodonts from Yaoxain and Fuping, Shaanzi Province, and their stratigraphic significance. Acta Geologica Sinica 59, 97-108 [in Chinese].

An Taixiang, Zhang Fang, Xiang Wei-da, Zhang Youqui, Xu Wen-hao, Zhang Hui-juan, Jiang De-biao, Yang Chang-sheng, Lin Lian-di, Cui Zhan-tang and Yang Xin-chang, 1983. Conodonts of North China and the adjacent regions. 223 p. Chinese Science Publishing House.

Anceñolaza, F.G. and Anceñolaza, G.F., 1992. The

genus *Jujuyaspis* as a world reference fossil for the Cambrian-Ordovician boundary. In: Webby, B.D. and Laurie, J.R. (eds.), Global Perspectives on Ordovician Geology. p. 115-122. A.A. Balkema, Rotterdam.

Bagnoli, G., Barnes, C.R. and Stevens, R.K., 1987. Tremadocian conodonts from Broom Point and Green Point, western Newfoundland. Bolletino della Societa Paleontologica Italiano 35, 145-158.

Barnes, C.R., 1984. Early Ordovician eustatic events in Canada. In: Brunton, D.L. (ed.), Aspects of the Ordovician System. Palaeontological Contributions from the University of Oslo 295, 51-64. Universitetsforlaget, Oslo.

Barnes, C.R., 1986. The faunal extinction event near the Ordovician-Silurian boundary: a climatically-induced crisis. In: Walliser, O.H. (ed.), Global Bio-Events. Lecture Notes in Earth Sciences 8, 121-126. Springer, Berlin Heidelberg New York.

Barnes, C.R., 1988. The proposed Cambrian-Ordovician global boundary stratotype and point (GSSP) in western Newfoundland, Canada. Geological Magazine 125, 381-414.

Barnes, C.R., 1989. Lower Silurian chronostratigraphy of Anticosti Island Quebec. In: Holland, H.D. (ed.), A global standard for the Silurian System. National Museum of Wales Geological Series 9, 101-106.

Barnes, C.R., 1992. The uppermost series of the Ordovician System. In: Webby, B.D. and Laurie, J.R. (eds.), Global Perspectives on Ordovician Geology. p. 185-194. A.A. Balkema, Rotterdam.

Barnes, C.R. and Bergström, S.M., 1988. Conodont biostratigraphy of the uppermost Ordovician and lowermost Silurian. Bulletin of the British Museum (Natural History) Geology 43, 325-343.

Barnes, C.R. and Fåhraeus, L.E., 1975. Province, communities, and the proposed nektobenthic habit of Ordovician conodontophorids. Lethaia 8, 133-149.

Barnes, C.R. and Williams, S. H., 1988. Conodonts from the Ordovician-Silurian Boundary Stratotype, Dob's Linn, Scotland. Bulletin of the British Museum (Natural History) Geology 43, 31-39.

Barnes, C.R. and Williams, S. H., 1990. Ordovician-Silurian. In: Briggs, D.E.G. and Crowther (eds.), Paleobiology -- A Synthesis. pp. 478-480. Oxford, Blackwell Scientific Publication.

Barnes, C.R., Rexroad, C.B. and Miller, J.F., 1973. Lower Paleozoic conodont provincialism. In: Rhodes, F.H.T. (ed.), Symposium on Conodont Paleozoology. Geological Society of America Special Paper 141, 156-190.

Bergström, J., 1976. Lower Palaeozoic trace fossils from eastern Newfoundland. Canadian Journal of Earth Sciences 13, 1613-1633.

Bergström, S.M., 1971. Conodont biostratigraphy of the Middle and Upper Ordovician of Europe and eastern North America. Geological Society of America Memoir 127, 83-161.

Bergström, S.M., 1986. Biostratigraphic integration of Ordovician graptolite and conodont zones - a regional review. In: Hughes, C.P. and Rickards, R.B. (eds.), Palaeoecology and biostratigraphy of graptolites. Geological Society Special Publication 20, 61-78.

Bergström, S.M., 1990. Relations between conodont provincialism and the changing palaeogeography during the Early Palaeozoic. In: McKerrow, W.S. and Scotese, C.R. (eds.), Palaeozoic Palaeogeography and Biogeography. Geological Society Memoir 12, 105-121.

Berner, R.A., 1990. Atmospheric carbon dioxide levels of Phanerozoic time. Science 249, 1382-1386.

Berner, R.A., 1991. A model for atmospheric CO_2 over Phanerozoic time. American Journal of Science 291, 339-376.

Berry, W.B.N., 1992. A base for the Arenig: the *Tetragraptus approximatus* Zone. In: Webby, B.D. and Laurie, J.R. (ed.), Global Perspectives on Ordovician geology. p. 123-133. A.A. Balkema, Rotterdam.

Braithwaite, L.F., 1976. Graptolites from the Lower Ordovician Pogonip Group of western Utah. Geological Society of America Special Paper 166, 1-106.

Brenchley, P.J., 1984. Late Ordovician extinctions and their relationships to the Gondwana glaciations. In: Brenchley, P.J. (ed.), Fossils and Climate. p. 291-316. Wiley, New York.

Brenchley, P.J., 1988. Environmental changes close to the Ordovician-Silurian boundary. In: Cocks, L.R.H. and Rickards, R.B. (eds.), A Global Analysis of the Ordovician-Silurian Boundary. Bulletin of the British Museum (Natural History) Geology 43, 377-385.

Brenchley, P.J., Romano, M., Young, T.P and Storch, P., 1991. Hirnantian glaciomarine diamictites--evidence for the spread of glaciation and its effect on Upper Ordovician faunas. In: Barnes, C.R. and Williams, S.H. (eds.), Advances in Ordovician Geology. Geological Survey of Canada Paper 90-9, 325-336.

Briggs, D.E.G., Clarkson, E.N.K. and Aldridge, R.J., 1983. The conodont animal. Lethaia 16, 1-14.

Bruton, D.L., Erdtmann, B.D. and Koch, L., 1982. The Naersnes section, Oslo Region, Norway: a candidate for the Cambrian-Ordovician boundary stratotype at the base of the Tremadoc Series. In: Bassett, M.G. and Dean, W.T. (eds.), The Cambrian-Ordovician boundary: sections, fossil distributions, and correlations. National Museum of Wales, Geological Series 3, 61-69. Cardiff.

Bull, E., 1991. What sort of a life did the benthonic graptolites have? In: Palmer, D. and Rickards, R.B. (eds.), Graptolites: writing in the rocks. p. 53-58. Boydell Press, Woodbridge, U.K.

Bulman, O.M.B., 1970. Graptolithina (2nd edition). In: Teichert, C. (ed.), Treatise on invertebrate paleontology, Part V. Geological Society of America Inc. and University of Kansas, x + 1-163.

Burke, W.H., Denison, R.E., Hetherington, E.A., Koepick, R.B., Nelson, H.F. and Otto, J.B., 1982. Variation of seawater $^{87}Sr/^{86}Sr$ throughout Phanerozoic time. Geology 10, 516-519.

Burrett, C., 1979. *Tasmanognathus*, a new Ordovician conodontophorida genus from Tasmania. Geologica et Palaeontologica 13, 31-38.

Burrett, C., Stait, B., Sharples, C. and Laurie, J., 1984. Middle-Upper Ordovician shallow platform to deep basin transect, southern Tasmania, Australia. In: Brunton, D.L. (ed.), Aspects of the Ordovician System. Palaeontological Contributions from the University of Oslo 295, 149-158. Universitetsforlaget, Oslo.

Caldeira, K. and Rampino, M.R., 1991. The Mid-Cretaceous super plume, carbon dioxide, and global warming. Geophysical Research Letters 18, 987-990.

Chatterton, B.D.E. and Speyer, S.E., 1989. Larval ecology, life history strategies, and patterns of extinction and survivorship among Ordovician trilobites. Paleobiology 15, 118-132.

Chen Jun-yuan and Gong We-li, 1986. Conodonts. In: Chen Jun-yuan (ed.), Aspects of Cambrian-Ordovician Boundary in Dayangcha, China. p. 93-223. Contribution to Dayangcha International Conference on Cambrian-Ordovician Boundary, Beijing. China Prospect Publishing House.

Chen Xu and Rong Jiayu, 1991. Concepts and analysis of mass extinction with the late Ordovician event as an example. Historical Biology 5, 107-121.

Chen Xu, Yang Da-quan, Han Nai-ren and Li Luozhao, 1983. Graptolites from the *Tetragraptus (Eotetragraptus) approximatus* Zone of the lowermost Ningkuo Formation in Yushan, N. Jiangxi. Acta Palaeontologica Sinica 22, 324-330 [in Chinese, with English summary].

Chen Jun-yuan, Qian Yi-yuan, Zhang Jun-ming, Lin Yao-kun, Yin Lei-ming, Wang Zhi-hao, Wang Zong-zhi, Yang Jie-dong and Wang Ying-xi, 1988. The recommended Cambrian-Ordovician global boundary stratotype of the Xiaoyangqiao section (Dayangcha, Jilin Province), China. Geological Magazine 125, 415-444.

Cisne, J.L. and Chandlee, G.O., 1982. Taconic foreland basin graptolites: age zonation, depth zonation, and use in ecostratigraphic correlation. Lethaia 15, 343-363.

Cisne, J.L., Chandlee, G.O., Rabe, B.D. and Cohen, J.A., 1982. Clinal variation, episodic evolution, and possible parapatric speciation: the trilobite *Flexicalymene senaria* along an Ordovician depth gradient. Lethaia 15, 325-341.

Cocks, L.R.M. and Fortey, R.A., 1990. Biogeography of Ordovician and Silurian faunas. In: McKerrow, W.S. and Scotese, C.R. (eds.), Palaeozoic palaeogeography and biogeography. Geological Society Memoir 12, 97-104.

Cocks, L.R.M. and Rickards, R.B., 1988. A global analysis of the Ordovician-Silurian boundary. Bulletin of the British Museum (Natural History) Geology 43, 1-394.

Cooper, R.A., 1973. Taxonomy and evolution of Isograptus Moberg in Australasia. Palaeontology 16, 45-115.

Cooper, R.A., 1979a. Ordovician geology and graptolite faunas of the Aorangi Mine area, north-west Nelson, New Zealand. New Zealand Geological Survey, Paleontological Bulletin 47, 1-127.

Cooper, R.A., 1979b. Sequence and correlation of Tremadoc graptolite assemblages. Alcheringa 3, 7-19.

Cooper, R.A., 1992. A relative timescale for the Early Ordovician derived from deposition rates of graptolite shales. In: Webby, B.D. and Laurie, J.R. (eds), Global Perspectives on Ordovician Geology. p. 3-21. A.A. Balkema, Rotterdam.

Cooper, R.A. and Fortey, R.A., 1982. The Ordovician Graptolites of Spitsbergen. Bulletin of the British Museum (Natural History) Geology 36, 157-302.

Cooper, R.A. and Fortey, R.A., 1983. Development of the graptoloid rhabdosome. Alcheringa 7, 201-221.

Cooper, R.A. and Lindholm, K., 1985. The phylogenetic relationships of the graptolites *Tetragraptus phyllograptoides* and *Pseudophylograptus cor.* Geologiska Föreningens i Stockholm Förhandlingar 106, 279-291.

Cooper, R.A. and Lindholm, K., 1990. A precise worldwide correlation of Early Ordovician graptolite sequences. Geological Magazine 127, 497-525.

Cooper, R.A. and Stewart, I., 1979. The Tremadoc graptolite sequence of Lancefield, Victoria. Palaeontology 22, 767-797.

Cooper, R.A., Fortey, R.A. and Lindholm, K., 1991. Latitudinal and depth zonation of early Ordovician graptolites. Lethaia 24, 199-218.

Dalla Salda, L.H., Cingolani, C. and Varela, R., 1992. Early Paleozoic orogenic belt of the Andes in southwestern South America: Result of Laurentia-Gondwana collision? Geology 20, 617-620.

Dalziel, I.W.D., Dalla Salda, L.H. and Gahagan, L.M., 1994. Paleozoic Laurentia-Gondwana interaction and the origin of the Appalachian-Andean mountain system. Geological Society of America Bulletin 106, 243-252.

DeMott, L.L., 1987. Platteville and Decorah trilobites from Illinois and Wisconsin. In: Sloan, R.E. (ed.), Middle and late Ordovician Lithostratigraphy and Biostratigraphy of the Upper Mississippi Valley. Minnesota Geologic Survey Report 35, 63-98.

Destombes, J., 1972. Les trilobites du sous-ordre des Phacopina de l'Ordovicien de l'Anti Atlas (Maroc). Notes et memoires. Service Géologique Maroc 32, 1-79.

Druce, E.C. and Jones, P.J., 1971. Cambrian-Ordovician conodonts from the Burke River structural belt, Queensland. Australian Bureau of Mineral Resources, Bulletin 110, 159 p.

Druce, E.C., Shergold, J.H. and Radke, B.M., 1982. A reassessment of the Cambrian-Ordovician boundary sections at Black Mountain, western Queensland, Australia. In: Bassett, M.G. and Dean, W.T. (eds.), The Cambrian-Ordovician boundary: sections, fossil distributions, and correlations. National Museum of Wales, Geological Series 3, 193-209. Cardiff.

Dzik, J., 1983. Relationships between Ordovician Baltic and North American Midcontinent conodont faunas. Fossils and Strata 15, 59-85.

Erdtmann, B.D., 1982. A reorganisation and proposed phylogenetic classification of planktic Tremadoc (early Ordovician) dendroid graptolites. Norsk Geologisk Tidsskrift 62, 121-144.

Erdtmann, B.D., 1988. The Early Ordovician nematophorid graptolites: taxonomy and correlation. Geological Magazine 125, 201-221.

Erdtmann, B.D. and VandenBerg, A.H.M., 1985. *Araneograptus* gen. nov. and its two species from the late Tremadocian (Lancefieldian, La2) of Victoria. Alcheringa 9, 49-63.

Ethington, R.L., 1972. Lower Ordovician (Arenigian) conodonts from the Pogonip Group, Central Nevada. Geologica et Palaeontologica SB 1, 17-28.

Ethington, R.L. and Clark, D.L., 1971. Lower Ordovician conodonts in North America. In: Sweet, W.C. and Bergström, S.M. (eds.), Symposium on Conodont Biostratigraphy. Geological Society of America Memoir 127, 63-82.

Ethington, R.L. and Clark, D.L., 1981. Lower and Middle Ordovician conodonts from the Ibex area, western Millard County, Utah. Brigham Young University Geology Studies 28 (part 2), 155.

Ethington, R.L. and Repetski, J.E., 1984. Paleobiogeographic distribution of Early Ordovician conodonts in central and western United States. Geological Society of America, Special Paper 196, 89-101.

Ethington, R.L., Engel, K.M. and Elliott, K.L., 1987. An abrupt change in conodont faunas in the Lower Ordovician of the Midcontinent Province. In: Aldridge, R.J. (ed.), Palaeobiology of conodonts. p. 111-127. Ellis Horwood Limited, Chichester.

Finney, S.C. and Bergström, S.M., 1986. Biostratigraphy of the *Nemagraptus gracilis* Zone. In: Hughes, C.P. and Rickards, R.B. (eds.), Palaeoecology and biostratigraphy of graptolites. Geological Society Special Publication 20, 47-59.

Finney, S.C. and Ethington, R.L., 1992. Whiterockian graptolites and conodonts from the Vinini Formation, Nevada: biostratgraphic implications. In: Webby, B.D. and Laurie, J.R. (eds.), Global perspectives on Ordovician geology. p. 153-169. A.A. Balkema, Rotterdam.

Flower, R.H., 1964. The nautiloid order Ellesmeroceratida (Cephalopoda). Bureau of Mines and Mineral Resources of New Mexico Institute of Mining Technology Memoir 12, 1-234.

Fortey, R.A., 1974, 1980. The Ordovician trilobites of Spitsbergen. Part 1, Olenidae, and part 3, Remaining trilobites of the Valhallfonna Formation. Skr. norsk Polarinstitutt 160 (129 pp) and 171 (163 pp).

Fortey, R.A., 1975. Early Ordovician trilobite communities. Fossils and Strata 4, 331-352.

Fortey, R.A., 1979. Early Ordovician trilobites from the Catoche Formation (St George Group), western Newfoundland. Geological Survey of Canada Bulletin 321, 61-114.

Fortey, R.A., 1984. Global earlier Ordovician transgressions and regressions and their biological implications. In: Brunton, D.L. (ed.), Aspects of the Ordovician System. Palaeontological Contributions from the University of Oslo 295, 37-50. Universitetsforlaget, Oslo.

Fortey, R.A., 1989. There are extinctions and extinctions: examples from the Lower Palaeozoic. Philosophical Transactions of the Royal Society of London Series B325, 327-355.

Fortey, R.A. and Barnes, C.R., 1977. Early Ordovician conodont and trilobite communities of Spitsbergen: influence on biogeography. Alcheringa 1, 297-309.

Fortey, R.A. and Chatterton, B.D.E., 1988. Classification of the trilobite Suborder Asaphina. Palaeontology 31, 165-222.

Fortey, R.A. and Cooper, R.A., 1986. A phylogenetic classification of the graptoloids. Palaeontology 29, 631-654.

Fortey, R.A. and Mellish, C.J.M., 1992. Are some fossils better than others for inferring palaeogeography? Terra Nova 4, 210-216.

Fortey, R.A. and Owens, R.M., 1978. Early Ordovician (Arenig) stratigraphy and faunas of the Carmarthen District, southwest Wales. Bulletin of the British Museum (Natural History) Geology 30, 225-294.

Fortey, R.A. and Owens, R.M., 1987. The Arenig Series in South Wales. Bulletin of the British Museum (Natural History) Geology 41, 69-307.

Fortey, R.A. and Peel, J.S., 1989. Stratigraphy and hystricurid trilobites of the Christian Elv Formation (Lower Ordovician) of western North Greenland. Rapport Grønlands Geologiske Undersøgelse 144, 5-15.

Fortey, R.A. and Shergold, J.H., 1984. Early Ordovician trilobites, Nora Formation, central Australia. Palaeontology 27, 315-366.

Fortey, R.A., Landing, E. and Skevington, D., 1982. Cambrian-Ordovician boundary sections in the Cow Head Group, western Newfoundland. In: Bassett, M.G. and Dean, W.T. (eds.), The Cambrian-Ordovician boundary: sections, fossil distributions, and correlations. National Museum of Wales, Geological Series 3, 95-129. Cardiff.

Fortey, R.A., Beckly, A.J. and Rushton, A.W.A., 1990. International correlation of the base of the Llanvirn Series, Ordovician System. Newsletters in Stratigraphy 22, 119-142.

Goldman, D., Mitchell, C.E., Bergström, S.M., Delano, J.W. and Tice, S., 1994. K-bentonites and Graptolite Biostratigraphy in the Middle Ordovician of New York State and Quebec: A New Chronostratigraphic Model. Palaios 9, 124-143.

Hammann, W., 1983. Calymenacea (Trilobita) aus dem Ordovizium von Spanien; ihre Biostratigraphie, Ökologie, und Systematik. Abh. Senck. Natur. Ges. 542, 1-177.

Henningsmoen, G., 1957. The trilobite family Olenidae. Skr. Norske Vid. Akad. Mat. nat. Kl. 1, 1-303.

Hintze, L.F., 1953. Lower Ordovician trilobites from western Utah and eastern Nevada. Utah Geological and Mineral Survey Bulletin 48, 1-249.

Hoffman, P.F., 1991. Did the breakout of Laurentia turn Gondwanaland inside out? Science 252, 1409-1412.

Holland, H.D., 1989. Principles, history, and classification. In: Holland, H.D. and Bassett, M.G. (eds.), A global standard for the Silurian System. National Museum of Wales Geological Series 9, 7-26. Cardiff.

Holland, S.M., 1993. Sequence stratigraphy of a carbonate-clastic ramp: The Cincinnatian Series (Upper Ordovician) in its type area. Geological Society of America Bulletin 105, 306-322.

Huff, W.D., Bergström, S.M. and Kolata D.R., 1992. Gigantic Ordovician volcanic ash fall in North America and Europe: Biological, tectomagmatic, and event-stratigraphy significance. Geology 20, 875-878.

Jaanusson, V., 1969. Ordovician. In: Robison, R.A. and Teichert, C. (eds), Treatise on Invertebrate Paleontology, Part A, Introduction- Biogeography and biostratigraphy. University of Kansas and Geological Society of America, A136-A166.

Jaanusson, V., 1984. Ordovician benthic macrofaunal associations. In: Bruton (ed.), Aspects of the Ordovician System. Paleont. Contr. Univ. Oslo 295, 127-139.

James, N.P. and Stevens, R.K., 1986. Stratigraphy and correlation of the Cambro-Ordovician Cow Head Group, western Newfoundland. Geological Survey of Canada Bulletin 366, 143 p.

James, N.P., Stevens, R.K., Barnes, C.R. and Knight, I., 1989. Evolution of a Lower Paleozoic continental margin carbonate platform, northern Canadian Appalachians. In: Crevelo, T., Sarg, R., Read, J.F. and Wilson, J.L. (eds.), Controls on Carbonate Platforms and Basin Development. Society of Economic Paleontologists and Mineralogists Special Publication 44, 123-146.

Jenkins, C.J., 1980. *Maeandrograptus schmalenseei* and its bearing on the origin of the diplograptids. Lethaia 13, 289-302.

Jeppsson, L., 1990. An oceanic model for lithological and faunal changes tested on the Silurian record. Journal of the Geological Society, London 147, 663-674.

Ji Zailiang and Barnes, C.R., 1990. Apparatus reconstructions of Lower Ordovician conodonts from the Midcontinent Province. In: Ziegler, W. (ed.), First International Senckenberg Conference and fifth European Conodont Symposium (ECOS V), Contributions IV. Courier Forschungs-Institut Senckenberg 118, 333-352.

Ji Zailiang and Barnes, C.R., 1993. A major conodont extinction event during the Early Ordovician within the Midcontinent Realm. Palaeogeography, Palaeoclimatology, Palaeoecology 104, 37-47.

Ji Zailiang and Barnes, C.R., 1994. Lower Ordovician conodonts of the St. George Group, Port au Port Peninsula, western Newfoundland, Canada. Palaeontographica Canadiana 11, 149 p.

Ji Zailiang and Barnes, C.R., (in press). Conodont paleoecology of the Lower Ordovician St. George Group, Port au Port Peninsula, western Newfoundland. Journal of Paleontology.

Johnson, M.E., Cocks, L.R.M. and Copper, P., 1981. Late Ordovician-Early Silurian fluctuations of sea level from eastern Anticosti Island, Quebec. Lethaia 14, 73-82.

Johnston, D.I., 1986. Early Ordovician (Arenig) conodonts from St. Pauls Inlet and Martin Point, Cow Head Group, western Newfoundland. M.Sc. thesis, Memorial University of Newfoundland, St. John's, Newfoundland, 226 p.

Keto, L.S. and Jacobsen, S.B., 1987. Nd and Sr isotopic variations of Early Paleozoic oceans. Earth and Planetary Science Letters 84, 27-41.

Knight, I., James, N.P. and Lane, T.E., 1991. The Ordovician St. George Unconformity, northern Appalachians: The relationship of plate convergence at the St. Lawrence Promontory to the Sauk/Tippecanoe sequence boundary. Geological Society of America Bulletin 103, 1200-1225.

Koren, T.N. and Sobolevskaya, R.F., 1983. Graptolites. In: Sokolov, B.S., Koren, T.N. and Nikitin, I.F. (eds.), The Ordovician-Silurian boundary in the northeast of the USSR. p. 1-208. Nauka Publishers, Leningrad [in Russian].

Landing, E., 1977. *Prooneotodus tenuis* (Müller, 1959) apparatuses from the Taconic allochthon, eastern New York: construction, taphonomy and the protoconodont "Supertooth" model. Journal of Paleontology 51, 1072-1084.

Landing, E. and Barnes, C.R., 1981. Conodonts from the Cape Clay Formation (Lower Ordovician), southern Devon Island, Arctic Archipelago. Canadian Journal of Earth Sciences 18, 1609-1628.

Landing, E., Barnes, C.R. and Stevens, R.K., 1986. Tempo of earliest Ordovician graptolite faunal succession: Conodont based correlation from the Tremadocian of Quebec. Canadian Journal of Earth Sciences 23, 1928-1949.

Lapworth, C., 1878. The Moffat Series. Quarterly Journal of the Geological Society of London 34, 240-346.

Lapworth, C., 1879. On the tripartite classification of the Lower Palaeozoic rocks. Geological Magazine 16, 1-15.

Larson, R.C., 1991. Latest pulse of the Earth: evidence for a mid-Cretaceous superplume. Geology 19, 549-550.

Leggett, J.K., 1978. Eustasy and pelagic regimes in the Iapetus Ocean during the Ordovician and Silurian. Earth and Planetary Sciences Letters 41, 163-169.

Lindholm, K., 1991. Hunnebergian graptolites and biostratigraphy in southern Scandinavia. Lund Publications in Geology 95, 1-36.

Lindholm, K. and Maletz, J., 1989. Intraspecific variation and relationships of some Lower Ordovician species of the dichograptid, *Clonograptus*. Palaeontology 32, 711-743.

Löfgren, A., 1978. Arenigian and Llanvirnian conodonts from Jämtland, northern Sweden. Fossils and Strata 13, 129.

Long, D.G.F., 1993. Oxygen and carbon isotopes and event stratigraphy near the Ordovician-Silurian boundary, Anticosti Island Quebec. Palaeogeography, Palaeoclimatology, Palaeoecology 104, 40-59.

Lu, Y.H., 1975. Ordovician trilobite faunas of central and south-western China. Palaeontologica Sinica, new series B 11, 1-463.

Maletz, J., 1992. The Arenig/Llanvirn boundary in the Quebec Appalachians. Newsletters in Stratigraphy 26, 49-64.

Magaritz, M., 1991. Carbon isotopes, time boundaries and evolution. Terra Nova 3, 251-256.

McCracken, A.D. and Barnes, C.R., 1981. Conodont biostratigraphy and paleoecology of the Ellis Bay Formation, Anticosti Island, Quebec, with special reference to Late Ordovician-Early Silurian chronostratigraphy and the systemic boundary. Geological

Survey of Canada Bulletin 329, 51-134.

McCracken, A.D., Nowlan, G.S. and Barnes, C.R., 1981. *Gamachignathus* a new multielement conodont genus from the latest Ordovician, Anticosti Island, Quebec. Geological Survey of Canada, Paper 80-1C, 103-112.

McLaren, D.J. and Goodfellow, W.D., 1990. Geological and Biological Consequences of Giant Impacts. Annual Review of Earth and Planetary Sciences 18, 123-172.

McKerrow, W.S., 1993. The development of Early Paleozoic global stratigraphy. Journal of the Geological Society, London 150, 21-28.

McKerrow, W.S. and Scotese, C.R., (eds.) (1990). Palaeozoic palaeogeography and biogeography. Geological Society Memoir 12, 435 p. London.

Melchin, M.J., 1987. Upper Ordovician graptolites from the Cape Phillips Formation, Canadian Arctic Islands. The Geological Society of Denmark Bulletin 35, 191-202.

Melchin, M.J., McCracken, A.D. and Oliff, F.J., 1991. The Ordovician-Silurian boundary on Cornwallis and Truro islands, Arctic Canada: preliminary data. Canadian Journal of Earth Sciences 28, 1854-1862.

Miall, A.D., 1992. The Exxon global cycle chart: An event for every occasion? Geology 20, 787-790.

Middleton, P.D., Marshall, J.D. and Brenchley, P.J., 1991. Evidence for isotopic change associated with Late Ordovician glaciation, from brachiopods and marine cements of central Sweden. In: Barnes, C.R. and Williams, S.H. (eds.), Advances in Ordovician Geology. Geological Survey of Canada Paper 90-9, 313-323.

Miller, J.F., 1969. Conodont fauna from the Notch Peak Limestone (Cambro-Ordovician), House Range, Utah. Journal of Paleontology 43, 413-439.

Miller, J.F., 1980. Taxonomic revisions of some Upper Cambrian and Lower Ordovician conodonts with comments on their evolution. University of Kansas Paleontological contributions Paper 99, 1-39.

Miller, J.F., 1984. Cambrian and earliest Ordovician conodont evolution, biofacies, and provincialism. Geological Society of America Special Paper 196, 43-68.

Miller, J.F., 1988. Conodonts as biostratigraphic tools for redefinition and correlation of the Cambrian-Ordovician boundary. Geological Magazine 125, 349-362.

Miller, J.F., 1992. The Lange Ranch Eustatic Event: A regressive-transgressive couplet near the base of the Ordovician System. In: Webby, B.D. and Laurie, J.R. (eds.), Global Perspectives on Ordovician Geology. p. 395-407. A.A. Balkema, Rotterdam.

Mitchell, C.E., 1987. Evolution and phylogenetic classification of the Diplograptacea. Palaeontology 30, 353-405.

Mitchell, C.E., 1992. Evolution of the Diplograptacea and the international correlation of the Arenig-Llanvirn boundary. In: Webby, B.D. and Laurie, J.R. (eds.), Global perspectives on Ordovician geology. p. 171-183. A.A. Balkema, Rotterdam.

Morris, W.G., 1988. A systematic survey of Lancefieldian graptolites from Victoria, Australia. Unpublished Ph.D. thesis, University of Cambridge. 180 p.

Moskalenko, T.A., 1967. Conodonts from the Chunsky Stage (Lower Ordovician) of the rivers Moiero and Podkamennaya Tunguska. In: Ivanovskii, A.B. and Sokolov, B.S. (eds.), New data on the biostratigraphy of the lower Paleozoic deposits of the Siberian platform. Akademiya Nauk SSSR, Sibirskoye Otdeleniye, Instituta Geologii i Geofiziki, 98-116.

Moskalenko, T.A., 1976. Environmental effects on the distribution of Ordovician conodonts in western Siberian platform. In: Barnes, C.R. (ed.), Conodont Paleoecology. Geological Association of Canada Special Paper 15, 59-68.

Moskalenko, T.A., 1983. Conodonts and biostratigraphy in the Ordovician of the Siberian Platform. Fossils and Strata 15, 87-94.

Mu En-zhi, Ge Me-yu, Chen Xu, Ni Yu-nan and Lin Yap-kun, 1979. Lower Ordovician graptolites of southwest China. Palaeontologica Sinica, 192 p. [in Chinese with English abstract].

Müller, K.J., 1973. Late Cambrian and early Ordovician conodonts from northern Iran. Geological Survey of Iran Report 30, 1-77.

Neilsen, A.T., 1992a. Ecostratigraphy and the recogniton of Arenigian (Early Ordovician) sea-level changes. In: Webby, B.D. and Laurie, J.R. (eds.), Global Perspectives on Ordovician Geology. p. 355-366. A.A. Balkema, Rotterdam.

Neilsen, A.T., 1992b. Intercontinental correlation of the Arenigian (Early Ordovician) based on sequence and ecostratigraphy. In: Webby, B.D. and Laurie, J.R. (eds.), Global Perspectives on Ordovician Geology. p. 367-379. A.A. Balkema, Rotterdam.

Newell, N.D., 1967. Revolutions in the history of life. Geological Society of America Special Paper

89, 63-91.

Nicoll, R.S., 1992. Evolution of the conodont genus *Cordylodus* and the Cambrian-Ordovician boundary. In: Webby, B.D. and Laurie, J.R. (eds.), Global Perspectives on Ordovician Geology. p. 105-113. A.A. Balkema, Rotterdam.

Nicoll, R.S., Nielsen, A.T., Laurie, J.R. and Shergold, J.H., 1992. Preliminary correlation of latest Cambrian to Early Ordovician sea level events in Australia and Scandinavia. In: Webby, B.D. and Laurie, J.R. (eds.), Global Perspectives on Ordovician Geology. p. 381-394. A.A. Balkema, Rotterdam.

Norford, B.S., 1991. The international working group on the Cambrian-Ordovician boundary: report of progress. In: Barnes, C.R. and Williams, S.H. (eds.), Advances in Ordovician Geology. Geological Survey of Canada Paper 90-9, 27-32.

Orth, C.J., Gilmore, J.S., Quintana, L.R. and Sheehan, P.M., 1986. Terminal Ordovician extinction: geochemical analysis of the Ordovician/Silurian boundary, Anticosti Island, Quebec. Geology 14, 433-436.

Owen, A.W., 1986. The uppermost Ordovician (Hirnantian) trilobites of Girvan, southest Scotland, with a review of coeval trilobite faunas. Transactions of the Royal Society of Edinburgh 77, 231-239.

Patzkowsky, M.E. and Holland, S.M., 1993. Biotic response to a Middle Ordovician paleoceanographic event in eastern North America. Geology 21, 619-622.

Paul, C.R.C., 1982. The adequacy of the fossil record. In: Joysey, K.A. and Friday, A.E. (eds.), Problems of phylogenetic reconstruction. Special volume of the Systematics Association 21, 75-118. Academic Press.

Pohler, S.M. and Barnes, C.R., 1990. Conceptual models in conodont paleoecology. Courier Forschungs-Institut Senckenberg 118, 409-440.

Read, J.F., 1980. Carbonate ramp-to-basin transitions and foreland basin evolution, Middle Ordovician, Virginia Appalachians. American Association of Petroleum Geologists Bulletin 64, 1575-1612.

Read, J.F., 1989. Controls on evolution of Cambro-Ordovician passive margin, U.S. Appalachians. In: Controls on Carbonate Platforms and Basin Development. Soc. Econ. Paleont. Mineral., Special Publ. 44, 147-165.

Repetski, J., 1982. Conodonts from El Paso Group (Lower Ordovician) of westernmost Texas and southern New Mexico. New Mexico Bureau of Mines and Mineral Resources Memoir 40, 121.

Rickards, R.B., 1978. Major aspects of evolution of the graptolites. Acta Palaeontologica Polonica 23, 585-594.

Ripperdan, R.L., Magaritz, M., Nicoll, R.S. and Shergold, J.H., 1992. Simultaneous changes in carbon isotopes, sea level, and conodont biozones within the Cambrian-Ordovician boundary interval at Black Mountain, Australia. Geology 20, 1039-1042.

Rong, J.Y. and Harper, D.T.A., 1988. A global synthesis of the latest Ordovician Hirnantian brachiopod faunas. Transactions of the Royal Society of Edinburgh 79, 383-402.

Ross, J.R.P. and Ross, C.A., 1992. Ordovician sea level fluctuations. In: Webby, B.D. and Laurie, J.R. (eds.), Global Perspectives on Ordovician Geology. p. 327-336. A.A. Balkema, Rotterdam.

Ross, R.J., Jr., 1975. Early Paleozoic trilobites, sedimentary facies, lithospheric plates and ocean currents. Fossils and Strata 4, 307-329.

Ross, R.J., Jr. and Ethington, R.L., 1992. North American Whiterock Series suited for global correlation. In: Webby, B.D. and Laurie, J.R. (eds.), Global Perspectives on Ordovician Geology. p. 135-152. A.A. Balkema, Rotterdam.

Ross, R.J., Jr. et al., 1982. The Ordovician System in the United States. International Union of Geological Sciences, Subcommission on Ordovician Stratigraphy, Publication 12.

Ross, R.J., Jr., Ethington, R.L. and Mitchell, C., 1991. Stratotype of Ordovician Whiterock Series. Palaios 6, 156-173.

Sansom, I.J., Smith, M.P., Armstrong, H.A. and Smith, M.M., 1992. Presence of earliest vertebrate hard tissues in conodonts. Science (Washington) 256, 1308-1311.

Savage, N.M., 1990. Conodonts of Caradocian (Late Ordovician) age from the Cliefden Caves Limestone, southeastern Australia. Journal of Paleontology 64, 821-831.

Sepkoski, J.J., 1981. A factor analytic description of the fossil record. Paleobiology 7, 36-53.

Serpagli, E., 1974. Lower Ordovician conodonts from Precordilleran Argentina (province of San Juan). Societa Palaeontologica Italiana Bolletino 13, 17-93.

Sheehan, P.M., 1988. Late Ordovician events and the terminal Ordovician extinction. New Mexico Bureau of Mines and Mineral Resources Memoir 44, 405-415.

Shergold, J.H., 1988. Review of trilobite biofacies

distributions at the Cambrian Ordovician boundary. Geological Magazine 125, 363-380.

Shergold, J.H. and Nicoll, R.S., 1992. Revised Cambrian-Ordovician boundary biostratigraphy, Black Mountain, western Queensland. In: Webby, B.D. and Laurie, J.R. (eds.), Global Perspectives on Ordovician Geology. p. 81-92. A.A. Balkema, Rotterdam.

Sleep, N.H., 1992. Hotspot volcanism and mantle plumes. Annual Reviews of Earth and Planetary Sciences 20, 19-44.

Stait, K.A. and Barnes, C.R., 1991. Conodont biostratigraphy of the upper St. George Group (Canadian to Whiterockian), western Newfoundland. In: Barnes, C.R. and Williams, S.H. (eds.), Advances in Ordovician geology. Geological Survey of Canada Paper 90-9, 125-134.

Stillman, C.J., 1984. Ordovician volcanicity. In: Brunton, D.L. (ed.), Aspects of the Ordovician System. Palaeontological Contributions from the University of Oslo 295, 183-194. Universitetsforlaget, Oslo.

Stitt, J.H., 1971. Late Cambrian and earliest Ordovician triloibtes, Timbered Hills and lower Arbuckle Groups, western Arbuckle Mountains, Murray County, Oklahoma. Oklahoma Geological Survey Bulletin 110, 1-83.

Stitt, J.H., 1975. Adaptive radiation, trilobite paleoecology and extinction, Ptychaspid Biomere, late Cambrian of Oklahoma. Fossils and Strata 4, 381-390.

Stitt, J.H., 1977. Late Cambrian and earliest ordovician trilobites, Wichita Mountains area, Oklahoma. Oklahoma Geological Survey Bulletin 124, 1-79.

Stockmal, G.S., Colman-Sadd, S.P., Keen, C.E., O'Brien, S.J. and Quinlan, G., 1987. Collision along an irregular margin: A regional plate tectonic interpretation of the Canadian Appalachians. Canadian Journal of Earth Sciences 24, 1098-1107.

Stouge, S.S., 1984. Conodonts of the Middle Ordovician Table Head Formation, western Newfoundland. Fossils and Strata 16, 1-145.

Stouge, S.S. and Bagnoli, B., 1988. Early Ordovician conodonts from Cow Head Peninsula, western Newfoundland. Palaeontographia Italica 75, 88-179.

Sweet, W.C., 1988. The Conodonta: morphology, taxonomy, paleoecology and evolutionary history of a long extinct animal phylum. Oxford Monographs on Geology and Geophysics 10, 212 p. Oxford University Press.

Sweet, W.C. and Bergström, S.M., 1974. Provincialism exhibited by Ordovician conodont faunas. In: Ross, C.A. (ed.), Paleogeographic provinces and provinciality. Society of Economic Paleontologists and Mineralogists Special Publication 21, 189-202.

Sweet, W.C., Ethington, R.L. and Barnes, C.R., 1971. North American Middle and Upper Ordovician conodont faunas. In: Sweet, W.C. and Bergström, S.M. (eds.), Symposium on conodont biostratigraphy. Geological Society of America Memoir 127, 163-193.

Szaniawski, H., 1987. Preliminary structural comparisons of protoconodont, paraconodont, and euconodont elements. In: Aldridge, R.J. (ed.), Palaeobiology of conodonts. p. 35-47. Ellis Horwood, Chichester.

Trench, A., Dentith, M.C., McKerrow, W.S. and Torsvik, T.H., 1992. The Ordovician magnetostratigraphic time scale: Reliability and correlation potential. In: Webby, B.D. and Laurie, J.R. (eds.), Global Perspectives on Ordovician Geology. p. 69-77. A.A. Balkema, Rotterdam.

Tzaj, D.T., 1974. Lower Ordovician graptolites of Kazakhstan. USSR Academy of Science, Moscow, 127 p. [in Russian].

VandenBerg, A.H.M. and Cooper, R.A., 1992. The Ordovician graptolite sequence of Australasia. Alcheringa 16, 33-85.

Wang Kun and Chatterton, B.D.E., 1992. Iridium abundance maxima at the latest Ordovician mass extinction horizon, Yangtze Basin, China: Terrestrial or extraterrestrial? Geology 20, 39-42.

Wang Xiaofeng and Erdtmann, B.D., 1986. The earliest Ordovician graptolite sequence from Hunjiang, Jilin Province, China. Acta Geological Sinica 3, 226-236 [in Chinese with English abstract].

Wang Xiaofeng and Erdtmann, B.D., 1987. Zonation and correlation of the earliest Ordovician graptolites from Hunjiang, Jilin province, China. The Geological Society of Denmark Bulletin 35, 245-257.

Wang Kun, Orth, C.J., Attrep, M., Jr., Chatterton, B.D.E., Wang Xiaofeng and Li Ji-jin, 1993. The great latest Ordovician extinction on the South China Plate: Chemostratigraphic studies of the Ordovician-Silurian boundary interval on the Yangtze Platform. Palaeogeography, Palaeoclimatology, Palaeoecology 104, 61-79.

Westrop, S.R. and Ludvigsen, R., 1987. Biogeographic control of trilobite mass extinction at the Upper Cambrian biomere boundary. Paleobiology

13, 84-99.

Whittington, H.B. and Hughes, C.P., 1972. Ordovician geography and faunal provinces deduced from trilobite distribution. Philosophical Transactions of the Royal Society of London Series B 263, 235-278.

Wilde, P., 1991. Oceanography in the Ordovician. In: Barnes, C.R. and Williams, S.H. (eds.), Advances in Ordovician Geology. Geological Survey of Canada Paper 90-9, 283-298.

Wilde, P. and Berry, W.B.N., 1984. Destabilisation of the oceanic density structure and its significance to marine "extinction" events. Palaeogeography, Palaeoclimatology, Palaeoecology 48, 143-162.

Wilde, P. and Berry, W.B.N., 1986. Role of oceanographic factors in the generation of global bio-events. In: Walliser, O.H. (ed.), Global Bio-Events. Lecture Notes in Earth Sciences 8, 75-91. Springer, Berlin Heidelberg New York.

Wilde, P., Hunt, M.Q., Berg, W.B. and Orth, C.J., 1986. Anoxic facies in the lower Paleozoic ocean. In: The Geol. Soc. of America, 97th annual meeting, Abstracts, p. 694.

Williams, S.H., 1982. The late Ordovician graptolite fauna of the Anceps Bands at Dob's Linn, southern Scotland. Geologica et Palaeontologica 16, 29-56.

Williams, S.H., 1987. Upper Ordovician graptolites from the *D. complanatus* Zone of the Moffat and Girvan districts and their significance for correlation. Scottish Journal of Geology 23, 65-92.

Williams, S.H., 1988. Dob's Linn - the Ordovician-Silurian boundary stratotype. In: Cocks, L.R.M. and Rickards, R.B. (eds.), A Global Analysis of the Ordovician-Silurian Boundary. Bulletin of the British Museum (Natural History) Geology 43, 17-30.

Williams, S.H. and Stevens, R.K., 1988. Early Ordovician (Arenig) graptolites of the Cow Head Group, western Newfoundland Palaeontographica Canadiana 5, 1-167.

Williams, S.H. and Stevens, R.K., 1991. Late Tremadoc graptolites from western Newfoundland. Palaeontology 34, 1-47.

Williams, S.H., Barnes, C.R., O'Brien, F.H.C and Boyce, W.D., 1994. A proposed global stratotype for the second series of the Ordovician System: Cow Head Peninsula, western Newfoundland. Bulletin of Canadian Petroleum Geology 42, 219-231.

Wright, J.W., Miller, J.F. and Holser, W.T., 1987. Conodont chemostratigraphy across the Cambrian-Ordovician boundary: western USA and southeast China. In: Austin, R.L. (ed.), Conodonts: Investigative Techniques and Applications. p. 256-283. Ellis Horwood Limited, Chichester.

Young, T.P., 1990. Ordovician sedimentary facies and faunas of southwest Europe: palaeogeographic and tectonic implications. In: McKerrow, W.S. and Scotese, C.R. (eds.), Palaeozoic palaeogeography and biogeography. Geological Society Memoir 12, 421-430.

Yu Jianhua, Fang Yiting and Liu Huabao, 1982. The Xinchangian graptolite-bearing strata of the early Ordovician in Wuning, Jiangxi. Journal of Nanjing University (Natural Sciences), Series B 2, 478-488 [in Chinese with English abstract].

Zhou, Z.Y. and Zhang, J.L., 1984. Uppermost Cambrian and Lowest Ordovician trilobites of North and Northeast China. In: Stratigraphy and Palaeontology of systemic boundaries in China. Cambrian-Ordovician Boundary 2. Nanjing Institute of Geology and Palaeontology. Anhui Science and Technology Publishing House, 63-194.

Manuscript received May 1993
Revision received March 1994

Authors' addresses:

Christopher R. Barnes, School of Earth and Ocean Sciences, University of Victoria, P.O. Box 1700, Victoria, V8W 2Y2, Canada

Richard A. Fortey, Department of Palaeontology, The Natural History Museum, Cromwell Road, London SW7 5BD, U.K.

S. Henry Williams, Department of Earth Sciences, Memorial University of Newfoundland, St. John's, Newfoundland, A1B 3X5, Canada

Silurian Bio-Events

Dimitri KALJO, Arthur J. BOUCOT, Richard M. CORFIELD, Alain LE HERISSE, Tatyana N. KOREN, Jiri KRIZ, Peep MÄNNIK, Tiiu MÄRSS, Viiu NESTOR, Robert H. SHAVER, Derek J. SIVETER, and Viive VIIRA

Abstract. In the history of Silurian biota and ecosystem as a whole no "big" catastrophes occurred like the one at the Ordovician-Silurian boundary. Yet it was not a quiet period either. There were established 15 more or less remarkable bio-events, among others the most severe extinction of conodonts and acritarchs in the very beginning of the Wenlock (Ireviken Event), the Great Crisis or *lundgreni* Event among graptolites in the Homerian and the middle Ludfordian Event comprising many lineages of vertebrates, graptolites, conodonts and corals. Most remarkable diversity rises of the Silurian biota were in the late Rhuddanian, in the Telychian and in the early Gorstian. Both extinctions and originations were in good correlation with the global sea-level curve, but the effect has to be interpreted as an integrated process partly triggered by development of early Silurian glaciation and climate in general.

Contents

Introduction	173
Environmental Background	175
Changes in Shelf and Basin Macrofaunas	178
Introductory Remarks	178
Corals	179
Brachiopods	181
Agnathans and Gnathostomes	182
Graptolites	184
Microfossil Evidence	187
Conodonts	187
Chitinozoans	191
Acritarchs	192
Regional Environmental Differences	196
Wenlock-Ludlow Depositional Events, Midwestern craton, U.S.A.	197
Carbon Isotope Changes during *ludensis-nilssoni* Time in Wales and the Welsh Borderland	202
Mid-Silurian Transition in the Pelagic Sequences of South Tien Shan	204
Major Facies and Faunal Changes in the Silurian of the Prague Basin, Bohemia	206
Conclusions on the Mid-Silurian Transition	214
Summary and Concluding Comments	217

1 Introduction

In the history of Silurian bio-event studies, the so-called Great Crisis by H. Jaeger (1991) or *lundgreni* Event by T. Koren (1987, 1991b), comprising drastic extinction among graptolites, was the first to attract the serious attention of researchers.

IGCP Project 216 initiated new interest in the topic and lately several bio-events of different ranks have been described. For example Boucot (1990) has analyzed diversity changes of brachiopods and listed a dispersal event in the late Aeronian, some minor extinctions and radiations in the early and late

Wenlock, in the late Ludlow and at the Silurian-Devonian Boundary. Jeppsson (1990) has suggested a climatic and oceanic model for the interpretation of events on the basis of very detailed data on the conodont distribution in the Silurian of Gotland, emphasizing the importance of the Ireviken Event at the very beginning of the Wenlock. Kaljo and Märss (1991) have discussed changes in graptolite, coral and vertebrate faunas concurrent to eustatic movements of sea level. Schönlaub (1986) has noted, besides the O/S Event, also a *Cardiola* Event - a short-term regression in the late Ludlow which caused an income of new taxa among conodonts, graptolites and bivalves.

All the above-mentioned bio-events and the others described below are listed in Fig. 1, giving also some idea about the relative importance of the events. The hierarchy used is certainly arbitrary. Considering that a first order event should be comparable with the O/S or K/T boundary events, we do not use this rank in the Silurian.

A second order event, according to our understanding, shows concurrent involvement of several groups of organisms in the extinction and radiation processes, with mass extinction in at least one group.

A third order event is characterized by prevailing extinction in several groups, mass extinction in one group or final extinction of a family level group.

A fourth order event corresponds to a profound diversity change (extinction) or a clear innovation (at least on the genus level) in one or two groups.

A fifth order event is based on distinct diversity changes without any high level innovation in one or two groups.

In the following paragraphs we attempt to show the above bio-events against the background of changing environment, demonstrating the course of diversity changes going on in several groups of organisms inhabiting different life zones of a marine ecosystem. From the shallow shelf area we have chosen corals and brachiopods (partly distributed also deeper), from deeper shelf sea conodonts (partly also from shallow sea), chitinozoans and acritarchs, and from the basin facies graptolites.

A special paragraph is devoted to the comparison of the upper Wenlock-lower Ludlow sequences comprising the mid-Silurian Great Crisis and other events, in different geological and environmental settings (Fig. 2). The idea is to compare diversity changes in different groups and regions and to interpret their responses to environmental or biotic effects in the sense of the pattern and causes of bio-events. Figure 2 shows the location of sections (areas) discussed below for bio- or environmental events.

The present chapter is a team work, but every author has contributed in his own way: A. Boucot provided the brachiopod chapter, R. Corfield and D. Siveter discussed carbon isotope data, A. Le Herisse acritarchs, T. Koren graptolites and pelagic sequences of South Tien Shan, J. Kriz changes in the Bohemian Silurian, P. Männik and V. Viira conodonts, T. Märss vertebrates, V. Nestor chitinozoans, R. Shaver summarized the geological history of the Mid-Western craton. All other parts were written by D. Kaljo, who was also responsible for planning and organizing the work.

The authors tried to keep the number of references low, quoting mainly the latest papers containing longer lists of useful publications.

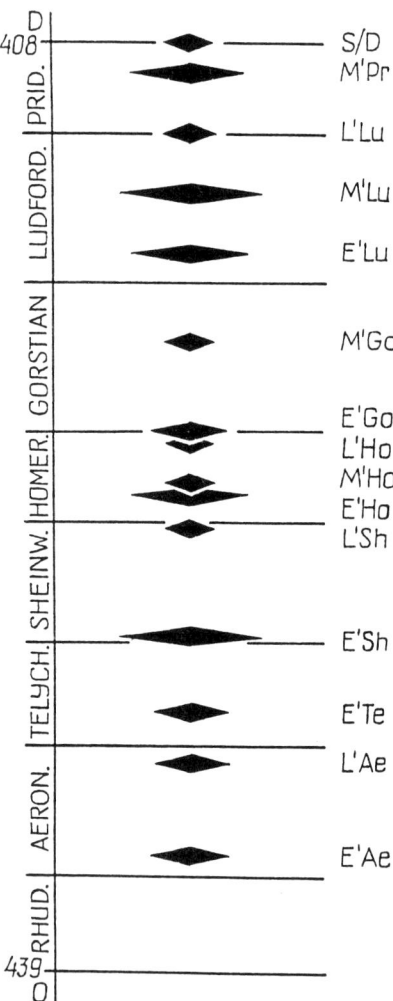

Fig. 1: Silurian bio-events. The size of the rhombs shows the relative importance of the event (see text)

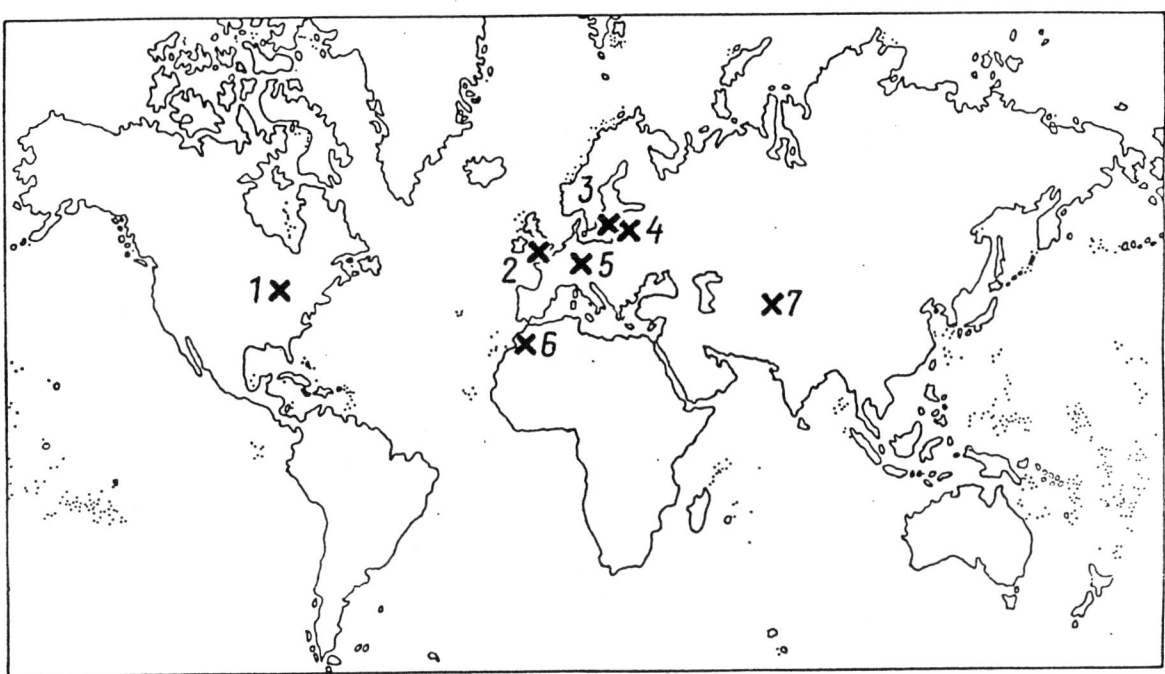

Fig. 2. Sites discussed for Silurian bio-events: 1 Midwestern craton; 2 Wales and Welsh Borderland; 3 Gotland; 4 East Baltic; 5 Thuringia and Bohemia; 6 West Morocco; 7 Alai Range, South Tien Shan

Environmental Background

Art Boucot (1991) has called the Silurian a unique period. Obviously, his conclusion is true, especially if one agrees that every period has its own specific nature. Both Boucot (1991) and Holland (1991) have listed several characteristic features of the Silurian, such as the content of biota, moderate provincialism in biogeography, dominating mild climate and the position of main landmasses, short duration of the period and detailed stratigraphy of the corresponding rocks etc. All these and other qualities, forming the environmental background for different bio-events, are also of notable importance for the following analysis.

During the Silurian many large continental masses and the connected sedimentary basins were located in the equatorial climatic belt and partly in the high latitude areas of the Southern Hemisphere (Fig. 3). Favourable climatic conditions combined with a generally low relief of the continents and consequently limited clastic influx resulted in predominantly carbonate sedimentation in the shallow shelf seas. Figure 3 shows clearly the general pattern of the distribution of the areas with carbonate and terrigenous types of sedimentation. There are some exceptions, but more often carbonate rocks were formed at low latitudes.

For our analysis it is important to emphasize that the Silurian cratonic, especially epicratonic seas (e.g. North American and Siberian), were usually shallow with a flat sea bottom, sloping only very gently towards the shelf margin. Consequently, wide tidal flat/lagoonal, shoal and shallow shelf facies areas with their characteristic communities of biota were also typical (Kaljo, Nestor and Einasto, 1991). The shallowness of the sea made it possible for relatively uniform tropical marine conditions valid over a wide belt to become easily affected by different, sometimes even low grade local factors. Nevertheless, the biogeographical provincialism is relatively moderate: warm water biota represents the Northern Realm subdivided into the North Atlantic and Uralian-Cordilleran regions; a few provinces and cool water areas of the Southern Hemisphere have been included into the Malvinokaffric Realm (Boucot, 1991). The latter is far less studied, but there are some indications of temperate climate faunas in East Siberia and in the Russian Far East (Kaljo, 1981).

Figure 3 illustrates the situation in the early Silurian characterized by a changeable (with warming and cooling episodes) climate after an end Ordovician glaciation period. The extent of seas was increasing until a turning point in the early Wenlock. Later times, espe-

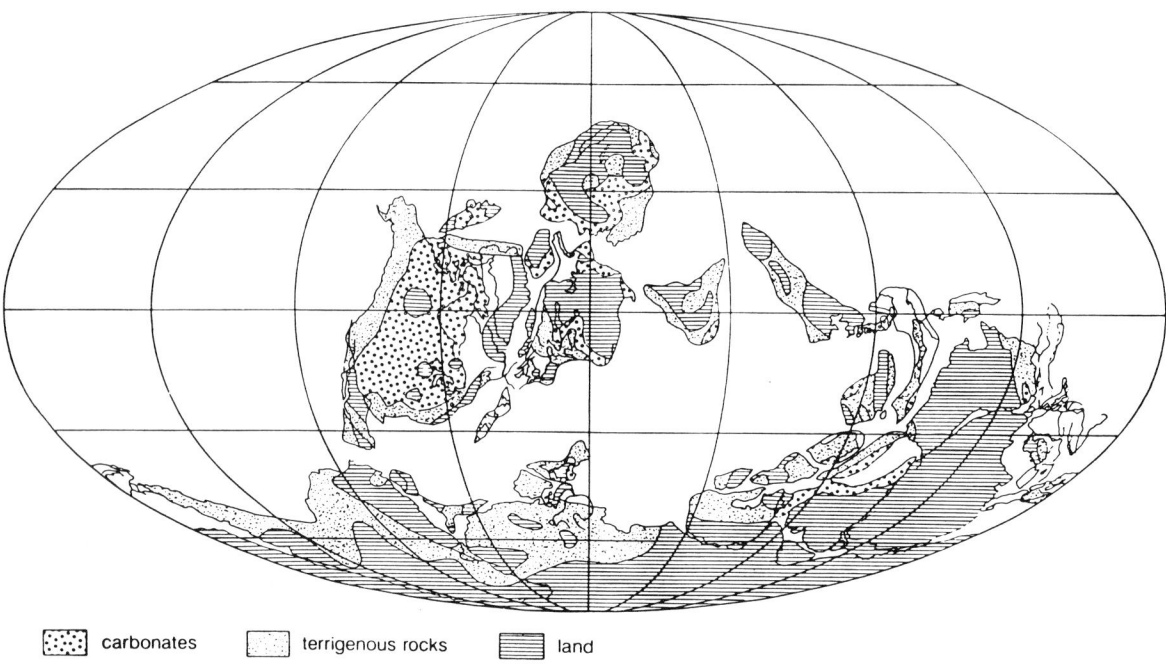

Fig. 3. Distribution of carbonate and terrigenous rocks and land areas in the early Silurian (after Kaljo, Nestor and Einasto, 1991)

cially the late Silurian, were dominated by the reduction of the seas. Also the temperature was rising, the weather became more hot and in many areas also drier, as evidenced by evaporite sedimentation (North American Michigan, Siberian Tunguska basin etc.; Shaver, 1992; Kaljo, Nestor and Einasto, 1991).

Wilde et al. (1991) have cited interesting data by Budyko et al. (1987) revealing several important parameters of the Silurian environment. Some of these are presented here together with the statements of the first authors. The mean global Silurian temperature was 20°C (present-day it is about 15°C), which is less than that established for the Ordovician and Devonian (Fig. 4).

The same pattern is also valid for atmospheric carbon dioxide - the Silurian value of about 900 ppm or about 300% of the modern one is less than the top-Ordovician 1200 ppm.

Atmospheric oxygen declines from about 65% PAL (Present Atmospheric Level) to about 35% PAL in the late Silurian. Other values have also been suggested but, in general, all authors agree that oxygen declined in the atmosphere until the Devonian when plants became more abundant on the continents and generated a significant amount of oxygen (Wilde et al., 1991).

The oxygen content of the warm surface waters in the tropics was also declining during the Silurian from about 3 ml/l (65% PAL) to about 2 ml/l (35% PAL). At a level of 50% PAL the oxygen content would reach zero at a depth of about 100 m. This means that the Silurian seas deeper than 100 m were, as a rule, anoxic. The expansion of the anoxic water zone during the Silurian was also favoured by the reduction of the area of the well oxygenated shelf seas (from about 40% to 30%) in the course of the regression (Wilde et al., 1991).

According to Wilde et al. (1991), the salinity of the Silurian ocean was similar to that of today.

The inventory data of volcanogenic rocks, carbonate carbon and organic carbon given by Wilde et al. (1991), based on Budyko et al. (1987), also show a declining tendency during the Silurian, particularly the first two of these in the late Silurian (Fig. 5). These authors see the reason in the shoaling and reduction of the shelf seas. Depletion of the organic carbon in the late Silurian was decelerating in comparison with earlier times, when increased anoxia in the ocean was having stronger inhibiting effect on primary production. It is important to remember the conclusion of Wilde et al. (1991) that due to a low oxygen content in the Silurian ocean far less organic matter is required

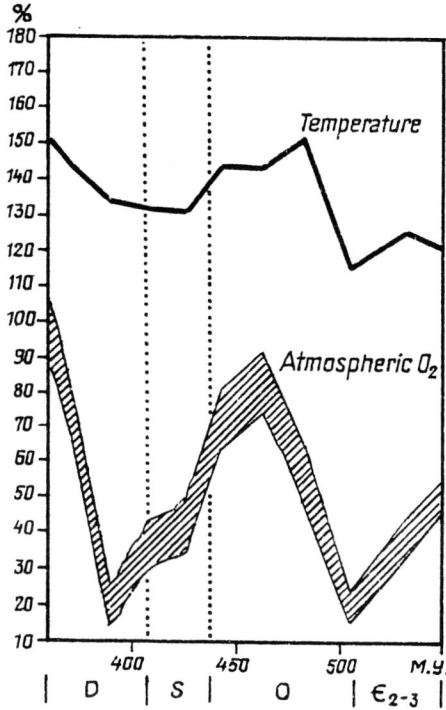

Fig. 4. Early-middle Palaeozoic atmosphere (temperature and oxygen in % to PAL) after Wilde et al. (1991)

to cause anoxic conditions in the sea. The same processes (oxygen content, anoxia, reduction of the shelf area) caused the general decrease in the primary production and consequently depletion of the $\partial^{13}C$ content.

The changes of the parameters just discussed were very general, showing only the main tendencies of the environmental evolution which cannot be applied for interpretation of detailed biodiversity curves. More promising in this respect seem to be sea-level curves demonstrating the Silurian eustasy pattern practically on an intercontinental scale (Johnson, Kaljo and Rong, 1991; McKerrow, 1979). Bathymetric changes are mainly based on stratigraphical replacement patterns of brachiopod, algal or coral-stromatoporoid dominated communities inhabiting different depth zones of the shelf seas, and the deeper water high diversity graptolite assemblages (Boucot, 1975; Johnson, Baarli et al., 1991; etc.). Several authors have also exploited lithofacies models and different lithological or facies characteristics reflecting changes in water depth through time (Nestor and Einasto, 1977; Tesakov et al., 1986; Droste and Shaver, 1987; etc.).

A generalized global sea-level curve, which is reproduced here with some additions (Fig. 6), was recently suggested by Johnson, Kaljo and Rong (1991). The curve summarizes data from 13 areas, therefore showing only the most important changes: the first sea-level rise during the earliest Silurian time started after the main glacioeustatic drop during the latest Ordovician, followed by four sea-level high stands and subsequent lowerings in the Llandovery with a maximum level at the end of it. Except for the very beginning of the Wenlock, the Sheinwoodian was in general characterized by sea-level high stand; in the Homerian the end-Silurian regression started interrupted by two deepening periods in the early and late Ludlow.

As shown by many authors (Johnson and McKerrow, 1991; Johnson, Baarli et al., 1991; etc.), curves from separate areas or even separate sections within the same area are much more individualized and therefore it is better to interpret local data sets from these. On the other hand, these might also be affected by some very local factors not having any remarkable influence on the composition of faunal-floral assemblages of the area. A good example is the cyclicity of sedimentary sequences and gaps appearing differently in separate parts of a palaeobasin. As a rule, gaps and

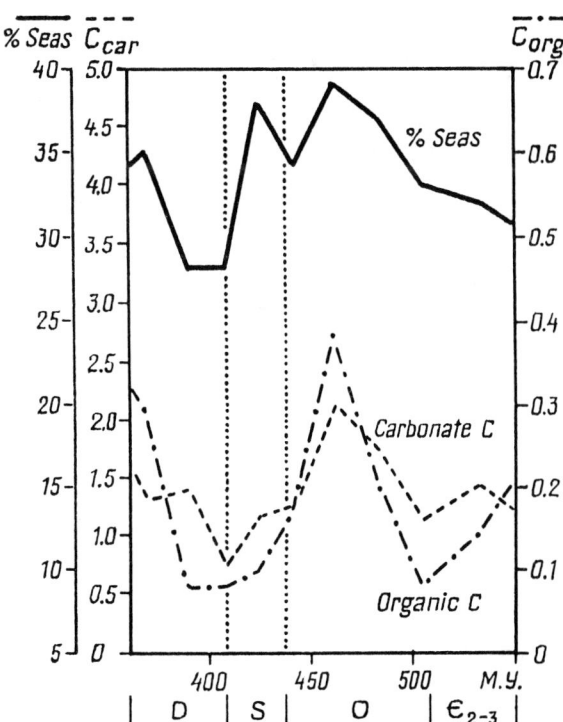

Fig. 5. Early-middle Palaeozoic inventory of carbonate and organic carbon (in 1021 grams) plotted against the areal percentage of seas overlying the continents (after Wilde et al., 1991)

clear cyclicity are well developed in the shallow shelf areas, but the most remarkable of them, however, reach the deep shelf and only a few the basinal area. Accordingly, the stratigraphical sequence of communities in the shallow sea is much more affected by cyclic environmental agents than the deep sea ones.

The geological time scale used in Fig. 6 and below for the compilation of some figures and calculation of total and per taxon rates is mostly based on the information published by Harland et al. (1989). Only the Ludlow was divided differently into two nearly equal parts (Gorstian and Ludfordian), in this way better fitting into the common understanding of the duration of these ages. The duration of standard graptolite zones was calculated proceeding from the above dates and taking into account also their relative extent as accepted by a special working group of the IUGS Subcommission on the Silurian System.

Changes in Shelf and Basin Macrofaunas

Introductory Remarks

Below we will study diversity changes in several groups of macrofossils, in the next section those concerning microfossils, with a goal of defining more or less widespread bio-events occurring in the course of the evolution of these groups and the whole ecosystems. Usually it is quite easy to establish a bio-event in a section or region, but finding out the reasons for it and tracing the event over the globe are a much more complicated task. A step in this direction is the comparison of changes going on during the Silurian in groups inhabiting different life zones. We proceed from the assumption that by the concurrent occurrence of bio-events in different facies belts they might have been caused rather by more general than local factors.

Corals are being analysed as belonging to the sessile benthos inhabiting the shallow shelf sea. They are most common in the high energy well aerated zone of the sea, called also the shoal facies belt (Bassett et al., 1989; Kaljo, Nestor and Einasto, 1991), but they occur also seaward from shoals in a lower energy zone corresponding to the inner belt of carbonate platform or the open shelf (op. cit.).

Brachiopods belong to the mobile or sessile benthos and were distributed over the whole shelf area forming specific communities adapted to different ecological conditions. The distribution patterns of brachiopod dominated communities are well studied (Boucot, 1975, 1992).

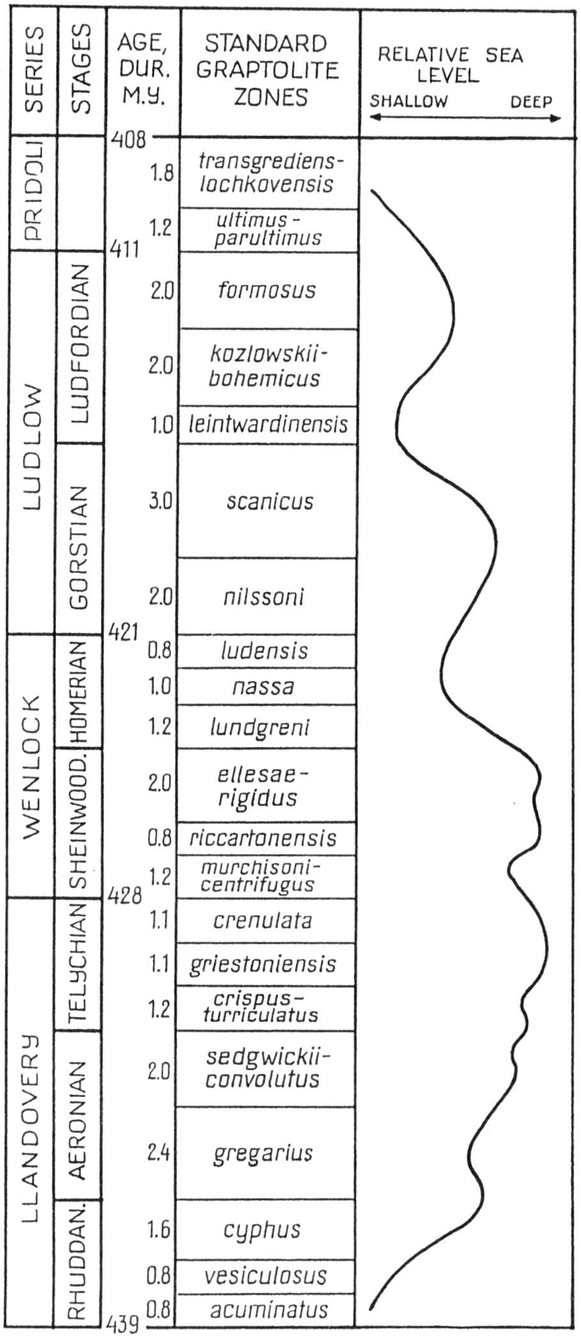

SERIES	STAGES	AGE, DUR. M.y.	STANDARD GRAPTOLITE ZONES	RELATIVE SEA LEVEL SHALLOW ←→ DEEP
PRIDOLI		408 1.8	transgrediens-lochkovensis	
PRIDOLI		1.2 411	ultimus-parultimus	
LUDLOW	LUDFORDIAN	2.0	formosus	
LUDLOW	LUDFORDIAN	2.0	kozlowskii-bohemicus	
LUDLOW	LUDFORDIAN	1.0	leintwardinensis	
LUDLOW	GORSTIAN	3.0	scanicus	
LUDLOW	GORSTIAN	2.0 421	nilssoni	
WENLOCK	HOMERIAN	0.8	ludensis	
WENLOCK	HOMERIAN	1.0	nassa	
WENLOCK	HOMERIAN	1.2	lundgreni	
WENLOCK	SHEINWOOD.	2.0	ellesae-rigidus	
WENLOCK	SHEINWOOD.	0.8	riccartonensis	
WENLOCK	SHEINWOOD.	1.2 428	murchisoni-centrifugus	
LLANDOVERY	TELYCHIAN	1.1	crenulata	
LLANDOVERY	TELYCHIAN	1.1	griestoniensis	
LLANDOVERY	TELYCHIAN	1.2	crispus-turriculatus	
LLANDOVERY	AERONIAN	2.0	sedgwickii-convolutus	
LLANDOVERY	AERONIAN	2.4	gregarius	
LLANDOVERY	RHUDDAN.	1.6	cyphus	
LLANDOVERY	RHUDDAN.	0.8	vesiculosus	
LLANDOVERY	RHUDDAN.	0.8 439	acuminatus	

Fig. 6. Silurian global sea-level curve (after Johnson, Kaljo and Rong, 1991) and stratigraphical classification with chronometric data (see text for explanation)

Silurian agnathans and gnathostomes belong to the nekton and nektobenthos dwelling in different zones of the shelf seas. It is important to emphasize that until the end of the Ludlow in the shallow shelf region, especially in the near-shore facies belts, thelodonts and osteostracans dominate. Beginning with the Pridoli, acanthodians became most numerous (Märss, 1986). The vertebrates discussed above were naturally animals of a "macro"size, but their remains (scales, tesserae, etc.) used for research are usually "microfossils".

Graptolites represent the planktic part of a marine ecosystem. Their habitat at some depth within the water column of a sea or ocean seems not to depend on the sea bottom but on some other factors of the environment (depth, chemistry, etc., Berry et al., 1987). In any case, graptolites are mostly connected with the so-called basin facies occurring only occasionally and in a very limited diversity in the carbonate platform facies.

The microfossils to be analysed include conodonts, chitinozoans and acritarchs, usually widely distributed over the whole shelf area. Acritarchs, less often chitinozoans and rarely conodonts occur also in the basin facies. In the Baltic, conodonts are the most diverse and abundant in the shallow shelf (carbonate platform) facies, in some other basins they are abundant also in pelagic sediments (Sweet, 1988). Chitinozoans, on the contrary, abound in the deeper shelf, especially in the transition to basin facies (Kaljo et al., 1986). Ecological control on acritarch distribution has not been discussed yet.

In general, considering the diversity and abundance patterns in the East Baltic Silurian given in Kaljo et al. (1986), we can conclude that the planktic groups (graptolites, chitinozoans, probably also acritarchs) prefer the deeper sea, nekton and nektobenthos (gnathostomes, agnathans, conodonts) favour the neritic part of the sea and benthos, especially the sessile forms, has a preference for the shallow sea.

Analyzing diversity changes of fossil groups in a section with the aim of finding out the bio-event levels, we have to consider also the above ecological preference pattern.

Corals

The Silurian coral fauna is rather rich and diverse, the total number of genera being very high (75 to 130 per series). The species level diversity in a number of genera is also high but unstable, depending on the author's views on taxonomy. On the other hand, species of corals such as sessile benthos are often strictly adapted to local environmental conditions and changes in the latter often caused nearly a total relay of the community. Due to this we will analyze only the generic level data, allowing us to understand more general tendencies in the coral fauna evolution.

The paper by Kaljo and Klaamann (1973) was used as a data base. The summary is a little out of date but the general picture is surely adequate enough.

Figure 7 shows the change of the total number of rugose, tabulate and heliolitid coral genera per series and tendencies in the coral diversity dynamics.

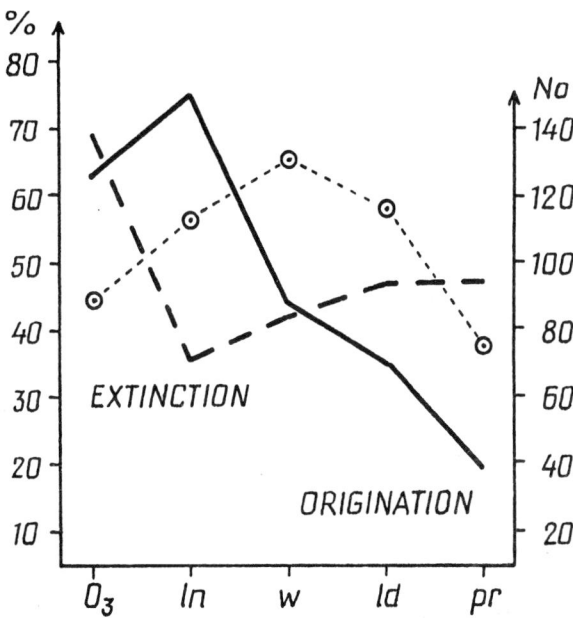

Fig. 7. Origination and extinction rates and the number (circle with a point) of coral genera per unit during the Silurian. Origination: percentage of appearing genera per unit; extinction - the same for disappearing genera. Units: O_3 late Ordovician; ln Llandovery; w Wenlock; ld Ludlow; pr Pridoli

In general, the composition of the fauna is changing step by step. The starting level in the late Ordovician was the time of a great crisis when 62 genera became extinct. A new Silurian coral fauna originated in the Llandovery when the innovation and radiation rates were the highest (84 new genera appeared). Later the curve falls quickly. In the Wenlock the extinction and origination rates were more or less balanced and as a result in the Wenlock the coral fauna reached its maximum diversity. Later, extinctions prevailed clearly over appearances and this caused quick decline of the coral fauna in the latest Silurian.

Figure 8 provides additional illustration to these processes on the basis of tabulate and heliolitid coral distribution in the Baltic Silurian according to Klaamann's (1970, 1986) data. The stratigraphic scale used is more detailed (stage level) but the general

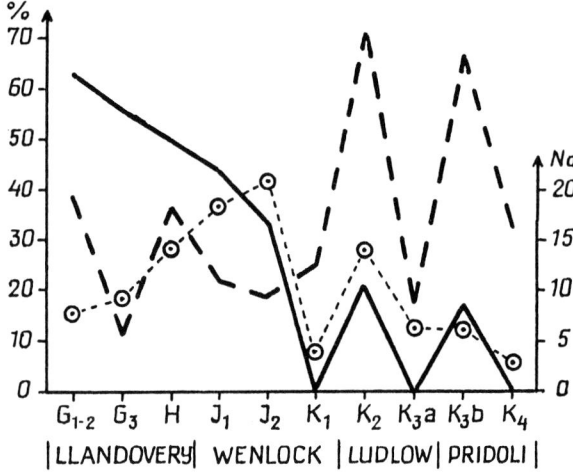

Fig. 8. Origination (solid line) and extinction (dashed line) rates and number (circle with a point) of tabulate coral species per unit in the Silurian of Estonia. Stage names: G_{1-2} Juuru; G_3 Raikkula; H Adavere; J_1 Jaani; J_2 Jaagarahu; K_1 Rootsikula; K_2 Paadla; K_3a Kuressaare; K_3b Kaugatuma; K_4 Ohesaare

tendencies are exactly the same - falling of the origination curve towards the younger strata and the opposite course of the extinction curve. The most important events among Baltic tabulates were the top-Paadla (approximately *bohemicus-kozlowskii* Zone) and top-Kaugatuma (approximately in the middle of the *lochkovensis-transgrediens* Zone) extinctions.

Continuing the above discussion about the local environmental influence on the abundance and diversity of corals, it should be emphasized that the big difference between the numbers of genera occurring in J_2 and K_1 (Fig. 8) did not cause any rise in extinction rate, because most of the genera involved occur again in K_2 and only then disappear finally. The situation seems to be the same in K_3a. However, here it is not so clearly expressed and the real disappearance time of the genera in question can be established only with difficulty.

Figure 9 shows the changing role of the endemic and widely distributed coral genera during the Silurian. There are some aspects in good accordance with the above general conclusions: (1) Rise of endemic genera (curve 1) in the Llandovery might be explained by quick innovations. (2) Wenlock corals were the most cosmopolitan ones. (3) Decline of the widely distributed genera (curves 4, 5, 6) in the late Silurian might be a result of the general decline of the group. (4) A considerable rise of endemic genera together with this

decline. The most relevant reason for this seems to be high specialization under conditions of a low level of innovation and the decline of cosmopolitan genera. On the other hand, geographical limitations due to the general late Silurian marine regression might have had some influence.

A series level analysis could show only general tendencies, not events in the strict sense. Having in mind also Baltic data (Fig. 8), we can note the following events among corals: an innovation and radiation event in the early Llandovery and a more energetic extinction event (or series of extinctions) in the late Ludlow (beginning with the *leintwardinensis* Zone). During the rest of the Silurian the diversity dynamics is more or less gradual or stepwise.

As to the causes of the process, some conclusions might be underlined. Strong diversification of the shallow shelf coral faunas during the early Silurian coincides with warming of the climate and transgression of the ocean. Late Wenlock regression seems not to affect the corals seriously but the late Silurian decline of the coral fauna coincides with the aridization period of the climate and with slow regression of the shelf seas. Together with this, the type of carbonate sedimentation was changing too, the shallow shelf became muddier, in some parts more salty. These may have been reasons for a general decline of the Silurian corals, they having been inhabitants of warm, "normal", clean waters.

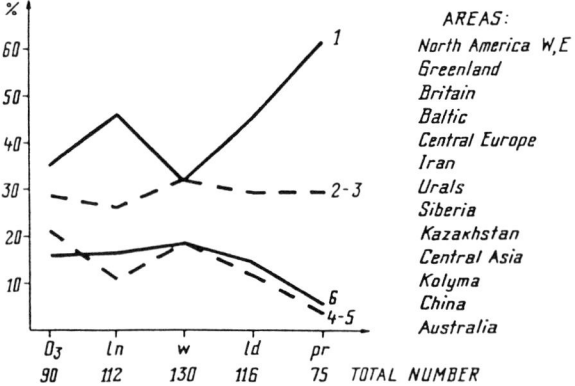

Fig. 9. Rate of endemic and cosmopolitan genera among Silurian corals. 1 endemics, distributed in one area; 2-3 genera of limited distribution (in 2-3 listed areas); 4-5 genera of wide distribution (in 4-4 listed areas); 6 cosmopolitan genera. Abbreviations of the unit names see Fig. 7

Brachiopods

Boucot (1992) discussed benthic brachiopod community changes reflecting Silurian bio-events in different habitats. A summary of this information will be given below.

During the entire Silurian there is a marked, high taxonomic contrast in diversity between the carbonate platform, level bottom communities (high diversity) and the siliciclastic, off-platform communities (low diversity). The carbonate platform communities tend to be rich in varied rugosans, tabulates, stromatoporoids, sponges, calcareous algae, bryozoans, nautiloids and pelmatozoans, whereas the reverse is true for the siliciclastic off-platform communities where brachiopods, trilobites, plus less abundant bivalves and gastropods are the dominant taxa in terms of both specimens and species - with most of these same taxa also commonly occurring on the carbonate platforms. The siliciclastic off-platform also lacks the reef and mound complexes that add so much to overall carbonate platform diversity.

Following the massive end-Ordovician global extinction event there is a relatively lengthy low diversity interval in the Rhuddanian and much of the Aeronian, up to the C_2-C_3, late Aeronian interval.

In the BA (= Benthic Assemblage, Boucot, 1975) 1 position this interval is characterized by the typical low diversity, high dominance, poorly studied orbiculoid-lingulid community complex, and by the equally poorly known rhynchonellid community complex. Both of these units persist through the entire period.

The interval BA 2 is characterized by several low diversity communities, including *Mendacella, Cryptothyrella, Dalmanella* Communities and BA 3 by a number of moderate diversity *Janius* Community Group units, in which the lack of the C_2-C_3 dispersalists such as *Atrypa "reticularis", Howelella, Eospirifer* and *Cyrtis* is notable. In the rough waters of the BA 3 there are also characteristic low diversity virgianid Community Group communities.

In BA 4 the low diversity, high dominance *Stricklandia* Community Group occurs and only BA 4-5 belt yields higher diversity, lower dominance representatives of the *Dicoelosia-Skenidioides* Community Group. These units lost much of their diversity during the Ashgill extinction interval. The still higher diversity so characteristic of this community group will not be regained until the C_2-C_3 dispersal event sees the addition of many taxa from the Uralian-Cordilleran Region into the North Atlantic Region.

The late Aeronian C_2-C_3 bio-event is the major, within Silurian benthic fauna, item, in many places corresponding to the beginning of the "late Llandovery" transgression. It does not reflect a great adaptive radiation, but rather, it appears to reflect a massive dispersal event of many taxa previously endemic within the Uralian-Cordilleran Region into the North Atlantic Region (Wang Yu et al., 1984).

The set of communities, those spanning up the late Aeronian to early Wenlock interval, mark the beginnings of what is commonly thought of as the Silurian benthic fauna, because of their globally far greater abundance and geographic distribution which correlates positively with overall Silurian transgression from a Rhuddanian low point.

During this second interval the virgianid communities are replaced by large, smooth pentamerid Community Group units. In BA 2 position a prominent feature is flourishing of *Eocoelia* communities.

The BA 3 *Janius* Community Group, the most areally widespread set of Silurian communities, now attains its maximum diversity with the addition of the immigrants from the Uralian-Cordilleran Region into the North Atlantic Region. This high diversity situation persists through the end of the period.

The BA 4-5 *Dicoelosia-Skenidioides* Community Group high diversity, low dominance communities show a marked diversity increase during C_2-C_3 that persists through the end of the period, with many of the additional taxa being immigrants from the Uralian-Cordilleran Region into the North Atlantic Region.

The early Wenlock, post-*amorphognathoides* Zone extinction is minor in terms of taxa, but notable in terms of shelly biomass and abundance relations for several previously important groups. The stricklandid Community Group now disappears except for Central North America where the *Microcardinalia-Plicostricklandia* lineage persists into the late Wenlock. The *Eocoelia* communities disappear except for a limited area of the European Subprovince where *E. angelini* persists a little longer in dominant condition.

By and large this interval merely sees, particularly in the high diversity units of the *Janius* and *Dicoelosia-Skenidioides* Community Groups, a continuous sequence of evolving communities featuring phyletic changes within varied generic lineages.

The next event is the late Wenlock adaptive radiation of the pentamerinids and subrianinids, especially the ribbed forms, all of them being low diversity, high dominance BA 3, rough water items. This late Wenlock interval also sees the adaptive radiation responsible for the Silurian reefs (not just mud mound type bioherms, which are present throughout the period) with their many unique taxa. These reef complexes persist through the Pridoli, but their community ecology is poorly known.

During the late Wenlock-Ludlow interval the high diversity community groups undergo little prominent alteration, as is also true through the Pridoli.

Beginning in this interval and continuing through the Pridoli, within the European Province of the North Atlantic Region, there is an extensive development of *Salopina* and *Protochonetes* dominated BA 2 communities, as well as the continuation, of course, of the rhynchonellid, nuculoid bivalve, linguloid-orbiculoid dominated BA 1 community types.

The most marked event at the end of the Ludlow is what Talent (oral comm.) has termed the Pentamerid Event, since it sees the extinction of almost all of the *Pentameridae* and the *Subrianidae*, together with the many communities they dominated.

Following the end-Ludlow Event in the Pridoli, the gypidulid-dominated communities take over the same BA 3, rough water role in a gypidulid Community Group that persists through the Frasnian.

Pridoli communities have much in common with those of both the Ludlow and the Gedinnian-Lochkovian, with the qualification that Pridoli faunas are most easily recognized by the absence of pentamerinid communities, presence of gypidulinid communities, and *Janius* Community Group faunas lacking key Devonian items such as *Cyrtina*, terebratuloids (*Podolella, Mutationella, Nanothyris*), and containing such forms as halysitid corals and *Merista*.

The end-Pridoli, a minor extinction at the Silurian-Devonian Boundary and subsequent adaptive radiation, saw a few lineages in the high diversity community groups eliminated, but by and large it was a minor, although recognizable event both taxonomically and ecologically.

Agnathans and Gnathostomes

The Agnatha and Gnathostomata have their first big radiation period in the Silurian, especially the Agnatha with 4 Llandovery, 21 Wenlock, 33 Ludlow and Pridoli genera occurring. The development of the Gnathostomata was slower but accelerated towards the Devonian, thus documenting another big vertebrate radiation. Comparing the origination and extinction rates shown for vertebrates (Fig. 10) and corals (Fig. 7), we can see a very similar character of curves with rapidly falling origination and slightly rising extinction rates. This coincidence is surprising because of great differences between these groups, bearing in mind their stage of evolution, mode of life and habitats. It seems logical to think that the reason is not internal but external and may be common for both groups.

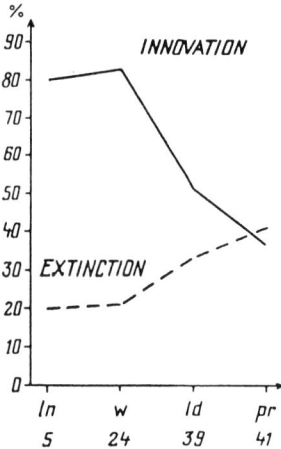

Fig. 10. Origination and extinction rates and number of vertebrate genera per unit. Unit names see Fig. 7

Recently Blieck and Janvier (1991), Kaljo and Märss (1991) and Märss (1992), discussing vertebrate history in the Silurian, noted several periods of energetic innovation and two clear extinction events, the first extinction in the late Wenlock and the other in the middle of the Ludlow. A gap in vertebrate history, perhaps due to late Ordovician glaciation (Blieck and Janvier, 1991), was succeeded by a series of originations during the Llandovery and early Wenlock. More significant among these are the appearances of vertebrates in the Llandovery of Scotland where the earliest anaspids have been identified from the Priesthill Group (upper Telychian). A thelodont *Loganellia* with a micromeric exoskeleton appears in the Llandovery (approximately on the level of the upper Aeronian) and spreads over a very extensive area: North Greenland, British Isles, Norway, Baltic, Timan-Pechora Region, Severnaya Zemlya, Tuva and locally possibly also in North America (Märss, 1986, 1989; Blieck and Janvier, 1991).

The most remarkable innovation event was the "burst" of Osteostraci in the latest Wenlock (*nassa* and *ludensis* zones) with a very quick radiation of several species of *Tremataspis, Saaremaaspis, Oeselaspis, Witaaspis* and *Thyestes* (Fig. 11), giving rise to the high diversity peak of agnathans.

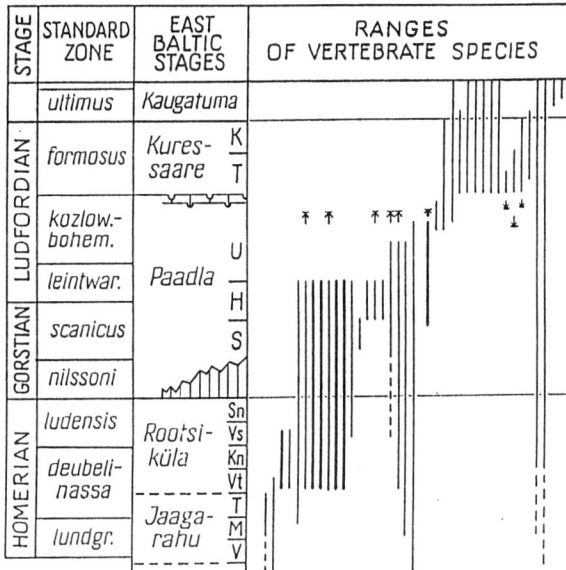

Fig. 11. Distribution of vertebrate species in the Homerian and Ludlow of Estonia. Solid line: certain occurrence; dashed line: doubtful identification; bold line: the last species and disappearing genus (from left to right): *Saaremaaspis* disappearing in Vs (= Vesiku Beds), *Oeselaspis, Vitaaspis, Birkeniida* gen. C and D disappearing in H (= Himmiste Beds), *Phlebolepis* disappearing in U (= Uduvere Beds). Small arrows show extinction or appearance levels of some species in other regions according to the current correlation

Many Wenlock vertebrates are more or less long-ranging and therefore the extinction rate is relatively low, especially in the *lundgreni* Zone (early late Wenlock), but it is quite remarkable at the species level at the very end of the Wenlock. The second diversity maximum in the mid-Ludlow occurred mostly at the species level with a few new genera of Osteostraci (*Procephalaspis, Dartmuthia*), and acanthodians (*Nostolepis* and *Gomphonchus*) becoming numerous. However, most of the vertebrates occurring in the Himmiste Beds of the Paadla Stage, disappear before the end of the same unit or in the lower part of the next one (Uduvere Beds of the Paadla Stage) (Fig. 11).

Thus, approximately at the level of the *leintwardinensis* Zone (mid-Ludlow) both innovation and subsequent extinction events occurred among vertebrates, but the latter seems to have been regionally restricted (Estonia), affecting some osteostracans, while several thelodonts became extinct only during the next *bohemicus-kozlowskii* Zone (Gotland, Central Urals; Märss, 1992).

A new profound innovation bringing about the long-ranging *Thelodus parvidens* Agassiz, the earliest cyathaspidid heterostracan *Archegonaspis* and the first osteichthyan *Andreolepis hedei* Gross started at the same level correlated with the *bohemicus-kozlowskii* Zone (partly it might be also the latest *leintwardinensis* Zone). In the whole Baltic area it corresponds to the topmost Paadla Stage of Estonia and Latvia, Hemse Beds e and Burgsvik Sandstone of Gotland, Long Quarry Beds of South Wales, the upper part of the Kuba Beds of the Central Urals and of the Gerdju Stage of the Timan-Pechora Region. It is the first step of the late Ludlow innovation called the *hedei* Event and followed by the next step (might be called the *sculptilis* Event) in the highest *bohemicus-kozlowskii* Zone characterized by the incoming of new anaspids of the order Birkeniida and acanthodians (genus *Poracanthodes*) in the Gotland and Central Ural sections. In Estonian sections the rich *Thelodus sculptilis* assemblage appeared in the lowermost Kuressaare Stage correlated with the *formosus* Zone (Fig. 11).

In the Pridoli (the Kaugatuma and Ohesaare stages of Estonia) three events might be distinguished which can also be traced outside the Palaeobaltic basin: (1) The practical predominance of acanthodians from above the Kuressaare-Kaugatuma boundary up to the top of the latter (acanthodians prevailed over other groups until the end of the Silurian). (2) The beginning of a new stage in the development of heterostracans and osteoichthyans (*Tolypelepis, Strosipherus, Lophosteus*) in late Kaugatuma times. (3) A pre-Devonian thelodont event in the latest Silurian, preceded by the extinction of the *Thelodus* species and the appearance of thelodonts transitional to the early Devonian (*Loganellia kummerowi, Katoporus timanicus*).

In summary, the most significant events in the evolution of the Silurian vertebrates studied were (see also Blieck and Janvier, 1991) as follows: appearance of thelodonts in the Llandovery; incoming of cyathaspidid heterostracans in the Ludlow (*leintwardinensis* Zone), reappearance of eriptychiid ones in the Pridoli after an interregnum beginning with the late Ordovician; emergence of anaspids in the Llandovery, of osteostracans in the Wenlock *(nassa* Zone) and their abundance in the late Wenlock and Ludlow; origination of acanthodians in the Llandovery, increase in their diversity during the Ludlow and prevalence in the Pridoli; entrance of osteichthyans in the Ludlow (*leintwardinensis* Zone) and their taxonomic diversification in the Pridoli.

Graptolites

The dynamics of morphological changes in the history of Silurian graptolites shows repeated successions of extinction, interval and radiation, followed by a period of normal evolution (nomismogenesis sensu Walliser, 1986).

Close to the Rawthean/Hirnantian boundary, the late Ashgill graptolites went through a major evolutionary crisis. All morphologically specialized taxa (families, genera, many species) became extinct within the short *pacificus* interval (Melchin and Mitchell, 1991; Koren, 1991a). It is supposed that the *pacificus* mass extinction event resulted from an oceanic cooling, oxygenation of water masses and probably from the contraction of the tropical plankton belts. Only some ecological generalists, such as *Normalograptus* and *Glyptograptus*, represented by a few species of very simple morphology, survived this event. Such taxa continued through the *extraordinarius* Zone interval and formed the ancestral stock for the quickly originating (beginning in the *persculptus* Zone) new morphological types of the Rhuddanian and Aeronian biserial graptoloids. The appearance of the monograptid colony (in the *persculptus* Zone) is the most striking morphological novelty ("the uniserial event" of Rickards, 1988) of graptolite evolution. The majority of the biserial and uniserial Silurian morphotypes are established in the *persculptus* to the early *acuminatus* zones, marking the beginning of stepwise radiation.

The Rhuddanian and Aeronian graptolite assemblages are the most diverse taxonomically and morphologically. The great increase in their diversity and numerical abundance (Fig. 12) results from several radiation events at the *acuminatus, vesiculosus, cyphus* and *gregarius* zones. The major taxonomic groups within diplograptaceans and monograptaceans reach their diversity maxima (an evolutionary equilibrium) within the Aeronian. They show the distinctive replacement of taxa, resulting from frequent episodes of extinctions and originations. The majority of Llandovery diplograptaceans go through the last peak of their diversity in the *cyphus* to *gregarius* zones. Later, at the end of the *convolutus-sedgwickii* Zone, diplograptaceans enter a short-term phase of decline and the last petalograptids finally disappear prior to the *griestoniensis* Zone.

The Aeronian and early Telychian (the *turriculatus* Zone) monograptid fauna display a great morphological variety of thecal and rhabdosomal shape and especially of the structure of apertural apparatuses.

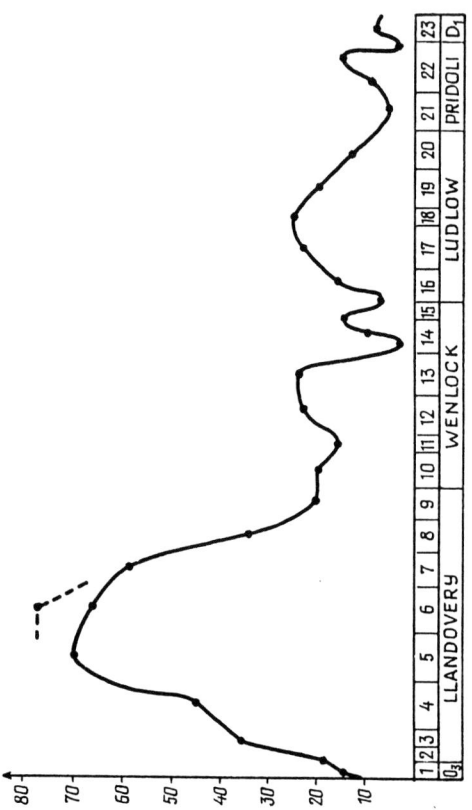

Fig. 12. Diversity of Silurian graptolites (number of species per zone). Data taken from Jaeger, 1991; Koren, 1983, 1987, 1991b, 1992; Koren et al., 1986; Kriz et al., 1986; Lenz and Melchin, 1991; Loydell, 1991; Melchin and Mitchell, 1991; Rickards, 1976; Storch, 1988. Standard graptolite zones shown by numbers: 1 *extraordinarius-persculptus*, 2 *acuminatus*, 3 *vesiculosus*, 4 *cyphus*, 5 *gregarius*, 6 *convolutus-sedgwickii*, 7 *turriculatus-crispus*, 8 *griestoniensis*, 9 *crenulata*, 10 *centrifugus-murchisoni*, 11 *riccartonensis*, 12 *rigidus-ellesae*, 13 *lundgreni*, 14 *nassa-deubeli*, 15 *ludensis*, 16 *nilssoni/colonus*, 17 *scanicus/chimaera*, 18 *leintwardinensis*, 19 *bohemicus-kozlowskii*, 20 *formosus/spineus*, 21 *parultimus-ultimus*, 22 *lochkovensis-transgrediens*, 23 *uniformis*. The species diversity in the turriculatus Zone, shown by the dotted line is taken from Loydell (1991)

Many specialized taxa belonging to the early monograptid lineages, based on the atavograptid stock, become extinct within the *convolutus-sedgwickii* to *turriculatus* zones ("*Demirastrites*", *Rastrites* and *Campograptus*). The simultaneous or stepwise appearance of more advanced morphotypes, including the new groups of lobate monograptids (*M. knockensis*,

M. sedgwickii, M. priodon, "*Streptograptus*", "*Globosograptus*" among others), takes place at the same time. At the beginning of the *turriculatus* Zone some monograptid genera undergo radiation events *(Rastrites,* Storch and Loydell, 1992; *Pristiograptus,* Loydell, 1991) and the introduction of a phylogenetically important morphological novelty, that is the appearance of thecal cladia *(Sinodiversograptus).*

The phase of decline of the morphological and taxonomic diversity starts prior to the beginning of the Telychian (Fig. 12), though locally up to 75 species may occur in the *turriculatus* Zone (Fig. 12; Loydell, 1991). An almost complete change of morphotypes takes place as a result of a radiation event within the *Monograptus priodon* group and *Monoclimacis*. The Telychian is characterized by the first diversification of retiolitids, a group which first appears in the *gregarius* Zone *(Pseudoretiolites).* The majority of important early Silurian genera, such as *Paraplectograptus, Plectograptus, Retiolites* and *Stomatograptus* appear for the first time in the *turriculatus-crispus* and *griestoniensis* zones. In view of the very extensive stratigraphical distribution of the above genera, a fact established by Lenz and Melchin (1987), the retiolitid phylogeny has proved to be different from that suggested earlier (Rickards et al., 1977). The earliest finds of cyrtograptids correspond to the *crenulata* Zone of the late Telychian. This typical Wenlock fauna is definitely of polyphyletic origin, deriving from at least three different monograptid lineages (Rickards et al., 1977; Lenz and Melchin, 1989).

Wenlock graptolites, embracing monograptid, cyrtograptid and retiolitid faunas display much smaller morphological variation compared to the Llandovery graptolites and, furthermore, show comparatively small taxonomic diversity (Fig. 12) resulting from phyletic evolution in several long-lived lineages (*M. priodon, M. vomerina, P. dubius).* Among monograptids, those with hooked thecae dominate assemblages. Monograptids differ mostly in the degree of astogenetic penetrance and expressivity of thecal apertural apparatuses. Some diversification events are distinguished among cyrtograptids, and occur at successive zonal levels: (1) within the *crenulata* to *centrifugus-murchisoni* zones, when multicladial colonies, with the main stipe tightly curved proximally, appear for the first time and become extinct prior to the *riccartonensis* Zone; (2) within the post-*riccartonensis* time, characterized by the origination of the *rigidus* group, showing thereafter a tendency of delay in the cladium formation and gracilization of the stipes (Rickards et al., 1977); and (3) at the beginning of the *lundgreni* Zone, showing the simultaneous reappearance of both colonies, with few and many cladia. The evolution of Wenlock retioliids is very slow, with no tendency to diversification until the late *lundgreni* Zone.

The terminal Wenlock mass extinction (the *lundgreni* Event) drastically changes the taxonomic composition and trends of morphological evolution of graptolites in general (Figs. 12, 13). The *lundgreni* Event stands out as a true crisis for all components of the early Homerian graptolite assemblages (Jaeger, 1991; Koren, 1991b). The pre-extinction (*lundgreni* Zone) fauna consists of morphologically diverse taxa, mostly belonging to long-lived and relatively unspecialised lineages. The crisis results in the final extinction of cyrtograptids and in the abrupt mass extinction of an equal magnitude for both monograptids and plectograptids. The *Pristiograptus dubius* stock is the only group among monograptids that survives the *lundgreni* extinction and gives origin to new lineages such as *praedeubeli-ludensis* and *idoneus-sherrardae.* Both become the ancestral stocks to the quickly diverging early Ludlow monograptids (Koren, 1991b, 1992). The Gorstian plectograptids show a diversity pattern contrary to monograptids and enter a phase of decline resulting in their final extinction in the *leintwardinensis* Zone (Fig. 13).

In general, late Silurian graptolites show a more fluctuating development, with frequent extinction, origination and specialization events. An overall decrease in the diversity combined with the high rates of species turnover is characteristic of their evolution during the Ludlow and most of the Pridoli. Some generalists persist through an extensive interval, providing a constant source stock for the majority of newly appearing taxa. Most Gorstian monograptids are represented by the newly originated *Neodiversograptus, Lobograptus, Cucullograptus, Neocucullograptus, Saetograptus, Colonograptus* and *Monograptus uncinatus* group. The morphological and taxonomical diversity of such taxa increases rapidly due to several radiation events at the *nilssoni, scanicus* and *leintwardinensis* zones (Fig. 13), each time followed by a period of directed phyletic evolution. Morphological changes are mostly concentrated on the apertural part of the thecae. This profound increase in generic diversity takes place at the beginning of the *nilssoni* Zone, when cucullograptid, neocucullograptid and linograptid faunas appear as a result of splitting of the *idoneus-sherrardae* lineage (Koren, 1992). Cucullograptid evolution is based on the specialized development of apertural apparatus, often connected with the

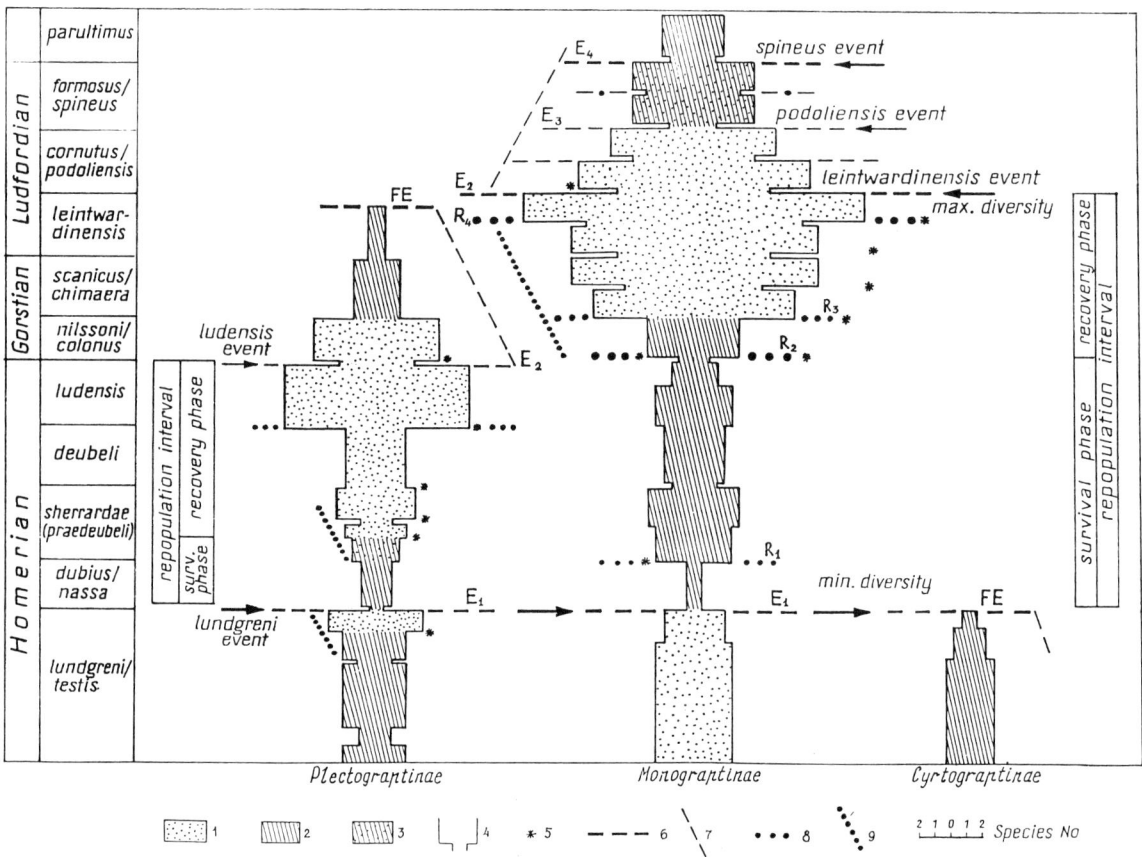

Fig. 13. Diversity of graptolites from the Homerian to Ludfordian sequences in the Kursala Formation, South Tien Shan: 1-3 morphological diversity: 1 high, 2 low, 3 average; 4 taxonomic diversity; 5 introduction of morphological novelty; 6 mass extinction event; 7 declining period; 8 radiation level; 9 diversification period; R radiation; E extinction; FE final extinction

elaboration of second assymetry of the paired apertural lobes (Urbanek, 1966). The neocucullograptid fauna, which develops parallel to the *Cucullograptinae*, is first of all represented by the slowly evolving *Bohemograptus* (Urbanek, 1970). The Gorstian linograptids are represented by *Neodiversograptus*, a genus which has an ability to develop a sicular cladium, a feature seen for the last time in the Silurian. The linograptids show no changes in thecal morphology during Gorstian to Pragian, when they become extinct.

As a result of radiation of the *praedeubeli-ludensis* lineage at the beginning of the *nilssoni* Zone the first saetograptids and colonograptids appear *(C. colonus* and *S. varians)*. Their further evolution is based on the progressive elongation and enrolling of lateral apertural processes (Hutt, 1969; Urbanek, 1970). Some other members of the Gorstian assemblages, for example the conservative *P. dubius* stock and the *M. uncinatus* group of a cryptogenic origin, develop slowly, displaying low taxonomic diversity.

Up to the beginning of the Ludfordian the amount of graptolite species origination exceeds the amount of their extinction (Fig. 13). All existing monograptid genera reach their diversity maxima within the *scanius* to *leintwardinensis* zones. After a short equilibrium interval graptolites begin to decline gradually until the end of the Ludlow. A prevalence of extinctions is characteristic of Ludfordian graptolite evolution. The extinction events share several common features: population bursts in some surviving and newly appeared taxa *(B. b. tenuis, P. fragmentalis, M. parultimus)*, as an evidence of changes at the population level, and a rapid diversification. The latter means that an extinction event is followed by a recovery phase with no post-crisis survival period (sensu Harries and Kauffman, 1990).

The *leintwardinensis* extinction event can be characterized as the most drastic reduction of taxonomic diversity (genera, species group) in the evolution of Ludlow graptolites (Urbanek, 1970; Koren, 1991b). It eliminated numerically abundant, morphologically diverse and highly specialized monograptids, belonging to different lineages. However, this extinction was selective in that it was coeval with the progressive development of some genera and species groups (*M. dalejensis* and *M. uncinatus* groups, *Cucullograptus, Polonograptus* and others) which successfully survived and radiated thereafter. The last few plectograptid species finally disappear (Fig. 13) within the *leintwardinensis* Zone prior to the main event and this can be qualified as a background extinction.

The *podoliensis* extinction event is comparable with the *leintwardinensis* event in the number of species which disappear, but shows a much smaller number of survivals. This event eliminates both conservative *(Bohemograptus)* and highly specialized taxa *(Neocucullograptus, Neolobograptus* and *Polonograptus)*. At the same time a number of important appearances take place within the different lineages, showing a distinctive faunal replacement. The newly appeared taxa are specialized and short-lived *(M. hamulosus, M. balticus, M. spineus, M. formosus* among the others) and become extinct prior to the end of the *formosus/ spineus* Zone. Even single relics *(M. formosus)*, which survive for a short time, became very rare.

By the end of the Ludlow graptolite diversity drops critically, to two or three species (Fig. 12). An important development is the appearance of a new morphotype, at the end of the *formosus/spineus* Zone, which has paired lateral lobes on the thecal apertures. This morphotype later dominates the Pridoli monograptids and serves as a basic structure for further specialization within several lineages. This pattern can be taken as an example of parallel evolution. The overall taxonomic diversity of Pridoli graptolites is the lowest in the Silurian. Furthermore morphological variety is limited to a few thecal structures, including: (1) the paired lateral lobes of different astogenetic expressivity and penetrance *(M. ultimus, M. branikensis, M. lochkovensis, M. transgrediens, M. nimius)*; (2) apertural dorso-lateral hoods *(M. bouceki, M. perneri, M. beatus, M. mironovi)*, sometimes with paired lateral processes in specialized taxa *(M. anerosus, M. supinus)*; and (3) sharply geniculate thecae with small hoods *(M. balaensis, M. microdon aksajensis)* or with even thecal apertures *(Psedomonoclimacis)* (Koren, 1983). A diversity equilibrium level is reached within the *bouceki* and *perneri* zones. The extinction event at the end of the *perneri* Zone eliminates all taxa except the conservative *M. transgrediens* and *L. posthumus* stocks. They form the latest Pridoli assemblage which finally disappears close to the *uniformis* boundary.

Microfossil Evidence

The next three sections of this chapter are dealing with microfossils - conodonts, chitinozoans and acritarchs. Detailed sampling and large numbers of both species and specimens per unit or sample have allowed statistical treatment of the data. The results obtained are used below besides the ordinary palaeontological information on morphological innovations and distribution. To describe the changes of the taxonomic diversity of the fossils, we have applied the number of taxa per stratigraphical unit, number and percentage (rate) of appearing and disappearing species, also the total and per taxon rates of these taxa. All these different rates harmonize well enough and therefore to save space in the following text and figures we use only the number of taxa and total rates of the appearing and disappearing species (species per Ma).

Conodonts

Clark (1972, 1981) was the first to observe several generic and species diversity lows and peaks in the history of conodonts, whereas the Silurian was noted as a period of a major evolutionary crisis.

According to Sweet (1985), the species diversity started to decrease in the late Ordovician and continued throughout the Silurian up to the early Devonian, with two diversity peaks in the late Llandovery and early Ludlow showing the prevalence of the appearing species over the disappearing ones.

In the Silurian, Aldridge (1988) documented only one clear generic origination event in the late Llandovery, preceded by relatively slow innovation in the earliest Silurian and followed by a rapid extinction in the early Wenlock. This event shows the appearance of many new genera with new types of apparatuses (e.g. *Pterospathodus, Apsidognathus*, etc.) in the late Llandovery. After the above-mentioned main extinction event only a few new genera were introduced in the late Wenlock and Ludlow.

Sweet (1988) recognized the cyclic pattern of conodont diversity changes and established 4 long-term and 20 second-order cycles. The Silurian history of conodonts corresponds to the IV and V second-order cycles with high diversity episode in the late Llandovery and a smaller one in the early Ludlow. These cycles constitute the upper part of the late Cambrian - Silurian long term cycle by Sweet (1988).

The results of the study of the Silurian conodonts from Estonia largely agree with the above conclusions. Männik and Viira (1993) recently discussed this topic in a special paper, therefore here only a summary of their data is presented.

The number of conodont species per stratigraphical unit in Estonia correlated with graptolite zones varies between 5 and 37, on the average 15, being twice higher in the pre-*riccartonensis* time (23) than in the Silurian after the Ireviken extinction (10) (Figs. 14, 15).

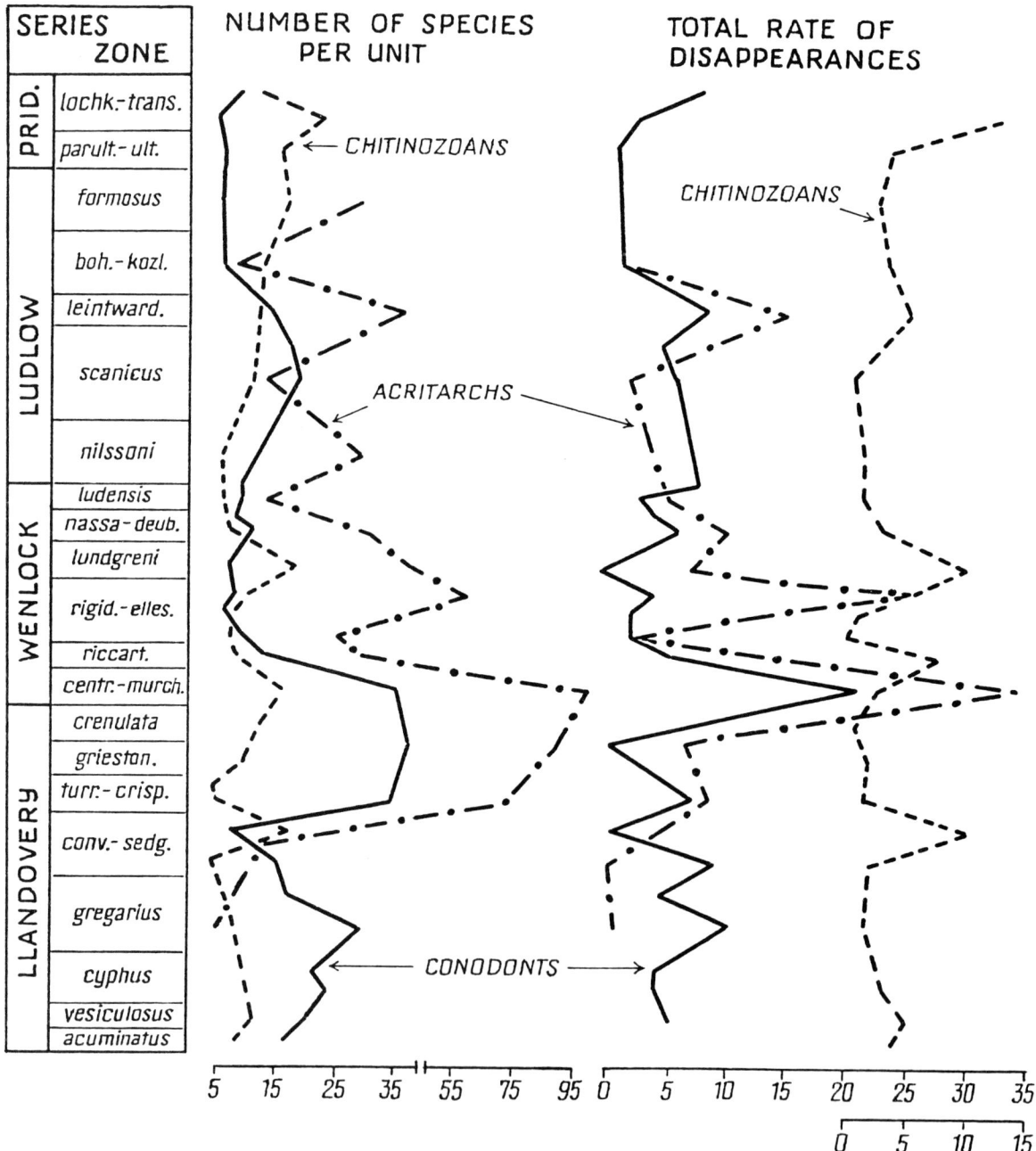

Fig. 14. Diversity dynamics of conodonts, chitinozoans and acritarchs in the Silurian of the Baltic area

There is certain cyclicity observable in the diversity dynamics of conodonts: low stand - diversity rise (radiation or innovation event) - extinction event, but the boundaries of the cycles are, as a rule, to some extent arbitrary.

The diversity curve (number of species per unit) shows three intervals of the high stand: top Rhuddanian-early Aeronian (maximum mid-*gregarius* Zone), Telychian until the very beginning of the Wenlock and early Ludlow (Fig. 14). In between there are low stands. The total rate curves are in accordance with the first one. In the beginning of the Rhuddanian there appear new genera *Distomodus* and *Oulodus*?, higher *Icriognathus*, *Kockelella*, *Rotundacodina*? and several new taxa, and *Pranognathus* in the early Aeronian. The first maximum of the appearance rate was reached at the mid-*gregarius* Zone level. At the same time also the first energetic extinction event began which continued and reached a maximum in the *convolutus* time and caused the first diversity minimum of conodonts at the end of the Aeronian.

The second peak of appearances was at the *turriculatus-crispus* Zone level which introduced characteristic so-called platform conodonts *Astropentagnathus*, *Aulacognathus*, *Apsidognathus*, and some others such as *Carniodus* and *Pterospathodus* to the Baltic basin, and also elsewhere.

This interval ends with the most serious extinction event in the history of Silurian conodonts, the so-called Ireviken Event of Jeppsson (1987).

In the course of this event within a short time interval (in general during the *murchisoni* time) more

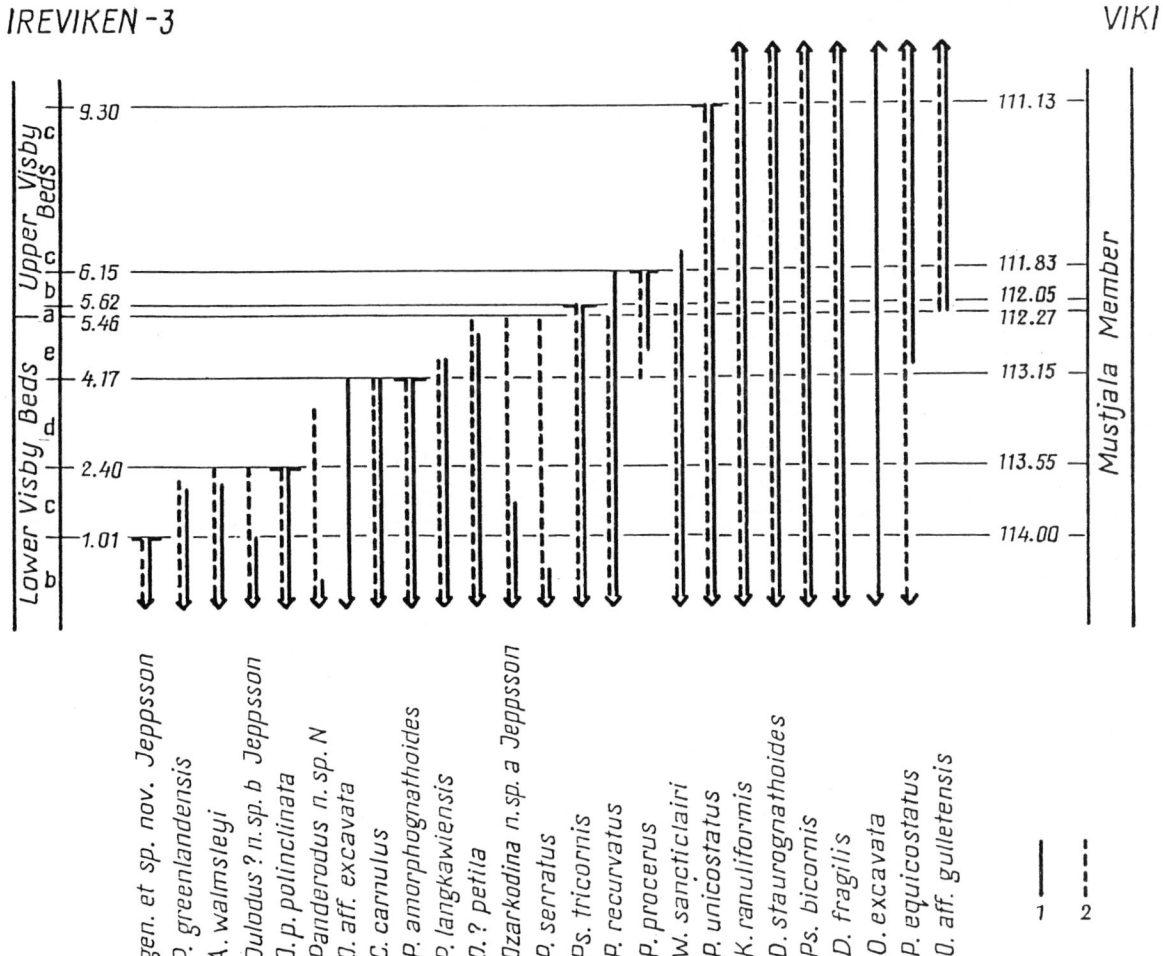

Fig. 15. Conodont ranges during the Ireviken or *amorphognathoides* Event (simplified after Jeppson and Männik, 1993). The distribution of the conodont species: 1 in Viki (Island Saaremaa, Estonia) and 2 in Ireviken; 3 (Island Gotland, Sweden) sections

than 70% of conodont species and 40% of genera disappeared.

The extinction event looks very sharp (Fig. 14) and it is so geologically speaking, but a detailed observation allows us to see also here the so-called stepwise development (Fig. 15).

Jeppsson and Männik (1993) made a very detailed study of two sections - Ireviken on Gotland and Viki on the Estonian Island Saaremaa (distance more than 200 km). These sections have revealed nearly exact coincidence of the order and levels of the extinction of taxa and the step by step character of the Ireviken Event. Noteworthy are also two cycles of the process. The most energetic extinction took place just below the boundary of the Lower and Upper Visby Beds, which in the Estonian sequence falls into the lower part of the Mustjala Formation (Wenlock).

In the course of this extinction there disappeared practically all abundant conodont assemblages, well known from the *celloni* and *amorphognathoides* conodont zones. The event was survived mostly by the simple-cone conodonts, such as *Dapsilodus, Panderodus, Walliserodus,* etc. From the ramiform taxa only *Kockelella, Oulodus?* and *Ozarkodina* have descendants also in the overlying strata.

During the following Wenlock a low diversity pattern prevailed - some new taxa appeared and disappeared. A somewhat more remarkable rise in both appearance and disappearance of total rates might be noted in the *nassa* and *ludensis* zones, and again in the very beginning of the Ludlow when the rate of appearances rose at the transition of the *nilssoni* and *scanicus* zones and that of disappearances at the *leintwardinensis* Zone level.

During the low diversity period in the Wenlock, most important were the index species *Ozarkodina sagitta rhenana, O. bohemica bohemica* and *Kockelella walliseri*. In the shallow shelf Rootsiküla Formation (upper Wenlock) there occurs an ecological specialist, *Ctenognathodus murchisoni*, this constituting a new apparatus type.

The early Ludlow rise in the appearances rate is also caused by a small radiation event where changes occurred on the species and subspecies levels. This pattern continued until the end of the Silurian. We can list such well-known Upper Silurian zonal et al. conodonts as *Ozarkodina snajdri, O. crispa* and *O.* aff. *tillmanni* showing different variations of the basal cavity or changes in denticulation, such as *O. c. cornidentata, O. c. densidentata, Oulodus siluricus,* etc.

The last remarkable extinction event occurred in the *leintwardinensis* Zone when more than 85% of the assemblage disappeared and only a few species, mostly different forms of *Ozarkodina eosteinhornensis*, continued until the end of the Silurian.

In summary, there are four main cycles in the diversity dynamics of conodonts (early-middle Llandovery, late Llandovery-early Wenlock, middle Wenlock-middle Ludlow, late Ludlow-early Devonian) established in Estonia.

In the studied stratigraphic interval three main extinction events, marking the ends of diversity cycles, are recognized (terminology according to Jeppsson, 1993): the prolonged Sandvika Event with its peak within the upper part of the *convolutus-sedgwickii* Zone, the *amorphognathoides* or Ireviken Event in the *centrifugus-murchisoni* Zone and the *siluricus* or Lau Event within the *leintwardinensis*-early *bohemicus-kozlowskii* Zone.

In the Baltic area (Gotland) Jeppsson (1993) recognized also several lower rank events: the Boge Event at the end of the *Kockelella patula* Zone (= *Cyrtograptus rigidus* Zone), the Valleviken Event in or below the *Ozarkodina s. sagitta* Zone (= *Monograptus testis* Zone); the Mulde Event from the beginning of the *Monograptus parvus* Zone to a part or all of the *M. praedeubeli* Zone; the Linde Event within the *Ancoradella ploeckensis* Zone (= the *Monograptus chimaera* Zone, Gorstian). These events and some others in the Pridoli need additional detailed studies.

According to Schönlaub (1986), and also Männik and Viira (1993) cyclicity of conodont diversity was to some extent influenced by sea-level changes. The high diversity intervals seem more or less coinciding with the transgressive phases and the three main extinctions with the regressive phases of basin development (Männik and Viira, 1993). These coincidences, however, cannot be regarded as simply unambiguous.

Jeppsson (1990) considered the explanation of faunal changes by only transgressions and regressions inadequate and elaborated a more complex model. In this model the oceanic cycles are expressed as changes between more humid low latitude and cooler high latitude climates (P episodes), and drier low latitude and warmer high latitude climates (S episodes), that caused also changes of different other environmental parameters, such as the water temperature and the content of dissolved gases, character of sedimentation, nutrient supply, etc. Typical P-S transitions causing the most severe extinctions were the Sandvika, Ireviken and Lau events (Jeppsson, 1987, 1990, 1993). All other events listed above are also P-S events, the only exception is the Mulde Event classified as a S-S event. Perhaps also the end-Silurian event is a S-S event.

Aldridge, Jeppson and Dorning (1993) concretized the above model proposing a succession of primo (P) and secundo (S) episodes basing on the sections of the Oslo area and Gotland (Fig. 28). The first is usually characterized by decline of carbonate deposition and increase in clayey material in the rocks. Planktic and connected organisms are abundant in primo episodes. During the secundo episodes vice versa pure carbonates, often with reefs, are widely distributed, rich benthic communities occur. Planktic communities show low diversity and abundance.

The above authors underline that the mentioned differences between P and S episodes are the results of climatic differences linked with changes in atmospheric carbon dioxide concentration and variation in storage capacity for CO_2 in the ocean depending on water temperature. Cold waters of P episodes have a higher capacity and therefore extract from the atmosphere a certain amount of carbon dioxide. The latter is released during the S episode.

This mechanism involves also other environmental aspects mentioned above causing abrupt (in case of transition from P to S state) or more gradual (in case of S-P or S-S) changes of ecosystems.

Identifying episodes and events (Fig. 28) in the Llandovery and early Wenlock, Aldridge, Jeppsson and Dorning (1993) used as guides long-ranging conodont lineages that reflect fluctuating environmental conditions. For example, *Panderodus* was very diverse during the primo episodes, but in the secundo episodes it was represented by only one species. Members of the *Ozarkodina* lineage are usually typical of secundo episodes. Some conodonts may occur abundantly during both episodes or may change their preferences.

Finalizing their discussion, the above authors conclude that not all sections are showing a typical set of changes as local environmental agents are influencing the process.

Anyhow, this model, as well as simple transgression-regression, is explaining the events in conodont history by ordinary terrestrial environmental agents despite the possible influence of Milankovitch cycles.

Chitinozoans

The evolution of Silurian Chitinozoa on the higher taxonomic level (formal genera and families) was rather slow. Most of the genera occurring in the Silurian are incomers from the Ordovician continuing in the Devonian. The species diversity of chitinozoans, however, considerably changed throughout the Silurian, reflecting general tendencies in their evolution, but also depending on environmental impact. Accordingly, several bio-events of different ranks were recently established by Nestor (1992).

The number of chitinozoans per zone is a little smaller than that of conodonts - 11 on average, but it is considerably different when comparing the pre-*scanicus* (9 on the average) and higher parts of the section (15).

The diversity curve of the East Baltic Silurian chitinozoans (Fig. 14) shows four well-pronounced peaks. In the early Llandovery the chitinozoan diversity was relatively high but was followed by a steady decline up to the short-time radiation in the *sedgwickii* Zone. This late Aeronian peak coincided with the beginning of the second main transgressive phase in the East Baltic Silurian and was accompanied by the first clear extinction event. The second diversity maximum occurred after a step by step radiation during the Telychian and in the very beginning of the Wenlock in the *murchisoni* Zone. A major extinction event followed in the *riccartonensis* Zone. A new short-time radiation maximum was attained in the early Homerian *lundgreni* Zone, followed by another mass extinction event at the same level. The last Silurian radiation period began in the early Ludlow *scanicus* Zone and, continuously increasing, extended up to the middle Pridoli *bouceki* Zone. The most significant late Silurian chitinozoan extinction event fell within the *bohemicus-kozlowskii* Zone. Similarly, with the *riccartonensis* Zone mass extinction it was characterized by the disappearance of more than 70% of the occurring species.

All these diversity changes were also accompanied by origination of new morphological features, e.g. in the Aeronian there appeared regular spinose *Gotlandochitina*, mucronate species of *Conochitina*, and widely distributed *Eisenackitina*. On the same level (*sedgwickii* Zone) there disappeared *Coronochitina* and *Cyathochitina*.

The radiation at the Llandovery-Wenlock boundary, especially in the *murchisoni* time, brought new *Anthochitina* and *Margachitina*, also several new species of *Ango-* and *Gotlandochitina* appear. At the *riccartonensis* level most of the earlier distributed species disappeared.

The next radiation occurred at the *ellesae* level when *Cingulochitina cingulata* etc. appeared, and was followed by energetic extinction at the *lundgreni* level when most of the typical Wenlock chitinozoans disappeared. The last of them disappeared together with many Ludlow species at the *bohemicus-kozlowskii* level.

A new graded radiation began at the early Ludlow *scanicus* Zone level, characterized by the appearance of *Conochitina latifrons* and *Angochitina elongata* at the end of this zone. The late Ludlow innovation brought along the reappearance and domination of *Eisenackitina* species *(E. lagenomorpha, E. philipi,* etc.), appearance of *Pterochitina* at the *formosus* Zone level and *Fungochitina* in the middle Pridoli. A final Silurian extinction started in the late Pridoli.

A comparison of the Baltic data with those of the Welsh Basin (Dorning, 1981) and North Spain (Schweineberg, 1987) shows relatively good coincidence of major event levels; e.g. according to total rate values well traceable levels seem to be *flexilis* (app.), *lundgreni* (app.+disapp.), *(leintwardinensis)* and *bohemicus-kozlowskii* (app.+disapp.). Other event levels are less concordant.

The coincidences in sections belonging to different regions is evidence that at least some of these levels are traceable over a large area. Of course, the exact correlation of carbonate rock sequences with the graptolite zonation used as a standard is problematic, as we cannot be sure which levels are in fact synchronous.

In general chitinozoan diversity dynamics the major innovation and extinction events are in good accordance with the transgression-regression cycles of the Baltic Silurian basin and the global sea-level curve (Johnson, Kaljo and Rong, 1991). Most of the high diversity peaks coincide with the rising phases of sea-level, extinction events with its lowering phases or low stands. Exceptions (e.g. late Ludlow extinction) show that the pattern is only tentative.

Acritarchs

Globally 90 acritarch genera are represented in the Silurian, whereas about 30% of the taxa are long ranging forms, directly derived from the Cambrian or the Ordovician.

Figure 16 shows general diversity changes in the acritarch assemblages from the late Ordovician to the early Devonian, including the number of genera present in each period, the levels of appearances of new forms and the levels of extinctions. In the Silurian acritarchs had distinct preferences in their regional distribution, so we have distinguished between the taxa locally restricted to low- and high-latitude provinces, and cosmopolitan taxa. Sometimes, we have indicated a number of genera changing their distribution area between two geographical regions, in this way reflecting also a biogeographical differentiation in diversity. Of course, at present it is possible to make

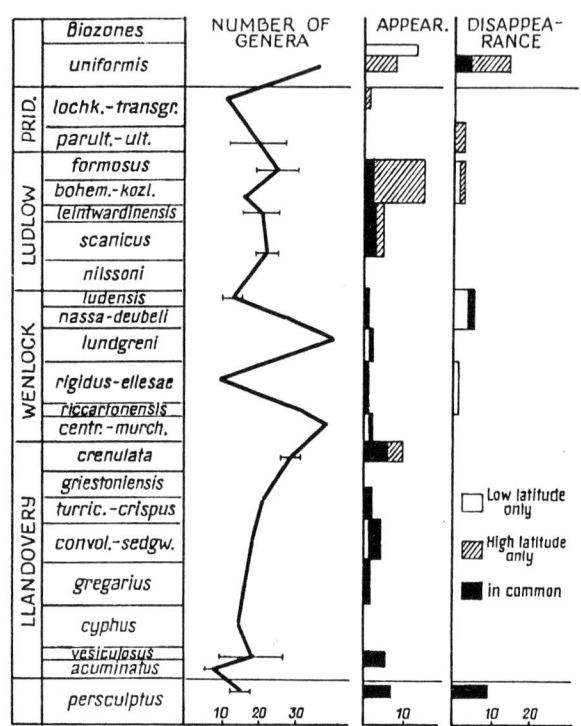

Fig. 16. Generic diversity, levels of appearance and disappearance of acritarch genera of different biogeographic origin from late Ordovician to early Devonian

only tentative statements on adaptative radiations and extinction events in Silurian acritarch assemblages, because of insufficiency of data from many parts of the world. Obviously, there is also a problem of precision in long-distance correlations, therefore much more taxonomic work will be necessary in continuous sections before we can draw certain conclusions.

The late Ordovician shows the disappearance of many characteristic genera, e.g. *Actinodissus, Aremoricanium, Ordovicidium, Orthosphaeridium, Peteinosphaeridium,* but also the first appearance of several genera with Silurian characters, such as *Dilatisphaera, Hoegklintia, Estiastra, Evittia (= Diexallophasis), Neoveryhachium, Moyeria,* or *Tunisphaeridium*. This epoch can be considered as a time of reorganization into assemblages, and there is no drastic extinction event among acritarchs. However, the comparison of Ordovician and Silurian assemblages has revealed profound differences in their composition and an important reduction in the number of genera and species taking place near the boundary between the two systems. The changes in the composition of the microflora are probably related to the influence of the late Ordovician glaciation centered on Gondwana, and its connected sea-level changes. Other major differences have been determined, for example in the size of some species assigned to the same genera (e.g. *Baltisphaeridium* or *Dilatisphaera)*, with abrupt size decrease across the Ordovician-Silurian boundary.

The earliest Silurian *acuminatus* Zone, which follows the maximum of the regressive phase of the late Ordovician, is a time of considerably reduced diversity, represented by only few genera that are generally long ranging in the Silurian.

This time is in sharp contrast with the preceding one and also with the succeeding middle Rhuddanian *vesiculosus* Zone, which shows rapid increase in the diversity of acritarch genera, with the first appearance of the very distinctive *Helosphaeridium, Salopidium, Tylotopalla,* or *Visbysphaera* and the complex acritarchs assigned to *Carminella* and *Geron*. The latter represent some important innovative morphological features among acritarchs.

Significant differences in the generic and specific diversity have been established in the Lower Silurian between the regions of low and high latitudes. The diversity is minimum in the regions of the Northern Hemisphere situated in the warm-water "province", and maximum in Gondwanan and Perigondwanan regions characterized by cool to cold environments.

The late early Llandoverian *cyphus* Zone does not reveal noticeable changes in the composition of the assemblages, but is marked, particularly in regions of high latitudes, by the relatively high abundance of diverse fusiform acritarchs, including *Dactylofusa, Eupoikilofusa* and *Leiofusa*.

The early Aeronian *gregarius* Zone shows the reappearance of *Domasia* and *Oppilatala,* and the beginning of abundant distribution of the genus *Multiplicisphaeridium*. The progressive increase in diversity is followed by the first occurrences of *Ammonidium, Cymbosphaeridium, Dilatisphaera* and *Duvernaysphaera* in the *convolutus-sedgwickii* Zone. The genus *Dilatisphaera* has an upper Ordovician ancestor *D. wimani*. This very distinctive genus, with a subspherical vesicle and a variable number of hollow distally open tubular processes, shows some similarities with chorate cysts of dinoflagellates of the genus *Hystrichosphaeridium*. However, as other morphological features distinctive to dinoflagellates are lacking, they cannot be considered as true dinoflagellates. In the Llandovery, *Dilatisphaera* spp. are restricted to low latitude regions of the equatorial zone.

A high degree of morphological diversification is attained in the Telychian, with a successive appearance of new taxa and a maximum variety of acritarch forms in the late Telychian (Fig. 16). In the microfloras of low and high latitude regions there is a persistent geographic differentiation. For example, the distribution of giant acritarchs belonging to *Estiastra, Hogklintia* and *Pulvinosphaeridium,* seems to be parallel to the carbonate belt and associated reefs of the warmer regions. By contrast, *Crassiangulina, Tyrannus* and several new genera recently discovered, are characteristic of the "Mediterranean province".

The Telychian is also characterized by the prevalence of the *Deunffia-Domasia* group, with many short ranging species of a particularly high stratigraphical value. Even if variants of the two taxa seem to be more abundant in the low latitude regions, many of them have recently been recognized also in the Gondwanan and Perigondwanan regions.

This time is also marked by the first occurrence of the very distinctive triangular *Onondagaella,* and other taxa such as *Gracilisphaeridium* or *Psenotopus,* limited only to the Llandovery-Wenlock boundary beds. These two latter genera have relatively short ranges compared to most of the other genera with a notably longer duration. The genus *Psenotopus* and some species of the genus *Visbysphaera* recorded on the Llandovery-Wenlock boundary show a particular arrangement of surficial ornaments (granules or processes) in patches or bands, forming some polygonal areas on the vesicle. These exterior features show some similarities with the arrangement of ornamentation in the Ordovician genus *Cymatiogalea,* and also a convergence with the paratabulation observed on dinoflagellate cysts.

The very beginning of the Wenlock is a phase of relatively high diversification and shows also the first appearance of new forms like *Florisphaeridium* belonging to acritarchs with florate terminations. This new morphological structure concerns other species in the Ludlow and in the lower Devonian.

The late Sheinwoodian is characterized by a low diversity of acritarch assemblages. Depending on the correlation of the Gotland sequence with the graptolite zonation, the diversity low might move to the middle of the Sheinwoodian.

The genus *Hapsidopalla*, which first appears in the basal Homerian, and is characterized by palmate processes and distinctive stellate or rosette-like arrangement of the ornamentation, is morphologically related to Devonian species. The Silurian species suggest a possible beginning of a long evolutionary lineage progressing through the entire Devonian with *H. sannemanii, H. chela, H. exornata, H. invenusta* and the succeeding *Naevisphaeridium* spp. and *Puteoscortum* spp.

A decrease in acritarch diversity is observed in the late Wenlock *nassa-deubeli* Zone and persists in the *ludensis* Zone. In this phase of decline, the extinctions, e.g. *Hogklintia, Psenotopus, Salopidium* and *Deunffia, Domasia* and *Wrensnestia* in the Northern Hemisphere, surpass the morphological innovations. However, in the late Homerian we can mention the first occurrence of the genus *Percultisphaera* that ranges up to the late Ludlow and is characterized by a complex ornamentation without an equivalent within the acritarch group.

The lowermost Ludlow seems to correspond to a period of stability in acritarch community, without marked changes in the composition of the assemblages. The Gorstian, however, shows rapid radiation of new genera like *Fimbriaglomerella* or *Leoniella* leading to an increased diversification that continued to expand through the Ludlow and lowermost Pridoli, particularly in high latitude regions. In the regions of the Northern Hemisphere a slight decrease in diversity can be noted in the basal Ludfordian, and only few appearances of new taxa are recorded up to the Ludlow-Pridoli boundary.

Within the Gondwanan realm, the late Ludfordian and early Pridoli microflora is well diversified and represented by a lot of new taxa with some triangular forms like *Antruejadina* or *Cantabrica*, some discoid or lenticular forms, e.g. *Ovnia, Pardaminella* or *Perforela*, a distinctive group with a coenobial organization such as *Cuatrifolia, Deflandrastum* (sometimes assigned to *Hydrodictyacea*) or *Morcoa*, and others. We can note also the good representation of some species like *Leprotolypa gordonense* and of some variants of *Neoveryhachium carminae* with three or five processes. Some similar "blooms" of particular genera, e.g. *Visbysphaera*, have also been described in the Upper Ludlow of Great Britain. Many of the Gondwanan endemic taxa cross the Silurian-Devonian boundary and disappear in the early Devonian, but most of them seem more ephemeral, dying out in the Pridoli.

The information available on Pridolia acritarch assemblages tends to suggest that the assemblages are often dominated by *Cymbosphaeridium* species and cysts of Prasinophycean algae (e.g. large *Cymatiosphaera*). The representatives of the early Devonian genus *Ozotobrachion* first appeared in the late Pridoli.

The data concerning the Silurian-Devonian boundary are still sparse and limited to few areas. This is partly due to the lack of marine sediments in many regions, in relation to the increasing marine to continental regression from the latest Silurian until the earliest Devonian. However, records of acritarchs from Eodevonian marine sequences in Gondwanan and Perigondwanan regions and North America, show that the basal Lochkov (not the lowermost part) is marked by abrupt changes with both extinctions of numerous genera and the appearance of new forms. Some of these new forms are cosmopolitan but many of them are endemic, indicating a high level of provincialism of the early Devonian acritarchs.

General Trends and Remarks

To sum up, Silurian acritarch assemblages appear to be quite different from the Ordovician and Devonian ones, although there is a regular progression of taxa, characterized by gradual incoming of new forms and extinction of others, and a continued reorganization of the assemblages with the appearance of taxa with Silurian characters in the upper Ordovician as well as some taxa with Devonian characters in the Silurian.

The evolution trends towards the increased diversification of taxa, and some morphological innovations concerning the general shape, the ornamentation or other characteristics, such as excystment openings which appear to be more diversified compared to the Ordovician.

The acritarch "acanthomorphs", i.e. the spiny forms, particularly the forms with ramified processes, but also *Veryhachium* and the fusiform acritarchs such as *Dactylofusa, Eupoikilofusa* and *Leiofusa*, greatly predominate in the Silurian assemblages.

Prasinophycean cysts such as *Cymatiosphaera* or *Pterospermopsis* but also *Dictyotidium* (by some authors included into prasinophycean algae), that were relatively rare in the subsequent periods, are also significant components of the Silurian assemblages.

The distribution of certain genera can be used to determine some biogeographical subdivisions. For example, we have established a high development of provincialism within acritarch assemblages during the late Ludlow and early Devonian. On the other hand, the known ranges of genera and their levels of appearance and disappearance may vary from one biogeographical realm to another. There is some evidence of that in the earliest Silurian, considering, for example, the level of appearance of *Oppilatala*, or in younger strata, the disappearance of *Deunffia* in low latitude regions, in South America or Florida.

The most significant changes in the composition of Silurian acritarchs took place in the middle Rhuddanian, the late Telychian and the upper Ludlow.

During these periods, the acritarchs evolved at a rate of about 4 to 6 genera per Ma. In other periods, the rate of evolution is much lower, only 1 genus per Ma. The basal Rhuddanian, the upper Sheinwoodian, the uppermost Homerian and upper Pridoli are particularly poorly diversified.

It is difficult to speculate about the possible implications of the morphological trends outlined above, due to the lack of precise ideas about the biological significance and palaeoecology of acritarchs and their possible morphological adaptations to varying environmental conditions. It is also difficult to judge about derivation of new taxa from earlier antecedents. We should not forget just what genera and species are in taxonomic practice, and that morphological differences between genera are sometimes relatively small, mostly concerning details or ornamentation, without real biological implications. In reality, very few evolutionary lineages can be recognized between acritarchs, except on the species level, and in most cases genera seem to be disjunct morphologically.

In order to detail the above discussion on the acritarch evolution and diversity patterns on the genus level presented by A. Le Herisse, there is added below an analysis of the diversity on the species level. As a data base the content and distribution of acritarchs in the Silurian of Gotland is used according to a monograph by Le Herisse (1989). The following analysis has to be considered as an example, more or less true, for one low latitude area in the Silurian Southern Hemisphere, but not obligatory for other regions.

Most of the species considered in Table 1 are so-called Lazarus taxa, missing in some intervals of their whole range. Ascertaining the number of species per zone, only taxa identified from the zone were counted. The missing Lazarus taxa were not taken into consideration, leaving aside also their repeated appearances-disappearances and tabulating only the first and the last of these.

Comparison with the data set where Lazarus taxa were counted for all zones during their whole range shows no principal differences, i.e. the most important diversity changes were established by both methods, but the usage of the real numbers of the occurring taxa allows us to understand better the ecological influence on the assemblage composition.

Plotting the data (Table 1, Fig. 14), the standard graptolite biozonation was used (Fig. 6), only the *rigidus-ellesae* Standard Zone was divided into two, the *rigidus-linnarssoni* and *ellesae* zones. Otherwise, the high number of species occurring in the upper part would overshadow a diversity low in the lower part.

For making comparisons between different regions, all local or topostratigraphic units were correlated with the graptolite biozonation. At many levels such a detailed correlation cannot be well grounded and has to be considered more or less as tentative.

In the case of Estonia and Gotland there were used all graptolite occurrences (see Hede, 1942; Jaeger, 1981; Kaljo, Paskevicius and Ulst, 1984; etc.) in carbonate rocks and possibilities to correlate by the assistance of conodonts (Jeppsson, 1983; Männik and Viira, 1990; etc.) and chitinozoans (Laufeld, 1974; Nestor, 1990). We believe the correlation is acceptable, but, of course, by no means fully exact in details.

Confining ourselves to these introductory remarks, below we shall comment briefly on the diversity data presented in Table 1 and Fig. 14.

Acritarchs were more diverse than other groups of microfossils in the Silurian of the Baltic area. The average number of taxa per unit reaches 35, maximum 99 in the *centrifugus-murchisoni* and minimum 5 in the *gregarius* zones. Figure 14 shows a very changing curve, but with in general the number of taxa per unit having its high peak at the Llandovery-Wenlock junction, and declining toward the end of the Silurian.

The appearance rate was very high during the Telychian, it dropped in the very beginning of the Wenlock nearly to the mean level and was below it onwards from the *riccartonensis* level (with one exception in the *ellesae* Zone). Considering both the total and per taxon rates, we can conclude that the most remarkable radiation event among Gotland acritarchs on the species level occurred at the very beginning of the Telychian (*turriculatus-crispus* Zone) when the total appearance rate was six times higher than the average.

Disappearance of acritarch taxa was most energetic in the earliest Wenlock (*centrifugus-murchisoni* and *riccartonensis* zones) and later, beginning with the *deubeli-nassa* Zone, especially in the Ludfordian. Such a more or less expected picture was obtained on the grounds of the percentage of the disappearing taxa. The total rate of disappearances shows only one clear extinction event in the very beginning of the Wenlock, when the rate exceeds the mean level nearly four times. This extinction level seems to coincide with the so-called Ireviken Event among conodonts and therefore deserves serious attention.

Later in the Silurian there is a smaller peak of the total rate of disappearances at the *ellesae* Zone level (= low in the percentage curve) and a general low stand of the curve at the high percentage of disappearances.

Here we can well see the direct influence of the

Table 1. Numerical data for Gotland acritarch species diversity

Gotland strat. units	Standard graptolite zones	Dur. m.y.	Species per unit					Total rate		Per taxon rate	
			Occurring No	Appearing No	%	Disappearing No	%	App.	Disapp.	App.	Disapp.
Sundre / Hamra	*formosus*	2.0	29	5	17	?	?	2.5	0	0.09	0
Burgsvik	*bohemic.-kozlowsk.*	2.0	8	1	13	4	50	0.5	2.0	0.05	0.21
Eke	*leintwardinensis*	1.0	37	6	16	15	41	6.0	15.0	0.14	0.34
Hemse	*scanicus*	3.0	13	1	8	5	38	0.3	1.7	0.03	0.09
	nilssoni	2.0	29	11	38	7	24	5.5	3.5	0.14	0.09
Klinteberg	*ludensis*	0.8	13	0	0	4	31	0	5.0	0	0.51
Mulde	*nassa-deubeli*	1.0	30	3	10	10	33	3.0	10.0	0.13	0.43
Halla / Slite	*lundgr.*	1.2	38	3	8	8	21	2.5	6.7	0.09	0.23
	ellesae	0.7	59	8	14	18	26	11.4	25.7	0.19	0.44
Tofta	*linn.-rigidus*	1.3	25	7	28	2	8	5.3	1.5	0.20	0.06
Högklint	*riccartonensis*	0.8	29	1	3	11	38	1.3	13.8	0.04	0.47
Up.Visby	*centrif.-murchis.*	1.2	99	19	18	41	41	15.8	34.2	0.16	0.35
Low. Visb.	*griest.-crenul.*	2.2	89	38	43	14	16	17.3	6.4	0.36	0.13
Beds 5 in När	*turric.-crispus*	2.0	71	61	86	10	14	50.8	8.3	0.86	0.14
borehole 6	*convol.-sedgw.*	2.0	13	9	69	0	0	4.5	0	0.35	0
7 350 m	*gregar.*	2.4	5	3	60	1	20	1.3	0.4	0.26	0.09
	Mean:		35	11	27	10	27	8.0	8.9	0.24	0.24

duration of the zonal units in terms of absolute geochronology discussed above. The high total rate of disappearances for the *ellesae* Zone is due to its short duration, and the low rate figures for the late Silurian zones are a result of very long durations given by Harland et al. (1989) for the Gorstian and Ludfordian. We cannot criticize the geochronological datings, but surely this uncertainty is a weak point in the usage of different age- (or duration) based rates.

Regional Environmental Differences

Another way to understand the causes of bio-events is the comparison of the corresponding intervals of sections in different geological settings. This makes it possible to sort out geological events connected or not with bio-events and to discriminate between local and wider (global, subglobal) causes of the latters.

The sections discussed below as examples are situated in different facies regions: Midwestern area - an epicratonic shallow sea (carbonate platform), Wales and Welsh Borderland - a transition from the carbonate platform to basin, South Tien Shan - passive margin of the Alai microcontinent with occasional turbidite sedimentation, Bohemia - a rift type trough with volcanic islands, and Thuringia - basinal area with euxinic deep sea and pelagic sedimentation. The treatment is confined to the upper Wenlock-lower Ludlow interval comprising a most striking and well known sequence of different events.

Wenlock-Ludlow Depositional Events, Midwestern craton, U.S.A.

The Wenlock and Ludlow of the midwestern part of the North American craton are especially strategic for testing two somewhat opposed schools of thought that tend to champion tectonism and global eustasy respectively to explain cyclicity in stratigraphic sequences. Although the cyclicity focus here is on sedimentary and biotic events and their interpretation, especially across the Wenlock-Ludlow transition, evidence for the whole of the Wenlock and of the Ludlow, and even evidence for the Llandovery and Pridoli must be examined. As suggested in large part by the data presented in Figs. 17 and 18, the following basic premises are thought to be paramount (see also Droste and Shaver, 1987; Shaver, 1992):

1. Carbonate sedimentation was overwhelmingly dominant in the proto-Illinois Basin and on the middle Paleozoic feature called the Wabash Platform, that is, the large area once separating the three basins noted in Fig. 17 and now named as broad arches. In contrast, evaporite sedimentation in deposits individually as thick as 125 m soon became dominant in the Michigan Basin, there to make up most of the 1,200 m of total Silurian thickness. Three major salts (in Salina A through B units and separated by carbonate rocks, Fig. 18) record late Wenlock and early Ludlow depositional events. Events were similar in the Appalachian Basin but differed by less total thickness (about 770 m) and by increasingly dominant siliciclastic sedimentation (section E, Fig. 17; W. VA. column, Fig. 18).

2. Two notable influxes of far-travelled siliciclastic sediments modified the otherwise totally carbonate dominated Wenlock and Ludlow sequences in the Illinois Basin and on the basin-intervening platform. The lesser of these influxes is represented by the Waldron Shale and equivalents, the larger, by the Mississinewa Shale Member and equivalents. Indeed, the Wenlock began with a siliciclastic influx as represented by the Osgood, Rochester, and other units (Figs. 17 and 18), and at or soon after the close of Ludlow time and D salt deposition, shale became a minor but noted component of the Salina E unit and equivalents.

3. An areally restricted unconformity apparently developed during the Wenlock-Ludlow transition along the western flank of the Cincinnati Arch in western Ohio (Fig. 18), but other intra-Wenlock and Ludlow unconformities appear to be minor and local, as over the tops of large reefs and carbonate banks. Therefore, the overall regional stratigraphy is a facies stratigraphy in the sense of both time and space, between carbonate-bearing rocks (reefs included), evaporite-bearing rocks, and siliciclastic infused rocks. Such units as the Louisville Limestone and equivalents are strongly illustrative. Environmentally it hosted normal-marine reef growth southward in the Ohio River country, but northward it and its equivalents became reef hostile and favoured evaporite deposition and production of algal laminates. Locally, the interaction of hydrodynamic factors resulted in modification of this generality, for example, along the Fort Wayne Bank that fringed the later evaporite basin in Michigan (section C, Fig. 17).

4. The thickness of Silurian rocks in the Michigan Basin is about five times greater than that of the chronologically equivalent platform-situated rocks (where not eroded). This generality holds well for both reef and nonreef sections, for example, the approximately age equivalent late Wenlock reefs of generations 3 and 4 as portrayed in Figs. 17 and 18 for thin Silurian sequences (25-30 m) on the platform and thick sequences containing 125 m-thick pinnacle reefs in the Michigan Basin.

5. Especially the several well-coordinated (for starts and some abortions) reef generations (Fig. 18) evince strong evidence of a regional-scale, or broader, depositional cyclicity. This evidence includes that of a reciprocal relationship with major evaporite episodes but in-phase relationships with the more significant siliciclastic influxes (Fig. 18); only in the Appalachian Basin did the detrital influx as of itself become so great, after Wenlock time, that it was detrimental to reef growth or prevented it altogether (W. VA. column, Fig. 18).

6. Although the question has been strongly debated, most of the rocks (and areas) denoted in Figure 18 were the products of (were dominated by) shallow-water regimes, even intratidal to supratidal regimes in places. The two evaporite basins may have repeatedly filled to and above sea-level while the intervening platform remained mostly awash.

7. As shown by Fig. 19, the palaeogeographic and chronologic transitions from one kind of regime to another were exceedingly interinvolved. And the sedimentational and biotic events were inextricably intertwined. Where a normal-marine biota was present, it was characterized by heavy-skeleton invertebrates.

8. Notable distinction existed between level-bottom and reef communities, both mostly of cosmopolitan composition except in the extreme examples of almost exclusively algae dominated communities.

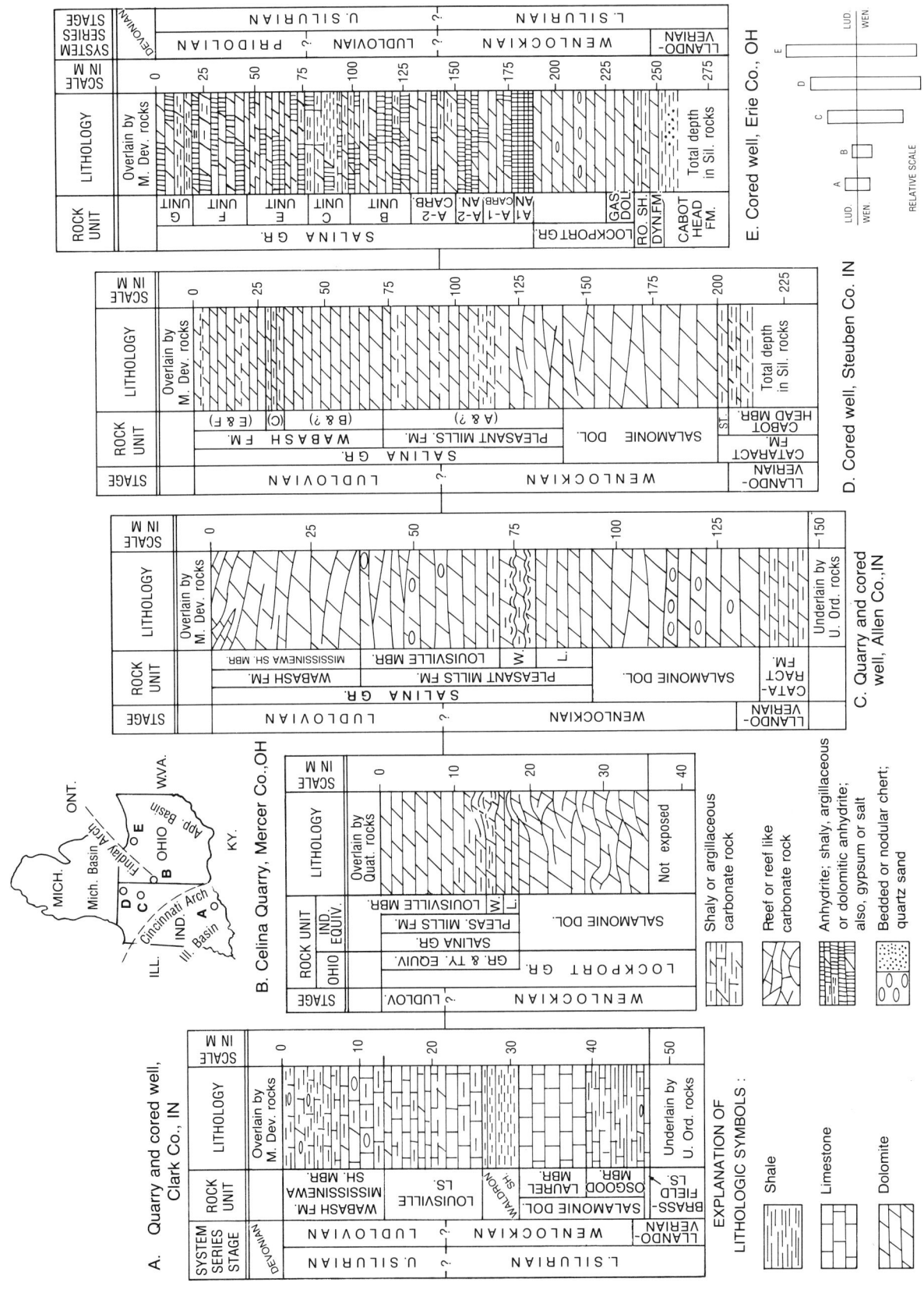

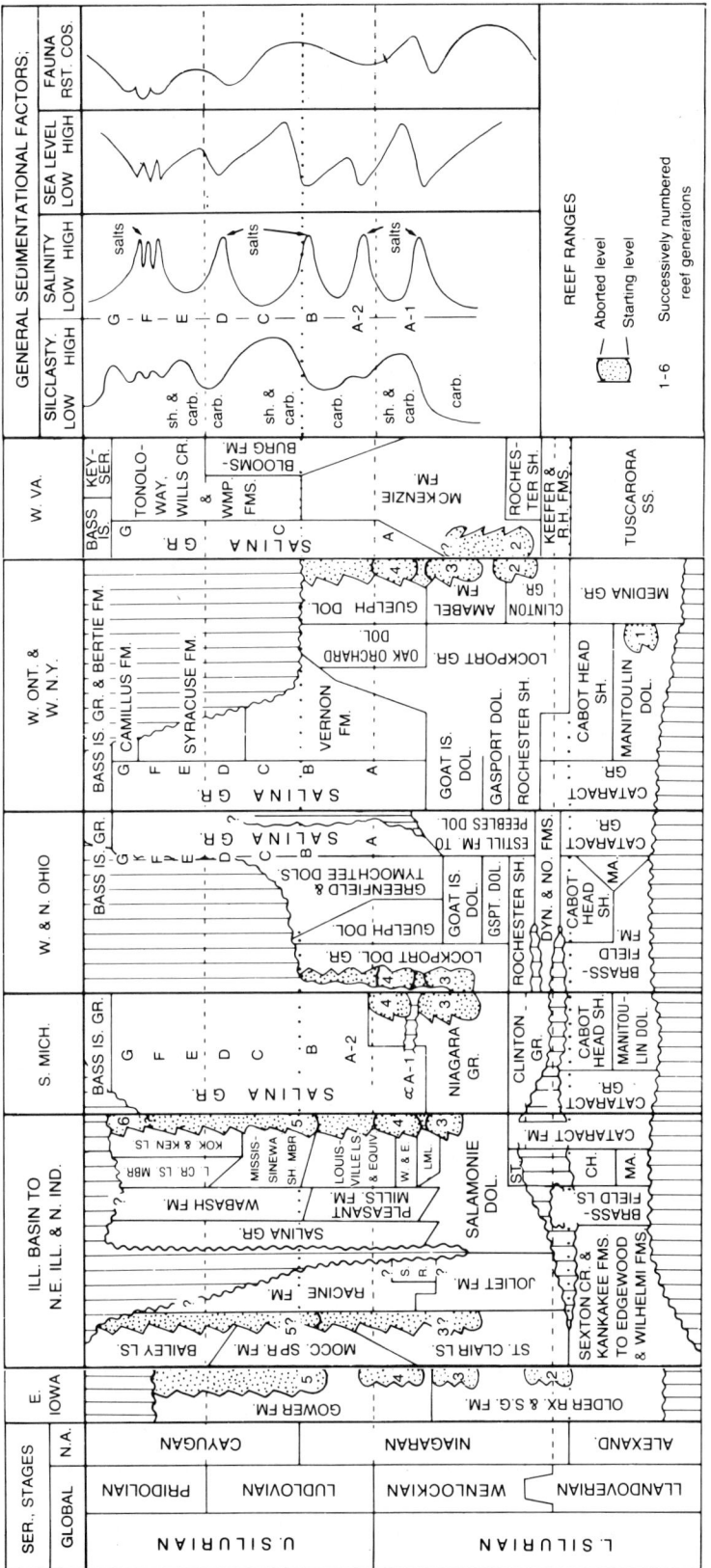

Fig. 17. Sections for three Silurian protobasins and the intervening platform of the Midwestern area. Note: Vertical scale of the sections changes. Abbreviations from the left: GR. & TY., Greenfield and Tymochtee Dols.; PLEAS., Pleasant; W., Waldron Mbr.; L, Limberlost Dol. Mbr.; (A & ?) and similar, equivalent of the A unit, Salina Gr. of Michigan; ST., Stroh Mbr.; AN., anhydrite; GAS., Gasport; RO., Rochester; and DYN., Dayton. Sources of all information are cited in Shaver (1991).

Fig. 18: Chart showing stratigraphic nomenclature for selected areas of Silurian rocks in the Midwestern craton, U.S.A.; ranges of six principal reef generations; and factors in sedimentation and biotic development. Note: a common starting level is the basis for a single generation; especially generation 1 is a collective generation. Abbreviations from the left: S. G., Scotch Grove; Mocc. Spr., Moccasin Springs; S. R., Sugar Run; L. CR., Liston Creek; KOK., Kokomo; KEN., Kenneth; W. & E., Waldron Sh. and equivalents; LML., Limberlost Dol. Mbr.; ST., Stroh Mbr.; C. H., Cabot Head Mbr.; MA., Manitoulin Dol. Mbr., BASS. IS., Bass Islands; GOAT IS., Goat Island; GSPT., Gasport; DYN. & NO., Dayton and Noland; WMP., Wiliamsport; R. H. Rose Hill; SILCLASTY, siliciclasticity; RST., restricted; COS, cosmopolitan. Adapted from Shaver (1991).

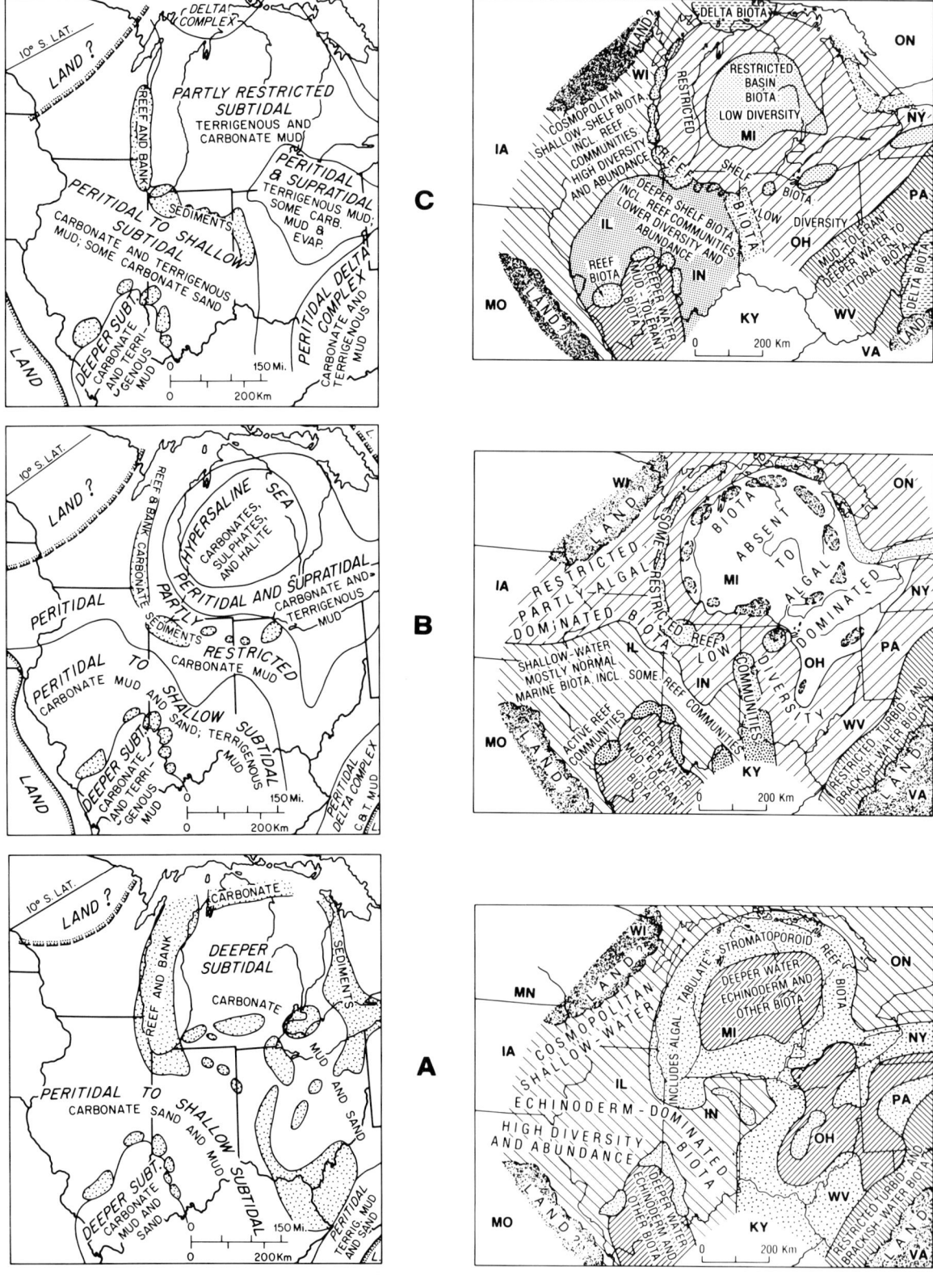

Fig. 19 (opposite page). Three maps showing dominant biotic character of the Midwestern area. A, Mid-Wenlock time; B, late Wenlock-early Ludlow time; and C, mid-Ludlow time. The maps are from Droste and Shaver (1987)

Discussion

Much evidence attests to an active tectonism: (1) disparate thicknesses for contemporaneously deposited rocks in the basins and on the adjacent platform; (2) generally shallow water lithologies that do not support the idea of initial Silurian depths to accommodate eventual thicknesses; (3) special interaction at the basin-platform transitions such as thickening in those areas into reeflike carbonate banks (e.g., the great northern Indiana carbonate bank depicted at the upper Louisville Mississinewa level in section C of Figure 17); and (4) the differential magnitude and the presence/absence of a late Wenlock unconformity (W. & N. Ohio column, Fig. 18).

Another part of the evidence strongly favours fluctuation of at least relative sea-level: (1) the Wenlock began with a short deepening episode and ended after a longer shallowing episode, and thus this stage is characterized by a thin deepening upward sequence followed by a thicker shallowing upward sequence; (2) the Ludlow, or approximately late Wenlock through Ludlow time, is characterized in the same way; (3) these two principal cycles compare best with the third-order cycles in global sea-level change set forth by Vail et al. (1977), the same cycles that were recognized globally by Johnson, Kaljo and Rong (1991)

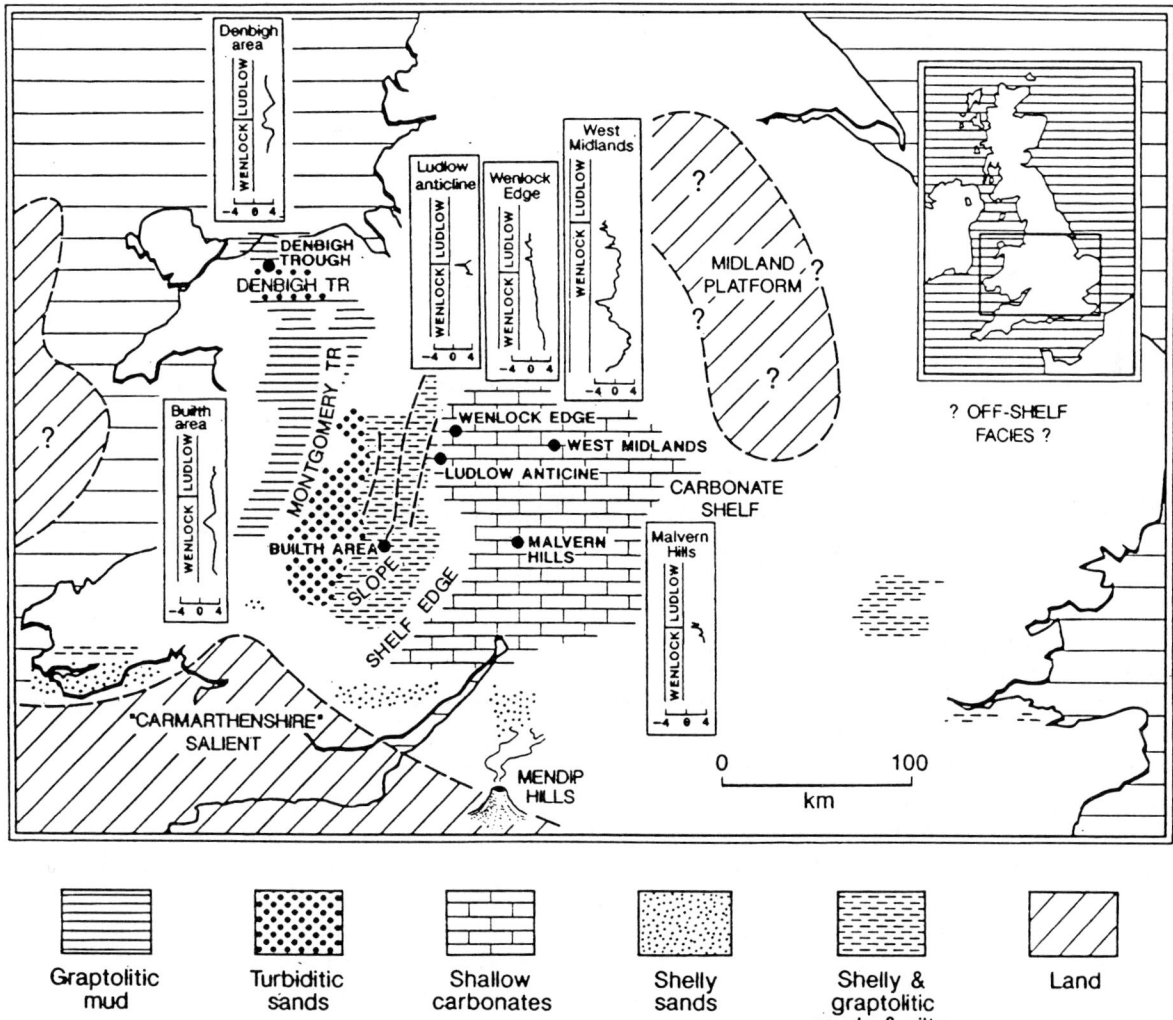

Fig. 20. Generalized palaeogeography and facies for southern Britain in the late Wenlock (Homerian), modified from Siveter, Owens and Thomas (1989, fig. 8), together with the location of sites sampled and their $\partial^{13}C$ data

among others; here, therefore, seems to be evidence that midwestern Silurian cratonic rocks reflect a global eustasy; (4) the smaller scale cyclicity noted with respect to evaporite-carbonate deposition beginning during late Wenlock is not well understood, although eustasy, local tectonism, and climate could be instrumental; (5) the longer cycles appear to have durations of about 4 to 5 million years; the next shorter ones relating to major salt depositions, less than 1 million years; and (6) the view much favoured here as to the magnitude of sea-level change is that only modest fluctuation occurred, say a few tens of feet or a few tens of metres at the most (Droste and Shaver, 1987; Johnson, 1987).

Both tectonism and relative sea-level change seem necessary, therefore, to explain the phenomena discussed here, and probably some global-scale eustasy was involved to account for the longer cycles. The purpose here does not extend to deeper explanation. Sloss (1988), however, eschewed glacio-eustatic explanations when hard evidence of glaciation is unknown but did favour eustatic theory in relation to sea-floor spreading and thermal changes within the earth. These causes cannot be showed to operate on a lesser scale than that assigned here to the two large cycles. But local tectonic, climatic, and possibly other causes could operate at the lesser scale.

Carbon Isotope Changes during *ludensis-nilssoni* Time in Wales and The Welsh Borderland

Carbon isotope stratigraphy is becoming a widely used correlation tool in marine sequences from a variety of time intervals (see Holser and Magaritz, this book).

The first data from strata of Silurian age were recently published by Corfield et al. (1992) and Corfield and Siveter (1992) and are summarized here with the aim of contributing to our understanding of the connection between biotic and environmental events. Hence our aim has been to investigate whether any systematic changes in $\partial^{13}C$ or $\partial^{18}O$ occur across Wenlock-Ludlow boundary sections in Wales and the Welsh Borderland, including the type areas for the Wenlock and Ludlow Series. These boundary sections include the *lundgreni* to *nilssoni* graptolite biozones interval.

The sites sampled (Fig. 20) occur over a transect which includes shallow marine facies on a carbonate platform area (the West Midlands of England; the Wenlock Edge to Ludlow area; the Malvern Hills), to more offshore shelf-edge/slope deposits (Builth area), to fully graptolitic shales deposited more nearly in the centre of the Welsh Basin (Denbighshire). Sampling intervals, methods of analysis, and checks for diagenetic effects are outlined in Corfield et al. (1992). Present evidence suggests that the isotope change recorded across the Wenlock-Ludlow boundary reflects palaeoceanographic change on at least a regional, sedimentary basin scale rather than reflecting local diagenesis.

Isotope Curves and Faunal Change

Our data (Fig. 21) show an approximately monotonic decline in the $\partial^{13}C$ of whole-rock samples from the uppermost Wenlock into the basal Ludlow. This depletion occurs in carbonate shelf environments (the West Midlands; Wenlock Edge; the Ludlow anticline; the Malvern Hills) as well as more offshore, deeper water facies (Builth area). Our data also show another $\partial^{13}C$ decline slightly lower in the upper Wenlock, in the West Midlands and Builth sections.

The Builth area is now regarded as the type area for the Wenlock graptolite sequence in the Welsh Basin (Elles, 1900). The recent study of Harris (1987) has shown that in the River Irfon section the *G. nassa* Biozone begins some 10 m below the base of the Ludlow and that here this biozone is 1.75-2 m thick. The stratigraphically lower $\partial^{13}C$ depletion coincides in the River Irfon section with the *G. nassa* Biozone.

A major, global crisis in pelagic graptolite diversity occurred in post-*lundgreni* time, at the level of the *nassa* Biozone, with just a few species surviving the *lundgreni* extinction Event. Diversity stayed very low through *M. ludensis* Biozone times until it recovered in the lower Ludlow *nilssoni* Biozone (Jaeger, 1991; Koren, 1987, 1992; Koren and

Fig. 21 (opposite page). $\partial^{13}C$ data and correlation ties for the localities sampled (see Corfield and Siveter, 1992). The basin dysaerobism data, including recognition of the two late Wenlock oxic events (indicated by lack of diagonal ornament), are taken from Kemp (1991). The position of the *G. nassa* Biozone at Builth is based on Harris (1987), and its position at Wenlock Edge on Bassett et al. (1975). The strata which possibly correspond to the *G. nassa* Biozone in the West Midlands are correlated with the Builth section on the basis of the $\partial^{13}C$ curves. To highlight $\partial^{13}C$ trends, the raw data have been filtered with a three point moving average

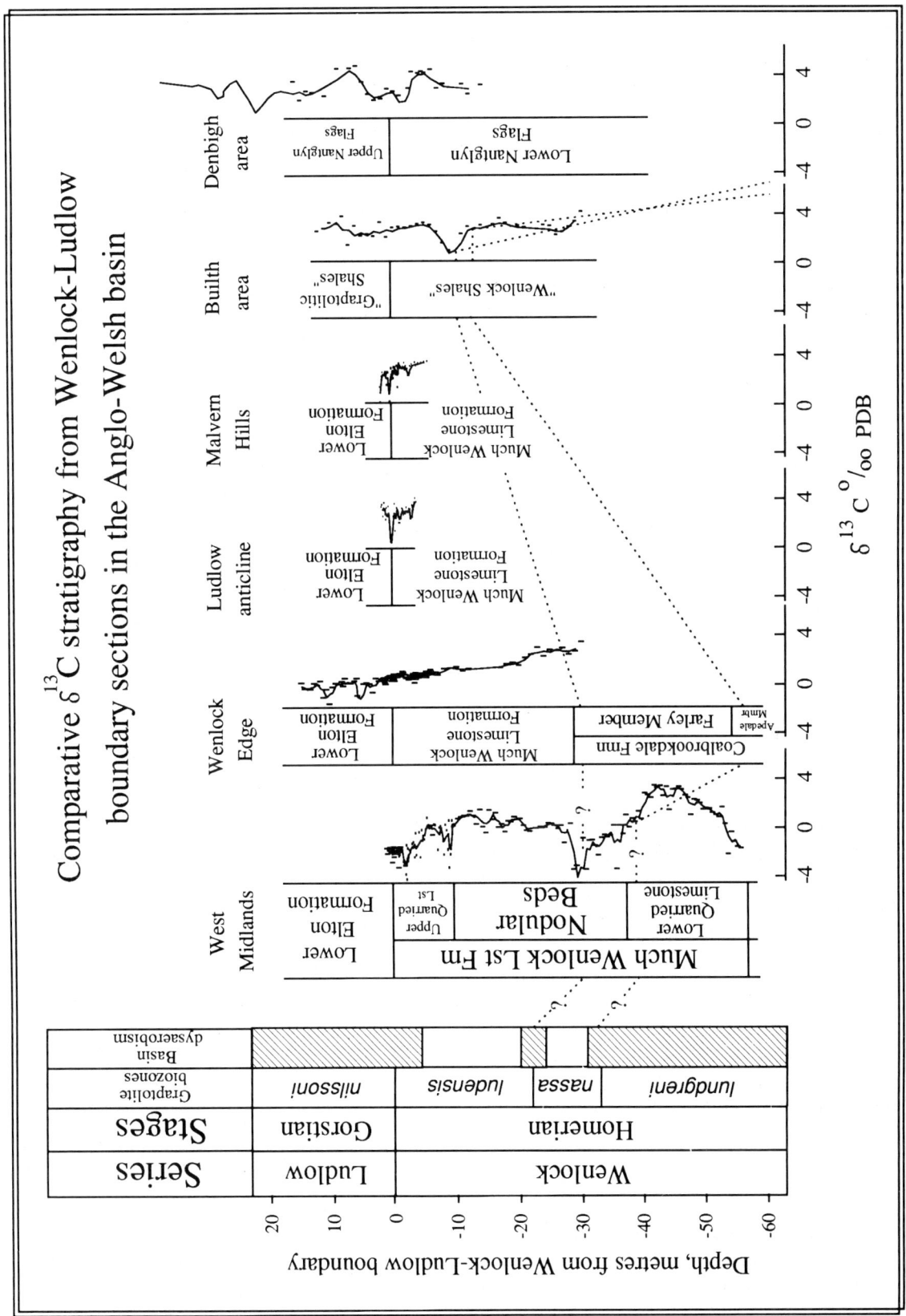

Rickards, 1980). Corfield et al. (1992) have suggested that the stratigraphically lower $\partial^{13}C$ depletion noted in the River Irfon section may be the expression in the whole-rock carbon isotope record of the *G. nassa* Biozone diversity decline. Higher in this section, $\partial^{13}C$ values recover to intermediate values before declining once more across the Wenlock-Ludlow boundary.

Given the very short time interval associated with the sections investigated (about 1-2 Ma; see Rickards, 1989; Zalasiewicz, 1990), it seems possible that the two episodes of carbon isotope depletion are in fact two aspects of the same phenomenon. The graptolite diversity decline may in fact be a stepped process, this being reflected in the shape of the $\partial^{13}C$ decline (Corfield et al., 1992).

In terms of amplitude the $\partial^{13}C$ depletion straddling the Wenlock-Ludlow boundary is approximately 4‰ in the sections analysed. Current estimates of the average length of Silurian graptolite biozones (see below) suggest that the section in the West Midlands is about 2 Ma or less in duration. The resulting minimum rate of $\partial^{13}C$ decline of about 8‰/Ma is therefore comparable with the rate of ^{13}C depletion at the Cretaceous-Tertiary boundary (between 15‰/Ma at DSDP holes on the Walvis Ridge (Shackleton and Hall, 1984) to 5‰/Ma in the Bottaccione Gorge, Italy (Corfield et al., 1991). As is well known, at the latter horizon there is also a major episode of marine and terrestrial extinction.

Extinction and reduced productivity in the marine realm have palaeoceanographic implications. Reduced productivity may lead to a decrease in the depth and intensity of the oxygen minimum zone in a manner analogous to that suggested for the other intervals of profound carbon isotope depletion. This hypothesis can be tested by using sedimentological data. Kemp (1991) has identified two major intervals of increased ocean basin ventilation in late Wenlock times. He suggested that this late Wenlock "oxic" event comprised two sub-events when at least the Iapetus Ocean basin was flushed with relatively more oxygenated waters. These oxic events coincided with intervals of eustatic sea-level low stand and increased carbonate deposition on the shelves and were suggested by him to be the result of enhanced deeper water circulation.

Both oxic events identified by Kemp appear to correlate with the two episodes of ^{13}C depletion documented in Fig. 21. Kemp (1991) was unable to distinguish between reduced primary productivity or increased deep water ventilation to account for his suggested oxic intervals. We suggest that the late Wenlock graptolite extinctions are implicated in the generation of the oxic events and that accordingly greater emphasis should be placed on the reduced productivity hypothesis.

In order to fully appreciate the significance of the $\partial^{13}C$ minima near the Wenlock-Ludlow boundary we need to establish the carbon isotope variability of adjacent strata and also that of sections outside the Welsh Basin. Finally, we regard our data as preliminary, and the $\partial^{18}O$ signal remains to be investigated further due to its susceptibility to diagenetic processes.

Mid-Silurian Transition in the Pelagic Sequences of South Tien Shan

Pelagic deposits of Wenlock and Ludlow age form a part of the allochthonous terrigenous suites in the Alai Range, South Tien Shan. They are assigned to the Kursala Formation and represented by the condensed sequences of black carbonaceous mudstones and siltstones with subordinate calcareous layers and lenses (Koren et al., 1986). The Kursala Formation is well exposed along the northern slope of the Alai Range to the south of Fergana city between the Sokh and Isfajramsai rivers (Fig. 22). Though the sections are often complicated as a result of folding and thrusting, tectonically undisturbed continuous sequences are quite often preserved within the area. Abundant flattened, but well preserved graptolites allow cm-scale strati-

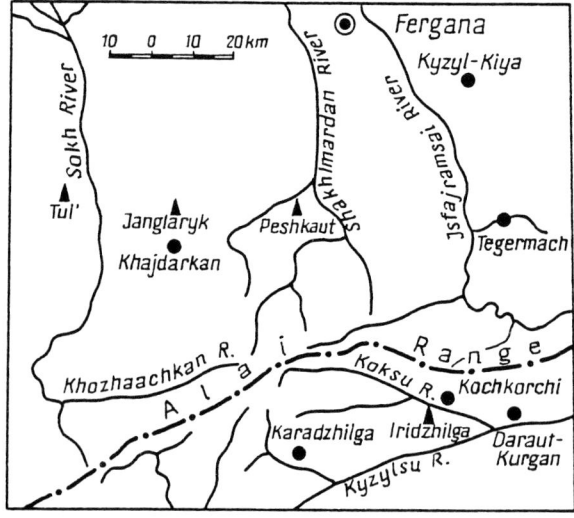

Fig. 22. Sketch map showing location of the sections studied (black triangles) within the Alai Range, South Tien Shan

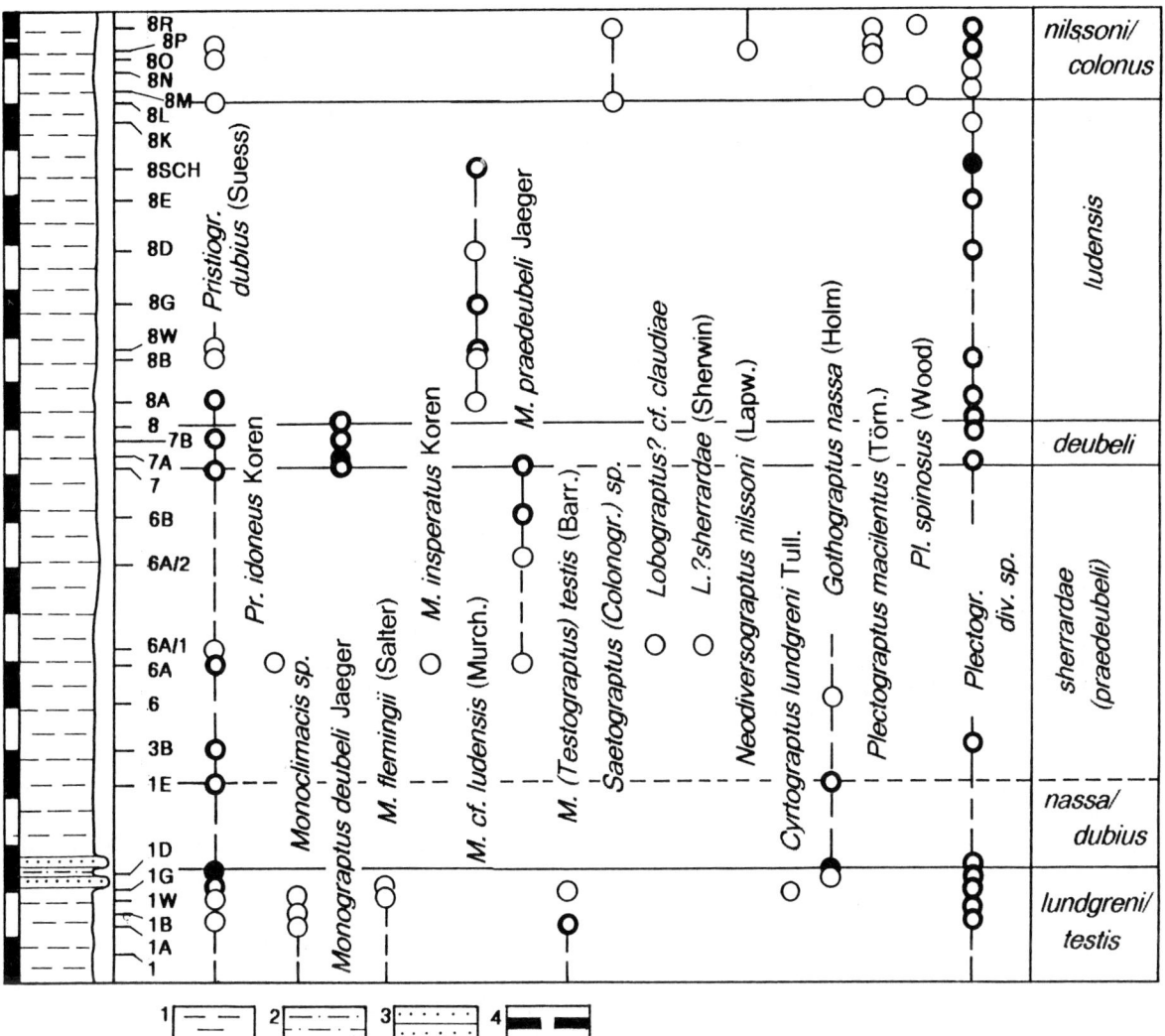

Fig. 23. Stratigraphic section and distribution of graptolites in the lower Kursala Formation near the Tul' Village. 1 mudstones; 2 siltstones; 3 sandstones; 4 a fault. Filled circles: abundant specimens; half-filled circles: a few specimens; blank circles: single specimens. Scale on the left in metres

graphical observations, providing the possibility for detailed zonal subdivision and studies of graptolite diversity dynamics within the Wenlock-Ludlow boundary interval.

The Homerian to Gorstian zonal sequence has been worked out resulting from the study of seven measured sections in the Peshkaut and Jangiaryk Valleys, and near the Tul village (Fig. 22). These studies revealed about 50 species of monograptids, plectograptids and cyrtograptids which form successive zonal assemblages (Koren, 1991b, 1992). Based on their stratigraphical ranges in the lower Kursala Formation, the following graptolite zones have been established: *lundgreni/testis, nassa/dubius, sherrardae/praedeubeli,* *deubeli, ludensis* and *nilssoni/colonus* (Fig. 23). Some zonal boundaries are based on the distinctive changes in the graptolite diversity expressed in extinctions, radiations and population bursts (the *nassa/dubius* and *nilssoni/colonus* zones). The others are defined by speciation events, namely by the first appearance of the diagnostic species (the *sherrardae, deubeli* and *ludensis* zones).

The general information has been presented earlier in this chapter where Fig. 13 demonstrates the graptolite diversity pattern in the Kursala Formation. In order not to repeat it here only one figure (Fig. 24) is added for characterizing graptolite diversity dynamics in the section.

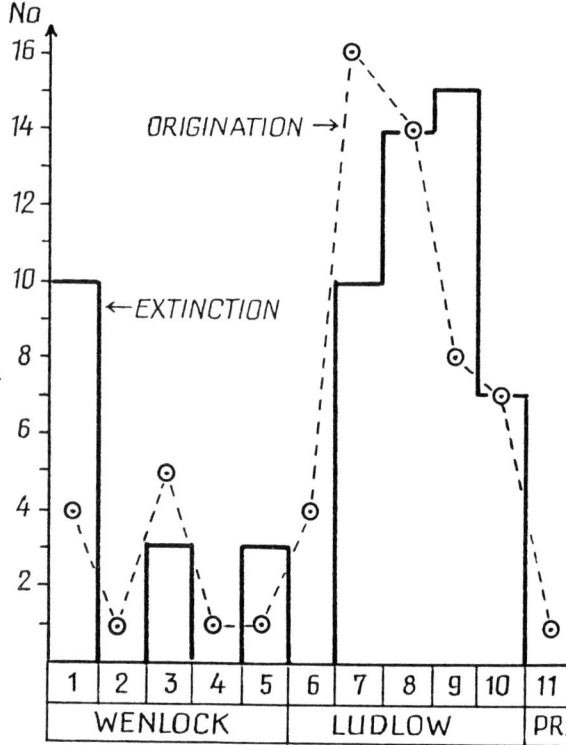

Fig. 24. The number of graptolite extinctions and originations in late Wenlock to early Pridoli in the Kursala Formation. Zones: 1: *lundgreni/testis;* 2: *dubius/nassa;* 3: *sherrardae;* 4: *deubeli;* 5: *ludensis;* 6: *nilssoni/colonus;* 7: *scanicus/chimaera;* 8: *leintwardinensis;* 9: *cornutus/podoliensis;* 10: *formosus/ spineus;* 11: *parultimus*

Distinctive changes of environmental factors are difficult to recognize in the sequences studied. The *lundgreni/testis* Zone, about 70 m thick, consists of dark coloured mudstones and siltstones bearing calcarenite layers and lenses which give evidence of the frequent episodes of turbidite sedimentation. The overlying Homerian is predominantly represented by dark-coloured mudstones with occasional thin layers of sandstones. These changes can be attributed to local phenomena. Another change takes place in the *ludensis* Zone, when the mudstones and siltstones become of lighter colour compared with the underlying and overlying strata. This could have resulted from the oxygenation of the water in the terminal phase of the Wenlock.

The post-*lundgreni* graptolite diversity dynamics affords an excellent opportunity to analyse the patterns of repopulation. In the sequences studied in the Kursala Formation, both survival and recovery phases (Harries and Kauffman, 1990) are easily defined. The post-extinction fauna, embracing a small number of taxonomically different groups, shares some common features such as extremely low morphological diversity (monograptids with simple thecae and those with small lateral incisions on the thecal aperture proximally), and survival of the long-lived ecological generalists like *Pristiograptus dubius* represented by taxa of a simple morphology from which a new and more specialized fauna emanates. Another common feature is the population bloom in single taxa (*P. dubius, G. nassa*).

The survival phase (Fig. 13) is defined by the alternation of episodes when origination exceeds extinction or origination rate and amount of the background extinctions are close to zero. The recovery phase begins with the first radiation within at least two lineages: *idoneus-sherrardae* and *praedeubeli-deubeli,* resulting in the appearance of morphologically diverse graptolite faunas.

During the repopulation interval plectograptids and monograptids, which suffered equally at the *lundgreni* crisis, show contrary tendencies in their extinction and origination rates (Fig. 13). This might be explained by their differential responses to the ecological crisis and new environmental opportunities. In one way or the other, plectograptids and monograptids might be taken as ecological antagonists. This conclusion highlights an advantage of the differential approach when analysing taxonomically or/and ecologically heterogeneous fauna.

Graptolite diversity dynamics at the Wenlock-Ludlow boundary observed in the pelagic facies of the South Tien Shan is based on the evolution of the indigenous faunas with only a small amount of immigration. The observed trends at the species-zonal level is considered to be a representative of the actual extinction and recovery patterns on a global scale due to the cosmopolitan distribution of the graptolites and completeness of the regional data-base. Causality of the *lundgreni* Event in pelagic facies is not yet properly understood when compared with climatically triggered mass extinction at the end of the Ordovician.

Major Facies and Faunal Changes in the Silurian of the Prague Basin, Bohemia

The Ordovician - Middle Devonian Prague Basin was defined as a linear depression of a rift type (Havlicek, 1981). It represents one of the troughs on the northern Gondwanaland of which history started in the lowermost Ordovician and lasted till the Middle

Devonian. The Prague Basin was in the early Ordovician narrow, 10-15 km, later up to 25 km wide depression which strikes 65° (Havlicek, 1992). It is a part of the Teplá - Barrandian block of the Bohemian Massif.

Tectonic, Eustatic and Volcanic Controls on Facies and Faunal Development in the Silurian of the Prague Basin

The early Silurian transgression flooded the peneplaned land adjacent to the Prague Basin, causing a marked extension of the marine areas on the craton. The Silurian in the Prague Basin and adjacent basins was characterized by a slow sedimentation rate because of the much greater distances away from the sources of terrestrial material and because of the major changes in water circulation. Anoxic conditions developed across the Gondwanan shelves, as documented in all known parts of the Prague Basin by the presence of black graptolite shale facies. These anoxic conditions persisted in the deeper parts of the basin at least until the end of the Silurian. Only the shallower parts of the basin were ventilated by wind action and currents. Oxygenation of the bottom permitted the accumulation of carbonate sediments and the development of diversified communities of benthic organisms (Kriz, 1991, 1992).

Movements of individual segments along the deep synsedimentary faults (Fig. 25) distinctly influenced the Silurian facies developments within the Prague Basin (Kriz, 1991, 1992).

In the Prague Basin the earliest Silurian sediments were pelitic, but during the Telychian (late Llandovery) and especially in the early Sheinwoodian (early Wenlock) the carbonate admixture increased continuously (Fig. 26). In the calcareous shales and rare limestone intercalations the brachiopod dominated, low diversity *Valdaria budili* Community occurs in the *Cyrtograptus murchisoni* Biozone (Havlicek and Storch, 1990; Kriz, 1992).

During the middle Llandovery (late Aeronian) and from the middle Wenlock to the middle Ludlow the sedimentation was influenced by volcanic activity, accompanied by the production of volcaniclastic sediments, magma intrusions and effusions. Several

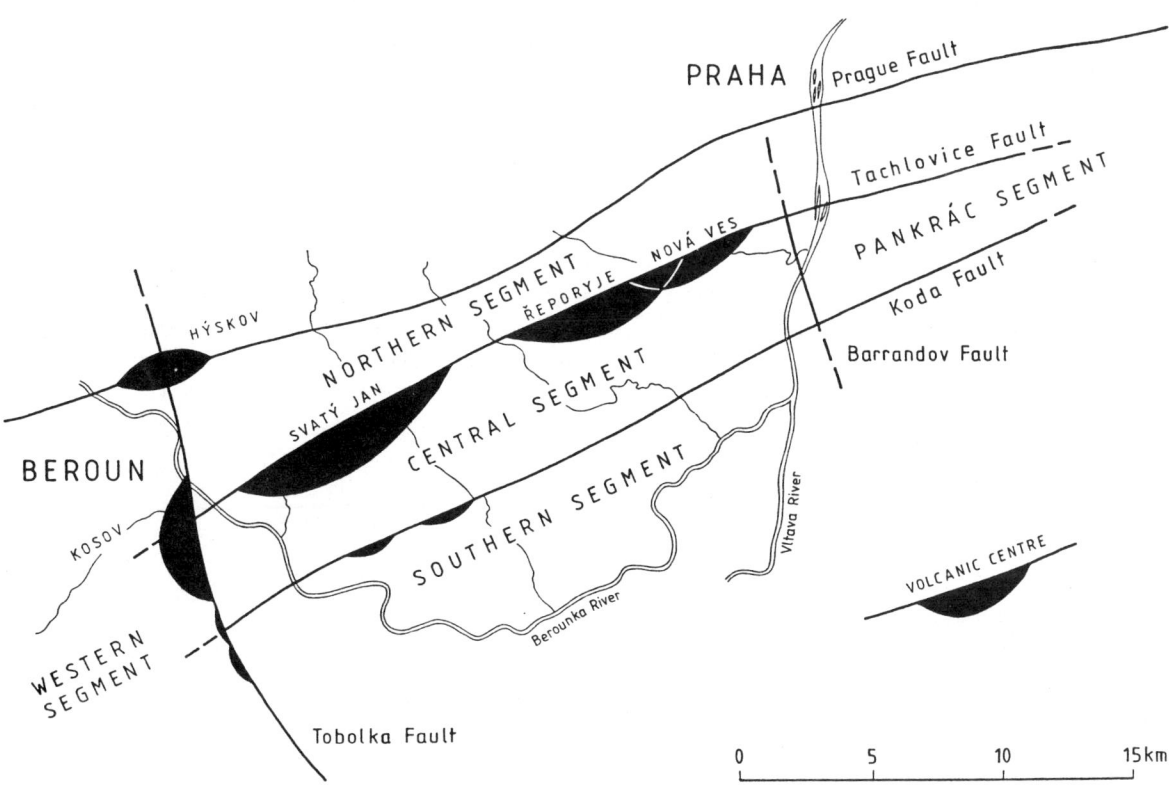

Fig. 25. Silurian synsedimentary tectonics, volcanic centers and segments (sub-basins) in preserved parts of the Prague Basin (after Kriz, 1991)

CHRONO-STRATIGRAPHY			BIOSTRATIGRAPHY		LITHOSTRATIGRAPHY	LITHOLOGY (shallow → deeper)
S I L U R I A N	PŘÍDOLÍ		M. transgrediens Interzone		Požáry (Přídolí) Formation	
			M. perneri			
			level with M. beatus			
			M. bouceki			
			M. lochkovensis			
			level with M. pridoliensis			
			M. ultimus			
			M. parultimus			
	LUDLOW	LUDFORDIAN	M. fragmentalis		Kopanina Formation	
			zones not yet established			
			M. fritschi linearis			
		GORSTIAN	M. chimaera			
			C. colonus (N. nilssoni)			
	WENLOCK	HOMERIAN	M. ludensis		Motol Formation	
			levels with G. nassa			
			M. dubius parvus			
			C. lundgreni	T. testis		
				C. radians		
		SHEINWOODIAN	C. perneri / C. ramosus			
			C. rigidus			
			M. belophorus			
			M. dubius dubius			
			M. riccartonensis			
			C. murchisoni			
			C. centrifugus			
			C. insectus			
	LLANDOVERY	TELYCHIAN	S. grandis		Litohlavy Formation	
			O. spiralis			
			M. tullbergi			
			M. griestoniensis			
			P. crispus			
			S. turriculatus			
			R. linnaei			
		AERONIAN	M. sedgwickii		Želkovice Formation	
			D. convolutus			
			D. simulans			
			D. triangulatus – D. pectinatus			
		RHUDDANIAN	C. cyphus			
			C. vesiculosus			
			A. ascensus – A. acuminatus			

Legend: SHALES; SILICITES; VOLCANITES; LIMESTONES; GREEN CALCAREOUS SHALES; CALCAREOUS SHALES; TUFFS / TUFFITES; TUFFACEOUS LIMESTONES; BIOCLASTIC LIMESTONES; LIMESTONES WITH CEPHALOPODS

Fig. 26 (opposite page). Silurian stratigraphy of the Prague Basin (after Kriz, 1992)

volcanic centres developed along the deep synsedimentary (growth) faults and on their zones of intersection (Kriz, 1991, 1992) and composed the Silurian volcanic archipelago on which the life evolution was like that on some virgin archipelagos (e.g. Galápagos).

Redeposition of volcaniclastics around the volcanic centres led to the origin of well-ventilated shallow-water flats, which were occupied by a rich benthos. Bioclasts were produced here in large quantities and formed a substantial proportion of the sediments. The bioclasts were also transported by bottom currents into the deeper parts of the basin where pelitic or tuffaceous - pelitic sedimentation was dominant. During periods of maximal volcanic activity and subsequent redeposition of volcaniclastics, the tuffaceous admixture in the sediments increased. In the region of the Svaty Jan Volcanic Centre, reworked limestone pebbles and fragments of corals, stromatoporoids and bryozoans document the presence of very shallow subtidal and intertidal environments during the Wenlock, and thus suggest the existence of an emerged volcanic island at that time at least. Local emergence related to the volcanic activity is also indicated in the Kosov Volcanic Centre by presence of reworked limestone pebbles in the middle Gorstian (Ludlow) sediments (Kriz, 1991).

In the late Wenlock (Homerian), eustatic movements were a primary cause of shallowing, for which there is evidence also from other Gondwanan basins. This is reflected locally in the presence of shallow-water bioclastic limestones with cephalopods (in the upper *Cyrtograptus lundgreni* Biozone), which are known also from the Massif Mouthoumet, Montagne Noire, Carnic Alps and Sardinia (Kriz, 1984, 1995b; Kriz and Serpagli, 1993). It can be assumed that the emergence of the Svaty Jan Volcano was contemporaneous with the late Wenlock low stand of the sea-level. There were other similar low stands documented by cephalopod limestone biofacies in the earliest Ludlow (lower part of the *Monograptus colonus* Biozone), middle Ludlow (early Ludfordian), and in the latest Ludlow (*Monograptus fragmentalis* Biozone; biozonation used above, see Fig. 26). Locally, the cephalopod limestone biofacies also existed on sea-floor elevations in the early Pridoli (Kriz, 1984, 1991, 1992).

Transgressions in the middle Ludfordian and in the earliest Pridoli caused the major change of the facies and the benthic communities.

Benthic Communities and Facies Development during the Llandovery and Wenlock; Important Role of the Volcanic Archipelago

When volcanic activity started in the Hyskov Volcanic Centre (late Aeronian, middle Llandovery), a shallow sea bottom on the top of volcanic accumulations became ventilated. This represented good conditions for existence of the first rich benthic fauna in the Silurian of the Prague Basin which formed the brachiopod, trilobite and dendroid dominated *Aegironetes - Aulacopleura* Community (Havlicek and Storch, 1990). It is composed mainly of brachiopod genera *Stricklandia, Aegironetes, Glyptorthis, Giraldibella, Rhaetorthis, Saukrodictya* and *Hirnantia* and trilobite genera *Stenopareia, Aulacopleura, Leonaspis, Balizoma, Diacalymene, Dicranopeltis, Trochurus, Bohemoharpes, Selenopeltis, Hadromeros* and *Youngia* (Havlicek and Kriz, 1973). Original species of these genera migrated with sea currents from the other basins and adapted, as did most of endemic species to new conditions developed in the Prague Basin.

A similar situation existed later during the Wenlock and early Ludlow in vicinity of the Svaty Jan Volcanic Centre, the Reporyje Volcanic Centre, the Kosov Volcanic Centre and in the Nová Ves Volcanic Centre. Many different benthic communities (Havlicek and Storch, 1990; Chlupac, 1987; Kriz, Dufka and Schönlaub, 1993) flourished on the bottom of the shallowed and well ventilated sea on the slopes of the volcanic islands which originated in the Prague Basin. They were dominated in various facies and stratigraphical levels especially by the brachiopods (*Niorhynx* Community, *Leptaena rugaurita* Community, *Strophoprion - Eoplectodonta* Community, *Plicocyrtia* Community, *Hircinisca - Ancylotoechia* Community, *Septatrypa lissodermis - Cyrtia maior* Community, *Bucequia obolina* Community and *Indaclor sulcicarens* Community) and by the trilobites (*Miraspis - Mezounia* and *Aulacopleura konincki* Communities).

These shallow water communities were composed mostly of endemic species developed by adaptive radiation in special local conditions on the virgin Silurian volcanic archipelago.

In slightly deeper environment, which was only temporarily ventilated by currents, cephalopod limestone biofacies with the Bivalvia dominated *Cardiola agna agna* Community developed (Kriz, 1995a). Very similar *Cardiola agna figusi* Community occurs also in Sardinia and Montagne Noire (Kriz and Serpagli,

Table 2. Graptolite zonation and distribution of benthic communities at the Wenlock-Ludlow boundary interval in the Prague Basin

GRAPTOLITE BIOZONES			BIVALVIA COMMUNITIES	TRILOBITE ASSEMBLAGES	BRACHIOPOD COMMUNITIES
M. fritschi linearis			Cardiola docens	Encrinuraspis-Eophacops	Smooth Atrypid
					Leptostrophiella nebulosa
M. chimaera (M. scanicus)			Cardiola donigala	Sphaeroxochus-Proetus	Atrypoidea renitens
C. colonus (N. nilssoni)			Cardiola gibbosa	Interproetus-Eophacops	Indaclor sulcicarens
					Indaclor sulcicarens
M. vulgaris-M. gerhardi			Cardiola gibbosa	Richterarges-Eophacops	Septatrypa lissodermis-Cyrtia maior
				Richterarges-Eophacops	Coral-Leptaenid
					Hircinisca-Ancylotoechia
M. deubeli					
M. praedeubeli					
M. dubius frequens-G. nassa					
M. dubius parvus				Raphiophorus-Delops	Ravozetina-Bracteoleptaena
C. lundgreni	T. testis		Cardiola agna	Aulacopleura konincki	Buceqia obolina
				Bumastus-Sphaerexochus-Cheirurus	Plicocyrtia
	C. radians			Bumastus-Sphaerexochus-Cheirurus	Strophoprion-Eoplectodonta

1993; Kriz, 1995a, 1993b). These communities developed from the larvae transported into the basin by currents. They are mostly composed of the cosmopolitan species adapted to similar conditions which existed also in the other basins.

Wenlock - Ludlow Boundary Development

During the summer of 1991 the complete sequence of the graptolite zones between the *Cyrtograptus lundgreni* and *Colonograptus colonus* biozones, identical with that described by Jaeger (1991) from Thuringia, was discovered in the Southern Segment of the Prague Basin. The *Testograptus testis* Subzone is overlain here by the *Monograptus dubius parvus* Biozone (Table 2). Together with this graptolite appeared in the deeper Western, Southern, Northern and Pankrác Segments of the Prague Basin (Fig. 25) a distinct new community of brachiopods and trilobites (*Ravozetina - Bracteoleptaena* Community) characterized by genera *Ravozetina, Strophochonetes, Bracteoleptaena, Rabuloproetus, Decoroproetus, Delops* and *Raphiophorus* accompanied by "*Plumulites*" *minimus*, "*Ortotheca*" *pulchra* and common thallophyts (Havlicek and Storch, 1990; Kriz, Dufka and Schönlaub, 1993). At the very end of Wenlock and beginning of Ludlow (corresponding to the *Monograptus vulgaris* and the *Colonograptus colonus* Biozones) in the western part of the Northern Segment a distinct uplift is documented by formation of limestones with the brachiopod dominated *Indaclor sulcicarens* Community (Kriz, Dufka and Schönlaub, 1993). High energy crinoidal limestones with corals and leptaenid brachiopods of the "Kozel" complex were formed in the shallow parts of the Central Segment of the Prague Basin on slopes of the Svaty Jan Volcano Island during the *Monograptus dubius parvus* (?) - *Monograptus vulgaris* Biozones (Fig. 27).

The crisis in graptolite evolution (Jaeger, 1991) was contemporary with a distinct change in benthic fauna in the Prague Basin when in the *Monograptus dubius parvus* Biozone time appeared new brachiopod and trilobite communities (Table 2). The graptolite crisis was also contemporary with a distinct change of *Chitinozoa* associations. The assemblage characterized in the *Testograptus testis* Subzone by *Conochitina tuba* was replaced at the base of the *Monograptus dubius parvus* Biozone by assemblage with *?Eisenackitina* sp. and *Eisenackitina* cf. *intermedia*.

Benthic Communities and Facies Development in the Lower Ludlow

The early Ludlow shallow and better ventilated parts of the sea were occupied in the Prague Basin by the coral, crinoid, brachiopod and trilobite dominated

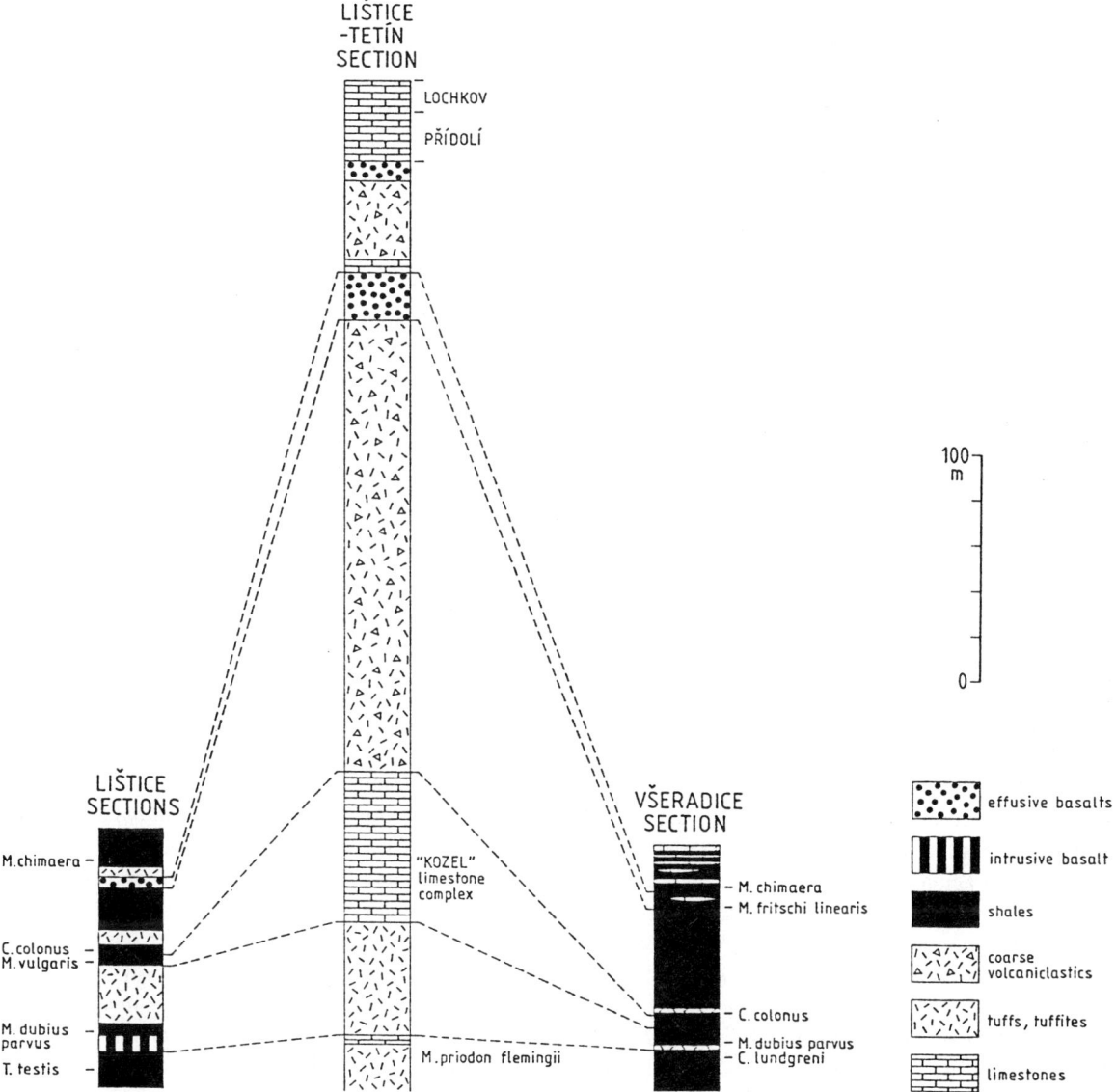

Fig. 27. Correlation of the Wenlock-Ludlow boundary in the Northern Segment (Listice Pipeline Section), the Central Segment (Listice-Tetin Section) and the Southern Segment (Vseradice Section) of the Prague Basin

Coral - Crinoid Community, *Atrypoidea renitens* Community, *Leptostrophiella nebulosa* Community and *Encrinuraspis beaumonti* - smooth Atrypoid Community (Chlupac, 1987; Havlicek and Storch, 1990; Kriz, Dufka and Schönlaub, 1993). Very characteristic for this period was the occurrence of the encrinurid trilobite fauna.

During the early Gorstian the deeper parts of the sea were temporarily ventilated by currents during low stands of the sea level. They were characterized by the development of the cephalopod limestone biofacies with the Bivalvia dominated *Cardiola gibbosa* Community in the *Monograptus colonus* Biozone and the *Cardiola donigala* and *Cardiola docens* Communities in the upper Gorstian and lower Ludfordian (Kriz, 1995a). In the deep parts of the basin, calcareous shales with ostracodes *Entomis migrans* and *Bolbozoe bohemica* accumulated. The *Cardiola gibbosa* Community and *Cardiola docens* Community and the other analogous and homologous contemporary communities were distributed worldwide (Australia: Yass Basin; Asia: Tajmyr, Caucasus; Africa: Morocco; Europe: Carnic

Alps, Germany, Montagne Noire, Poland, Sardinia, Spain, Serbia, Welsh Borderland; North America: Alaska).

Middle Ludfordian Change in Facies and Fauna

In the middle Ludfordian (late Ludlow), the general development of the communities which started in the Wenlock and lasted through the early Ludlow was stopped by a major change in the facies development. It is best documented in the shallow environment just above the horizon characterized by occurrence of the cosmopolitan ostracodes *Entomis migrans* and *Bolbozoe bohemica* (Boucek and Pribyl, 1955; Kriz, 1991, 1992). In relatively shallow parts of the Prague Basin (regions of the former Kosov Volcanic Center and southeastern part of the former Svaty Jan Volcanic Center), a distinct level of the calcareous shales developed. It is dated by the occurrence of *Bohemograptus bohemicus bohemicus, Bohemograptus bohemicus tenuis, Neocucullograptus inexpectatus* and *Polonograptus podoliensis* (Pribyl, 1983) and by the trilobite dominated *Acanthalomina minuta* Community (Chlupac, 1987; Havlicek and Storch, 1990; Kriz, 1992). Contemporaneous bioclastic limestone facies with the almost monospecific brachiopod *Atrypoidea linguata* Community (Havlicek and Storch, 1990; Kriz, 1992) was developed in shallowest parts of the basin, in the regions of the Wenlock - lower Ludlow volcanic centers.

In relatively shallow parts of the basin the facies is overlain by the endostratic breccia level (Kriz, 1992) and facies of the dark grey calcareous, often laminated shales with alternating dark micritic limestones with crinoidal clasts. It is first characterized by the occurrence of the brachiopod and trilobite dominated *Kosovopeltis - Scharyia - Metaplasia* Community and later by the *Ananaspis fecunda - Cyrtia postera* Community (Havlicek and Storch, 1990; Kriz, 1991, 1992).

In the deeper parts of the Prague Basin (Vseradice, Lounín) the facies is developed as the thin accumulation of the bioclastic limestone with endostratic breccia of micritic limestone (Horny, 1960; Kriz, 1992). In the bioclastic limestone the brachiopod dominated *Ananaspis fecunda - Cyrtia postera* Community occurs. The micritic pebbles and fragments contain just the trilobite *Ananaspis fecunda,* brachiopod *Metaplasia hemicona*, inarticulates, rare gastropods and *Polonograptus* aff. *podoliensis* (Kriz, 1992).

The major change in sedimentation and fauna in the middle Ludfordian was related to a high stand of the sea-level or transgression which may be detected also in the other Gondwanaland regions (Sardinia, Montagne Noire) above the level of cephalopod limestones with the Bivalvia dominated *Cardiola docens* Community and with the cosmopolitan ostracodes *Entomis migrans* and *Bolbozoe bohemica* (Kriz, 1991, 1992; Kriz and Serpagli, 1993; Kriz, 1995b). The change caused in the Prague Basin extinction of all shallow water benthic communities characterized by occurrence of the trilobite *Encrinuraspis beaumonti.*

Ludlow-Pridoli Boundary Change in Facies and Fauna

The latest Ludlow was characterized by continuous shallowing of the basin. In the shallow parts (former Wenlock-early Ludlow volcanic centers) the bioclastic limestone accumulations with the brachiopod and trilobite dominated *Prionopeltis archiaci - Atrypoidea modesta* Community (Chlupac, 1987; Havlicek and Storch, 1990) were developed. In the slightly deeper areas, ventilated only temporarily by surface currents, the cephalopod limestone biofacies characterized by the Bivalvia dominated *Cardiola conformis* Community originated (Kriz, 1995a). This community is analogous and homologous with the upper Wenlock and lower Ludlow communities of the *Cardiola* Community Group (Kriz, 1995a) and most of its representatives are closely related to the species which occurred in the Prague Basin before the major Ludfordian facies and faunal change. The deeper parts of the basin are characterized by accumulations of micritic limestones with trilobites and brachiopods while in the deepest parts of the basin calcareous shales accumulated.

In the earliest Pridoli, the sea transgression caused another major change in the development of facies and fauna (Kriz, Jaeger, Paris and Schönlaub, 1986). Deeper facies of the Pridoli are in the Prague Basin characterized by accumulation of micritic limestones and laminites alternating with calcareous shales. The facies was occupied mainly by the Bivalvia dominated communities of the *Snoopyia* Community Group (Kriz, 1995a). These communites are characterized by infaunal or reclining forms. The most favourable parts of the environment were occupied by relatively higher diversity communities with relatively low population densities (Kriz, 1995a). In the parts of the bottom with restricted living conditions (limited current activity and low oxygen content) the Bivalvia dominated communities were characterized by a low diversity and by very high population density of the dominant species (e.g. the *Cheiopteria bridgei* Community).

Deeper water biofacies of the same type and with almost the same Bivalvia dominated communities were also developed in numerous other regions (e.g. Montagne Noire, Massif Armoricain, Sardinia, Florida, Poland, Serbia and Morocco).

Shallower facies are characterized by the presence of bioclastic limestone levels with the Bivalvia dominated *Patrocardia* Community Group (Kriz, 1995a). In the lowermost Pridoli, the brachiopod dominated *Gracianella graciosa* Community (Havlicek and Storch, 1990) occurred in the shallow water facies on the Ludlow Kosov Volcanic Center.

The activity of the Svaty Jan Volcanic Center terminated during the latest Wenlock and early Ludlow (Gorstian) with repeated basalt lava flows into the shallow-water environment (Fiala, 1970, 1982). The thick lava sheet (up to 60 m) was formed and covered earlier volcaniclastic deposits. The lava sheet was overlain by bioclastic limestones in some areas only in the latest Pridoli. This indicates that during the Ludfordian (late Ludlow) and almost the whole of the Pridoli the top of the volcano was above sea level. The sheet also protected earlier volcaniclastics from erosion and redeposition, as shown by the fact that there is no tuffaceous admixture in the Pridoli limestones and shales that were deposited in the vicinity of the Svaty Jan Volcanic Center while the emerged volcanic island existed (Kriz, 1991).

Shallow bioclastic limestones accumulated on the top of the Ludlow Svaty Jan Volcano are characterized by occurrence of the brachiopod dominated *Dubaria* Community (Havlicek and Storch, 1990).

In the upper Pridoli between the relatively deep facies of micritic limestones alternating with shales and the shallow facies of bioclastic limestones occur the accumulations with the *Scyphocrinites* Community (Havlicek and Storch, 1990). These accumulations protected against high energy environment the shallow water plains where bioclastic limestones with the *Dayia bohemica* Community originated (Havlicek and Storch, 1990). In the *Scyphocrinites* Community the sessile benthos is strongly restricted or almost missing.

Conclusions

The general facies and faunal development of the Prague Basin was influenced during the Silurian by eustatic movements, water currents, volcanism and by synsedimentary tectonics.

After the early Silurian transgression, anoxic conditions in the basin persisted up to the late Aeronian (middle Llandovery) when ventilated, shallow sea conditions developed due to the volcanic activity. This caused development of the shallow water benthic communities formed of the new, most probably endemic species which originated by very quick adaptive radiation on the slopes of virgin volcanic island.

In the late Wenlock and in the early Ludlow, a volcanic archipelago originated in the Prague Basin and the situation was analogous with that in the middle Llandovery. New shallow water, high diversity and high density brachiopod and trilobite dominated benthic communities developed by very quick adaptive radiation of first immigrant species.

A different situation was in deeper parts of the basin which were ventilated only temporarily by surface currents. The currents provided not only ventilation of deeper parts of the basin but also transport of larvae and cephalopods. Here, during low stands of the sea level when currents were reaching the sea bottom, the cephalopod limestone biofacies developed with the Bivalvia dominated *Cardiola* Community Group (Kriz, 1995a). This group is composed of the recurring *Cardiola* dominated analogous and homologous communities. They are developed in several horizons which indicate low stands of the sea level stratigraphically. The communities are formed by cosmopolitan species which have relatively long larval life favourable for long transport of larvae and which are known from many other basins within the reach of the South Tropical Current (Wilde, Berry and Quinby-Hunt, 1991) - e.g. Morocco, Spain, Montagne Noire, Massif Armoricain, Welsh Borderland, Scania, Carnic Alps, Moesian Platform, Serbia, Sardinia, Australia, Caucasus and Tajmyr.

The first major change in facies and fauna was in the middle Ludfordian (late Ludlow) caused by high stand of the sea-level which influenced the whole basin, even the shallowest parts of it. In the basin new shallow water communities developed while old shallow water communities characterized by encrinurid trilobites became extinct; there was neither time nor the right conditions for the deeper water communities to recover.

Subsequent shallowing of the basin during the latest Ludlow caused development of similar conditions in the basin which existed here before the middle Ludfordian change. In the shallow parts of the basin (in the regions of former volcanic centers) the brachiopod and trilobite dominated communities composed mainly of endemic species flourished. The deeper parts of the basin ventilated by currents were characterized by the Bivalvia dominated communities in which majority of the species was cosmopolitan.

The second major facies and faunal change was at the Ludlow - Pridoli boundary, caused by another transgression of the sea. High stand of the sea level killed all the Ludlow benthic communities and just a few species have survived. The new Bivalvia dominated communities, composed mainly of the infaunal and cosmopolitan species distributed as larvae together with cephalopods by surface currents, developed in deeper water under somewhat restricted bottom conditions (mostly only little and temporarily ventilated). These communities are known also from other regions (e.g. Morocco, Montagne Noire, Poland, Moesian Platform, Carnic Alps, Sardinia, Serbia). In the shallow bottom which persisted in the regions of former volcanic centers flourished the brachiopod dominated communities with majority of endemic species. Parts of these shallow flats were protected against high energy by barriers of crinoid sand composed mainly of *Scyphocrinites* disarticulated stem plates and numerous loboliths.

Conclusions on the Mid-Silurian Transition

The global character of the latest Wenlock sea-level lowering has been advocated by many authors (Johnson, Kaljo and Rong, 1991; Johnson and McKerrow, 1991; Vail et al., 1977; etc.) and can be taken as proved. In the shelf seas, in shallow ones in particular, this regression event is usually a complicated process.

The above data on the Midwestern craton history (see also Droste and Shaver, 1987; Shaver, 1992) show two larger (series) and several smaller scale cycles during the Wenlock and Ludlow. The upper part of the Wenlock shallowing - upward sequence was complicated by comparatively rapid relative deepenings and shallowings, facilitating the alternating deposition of evaporite and carbonate rocks.

Three principal reef generations (nos. 3-5, Fig. 18) and two widespread reef abortions ensued during late Wenlock and early Ludlow. Many of the reefs exhibit a cosmopolitan fauna below and a restricted, algae-dominated community above (Shaver, 1992).

The salts A-1, A-2 and B of the Salina group (Fig. 18) are confined to the coinciding high-salinity and sea-level lowering phases of the general mid-Silurian eustatic drop. Insufficiency of palaeontological datings of these salt or reef levels do not allow their exact correlation with analogous ones elsewhere. With a working accuracy, the salt A-2 is placed into the very beginning of the Ludlow (Fig. 18) or a little earlier (Shaver, 1992).

A similar general pattern of cyclicity is valid also for the Baltic carbonate platform area. The late Llandovery-Wenlock macrocycle is complicated with several lower rank cycles. The sedimentary cyclicity is interpreted as a reflection of sea-level excursion. Nestor and Nestor (1991) correlated the upper Wenlock carbonate sequences of Estonia with deeper shelf and marginal basinal sequences in western Latvia and Gotland using occurrences of different groups of fossils, especially chitinozoans and graptolites. They concluded that against the background of a general regressive development of the Baltic and adjacent platform seas in the Wenlock more abrupt shallowing took place at the time of the *lundgreni* to *nassa* graptolite zones. The shallowing event culminated according to the Nestors (op. cit.) in the early *nassa* time when the Ancia Member (see below) was formed. The slight difference in the dating of the event is not significant and in fact not ascertainable according to the data available. A short transgression episode was established (Einasto in Märss, 1986; Nestor and Nestor, 1991) in the second half of the *nassa* time coinciding with the basal part of the Rootsikula Stage and a couple of smaller ones higher in the stage.

In Britain the first indications of the shallowing in the type Wenlock area appeared in the *nassa* Zone together with the beginning of the alternating limestone and mudstone succession of the Farley Member of the Coalbrookdale Formation. The sea-level lowering culminated in the topmost beds of the following Much Wenlock Limestone Formation (Bassett, 1989). Corfield and Siveter showed (1992, Sect. 5.1.) that the *nassa* Zone is marked with a clear $\partial^{13}C$ depletion event, which might be caused by a serious drop in bioproduction generating also better oxygenation of the bottom waters. At the same time we have to remember that $\partial^{13}C$ depletion has been recorded in the West Midlands section, this representing the carbonate platform area without any serious amount of graptolite biomass but having a rich benthic fauna and flora. It seems logical to consider here also the effects of better oxygenation due to better ventilation of bottom waters. The latter is to a certain degree connected with the sea-level lowering. On the other hand, interpreting the $\partial^{13}C$ curve we have to bear in mind bioproduction in a wider area, not only that at the site of the studied section. The curves (Fig. 21) show at about the *nassa* Zone a sequence of depletion, followed by a recovering phase. As is known (see earlier in this chapter), the drastic drop in bioproduction in the pelagic area occurred at the very end of the *lundgreni* Zone and remained on a very low level (only a few species survived) during the next 1.5 - 2 Ma. Therefore it seems that at this locality the above recovering phase of the curve was contributed by biota from a wider area.

The above discussion leads us to the conclusion that the carbon isotope excursion is a complex phenomenon integrating the effects of different factors and processes.

There was no clear indication of any environmental change at the level of the *lundgreni* Event in the studied sections of South Tien Shan, but in some other regions such markers have been described. Jaeger (1991) presented new data about a pelagic sequence at Gräfenwarth in Thuringia. The whole section consists of black shales with cherts (so-called alum slates with lydites), whereas the latter are more numerous above the *dubius-nassa* Interregnum. The roof of the *lundgreni* Zone is made up of a 10 cm thick "boundary coal seam" (Grenz-Kohlenflözchen by Jaeger, 1991; an alum shale with very high contents [ca. 20%] of coaly matter). Just below this seam, on the level corresponding to the above described *lundgreni* Event, many typical Wenlock graptolites became extinct.

Detailed geochemical studies could not find anything anomalous in the section (28 elements were measured, the Ir content varies between 0.10-0.30 ppb, but was low in the boundary interval; Jaeger, 1991).

In conclusion, there was no "big" change in lithology, but the boundary coal seam marks a brief environmental change during a short episode in the Homerian. The meaning of the boundary coal seam is not clear yet.

In western Latvia the same position as the Grenz-Kohlenflözchen in Thuringia shows a very thin-bedded intercalation (one metre thick) of limestones and marls. The band belonging to the Ancia Member of the Riga Formation constitutes the very top of the *lundgreni* Zone and just above it Ulst (1974) has identified *Monograptus dubius parvus, Gothograptus nassa,* etc. The band occurring in the monotonous sequence of graptolitic mudstones is usually interpreted as a reflection of a short-term sea-level lowering.

In Morocco analogous marker beds were established in the same position by Hollard and Willefert (Destombes et al., 1985). In the Ain-Deliouine section (Anti Atlas) just below *G. nassa, M. dubius* and *M. ludensis,* they have recorded a thin horizon of black or rusty ferruginous limestone. The rocks below and above the marker are usual graptolitic shales. In another section at Ain Sidi Larbi (western Morocco) an ochreous ferruginous limestone bed (one metre thick) occurs just above the disappearance of *M. flemingii, C. lundgreni,* etc. and above the marker "*Pristiograptus vulgaris*" was recorded.

The end Wenlock shallowing was well marked also on the northern Gondwanaland in the narrow trough-like Prague Basin, where the late Wenlock-early Ludlow history was complicated with strong volcanic activity. These environmental causes seem to be responsible for biotic changes observed in the sequence (see above). The changes of the benthic communities are contemporaneous with the graptolite "crisis" of Jaeger (1991).

In pelagic sequences markers of the environmental events on the level of the transition of the *lundgreni* and *nassa* zones have not been found everywhere. It might be due to different local tectonical or/and lithological but also subjective reasons. The interpretation of the markers described in Thuringia and Morocco is not clear yet, but we do think that differences in responses to the influence of some global factors depend on the character of the geological setting and location of the site in the context of climatic belts. In this respect both Thuringia and Morocco were situated in the highest latitude belt in comparison with other areas.

Johnson and McKerrow (1991) discussed a series of sequences from most of the world's well-known Silurian outcrop areas showing late Wenlock shallowing and early Ludlow deepening events. They concluded that the 30-50 m deepening event was between 1-2 Ma in duration and most probably caused by melting of the continental ice cap in South America.

Discussing here the possible causes of the early Homerian extinction and environmental interpretation of a thin band marking the *lundgreni* Event in some sections, we have to concentrate on a by far briefer episode than 1-2 Ma of glaciation. The latter looks in general well-founded, especially when the evidence of Silurian glacial deposits near the contemporary South Pole (see Johnson and McKerrow, 1991) is sufficiently proved also in the stratigraphical sense (Grahn and Caputo [1992] gave evidence for three older glaciations, see below and Fig. 28).

The rapid sea-level lowering during end-*lundgreni*/early *nassa* time could be connected with the beginning (or probably with an advanced stadium) of the Homerian glaciation causing also substantial changes in ocean circulation, increase in the content of dissolved gases in the ocean leading to net lessening of

SERIES	STAGES	AGE, DUR. M.Y.	STANDARD GRAPTOLITE ZONES	BIO-EVENTS AND RELATIVE SEA-LEVEL CURVE (SHALLOW ↔ DEEP)		OCEANIC (P,S) EPISODES AND EVENTS	GLACIAL EVENTS
PRIDOLI		408 — 1.8	transgrediens-lochkovensis		S/D M'Pr	Mid-Pridoli S/P	
		1.2 411	ultimus-parultimus		L'Lu		
LUDLOW	LUDFORDIAN	2.0	formosus				
		2.0	kozlowskii-bohemicus		M'Lu		
		1.0	leintwardinensis		E'Lu	Lau S/P	
	GORSTIAN	3.0	scanicus		M'Go	Linde S/P	
		2.0 421	nilssoni		E'Go		
WENLOCK	HOMER.	0.8	ludensis		L'Ho		
		1.0	nassa		M'Ho	Mulde. S-S	LW
		1.2	lundgreni		E'Ho	Valleviken S/P	
	SHEINW.	2.0	ellesae-rigidus		L'Sh	Boge S/P	
		0.8	riccartonensis				
		1.2 428	murchisoni-centrifugus		E'Sh	Ireviken Ev.	BR 3
LLANDOVERY	TELYCH.	1.1	crenulata			Snipklint Primo Epis.	
		1.1	griestoniensis				
		1.2	crispus-turriculatus		E'Te	Malmoykalven Secundo Epis.	?BR 2
	AERONIAN	2.0	sedgwickii-convolutus		L'Ae	Sandvika Ev. Jong Primo Ep.	
		2.4	gregarius		E'Ae	Spirodden Secundo Episode	BR 1
	RHUDD.	1.6	cyphus				
		0.8	vesiculosus				
		0.8 439	acuminatus				

Fig. 28. Correlation of Silurian bio-events with the global sea-level curve (after Johnson, Kaljo and Rong, 1991), oceanic (P.S) episodes by Jeppsson (after Aldridge, Jeppson and Dorning, 1993; Jeppsson, 1993) and glacial events (after Grahn and Caputo, 1992; Johnson and McKerrow, 1991). Black rhombs: extinction events; white ones: radiation events; rimmed rhombs: extinction and radiation events closely following

carbonate deposition (cf. Berry and Wilde, 1978; Jeppsson, 1990; Aldridge, Jeppsson and Dorning, 1993). Opposite influence of the Homerian regression has also to be kept in mind. The summary effect of the above factors brought about the complicated extinction-origination pattern documented in chapters 3 - 5 above. Step by step extinctions, like the Ireviken Event among conodonts (see above), are in good accordance with such a scenario. Interpretation of thin bands, like Grenz-Kohlenflözchen of Thuringia, should include also interaction of climatic, oceanographic and regional tectonic factors making the episode very condensed.

Summary and Concluding Comments

In the history of Silurian biota and/or ecosystem as a whole no "big" catastrophes occurred like the first order bio-events at the Ordovician-Silurian or Cretaceous-Tertiary boundaries. Yet, it was not a quiet period either - 15 more or less remarkable bio-events were established (Fig. 28).

The environmental background for these events was relatively stable due to the location of most cratonic seas and connected basinal areas in the tropical climatic belt (Fig. 3) and in general the sea- level was standing high (Vail et al., 1977; Fig. 6). Nevertheless, having in mind the changes of the environment, some clear tendencies can be observed (Figs. 4 - 6). According to Wilde et al. (1991), many parameters, connected in some way with biotic productivity, were declining during the Silurian Period, e.g. the contents of atmospheric oxygen, organic carbon in the ocean (see Figs. 4, 5) and $\partial^{13}C$. Together with a relatively high temperature (20°C), these caused increased anoxia in the water column. Sea-level rise also brought about the spreading of anoxic waters.

Some of these parameters are primary, the others secondary, but surely they are all interacting and therefore important for biotic analysis.

The Silurian sea-level curve (Fig. 28) shows the following important tendencies and intervals: (1) rapid deepening in the Rhuddanian, (2) continuation of sea-level rise in the Aeronian until the late Telychian interrupted by three more or less significant shallowings (mid-*gregarius*, *sedgwickii* and *crispus* zones), (3) sea-level high stand in the Sheinwoodian, complicated by two short-term lowerings (*murchisoni* and *rigidus-linnarssoni* zones), (4) the general regression of seas during the Homerian and the whole late Silurian, divided by two deepening periods (*nilssoni* + *scanicus* and top-*bohemicus-kozlowskii* + *formosus* zones) into the Homerian, Gorstian-early Ludfordian and late Ludfordian Pridoli declining cycles.

The above general tendencies of sea-level changes are in good correlation with the picture shown in Fig. 5 after Wilde et al. (1991).

In the context of the above-mentioned environmental parameters the pattern of sea-level changes might be considered as a good instructive feature for bio-event studies.

As mentioned before, the Silurian temperature was higher than the modern mean value, but less than that suggested for the Ordovician and Devonian (Wilde et al., 1991; Fig. 4 here). The Rhuddanian deepening of the world ocean is usually interpreted as a result of melting of the ice caps of the late Ordovician glaciation (Brenchley and Newall, 1984). Johnson and McKerrow (1991) suggested the same explanation also for the late Telychian and early Gorstian deepening events. It means that before the melting there had to be a glaciation period when increasing ice caps caused sea-level lowering. This was the case in the Homerian but not so clearly expressed in the Llandovery. The late Telychian melting period was preceded by several (see above) lower rank regressions, which might be caused by the increased glacial process. Another speculation might be that the melting of the late Ordovician ice caps lasted with periodic interruptions until the Wenlock. One of these glaciation restoration intervals was the latest Aeronian, marked by rapid sea-level drop and better ventilation of the bottom waters (Kaljo, Nestor, Põlma and Einasto, 1991b).

Lately Grahn and Caputo (1992) reported early Silurian glaciations in Brazil that might serve as a proof of the above discussion. The authors presented evidence (tillites) for four glaciations dated as the latest Ashgill and/or earliest Llandovery, the *gregarius* Zone of early middle Llandovery, early late Llandovery (possibly starting already in the late middle Llandovery), and the latest Llandovery to earliest Wenlock. These levels of glacial sediments are relatively well coinciding with sea-level lowerings according to

Johnson, Kaljo and Rong (1991; Fig. 28 here) and corresponding to the extinction events described above.

The long-lasting melting period is in accordance with the temperature curve in Fig. 4 and data presented by Grahn and Caputo (1992) support the idea. As to eustatic changes of sea-lev l, these are obviously affected by different agents including processes in the upper mantle, therefore a thorough analysis is needed.

Considering the biota, we have to underline the most striking pattern - many groups of organisms, not depending on their habitats or mode of life, have experienced the same pattern of evolution during the Silurian. The period usually began with a low but rising diversity phase (Rhuddanian or longer), followed by a diversity maximum (most clearly pelagic-planktic graptolites in the Aeronian, nektobenthic conodonts and planktic acritarchs in the Telychian and earliest Wenlock, corals in the Wenlock) and a long declining phase, of course with several diversity rises, but the latter did not reach the previous level.

Brachiopods (see above) and trilobites (Männil, 1992) do not show the pattern so clearly, chitinozoans (Fig. 14) have several nearly equal diversity peaks, vertebrates as an emerging new group have clear diversity rise, but surprisingly in the same time their origination rate (Fig. 10) was falling in the late Silurian.

The causes of the graptolite events documented earlier are not yet clearly understood. However, the rapid increase in graptolite diversity in the Llandovery correlates with the climatic amelioration and coeval rise in sea level which followed the late Ordovician glaciation and associated extinction event. The Llandovery eustatic rise in sea level possibly provided new ecological opportunities for graptolites, in which morphological innovations and diversification took place. The coincidence of a general fall of sea level in the late Wenlock with a decline of graptolite diversity seems to support the importance of ecological factors in influencing diversity. A similar case could be made for the late Ludlow (post-*scanicus*) graptolites. On the other hand, surprisingly good coincidence of diversity peaks and minima with the phases of sea level rises and lowerings reported by Kaljo and Märss (1991) basing on the Baltic data, demonstrates the role of local factors and the need to keep these apart from the global ones.

Evidently every group has its own main reason affecting its evolution but there was also a common cause, the general environmental stress, discussed above, influencing the life of at least many groups of organisms.

Walliser (1986) suggested a sequence of evolutionary events and processes comprising the following stages: the extinction event - interval with low selectional stress - radiation with or without innovation, and nomismogenesis with the decreasing evolutionary rate. This scheme is instructive also for the interpretation of the evolutionary process in the Silurian, summarized above.

The rise and subsequent decline in general diversity were influenced by long-term changes of the environment, the bio-events as stages of the general process are usually the results of concrete short-term effects, sometimes not so easy to identify.

In the last context a complication should be noted. For the sake of correct correlation of shelly and graptolite sequences in this chapter a set of the so-called standard graptolite zones, not a detailed zonation, was used for scale. Now, on the other hand, it is difficult to be sure about the exact dating of usually short events in the limits of several relatively long-lasting zones (2-3 Ma, Fig. 28). In some cases the dating of events is not as good as is needed for the purpose of this study.

The preceding paragraphs concentrate on different kinds of events (the hierarchy was defined in the introduction, see also Fig. 1 for a list of events and their short descriptions in the chapter on the event-stratigraphy), below these are correlated with the global sea-level curve (Fig. 28).

Most clear extinctions occur at the following levels of sea-level lowerings: mid-*gregarius* (conodonts, graptolites), *sedgwickii* (conodonts, chitinozoans, graptolites), *centrifugus-murchisoni* (graptolites, mass extinction of conodonts and acritarchs, also trilobites according to data by Chatterton et al., 1990), *lundgreni* (chitinozoans, mass extinction of graptolites), *ludensis* (agnathans), *leintwardinensis* (agnathans, conodonts, graptolites), *lochkovensis-transgrediens* (corals, conodonts, chitinozoans and graptolites) zones. The conclusion is that all but one-two (late Ludfordian *spineus* Event and perhaps the mid-Ludfordian one) remarkable extinctions occurred at periods of the shallowing of the sea.

On the contrary, the most important origination events were tied to the deepening periods e.g. to the *turriculatus* (conodonts and acritarchs), *ellesae* (chitinozoans and acritarchs), *nassa* (agnathans), *nilssoni* (graptolites), *scanicus* (conodonts, chitinozoans) zones. Of course, step by step radiations and innovations occurred during the whole Rhuddanian. An exception from the last pattern was the *centrifugus-murchisoni* Zone when both originations and extinctions occurred during a short sea-level lowering.

The last statement needs a comment. A few paragraphs above a warning was made about correlation, another note is that we are discussing extinctions-appearances basing on the global sea-level curve (Fig. 28), but local curves are different in details (McKerrow, 1979; Chatterton et al., 1990; Johnson, Baarli et al., 1991; etc.). Therefore we can define the exact position of an event against the sea-level curve in a concrete section, but not so clearly in a generalized sequence.

From this position we could explain a few differences in levels of events among Silurian trilobites and conodonts of northwestern Canada (Chatterton et al., 1990) and those presented here. But in general both these are well concordant.

In general, the pattern looks rather unambiguous but we have to remember the above discussion on the interacting manner of the environmental processes.

Figure 28 shows the above events correlated with those defined by Jeppsson (1990, 1993; Aldridge, Jeppsson and Dorning, 1993) basing on climatically triggered changes in oceanic situation. Cyclic alternation of primo and secundo oceanic episodes and their good coincidence with sea-level curve (P/S events mostly coincide with regressive phases) and glacial events prove that the primary cause of the Silurian bio-events has been changing climate. In this context it would be important to make reference to a statement made by Erwin (1994) which emphasizes the interacting manner of environmental processes, especially the influence of the regression of the sea as a factor increasing climatic variability, triggering also changes in gas and carbon content in waters and reducing the habitat area for several groups of organisms. Therefore we tend to understand the above conclusion of climate as a general one, but every concrete event and the way it is expressed in different groups of organisms and different ecological situations needs specialized analysis. In this sense we would like to return to the principle quoted in the introductory remarks that by the analysis of bio-events ecological preferences of organisms and their distribution pattern in a basin have to be considered. In other words, the well-known Walther's law of facies distribution has to be kept in mind when interpreting diversity changes in a section. Therefore sometimes a diversity rise or lowering in a section is more realistic to be explained through changes in the onshore-offshore direction, not so much on the time axis.

Having in mind some discrepancies between the list of bio-events, sea-level curve, oceanic episodes and glacial events shown in Fig. 28, we think that for better correlation more detailed and complex study of concrete sections is needed.

Johnson and McKerrow (1991), analysing faunal changes connected with the late Llandovery and the basal Ludlow transgressions, concluded that extinctions and radiations were more prevalent in pelagic and nektic groups than in the benthos. Our data support this conclusion in part of radiation during the mentioned deepening phases, but main extinction events (in the *centrifugus-murchisoni, bohemicus-kozlowskii* zones and the end-Silurian one) affected all types of biota including sessile benthos.

One more comment is needed on the so-called Lazarus phenomenon. We consider it a natural effect of the interaction of often rhythmically changing environment and organisms. Depending on the sampling scale etc., the result may be different, but the Lazarus-type distribution seems to be a general pattern. A good example is the acritarch distribution in the Silurian of Gotland (see Le Herisse, 1989).

The last comment is devoted to acknowledgement that we have been explaining in this chapter biotic changes by influence of interacting abiotic components of an ecosystem. Plotnick and McKinney (1993) have recently shown that biotic control and internal structure of an ecosystem were among the major factors in past extinctions and diversity changes. We agree that this part of the Silurian bio-event analysis remains to be done.

Acknowledgments

As the organizer of this team work, the first author is glad to express the thanks of the whole party to all our supporters and helpers assisting in various ways. By name they are listed in papers of the authors published in Proceedings, Estonian Acad. Sci. Geology, v. 41, No 4, 1992. Here gratitude is expressed also to Mrs Anne Noor for linguistic and technical help, Mrs Kaie Ronk for drawings and Mrs Alla Shogenova for computer work. Last but not least, Prof. O.H. Walliser and his team are thanked for the invitation to participate in the project, for many fruitful discussions, and for kind support during the work.

References

Aldridge, R.J., 1988. Extinction and survival in the Conodonts. In: Larwood, G.P. (ed.), Extinction and survival in the fossil record. Systematic Assoc.

Spec. Vol. 34, 231-256.

Aldridge, R.J., Jeppsson, L. and Dorning, K.J., 1993. Early Silurian oceanic episodes and events. J. Geol. Soc. London 150, 501-513.

Bassett, M.G., 1989. The Wenlock Series in the Wenlock area. In: Holland, C.H. and Bassett, M.G. (eds.), A global standard for the Silurian System. Nat. Mus. Wales, Geol. Ser. 9, 51-73. Cardiff.

Bassett, M.G., Cocks, L.R.M., Holland, C.H., Rickards, R.B. and Warren, P.T., 1975. The type Wenlock Series. Rep. Inst. geol. Sci. 75/13, 1-19.

Bassett, M.G., Kaljo, D. and Teller, L., 1989. The Baltic region. In: Holland, C.H. and Bassett, M.G. (eds.), A global standard for the Silurian System. Nat. Mus. Wales, Geol. Ser. 9, 158-170. Cardiff.

Berry, W.B.N. and Wilde, P., 1978. Progressive ventilation of the oceans - an explanation for the distribution of the Lower Paleozoic black shales. Amer. J. Sci. 278, 257-275.

Berry, W.B.N., Wilde, P. and Quinby-Hunt, M.S., 1987. The oceanic and non-sulfide oxygen minimum zone: a habitat for graptolites? Bull. Geol. Soc. of Denmark 35, 103-113.

Blieck, A. and Janvier, Ph., 1991. Silurian vertebrates. In: The Murchison Symposium. Spec. Papers Palaeont. 44, 345-389. London.

Boucek, B. and Pribyl, A., 1955. On the Silurian ostracodes and the stratigraphy of the Budnany Beds (eß) from the immediate vicinity of the Kosov and the Kolednik near Beroun. Sbornik Ustredniho Ustavu Geologickeho 21-1954, 577-662.

Boucot, A.J., 1975. Evolution and extinction rate controls. XV + 427 pp. Elsevier.

Boucot, A.J., 1990. Silurian and pre-Upper Devonian bio-events. In: Kauffman, E.G. and Walliser, O.H. (eds.), Extinction Events in Earth History. Lecture Notes in Earth Sci. 30, 125-132. Springer, Berlin Heidelberg New York.

Boucot, A.J., 1991. Developments in Silurian studies since 1839. In: Bassett, M.G., Lane, P.D. and Edwards, D. (eds.), The Murchison Symposium. Spec. Papers Palaeont. 44, 91-107.

Boucot, A.J., 1992. Benthic brachiopod community changes that reflect Silurian bioevents. Proc. Estonian Acad. Sci., Geol. 41, 193-197.

Brenchley, P.J. and Newall, G.A., 1984. Late Ordovician environmental changes and their effect on faunas. In: Bruton, D.L. (ed.), Aspects of the Ordovician System. Palaeont. Contrib. 295, 65-79. Univ. of Oslo.

Budyko, M.I., Ronov, A.B. and Yanshin, A.L., 1987. History of the earth's atmosphere. 139 pp. Springer, Berlin Heidelberg New York.

Chatterton, B.D.E., Edgecombe, G.D. and Tuffnell, P.A., 1990. Extinction and migration in Silurian trilobites and conodonts of northwestern Canada. J. Geol. Soc. London 14, 703-715.

Chlupac, I., 1987. Ecostratigraphy of Silurian trilobite assemblages of the Barrandian area, Czechoslovakia. Newsletters on Stratigraphy 17, 169-186.

Clark, D.L., 1972. Early Permian crisis and its bearing on Permo-Triassic conodont taxonomy. Geologica et Palaeontologica, SB 1, 147-158.

Clark, D.L., 1981. Biological considerations and extinction. In: Robison, R.A. (ed.), Treatise on Invertebrate Paleontology, Part W, Miscellanea, Supplement 2, Conodonta. W83-W87.

Corfield, R.M., Cartlidge, J.E., Premoli-Silva, I. and Housley, R.A., 1991. Oxygen and carbon isotope stratigraphy of the Palaeogene and Cretaceous limestones in the Bottacione Gorge and the Contessa Highway sections, Umbria, Italy. Terra Nova 3, 414-422.

Corfield, R.M. and Siveter, D.J., 1992. Carbon isotope change as an indicator of biomass flux and an aid to correlation during *ludensis-nilssoni* (Silurian) time. Proc. Estonian Acad. Sci., Geol. 41, 173-181.

Corfield, R.M., Siveter, D.J., Cartlidge, J.E. and McKerrow, S., 1992. Carbon isotope excursion near the Wenlock-Ludlow (Silurian) Boundary in the Anglo-Welsh Area. Geology 20, 371-374.

Destombes, J., Hollard, H. and Willefert, S., 1985. Lower Palaeozoic rocks of Morocco. In: Holland, C.H. (ed.), Lower Palaeozoic of north-western and west central Africa. pp. 91-336. John Wiley & Sons Ltd.

Dorning, K.J., 1981. Silurian Chitinozoa from the type Wenlock and Ludlow of Shropshire, England. Rev. Palaeobot. Palynol. 34, 205-208.

Droste, J.B. and Shaver, R.H., 1987. Paleoceanography of Silurian seaways in the midwestern basins and arches region. Paleoceanography 2, 213-227.

Elles, G.L., 1900. The zonal classification of the Wenlock Shales of the Welsh Borderland. Quart. J. geol. Soc. London 56, 370-414.

Erwin, D.H., 1994. The Permo-Triassic extinction. Nature 367, 231-236.

Fiala, F., 1970. Silurian and Devonian diabase of the Barrandian Basin. Sbornik Geologickych Ved, Geologie 17, 7-71.

Fiala, F., 1982. Basaltoid diabase from Hostim with indications of hematite mineralization. Casopis pro Mineralogii a Geologii 27, 173-185.

Grahn, Y. and Caputo, M.V., 1992. Early Silurian glaciations in Brazil. Palaeogeogr., Palaeoclim., Palaeoecol. 99, 9-15.

Harland, W.B., Armstrong, R.L., Cox, A.V., Craig, L.E., Smith, A.G. and Smith, D.G., 1989. A Geologic Time Scale. 263 pp. Cambridge Univ. Press.

Harries, P.J. and Kauffman, E.G., 1990. Patterns of survival and recovery following the Cenomanian-Turonian (Late Cretaceous) mass extinction in the Western Interior Basin, United States. In: Kauffman, E.G. and Walliser, O.H. (eds.), Extinction Events in Earth History. Lecture Notes in Earth Sci. 30, 277-298. Springer, Berlin Heidelberg New York.

Harris, J.H., 1987. The geology of the Wenlock Shales around Builth Wells. Unpubl. Ph.D. thesis. Univ. of Cambridge.

Havlicek, V., 1981. Development of a linear sedimentary depression exemplified by the Prague Basin (Ordovician - Middle Devonian; Barrandian area - Central Bohemia). Sbornik Geologickykh Ved, Geologie 35, 7-48.

Havlicek, V., 1992. Prazska panev. In: Chlupac, I. (ed.), Paleozoikum Barrandienu. Cesky Geologicky Ustav, 56-59. Praha.

Havlicek, V. and Kriz, J., 1973. Upper Llandovery and Lower Devonian near Hyskow (Barrandian). Vestnik Ustredniho Ustavu Geologickeho 48, 103-107.

Havlicek, V. and Storch, P., 1990. Silurian brachiopods and benthic communities in the Prague Basin (Czechoslovakia). Rozpravy Ustredniho Ustavu Geologickeho 48, 1-275.

Hede, J.E., 1942. On the correlation of the Silurian of Gotland. Meddelanden Lunds Geologisk-Mineral. Instn. 101, 1-25.

Holland, C.H., 1991. What is so very special about the Silurian. In: Bassett, M.G., Lane, P.D. and Edwards, D. (eds.), The Murchison Symposium. Spec. Papers Palaeont. 44, 391-397.

Holland, C.H. and Bassett, M.G. (eds.), 1989. A global standard for the Silurian System. Nat. Mus. Wales, Geol. Ser. 9, 1-325. Cardiff.

Horny, R., 1960. Stratigraphy and tectonics of the western closures of the Silurian-Devonian synclinorium in the Barrandian area. Sbornik Ustredniho Ustavu Geologickeho, odd. geol. 1, 495-524.

Hutt, J., 1969. The development of the Ludlovian graptolite *Saetograptus varians*. Lethaia 2, 361-368.

Jaeger, H., 1981. Comments on the graptolite chronology of Gotland. In: Laufeld, S. (ed.), Proceedings of Project Ecostratigraphy Plenary Meeting. p. 12. Gotland, Uppsala.

Jaeger, H., 1991. Neue Standard-Graptolithenzonenfolge nach der "Grossen Krise" an der Wenlock/Ludlow-Grenze (Silur). N. Jb. Geol. Paläont., Abh. 182, 303-354.

Jeppsson, L., 1983. Silurian conodont faunas from Gotland. Fossils and Strata 15, 121-144.

Jeppsson, L., 1987. Lithological and conodont distributional evidence for episodes of anomalous oceanic conditions during the Silurian. In: Aldridge, R.J. (ed.), Palaeobiology of conodonts. pp. 129-145. Ellis Horwood Ltd, Chichester.

Jeppsson, L., 1990. An oceanic model for lithological and faunal changes, tested on the Silurian record. J. Geol. Soc. London 147, 663-674.

Jeppsson, L., 1993. Silurian events, the theory and the conodonts. Proc. Estonian Acad. Sci., Geol. 42, 23-27.

Jeppsson, L. and Männik, P., 1993. High-resolution correlations between Gotland and Estonia near the base of the Wenlock. Terra Nova 5, 348-358.

Johnson, M.E., 1987. Extent and bathymetry of North American platform seas in the Early Silurian. Paleoceanography 2, 185-211.

Johnson, M.E., Baarli, B.G., Nestor, H., Rubel, M. and Worsley, D., 1991. Eustatic sea-level patterns from the Lower Silurian (Llandovery Series) of Southern Norway and Estonia. Geol. Soc. America Bull. 103, 315-335.

Johnson, M.E., Kaljo, D. and Rong, J.-Y., 1991. Silurian eustasy. In: Bassett, M.G., Lane, P.D. and Edwards, D. (eds.), The Murchison Symposium. Spec. Papers Palaeont. 44, 145-163.

Johnson, M.E. and McKerrow, W.S., 1991. Sea level and faunal changes during the latest Llandovery and earliest Ludlow (Silurian). Hist. Biol. 5, 153-169.

Kaljo, D.L., 1981. Biogeografiya i klimaticheskaya zonal'nost' silura. In: Paleontologiya, paleobiogeografiya i mobilizm. Trudy XXI sess. Vsesojuznogo paleontologicheskogo obshchestva, 64-71. Magadan.

Kaljo, D. and Klaamann, E., 1973. Ordovician and Silurian Corals. In: Hallam, A. (ed.), Atlas of Palaeobiogeography. pp. 37-45. Elsevier.

Kaljo, D. and Märss, T., 1991. Pattern of some Silurian bioevents. Hist. Biol. 5, 145-152.

Kaljo, D., Nestor, H. and Einasto, R., 1991. Aspects of Silurian carbonate platform sedimentation. In: Bassett, M.G., Lane, P.D. and Edwards, D. (eds.),

The Murchison Symposium. Spec. Papers Palaeont. 44, 205-224. London.

Kaljo, D.L., Nestor, H.E., Põlma, L.J. and Einasto, R.E., 1991. Pozdneordovikskoye oledeneniye i ego otrazheniye v osadkonakoplenii Paleobaltiiskogo basseina. In: Kaljo, D., Modzalevskaya, T. and Bogdanova, T. (eds.), Major biological events in Earth history. pp. 68-78. Tallinn.

Kaljo, D., Paskevicius, J. and Ulst, R., 1984. Graptolitovye zony silura Pribaltiki. In: Männil, R. and Mens, K. (eds.), Stratigrafiya drevnepaleozoiskikh otlozhenii Pribaltiki. pp. 94-118. Acad. Sci. Est. SSR, Tallinn [Engl. summ. Graptolite zones in the East Baltic Silurian].

Kaljo, D., Viira, V., Märss, T. and Nestor, V., 1986. Soobchestva nektona, nektobentosa, planktona (ryb, bescheljustnykh, konodontonositelei, graptolitov, khitinozoi) silura Vostochnoi Pribaltiki. In: Kaljo, D. and Klaamann, E. (eds.), Theory and practice of ecostratigraphy. pp. 127-136. Tallinn, Valgus [Engl. summ. The nektic, nektobenthic and planktonic communities (fishes, agnathans, conodonts, graptolites, chitinozoans of the East Baltic Silurian)].

Kemp, A.E.S., 1991. Mid-Silurian pelagic and hemipelagic sedimentation and palaeoceanography. In: Bassett, M.G., Lane, P.D. and Edwards, D. (eds.), The Murchison Symposium. Spec. Papers Palaeont. 44, 261-299.

Klaamann, E., 1970. Tabulaty. In: Kaljo, D. (ed.), Silur Estonii. pp. 114-125. Tallinn, Valgus.

Klaamann, E., 1986. Soobchestva i biozonalnost Tabulatomorfnykh korallov silura Pribaltiki. In: Kaljo, D. and Klaamann, E. (eds.), Theory and practice of ecostratigraphy. pp. 80-98. Tallinn, Valgus [Engl. summ. The tabulate communities and biozones of the East Baltic Silurian].

Koren, T.N., 1983. New late Silurian Monograptids from Kasakhstan. Palaeontology 26, 407-434.

Koren, T.N., 1987. Graptolite dynamics in Silurian and Devonian time. Bull. Geol. Soc. Denmark 35, 149-159.

Koren, T.N., 1991a. Evolutionary crisis of the Ashgill graptolites. Advances in Ordovician Geology, Geol. Surv. of Canada pap. 909, 157-164.

Koren, T.N., 1991b. The *lundgreni* extinction event in Central Asia and its bearing on graptolite biochronology within the Homerian. Proc. Estonian Acad. Sci., Geol. 40, 74-78.

Koren, T.N., 1992. Novye pozdnevenlokskie monograptidy Alaiskogo khrepta. Paleont. Zhurnal 2, 21-23.

Koren, T.N., Klishevich, V.L. and Rinenberg, R.E., 1986. Opornyi razrez siluriiskikh i nizhne devonskikh pelagicheskikh otlozhenii Yuzhnoi fergany. Sovetskaya Geologiya 11, 62-74.

Koren, T.N. and Rickards, R.B., 1980. Extinction of the graptolites. In: Harris, A.L., Holland, C.H. and Leake, B.E. (eds.), The Caledonides of the British Isles reviewed. Spec. Publ. Geol. Soc. London 8, for 1979, 457-466.

Kriz, J., 1984. Autecology and ecogeny of the Silurian Bivalvia. In: Bassett, M.G. and Lawson, J.D. (eds.), Autecology of Silurian organisms. Spec. Papers Palaeontology 32, 183-195.

Kriz, J., 1991. The Silurian of the Prague Basin (Bohemia) - tectonic, eustatic and volcanic controls on facies and faunal development. In: Bassett, M.G., Lane, P.D. and Edwards, D. (eds.), The Murchison Symposium. Spec. Papers Palaeontology 44, 179-203.

Kriz, J., 1992. Silurian Field Excursions: Prague Basin (Barrandian), Bohemia. National Museum of Wales, Geol. Ser. 13, 1-111.

Kriz, J., 1995a MS. Bivalvia dominated communities of Bohemian type from the Silurian and Lower Devonian carbonate facies. In: Boucot, A.J. and Lawson, J.D. (eds.), Final Report, Proj. Ecostratigraphy. Cambridge Univ. Press.

Kriz, J., 1995b MS. Silurian Bivalvia of Bohemian type from Montagne Noire and the Massif Mouthoumet, France. Palaeontogaphica, Bonn.

Kriz, J., Dufka, P. and Schönlaub, H.-P., 1993. Wenlock-Ludlow boundary in the Prague Basin (Bohemia). Jb. Geol. Bundesanstalt Wien 136 (4), 809-839.

Kriz, J., Jaeger, H., Paris, F. and Schönlaub, H.P., 1986. Pridoli - the fourth subdivision of the Silurian. Jb. Geol. Bundesanstalt Wien 129, 291-360.

Kriz, J. and Serpagli, E., 1993. Upper Silurian and lowermost Devonian Bivalvia of Bohemian type from South-Western Sardinia. Boll. della Soc. Paleontologica Italiana 32, 289-347.

Laufeld, S., 1974. Silurian Chitinozoa from Gotland. Fossils and Strata 5, 1-130. Oslo.

Le Herisse, A., 1989. Acritarchs et kyster d'algues Prasinophycees du Silurien de Gotland, Suede. Paleontographia Italica 76, 57-302. Pisa.

Lenz, A.C. and Melchin, M.J., 1987. Silurian retiolitids from the Cape Phillips Formation, Arctic Islands, Canada. Bull. Geol. Soc. Denmark 35, 161-170.

Lenz, A.C. and Melchin, M.J., 1989. *Monograptus spiralis* and its phylogenetic relationship to early

cyrtograptid. J. Paleont. 63, 341-348.
Lenz, A.C and Melchin, M.J., 1991. Wenlock (Silurian) graptolites, Cape Phillips Formation, Canadian Arctic Islands. Trans. Roy. Soc. Edinburgh, Earth Sci. 82, 211-237.
Loydell, D.K., 1991. The biostratigraphy and formational relationships of the upper Aeronian and lower Telychian (Llandovery, Silurian) formations of western mid-Wales. Geol. J. 26, 209-244.
Männik, P. and Viira, V., 1990. Conodonts. In: Kaljo, D. and Nestor, H. (eds.), Field Meeting Estonia 1990. An excursion guidebook. pp. 84-89. Tallinn.
Männik, P. and Viira, V., 1993. Events in the conodont history during the Silurian in Estonia. Proc. Estonian Acad. Sci., Geol. 42, 58-69.
Männil, R., 1992. Trilobite faunal changes in the East Baltic Silurian. Proc. Estonian Acad. Sci., Geol. 41, 198-204.
Märss, T., 1986. Pozvonochnye silura Estonii i zapadnoi Latvii. Fossilia Baltica 1. 104 pp. Acad. Sci., Estonian SSR, Inst. Geol., Tallinn [Engl. summ. Silurian Vertebrates of Estonia and West Latvia].
Märss, T., 1989. Vertebrates. In: Holland, C.H. and Bassett, M.G. (eds.), A global standard for the Silurian System. Nat. Mus. Wales, Geol. Ser. 9, 284-289. Cardiff.
Märss, T., 1992. Vertebrate history in the Late Silurian. Proc. Estonian Acad. Sci., Geol. 41, 205-214.
McKerrow, W.S., 1979. Ordovician and Silurian changes in sea-level. J. Geol. Soc. London 136, 137-145.
Melchin, M.J. and Mitchell, C.E., 1991. Late Ordovician extinction in the Graptoloidea. Advances in Ordovician Geology, Geol. Surv. of Canada pap. 90-9, 143-156.
Nestor, V., 1990. Silurian chitinozoans. In: Kaljo, D. and Nestor, H. (eds.), An Excursion Guidebook. pp. 80-83. Tallinn.
Nestor, V., 1992. Chitinozoan diversity dynamics in the East Baltic Silurian. Proc. Estonian Acad. Sci., Geol. 41, 215-224.
Nestor, H. and Einasto, R., 1977. Facialno-sedimentologicheskaya model siluriiskogo Paleobaltiiskogo perikontinentalnogo basseina. In: Kaljo, D. (ed.), Facies and fauna of the Baltic Silurian. pp. 89-121. Tallinn. [Engl. summ. Facies-sedimentary model of the Silurian Paleobaltic pericontinental basin].
Nestor, V. and Nestor, H., 1991. Dating of the Wenlock carbonate sequences in Estonia and stratigraphic breaks. Proc. Estonian Acad. Sci., Geol. 40, 50-60.
Plotnick, R.E. and McKinney, M.L., 1993. Ecosystem organization and extinction dynamics. Palaios 8, 202-212.
Pribyl, A., 1983. Graptolite biozones of the Kopanina and Pridoli Formations in the Upper Silurian of central Bohemia. Casopis pro Mineralogii a Geologii 28, 149-167.
Rickards, R.B., 1976. The sequence of Silurian graptolite zones in the British Isles. Geol. J. 11, 153-188.
Rickards, R.B., 1988. Anachronistic, heraldic and echoic evolution: new patterns revealed by extinct planktonic hemichordates. In: Larwood, G.P. (ed.), Extinction and Survival in the Fossil Record. Systematics Assoc. Spec. 34, 211-230.
Rickards, R.B., 1989. Exploitation of graptolite cladogenesis in Silurian stratigraphy. In: Holland, C.H. and Bassett, M.G. (eds.), A global standard for the Silurian System. Nat. Mus. Wales, Geol. Ser. 9, 267-274. Cardiff.
Rickards, R.B., Hutt, J.E. and Berry, W.B.N., 1977. Evolution of the Silurian and Devonian graptoloids. Bull. Brit. Mus. (Nat. Hist.) Geol. 28, 120 pp.
Schönlaub, H.P., 1986. Significant geological events in the Paleozoic record of the Southern Alps (Austrian part). In: Walliser, O.H. (ed.), Global bio-events. pp. 163-167. Springer, Berlin Heidelberg New York.
Schweineberg, J., 1987. Silurische Chitinozoen aus der Provinz Palencia (Kantabrisches Gebirge, N-Spanien). Göttinger Arb. Geol. Paläont. 33, 1-94.
Shackleton, N.J. and Hall, M.A., 1984. Carbon isotope data from Leg 74 sediments. In: Moore, T.C., Jr., Rabinowitz, P.D. et al.: Initial Reports of the Deep Sea Drilling Project 74. pp. 613-619. Washington D.C.
Shaver, R.H., 1991. A history of study of Silurian reefs in the Michigan Basin environs. In: Catacosinos, P.A. and Daniels, P.A., Jr. (eds.), Early sedimentary evolution of the Michigan Basin. Geol. Soc. Amer. Spec. Paper 256, 101-138.
Shaver, R.H., 1992. Wenlockian and Ludlovian sedimentary and biotic events in the midwestern cratonic area, U.S.A.. Proc. Estonian Acad. Sci., Geol. 41, 182-192.
Siveter, D.J., Owens, R.M. and Thomas, A.T., 1989. Silurian field excursions: A geotraverse across Wales and the Welsh Borderland. Nat. Mus. Wales, Geol. Ser. 10, 1-133. Cardiff.

Sloss, L.L., 1988. Forty years of sequence stratigraphy. Geol. Soc. Amer. Bull. 100, 1661-1665.

Storch, P., 1988. Earliest Monograptidae (Graptolithina) in the lower Llandovery sequence of the Prague Basin (Bohemia). Sbor. geol. ved., Paleont. 29, 9-48.

Storch, P. and Loydell, D., 1992. Graptolites of the *Rastrites linnaei* group from the European Llandovery (lower Silurian). Neues Jahrbuch für Geologie und Paläontologie 184, 63-86.

Sweet, W.C., 1985. Conodonts: those fascinating little whatzits. J. Paleont. 59, 485-494.

Sweet, W.C., 1988. The Conodonta. Morphology, taxonomy, paleoecology and evolutionary history of a long-extinct animal phylum. 212 pp. Clarendon Press, Oxford.

Tesakov, J.I., Predtechenskii, N.N., Khromykh, V.G. and Berger, A., 1986. Siluriiskie biocenozy severa Sibirskoi platformy (bassein r. Moiero). In: Sokolov, B.S. (ed.), Fauna i flora Zapolarya Sibirskoi platformy. pp. 5-84. Nauka, Novosibirsk.

Ulst, R.Z., 1974. Posledovatel'nost' pristiograptov v pogranichnykh otlozheniyakh venloka i ludlova v Srednei Pribaltike. In: Obut, A.M. (ed.), Graptolites of the USSR. pp. 105-122. Nauka, Siberian Branch, Novosibirsk.

Urbanek, A., 1966. On the morphology and evolution of the Cucullograptinae (Monograptinae, Graptolithina). Acta Palaeont. Polonica 11, 1-540.

Urbanek, A., 1970. Neocucullograptinae n. subfam. (Graptolithina) - their evolutionary and stratigraphic bearing. Acta Palaeont. Pol. 15, 1-393.

Vail, P.R., Mitchum, R.M., Jr. and Thompson, III, S., 1977. Seismic stratigraphy and global changes of sea level, 4, Global cycles of relative changes of sea level. In: Payton, C.E. (ed.), Seismic stratigraphy - applications to hydrocarbon exploration. Amer. Assoc. Petrol. Geologists, 83-97. Tulsa.

Walliser, O.H., 1986. Towards a more critical approach to bio-events. In: Walliser, O.H. (ed.), Global Bio-events. Lecture Notes in Earth Sci. 8, 5-16. Springer, Berlin Heidelberg New York.

Wang, Yu, Boucot, A.J., Rong, Jia-yu and Yang, Xue-chang, 1984. Silurian and Devonian biogeography of China. Geol. Soc. Amer. Bull. 9, 265-279.

Wilde, P., Berry, W.B.N. and Quinby-Hunt, M.S., 1991. Silurian oceanic and atmospheric circulation and chemistry. In: Bassett, M.G., Lane, P.D. and Edwards, D. (eds.), The Murchison Symposium. Spec. Papers Palaeont. 44, 123-143.

Zalasiewicz, J.A., 1990. Silurian graptolite biostratigraphy in the Welsh Basin. J. Geol. Soc. London 147, 619-622.

Manuscript received February 1993
Revision received March 1994

Authors' addresses:

Dimitri Kaljo, Peep Männik, Tiiu Märss, Viiu Nestor and Viive Viira, Institute of Geology, Estonian Academy of Sciences, 7 Estonia Ave., EE0001 Tallinn, Estonia

Arthur J. Boucot, Department of Zoology, Oregon State University, Corvallis, OR 97331-2914, U.S.A.

Richard M. Corfield, Department of Earth Sciences, University of Oxford, Parks Road, Oxford OX1 3PR, U.K.

Alain Le Herisse, Laboratoire de Paléontologie et de Stratigraphie du Paléozoique, Université de Bretagne Occidentale, 6 avenue Le Gorgeu, 29287 Brest, France

Jiri Kriz, Czech Geological Survey, P.O. Box 85, Praha 011, 11821 Czech Republic

Tatyana N. Koren, Geological Institute, Sredni prosp. 74, St. Petersburg, 199026, Russia

Robert H. Shaver, Indiana Geological Survey, 611 North Walnut Grove, Bloomington, Indiana 47405, U.S.A., and Indiana University Department of Geological Sciences, 1005 East Tenth Street, Bloomington, Indiana 47405, U.S.A.

Derek J. Siveter, Geological Collections, University Museum, University of Oxford, Parks Road, Oxford OX1 3PW, U.K.

Global Events in the Devonian and Carboniferous

Otto H. WALLISER

Abstract. Devonian and Carboniferous global events are described. Their magnitude ranges from short-termed geo-events without major biotic changes up to highest order mass-extinctin events. All bio-events are proximately caused by environmental changes such as anoxic episodes and/or sea-level changes.

Contents

Introduction	225
Devonian and Carboniferous: A Time of Paramount Importance in Earth History	225
Palaeogeographic Setting	225
Climate	226
Colonization of Land Areas	226
Time-Specific Facies (TSF)	227
State-of-The-Art	228
Devonian Global Events	228
Carboniferous Global Events	241

Introduction

Devonian and Carboniferous: A Time of Paramount Importance in Earth History

In the course of Earth history we recognize 5 episodes of genuine revolution in evolution: (1) the origin of life, more than 3.500 Ma ago; (2) formation of metazoan organisms, less than 1.000 Ma before present; (3) skeletal mineralization at the turn of the Precambrian to Phanerozoic; (4) colonization of land areas with the decisive steps in the Devonian; and (5) spreading of grasses and mammals in the early Tertiary. In so far, the Devonian to Carboniferous Eras take an eminent place in Earth history. In addition, the Variscan orogenesis terminates the plate-tectonic supercycle which started in the terminal Precambrian and had its first culmination during the Caledonian orogenesis.

Palaeogeographic Setting

The late Silurian Caledonian orogenesis, which was caused by the collision of Laurentia and Fennosarmatia, resulted in the formation of the Old Red Continent in the northern hemisphere (Fig. 1). The distance of that to the Gondwanan continent in the southern hemisphere is not yet known with certainty but the distribution of faunas and facies indicates a relatively close position. Palaeomagnetic and biogeographic data indicate that in the Middle Devonian Europe and middle to northern parts of North America were crossed by the equator, whereas the South Pole was situated in South Africa.

In the further course of the Devonian and Carboniferous, the continents shifted slightly in northern direction. Thus the equator reached a position close to southern Europe and southern North America, and the South Pole was situated in Antarctica. By the Variscan

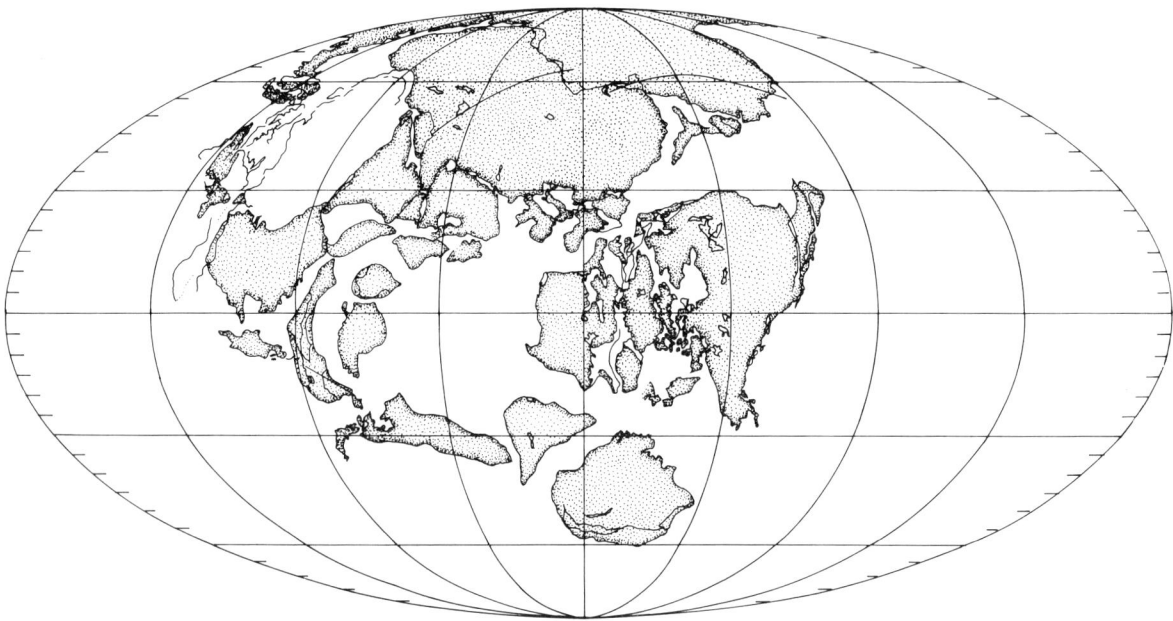

Fig. 1. Early Devonian geographic setting (From Scotese and McKerrow, 1990)

orogenesis, resulting from late Carboniferous and Early Permian plate movements, the continents in the northern hemisphere merged finally forming together with the former Gondwana the supercontinent Pangaea.

Climate

In the Early Devonian, southern parts of the southern hemisphere were characterized by a cold water fauna and the lack of extensive carbonate facies. This Malinokaffric Province was in contrast to the northern hemisphere, where warm-water faunas prevailed. This temperature distribution is in good accordance with the inferred position of the continents at that time. Lacking indications of an extended Gondwanan glaciation permits the assumption that the mean global temperature was relatively high.

Regarding widespread occurrences of clastic sediments in the Early Devonian, carbonates in the Middle Devonian, and even evaporites in the Late Devonian, a general increase in temperature during the Devonian can be assumed at least for the northern hemisphere.

From NE Brasil, i.e. within the southern hemisphere, Caputo and Crowell (1985) and Niklas et al. (1976) described early Famennian floras from a sequence below glacigenic sediments. This may be taken as an indication of a first fluctuation in direction towards a cooler climate in the southern hemisphere, which reached its climax in the Permocarboniferous glaciation. This was restricted to the southern parts of Gondwana, with the center at Antarctia.

Colonization of Land Areas

There are many indications that terrestrial areas were occupied to some extent by Thallophyta long before the Devonian. Even vascular plants – a Silurian biological innovation of paramount importance – may have advanced to marginal land areas in the late Silurian. Certainly, the main and decisive steps of colonization took place in the Early and Middle Devonian. Thereby, the pioneering plants were immediately followed by animals.

Geological Aspects of Land Colonization

In contrast to former times, the cover of terrestrial areas by plants caused principal changes with respect to processes in the geo-, hydro-, and atmosphere (Fig. 2). Thus, e.g., weathering became more strongly influenced by chemical processes, thereby causing changes in the chemical composition of the surface water and of the final sediments. Because a much higher percentage of the rainfall was bound by the vegetation and the soil, physical conditions such as the energy for erosion and transportation of debris also changed.

The storage of water by plants and soil led to strong changes in evaporation on land areas with respect to magnitude and time distribution. This in turn, together with changes in albedo as well as in the O_2 - CO_2 cycle, had a strong influence on the atmosphere and climate. Feed-back reactions and interactions may have increased the complexity of this new environmental system.

Evolutionary Aspects of Land Colonization

There is a great coincidence not only in the colonization of terrestrial areas and the diversification of plants and terrestrial animals, but also of primitive and advanced fishes. For such an unique increase in taxonomic and ecologic diversity two pre-conditions must be realized: (1) Presence of such biological bauplans or archetypes which are qualified and promising for further successful evolution, i.e. for opening new pathways in the evolution of ecosystems. This precondition has been fulfilled not only by the evolution of vascular plants, but also by many constructions which have long existed within both the invertebrates and vertebrates. (2) Presence of appropriate environmental conditions. These have been created as a result of the Caledonian orogenesis. Thus, e.g., on the periphery of the recently formed Old Red Continent, there developed large deltaic systems which offered a great amount of different and closely spaced ecological niches in a favorable climate realm, thus providing optimal conditions for a rapid diversification of plants, invertebrates and fishes (Walliser and Michels, 1983).

In addition to the optimal setting at the sea-land transition, the land areas themselves offered a great variety of un-occupied terrestrial environments or niches, respectively, which provided a completely new opportunity for co-evolution of plants and animals.

With all these unique and optimal parameters it is quite understandable that the Devonian and early Carboniferous, especially with respect to evolution of terrestrial ecosystems, is a time of rapid diversification and of evolving new bauplans or archetypes such as the diverse advanced classes in plants and also, e.g., in Tracheata and Tetrapoda. One even can state that with the colonization of land areas in the Devonian and Carboniferous all important steps in the evolution of terrestrial taxa and ecosystems had been completed and the evolution in the subsequent 280 Ma was mainly restricted to optimalization and variation of the bauplans which had already evolved.

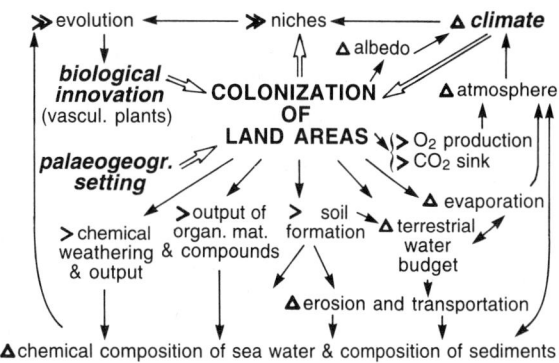

Fig. 2. Flow-chart indicating changes and interactions caused by the colonization of land areas by vascular plants

Time-Specific Facies (TSF)

Facies analysis is one of the most important tools in reconstructing the environmental conditions. Thus, facies changes connected with global bio-events may indicate the causation of the latter. This is especially valid for changes that are obviously caused by sea level changes or by fluctuation in oxygen content of the water masses. Later, in certain time intervals certain sediments may predominate, as e.g., reefs in the Middle Devonian and Frasnian, or black shales and bedded cherts in the Dinantian. In addition, there sometimes exist quite special features of the facies which characterize certain time intervals. As an example the Famennian nodular cephalopod limestones may be mentioned. Of course, nodular cephalopod limestones occur throughout the entire Devonian and even often in the entire Phanerozoic. But in most cases, the Famennian nodular cephalopod limestones are significantly different from the others with respect to formation and arrangement of the nodules, color, weathering, etc. Even within the Famennian characteristic differences exist: in the lower part, beneath the *annulata* horizon, the calcareous nodules have a mean diameter in the order of 1 to 2 cm, whereas in the upper part they normally are much larger. This pattern, called time-specific facies by Walliser (1984a), has been observed in Europe, North Africa, Iran, South China, and northeastern USA. Certainly it is the documentation of quite special conditions with respect to sedimentation and diagenesis, but we are still quite far from being able to give a sufficient explanation.

State-of-The-Art

With respect to global events the Devonian has received relatively strong attention. This is due to the fact that the biostratigraphy of the Devonian is well advanced, and that many of the biostratigraphers are involved in the research of the International Subcommission on Devonian Stratigraphy as well as in the IGCP Project "Global Biological Events in Earth History". However, there exist great differences in the intensity and therefore also in the knowledge of Devonian events. A first order bio-event such as the Kellwasser Event has been studied in many places around the world in greatest detail and by many different methods. In contrast, smaller overturns in fauna and/or facies still need more detailed investigations or even a verification of their global extent.

Most global events in the Devonian and Carboniferous have been initially recognized and then investigated in the pelagic facies, because the latter has the highest potential of being globally distributed and traceable. Furthermore, the faunal distribution in space and time is well known within pelagic groups, especially with regard to goniatites, conodonts, and dacryoconarids, because they serve as the main tools for detailed stratigraphic subdivision and correlation. With a few exceptions, much less is known about the influence of global events on the neritic facies realms. One reason for this may be the fact that the influence and result of local to regional processes can be quite strong, making it very difficult to distinguish them from global-scale changes and to correlate these changes over great distances.

The investigations of Carboniferous global events are not as advanced as those for the Devonian. In spite of the Mid-Carboniferous bio-event, the other stronger changes are assumed to be global, as correlative changes, e.g. in pelagic facies, are known from many areas. The event character of those changes is sometimes supported by minor turnovers in goniatites and/or conodonts.

Devonian Global Events

Silurian-Devonian Boundary Event (S/D-Event)

S/D Event (Walliser, 1985). The S/D boundary at the type locality (Klonk near Suchomasty, Barrandian) is fixed within a monotonous sequence of dark-gray platy limestones and interbedded calcareous shales (Hladil, 1992). Because no facies change occurs around the boundary, the Devonian base can only be recognized by the onset of *Monograptus uniformis*. Among the other index fossils, *Ozarkodina remsch. remscheidensis* is already present in a layer 40-50 cm below the boundary, whereas the first *Warburgella rugulosa rugosa* has been found about 190 cm above the boundary (Chlupac, 1980). This relative succession of index species also occurs in sections which contain different lithofacies. There, under more shallow conditions, *Icriodus woschmidti* occurs already below the boundary, together with *O. r. remscheidensis*.

In some other sections of the Bohemian facies realm, traceable from the Barrandian through Carnic Alps and Sardinia to the Moroccan Meseta near Rabat, the boundary is lithologically characterized by the occurrence of dark gray, finely organodetrital and micritic platy Lochkovian ("e γ") limestones. This change in facies is due to a sea-level rise that can also be observed in other regions. In the Rhenohercynian and Saxothuringian zones of the European Variscides a deepening effect and an onlap at the northern geosynclinal margin is obvious (Ochre Lmst. at Hüinghausen, Gedinnian transgression). In Podolia, shallow-water limestones with a neritic fauna successively change to pelagic, platy limestones where the first graptolite-shale layer is intercalated, containing *M. uniformis augustidens*. This transgressional tendency has also been recognized in Australia and SW Siberia (Talent and Yolkin, 1987).

Also characteristic for the S/D boundary interval is the occurrence of two blooming events within Scyphocrinitidae. This fossil group, short-lived but crossing the boundary, floated at the water surface due to their balloon-like loboliths (Haude, 1972). The lower horizon characterized by a more primitive cirrus lobolith as well as the upper horizon showing a more advanced plate lobolith are known from many regions, such as Northwest Africa, Spain, Germany, Bohemia, Podolia, and Yunnan (see Jahnke and Shi, 1989).

Faunistically there is no strong overturn at the event although the Pridolian and Lochkovian associations can be readily distinguished (see, e.g., Chlupac, 1980). Thus, the S/D Event appears to be a minor, relatively gradual but globally traceable event.

Lochkovian-Pragian Boundary Event (Lo/Pr Event)

G/S (Gedinnian-Siegenian boundary Event, Walliser, 1985); Lochkovian-Pragian boundary Event (Chlupac and Kukal, 1988). In Bohemia, i.e. the region of the stratotype, the boundary between the two

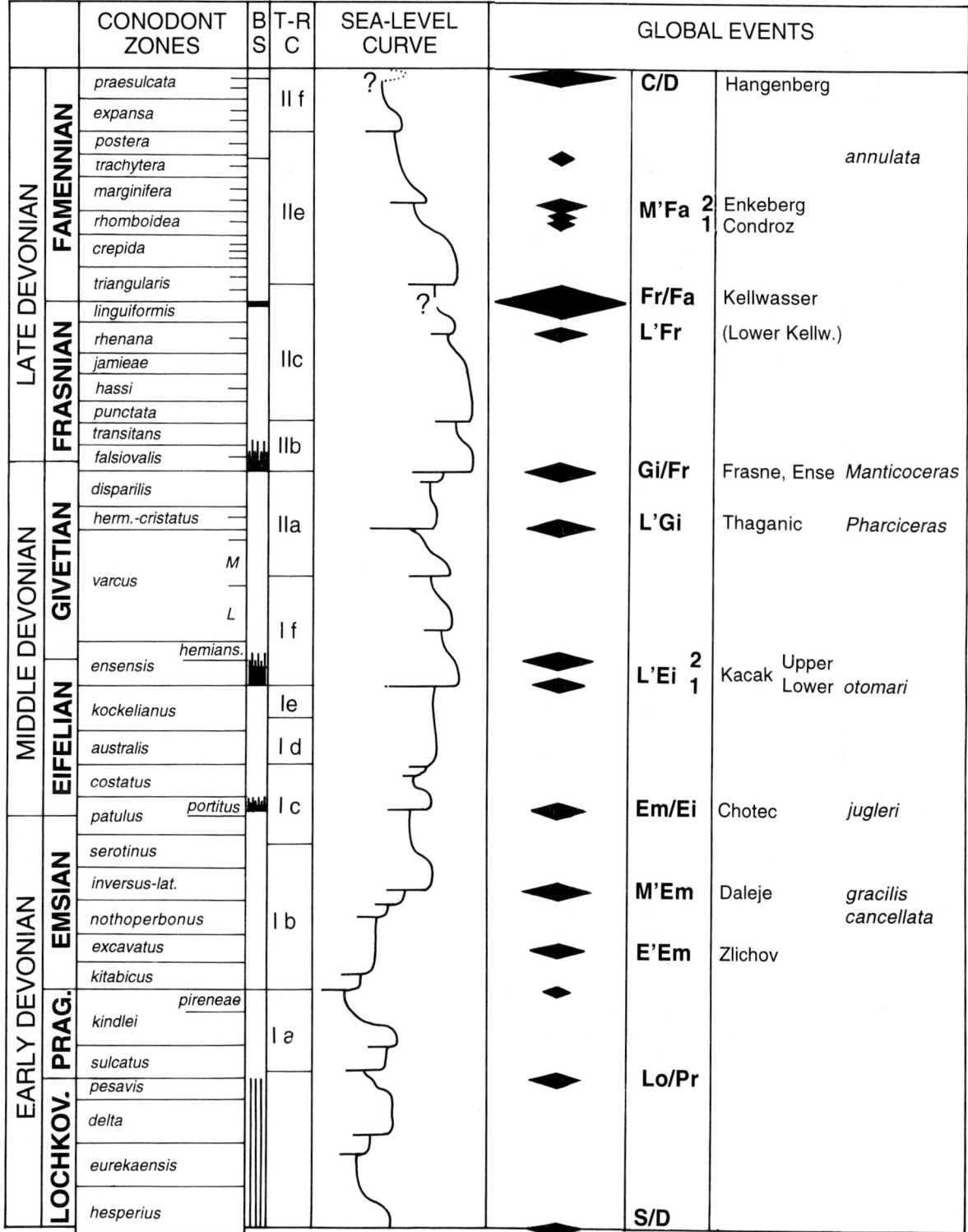

Fig. 3. Devonian global events and sea-level changes. Black rhombs: global events. Sea-level curve modified after Johnson et al. (1985)

stages is marked by a change in fauna and facies. The dark Lochkovian platy limestones with intercalated dark, partly graptolite-bearing calcareous shales are succeeded by gray or red micritic and nodular limestones (Dvorce-Prokop and Reporyje Lmsts.), or reef and organodetrital limestones (Upper Koneprusy and Slivenec Lmsts.). The change from dark to light sediments is not abrupt. Light color, as well as an increase in grain size of biogenic detritus can be observed already in the topmost Pragian.

This facies change clearly reflects "a fairly rapid but not very large lowering of the sea level" (Chlupac and Kukal, 1988). The mentioned authors demonstrate by a short compilation of concerned literature that the Lower Pragian regressive Event (Johnson and Murphy, 1984) is globally well represented (for details see lit. cit.). With respect to palaeontological features, the worldwide lowering of sea level led to well-recognized differences between the Lochkovian and Pragian faunal associations.

Basal Zlichov Event (B'Zl Event)

Basal Zlichovian Event (Chlupac and Kukal, 1988). When Walliser (1985) mentioned a Siegen-Emsian boundary Event, he presumed that the faunal overturn at the Siegenian-Emsian boundary coincides with the onset of *Polygnathus dehiscens* and approximately with the base of the topmost limestone horizon of the Praha Formation, characterized by *Guerichina strangulata*. In the meantime the Emsian base has been fixed in the Zinzilban area (Zeravshan Range, Central Asia) by the first occurrence of the *Po. dehiscens* lineage which is represented by the newly erected *Po. kitabicus* (see Yolkin et al., 1989, 1994). The onset of this species is correlated with the beginning of a prominent sea-level rise in the type area, marked by a quick change from the shallow water carbonate complex of the Madmon Formation to the Khodzha-Kurgan Formation that starts with well-bedded, fine-grained, dark limestones of the Zinzilban Member. Up to now it is not clear whether this transgressive Zinzilban Event is only regionally traceable and to which horizon in Bohemia it can be correlated. I assume that it is considerably lower in the Praha Formation than the known occurrence of *Po. dehiscens* in that region. Moreover, the correlation with the Siegenian-Emsian boundary is again uncertain. In any case, the B'Zl Event is significantly younger than the Zinzilban and Siegen-Ems Events.

The Basal Zlichov Event has been defined and described by Chlupac and Kukal (1988). For details the reader is referred to this publication. Summarizing their results it should be emphasized that the faunal change is not very distinct. According to the authors, sedimentological peculiarities above the boundary (e.g. Kaplicka Horizon, a breccia that is mostly deformed by slumping) is due to local tectonics.

Mid-Emsian Event (M'Em Event)

Unnamed global event (Walliser, 1984a, Fig. 3), Daleje Event (House, 1985), *gracilis* or *cancellata* Event (Walliser, 1985). The Daleje Event is named after the changes in lithology and fauna near the base of the Daleje Formation in Bohemia. This level is indicated by the first occurrence of *Nowakia cancellata* and *Gyroceratites gracilis*. In the pelagic facies realm this *gracilis* boundary has been formerly taken as the base of the Middle Devonian. As Carls et al. (1972) demonstrated, this level has to be correlated with about the Lower-Upper Emsian boundary.

The Daleje Event is connected with a transgressive period that had already started early in the Early Emsian or Zlichovian, respectively (T-R cycle I b in Johnson et al., 1985). Therefore the change in sedimentation, i.e. from carbonates to shales, increases gradually within the late Zlichovian (*N. barrandei* Zone: biomicritic limestones prevail; *N. elegans* Zone: limestones are reduced to thin intercalations within dark-gray calcareous shales; *N. cancellata* Zone: typical Daleje Shale). In more shallow positions where biosparitic limestones dominate during late Zlichovian, a sudden change to calcareous shales with micritic nodules or micritic nodular limestones occurs at the base of Dalejan. The late Zlichovian to early Dalejan transgression can be traced globally, as Chlupac and Kukal (1988) showed in a compilation work (for details see lit. cit.).

According to the gradual faunal change, the faunal turnover also spans a certain time interval, but mainly from base to top of the *N. elegans* Zone, i.e. from late *Po. gronbergi* to early *Po. laticostatus* Zones (correlation of dacryoconarid and conodont zones according to Klapper et al., 1978, and Chlupac et al., 1979). Especially significant is the faunal change in goniatites, which had their origination and first radiation within the Zlichovian. Most species of the more or less loosely coiled Mimosphinctidae vanished between the base and top of the *N. elegans* Zone, and the surviving species were already gone within the lowermost Dalejan. Including representatives of all families, about 75 percent of all species became extinct.

In higher parts of the Dalejan or Late Emsian,

respectively, the surviving goniatite genera, *Gyroceratites* and *Mimagoniatites*, were replaced by Agoniatitidae and Anarcestidae, which subsequently increased rapidly in diversity. Strong diversity increases also occurred in other fossil groups, as e.g., the Asteropygidae (Morzadek, 1992), i.e. a group of trilobites that is not represented in Bohemia.

Emsian-Eifelian Boundary Event (Em/Ei Event)

Jugleri Event (Walliser, 1995), Basal Chotec Event (Chlupac and Kukal, 1988). In addition to a gentle faunal turnover, this is an important geo-event documented by a distinct facies change. In Bohemia it is the change from the Trebotov to Chotec Limestones. The base of the latter "is characterized by abrupt onset of darker colored biomicritic and biosparitic limestone layers within lighter micritic limestones" (Chlupac and Kukal, 1988). A facies change is also recognizable in sequences of shallow water carbonates (boundary Suchomasty-Acanthopyge Limestones).

This facies change is due to a sea-level rise that simultaneously caused a decrease of oxygen content at the sea floor, even within shallow basins. The corresponding dark to black sediments are globally detectable. Often, as e.g. in South Morocco, the transgression and color change coincide with a change from calcareous shales with interbedded marlstones to the onset of dark platy limestones with *Pinacites jugleri* occurring shortly above the base. The change of environmental conditions is also indicated by a Styliolina and Nowakia coquina ("styliolinit") that occurs in many sections just before the first occurrence of *Pinacites*.

The event level is not far above the newly defined Early-Middle Devonian boundary, indicated by the first occurrence of *Po. costatus partitus*. The term "Chotec" Event is preferred herein, as it is more exact than the other mentioned terms for this event.

The Chotec Event comprises an important faunal change as Chlupac (1982) and Chlupac et al. (1979) pointed out in the Barrandian area. Extinctions during the Chotec Event occurred in nearly all fossil groups of both neritic and pelagic facies. Shortly after the event a new diversification took place in many of the surviving groups. This is especially distinct in goniatites. After the extinction of all *Mimagoniatites* and *Sellanarcestes* species, a diversity increase in agoniatitids and anarcestids occurred. A great loss in trilobites is also recognized. Within Asteropyginae about 90 % of species (57 of 63) disappeared (Morzadek, 1992).

In addition to the faunal change an important turnover in the spore flora occurred, most probably shortly before the Chotec Event, namely within the uppermost Wetteldorf Formation of the Eifel Mountains in Germany (Riegel, 1975a, b). Near the top of this formation the first calcareous intercalations occur within the sandstone sequence, indicating the change to prevailing carbonate sedimentation in the overlying uppermost Emsian Heisdorf Formation and the subsequent younger sequences.

Late Eifelian Events (L'Ei Events)

Otomari Event, *rouvillei* Event (Walliser, 1983), Kacak Event (House, 1985). The *otomari* Event is lithologically characterized by a sharp onset of black shale deposition, especially within basinal position but also partly extending to the hitherto neritic facies realm. This lithological change coincides with the boundary between the *Tortodus k. kockelianus* and *Po. ensensis* Zones. Thus it is well below the recently defined Eifelian-Givetian boundary, indicated by the first occurrence of *Po. hemiansatus* in the upper part of the *Po. ensensis* Zone. Together with the black shale occurs an early form of *Nowakia otomari*.

Because there was some confusion with respect to terminology and duration of this event, it should be emphasized once more that the present author restricts the term "*otomari* event" to the onset of the black shale. In contrast, House (1985) implied in his "Kacak Event" the entire black-shale horizon as, e.g., the Kacak Member of the Srbsko Formation in Bohemia or the Odershausen Formation in Germany. This could be accepted if the discussed black-shale intercalation would have been as short-termed as, e.g., the two Kellwasser Horizons or the *annulata* black shale. However, the interval of the Kacak Member certainly is several times longer and in addition it has two very distinct extinction events, one at the base, and a second at the top. In order not to use too many terms, in this chapter the lower event, i.e. the *otomari* Event, is called "Lower Kacak Event" or L'Ei 1 Event in contrast to the event at the top of the black-shale period, the "Upper Kacak Event" or L'Ei 2 Event.

With the Lower Kacak Event vanished a great number of typical Eifelian conodont species, such as *Tortodus k. kockelianus*, *Polygnathus trigonicus*, *Po. angustipennatus* and *Po. robusticostatus*. They were replaced by a quick evolution of the *ensensis* and *pseudofoliatus* group within polygnathids, and also by that *Icriodus* lineage which leads to *Icriodus obliqumarginatus* near the Eifelian-Givetian boundary. In goniatites all genera with the exception of *Exopinacites* crossed the base of the Kacak Event. *Foordites*, *Chlupacites* and *Pinacites* disappeared already prior to the event. Benthic communities, of course, disappeared in localities where the anoxic event occurred.

The Upper Kacak Event coincides widely with the newly defined Eifelian-Givetian boundary. Lithologically above this level a quick but transitional change to light colored limestones can be observed in the stratotype and other sections. Already prior to the boundary occurs *Maenioceras*, but *Fidelites* and *Subanarcestes* became extinct. *Mitraxites* and *Parodiceras* then vanish at the boundary. In contrast, conodont do not show a major turnover but rather are characterized by evolutionary transitions at the species level.

The black shales of the L'Ei Events indicate primarily an anoxic event that probably results from a transgressive pulse (T-R cycle I f in Johnson et al., 1985) which not necessarily was as abrupt as the onset of the black shales. Further investigations should examine whether there is any direct or indirect causal connection between this important anoxic *otomari* Event and the termination of the Malvinokaffric Province.

Late Givetian Event (L'Gi Event)

Pharciceras Event (Walliser, 1983); Thaganic Event (House, 1985). The *Pharciceras* Event coincides with the former Middle-Upper Devonian boundary in the pelagic facies realm. It is the level at or prior to which the typical Givetian goniatites became extinct and at which multilobate pharciceratids occurred. This is at the base of the *Schmidtognathus hermanni - Po. cristatus* Zone. House (1985) chose the name Thaganic Event because he considers the Thaganic Onlap (T-R cycle II a in Johnson et al., 1985) as the initiation of the event. Indeed, since that time within the Middle *Po. varcus* Zone a reduction in diversity can be observed in several fossil groups including the goniatites. However, the most important extinctions occurred about the top of the *Po. varcus* Zone. Within goniatites disappeared the last Agoniatitidae, Pinacitidae and most Agoniatitidae. Immediately after this extinction event radiated the multilobate Pharciceratidae evolving from a more primitive species that already occurred in the latest Givetian.

The Thaganic Event also decimated many benthonic groups. However, as Ebert (1993) showed in a compilation of then available data, the maxima of extinctions did not occur simultaneously in the different groups. Thus, the highest extinction rate in corals and stromatopors around the event occurred within the late *Po. varcus* Zone, certainly due to the main pulse of the Thaganic transgression. In contrast, the brachiopods had their main loss in diversity between Early and Late *hermanni-cristata* Zone.

Givetian-Frasnian Boundary Event (Gi/Fr Event)

Frasne Event (House, 1985); *Manticoceras* Event (Walliser, 1985); Ense Event (Ebert, 1993). House (1985) clearly characterized the Frasne Event by the extinction of most pharciceratids and by the onset of a major worldwide transgression. He mentioned in this connection the Bou Tchrafin section in southern Morocco, where the pharciceratid fauna suddenly is lost just beneath *Styliolina* shales that suggest a deepening pulse. This level of a major transgression-induced facies change and of the disappearance of most pharciceratids is the level of the *Manticoceras* Event (Walliser, 1985). It also characterizes the base of the *Manticoceras* Stufe, although the genus *Manticoceras* occurs only a certain time interval after the event (see Walliser in Bensaid et al., 1985). However, Becker (1993b) distinguishes an "End-pharciceratid extinction", followed by an "Earliest Frasnian radiation" and only then by the "Frasne Event (*Manticoceras* Event)". It might well be that the here discussed event occurs in several steps but it appears to me that the main bio-event coincides with the main geo-event, i.e. the transgressive facies change at the base of the *Mesotaxis falsiovalis* Zone (base of the former lowermost *Po. asymmetricus* Zone), i.e. very short below the newly defined Middle-Upper Devonian boundary.

Prior to the event-causing transgression occurred a regression period that is documented by the particular facies development of the upper *Pharciceras* limestones. It appears to me that as in many other events the rapid change from shallowing to deepening caused the main step of the extinction event.

Besides of the faunal overturn in goniatites, also other groups, especially the benthonic ones, were heavily affected by the event. Very high extinction rates occurred in brachiopods, corals and stromatopors (for details see the compilation in Ebert, 1993). Obviously this event caused not only a first heavy crisis in biohermal reef habitats but even the commencement of their disappearance that finally was caused by the Kellwasser Event at the close of Frasnian.

Late Frasnian Event (L'Fr Event)

Lower Kellwasser Event (Becker, 1993b). This black-shale event occurred short above the base of the Upper *Pa. rhenana* Zone. Since the 80th of the last century the Lower Kellwasser (L'KW) Horizon at the Schmidt Quarry (Rhenish Schiefergebirge) is famous for its excellently preserved Arthrodira.

Schindler (1990) emphasized that with the onset of the L'KW black-shale the late Frasnian Crises started (Schindler used the term "Kellwasser Crisis"). According to Feist and Schindler (1994), the trilobite families Proetidae, Tropidocoryphidae, Scutellidae and Phacopidae lost in diversity at the base of the L'KW Horizon. Schindler (1990) as well as Becker (1993b) mentioned further groups that became diminished by the event. Thus, e.g. goniatites and styliolinids (which disappeared, according to Schindler, 1990, already shortly prior to the event), but especially benthonic taxa such as reef builders and dwellers, and also shallow water brachiopods. However, an exact time correlation of these decimations with the L'KW Horizon is still missing in most cases. Therefore further investigations are needed to specify the role of this Event within the late Frasnian crisis.

The L'KW Horizon is an anoxic event intercalated into "normal" Frasnian sediments, either basinal shales or cephalopod limestones. This pattern resembles that of the Upper KW Horizon and other thin black-shale intercalations that occur in several sections between the two KW horizons. This supports the assumption that the black shales are caused by oxygen-depleted water masses that flooded over large parts of the relatively shallow sedimentary basins. On account of the repeated occurrence of the black shales, it appears more probable that this flooding is due to an episodic oscillation of the oceanic oxygen minimum zone rather than to an oceanic overturn in the sense of Wilde and Berry (1984, 1988).

Frasnian-Famennian Boundary Event

Fr/Fa Event; F/F Event (McLaren, 1970), Kellwasser Event (Eder et al., 1977; Walliser, 1980, 1984a).

The Kellwasser Event is one of the 7 strongest Phanerozoic faunal turnovers. This was already recognized when, within late Devonian, the Frasnian and subsequent Famennian stages were introduced more than 130 years ago. This subdivision was based on the strong changes in the communities of the neritic facies realm. But also the pelagic fauna, especially the goniatites, showed important changes and gave reason to distinguish the Adorfian (Adorf-Stufe, *Manticoceras*-Stufe, do I = Upper Devonian I) from the younger Nehdenian (Nehden-Stufe, *Cheiloceras*-Stufe, do II = Upper Devonian II) and following stages (Denckmann, 1895; Paeckelmann, 1924; Kayser, 1873; Wedekind, 1913).

Later, the exceptional magnitude of the Fr/Fa Bio-Event was emphasized by McLaren (1970). His argumentation was mainly based on the rapid turnover in fossil communities which was combined with a high extinction rate at the Frasnian-Famennian boundary. As the ultimate cause for this event he proposed the impact of a large extraterrestric body on Earth.

About at the Frasnian-Famennian boundary, in the Mid-European Variscides the so-called Upper Kellwasser (U'KW) Horizon is intercalated in light-grey pelagic cephalopod limestones or in basinal shales. This horizon, well known since the time of F.A. Roemer (1850), consists of a few dm of more or less black limestones in connection with black shales, both yielding a characteristic fauna of goniatites, orthocone cephalopods, conodonts, homoctenids, entomozoan ostracodes, and *Buchiola*, besides trilobites and large, thin-shelled pelecypods. A very similar black-shale horizon already occurs within the early Late *Pa. rhenana* conodont Zone, i.e. the Lower Kellwasser (L'KW) Horizon. Sometimes both black-shale horizons, but in any case the upper one, have been observed by the author as early as 30 years ago also in the state of New York, in Morocco, and in Iran, indicating a worldwide occurrence. Therefore, and on account of its position near the Frasnian-Famennian boundary, the U'KW horizon has been classified as a global bio-event which presumably was causatively connected with the disappearance of the biohermal reefs (Eder et al., 1977; Walliser, 1980) and the Frasnian/Famennian faunal overturn.

Intensive and detailed investigations were then carried out in the framework of the IGCP Bio-Event Project 216. The first important results were presented by Sandberg et al. (1987a, b, 1988a, b). They proved the worldwide synchronous character of the Kellwasser Event.

The importance of this first order global bio-event was taken into consideration when the International Subcommission on Devonian Stratigraphy chose the stratotype for the Frasnian/Famennian boundary. In the boundary type section at the abandoned quarry Coumiac (Montagne Noire/France), the boundary has been fixed between the top of the Upper Kellwasser equivalent (top of the *Pa. linguiformis* Zone) and the *Pa. triangularis* Zone.

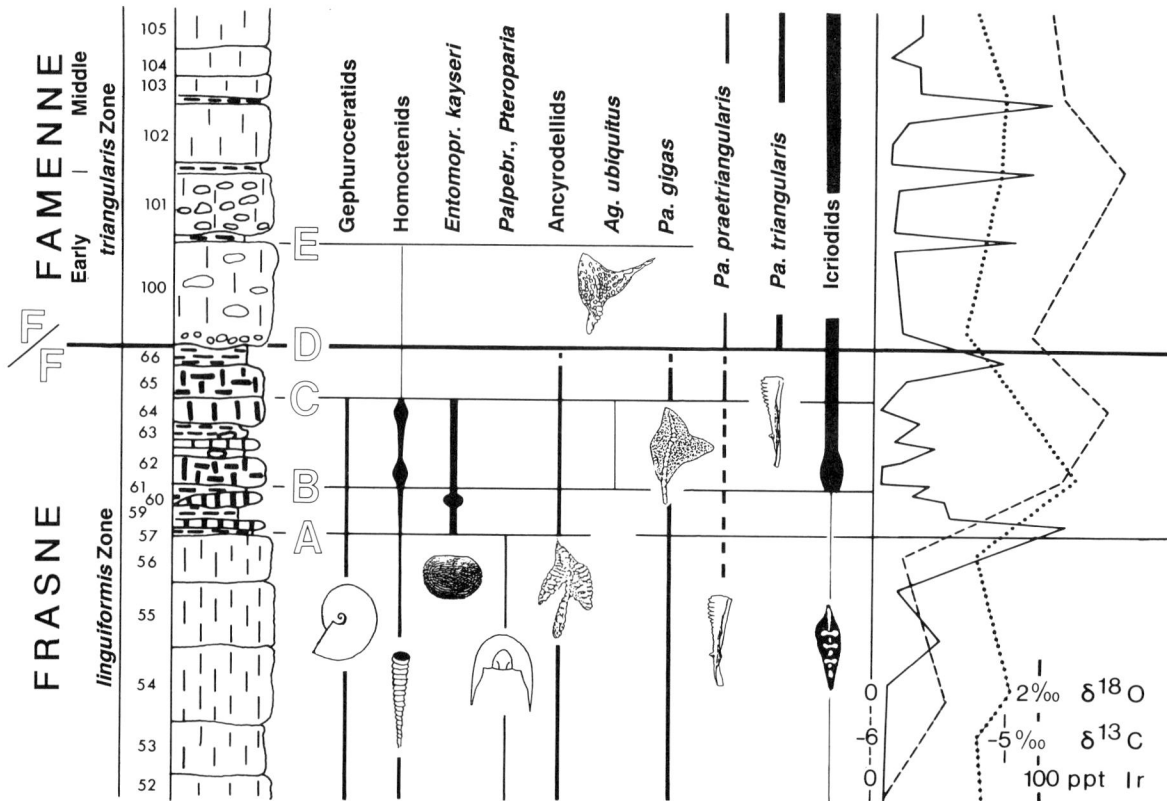

Fig. 4. The Kellwasser Event. Compilation based on Schindler (1990)

The Faunal Overturn

The Fr/Fa Bio-Event affected numerous animal groups. Thus the p h y t o p l a n k t o n shows a strong reduction in diversity, as pointed out by Tappan (1981), Streel et al. (1987), and Streel and Vanguestaine (1988). As Vachard (1994) showed, among foraminifers the Eogeinitzinidae and Semitextulariidae became extinct. In general, the calcareous forams became significantly reduced and then recovered only in the later parts of the Famennian (Marchant, 1987).

Among the hermatype organisms the S t r o - m a t o p o r a suffered from a drastic diversity reduction (Stearn, 1987; Cockbain, 1988). The same happened to framework building r u g o s e c o r a l s (Pedder, 1982; Sorauf and Pedder, 1986; Scrutton, 1988). Oliver and Pedder (1994) demonstrated that the extinction curves of the Old World Realm and Eastern American Realm parallel each other, i.e. that controls were worldwide rather than regional. According to these authors the extinction of the coral fauna was prolonged, with an abrupt final phase in the terminal Frasnian, which did not overlap with the origination of the Carboniferous-like Famennian coral fauna.

The extinction of the groups of organisms that provided the framework of biohermal build-ups is tantamount to the disappearance of these reefs. Certainly, Schindler (1990) showed in a compilation of then available data that the end of reef growth in different regions and even within a certain area obviously spanned a longer interval of late Frasnian time, primarily within the former *Pa. gigas* conodont Zone. But in most cases, the virtual end of reef growth with respect to conodont subzones is not known. The same is valid for the extinction of the framework building stromatopors and corals. In addition, the classification of the concerned "reefs" is not yet sufficiently investigated in many cases. In order to really understand the causes and processes, we need to decipher in detail the early Late Devonian turnover in reef growth. Thereby one should distinguish the different kinds of "reefs", such as bioherms and biostroms, coral-, stromatopor-, stromatolite-dominated, etc. With respect to the involved taxa we need the exact time of their extinction, as well as their ecological classification or valency, respectively. These data then have to be correlated with the extinction data of other groups which inhabit reefs and reef-like environments. Being far from this necessary knowledge, at the time being we can only summarize as follows: before the Kellwasser Event, true biohermal reefs had already suffered some kind of crisis that began with the early Frasnian transgression and led in many cases to an end of reef growth within the late Frasnian (former *Pa. gigas* Zone), followed by the final extinction at about the Kellwasser Event.

Among the o s t r a c o d e s the planktonic entomozoans were apparently less affected (Groos-Uffenorde and Schindler, 1990) than the benthonic groups, which show a stepwise but significant diversity reduction during the late part of the *Pa. linguiformis* conodont Zone, reaching the lowest diversity at the Fr/Fa boundary with a 65 % total loss of species (Lethiers and Feist, 1991).

Particularly interesting is the diversity pattern in t r i l o b i t e s . The detailed investigations of Schindler (1990), in cooperation with R. Feist/Montpellier (see also Feist in House et al., 1988; Feist and Schindler, 1994) proved that numerous trilobite groups became extinct at the base of the U'KW Horizon. Certainly, the main extinction interval is obviously preceded by a severe crisis which had already started at least by the times of the Lower KW Horizon. At the base of this black-shale event species of the Proetidae, Scutellidae and Phacopidae vanish, but the Aulacopleuridae, Harpinae, Dalmanitidae, Tropidocoryphidae, Asteropyginae, the phacopid genus *Cryphops* and the proetid genus *Palpebralia* are still present in the interval between the two KW Horizons. For the Asteropyginae, Morzadek (1992) showed that the main extinction phase already occurred at the *asymmetricus/gigas* Zones boundary, reducing the diversity from 35 to 4 species. These survivors disappeared in a stepwise manner prior to the Fr/Fa boundary.

According to Johnson (1971, 1974), Johnson and Boucot (1973), and McGhee (1981), the b r a c h i o p o d a suffered a great loss in diversity, estimated up to 86 %. But it has to be noted that until now reliable data are missing regarding the exact time of extinction within different groups. Thus, e.g., in contrast to former assumptions, the *atrypids* survived into the lowermost Famennian, at least into the Late *crepida* conodont Zone (Wang and Bai, 1988; Hou et al., 1988; Schindler, 1990).

The diversity pattern in b r y o z o a n s is interpreted by Bigey and Curie (1987) as a crisis with gradual reduction before the Fr/Fa boundary.

T e n t a c u l i t e s also became a victim of the Kellwasser Event. The large, thick-shelled, benthonic tentaculites s. str. did not survive (Lardeux, 1969; Farsan, 1986, 1994). The pelagic dacryoconarids had undergone a diversity reduction since early Frasnian time. In spite of some very questionable findings, nowakiids are known only as late as the early Frasnian. Styliolinids then disappear within the Early *Pa. rhenana* conodont Zone (Schindler, 1990), i.e. still beneath the L'KW Horizon. In contrast to that, the homoctenids are abundant in the late Frasnian but suffer a diversity reduction up to one species, *Homoctenus ultimus*. This species occurs in great abundance in certain layers of the U'KW Horizon and even oversteps the Fr/Fa boundary, extends no higher than within the Early *Pa. triangularis* Zone.

A m m o n o i d s : As early as the late 19th and early 20th century, the overturn in goniatitic ammonoids had been recognized. Thus, in the pelagic facies realm the *Manticoceras* Stufe (= Adorf-Stufe, U'Dev. I) has been distinguished from the younger *Cheiloceras* Stufe (= Nehden-Stufe, U'Dev. II). Some discrepancies in the last occurrences of gephuroceratid taxa such as *Crickites* were cleared up in the last decade due to advanced precision in the correlation of goniatite and conodont ranges (see House, 1985; Schindler, 1990; Becker, 1993b). It appears now that both gephuroceratids and beloceratids as well as many other tornoceratids and anarcestids became extinct. According to Becker (1993b) it is an 88 % extinction at the species level.

C o n o d o n t s are considered as excellent indicators of the Fr/Fa boundary as well as of the event pattern. Sandberg et al. (1987a, b, 1988a, b) recognized within the *Pa. linguiformis* Zone in North America and Europe a coinciding stepwise change in the ratio between the conodont genera *Palmatolepis*, *Polygnathus* and *Icriodus*, as well as a particularly high extinction rate at the top of the U'KW Horizon. These results have been confirmed with some variations by subsequent authors (e.g. Schindler, 1990; Matyja and Narkiewicz, 1992; Girard, 1994b; Schülke, 1995).

Up to now not many data on v e r t e b r a t e s are available. Placoderms and acanthodians suffered continuous diversity reduction already during the late Frasnian, but succeeded into the Famennian. In contrast, the jawless Agnatha became extinct at the close of the Frasnian (Halstead, 1988). Among them were the psammosteiform Heterostraci, which flourished in the Middle Devonian and Frasnian, as well as the last representatives of all other agnathian groups which had already suffered their biggest losses within or at the end of the Early Devonian.

Abiotic Parameters and Changes

The geochemical data from all sites with typical Kellwasser facies show values that are characteristic for black shales, i.e. sediments deposited under euxenic or near-euxenic diagenetic conditions. This was already shown by Buggisch (1972). Since Alvarez et al. (1980) found extremely high concentrations of iridium in the Cretaceous/Tertiary boundary layer, attributed to the impact of an extraterrestric body, the search for such an indicator at the Fr/Fa boundary has intensified. Only from a few Late Frasnian and Early Famennian sites and horizons has an enrichment of Ir been reported, but none of them reached the high values of the K/T boundary (e.g. Playford et al., 1984; Hou, 1987; Hou et al., 1988; Wang et al., 1991). These enrichments can be explained rather by an extremely low sedimentation rate or by biological activity.

Significant fluctuations in stable isotopes, such as C, O, and S, have been reported from all KW sites. Their meaning and causation are still under discussion, although it appears to be clear that they do not indicate or support the impact hypothesis. Interesting in this connection is that in a few cases a significant excursion of $\partial^{13}C$ was across the KW level in sequences without black shales. This is taken by the authors (Joachimski and Buggisch, 1993; Geldsetzer et al., 1993) as proof that the anomalies are due to a severe change in oceanic chemistry rather than to special diagenetic conditions in black-shale sediments, although in one of the cases (Joachimski and Buggisch, 1993) the $\partial^{13}C$ excursion is positive, and in the other (Geldsetzer et al., 1993) negative.

Further indications of an impacting asteroid include the occurrence of shocked quartz grains and microtectites. Only microtectites have been reported at the Fr/Fa boundary, but it has to be emphasized that microtectites are found in various quantities in Palaeozoic conodont samples from many layers lacking any indication of an unusual geological or biological change.

Aspects of facies and sedimentology are of great importance for deciphering the event processes. In many parts of the globe, the terminal Frasnian Kellwasser Horizon (U'KW) is represented in the pelagic facies realm as an intercalation of dark sediments within lighter limestones or shales. Depending on the palaeogeographic position and conditions, the U'KW often consists of black limestones and/or shales, a few cm or dm in thickness. A certain silt content, especially in the upper part of the shales, is frequent (for more details see Schindler, 1990). *Buchiola*, orthocone nautiloids, goniatites, *Homoctenus*, entomozoan ostracodes, and large pelecypods characterize this biofacies.

In the neritic realm the Kellwasser Event has also been recognized. In most cases there occurs about the time of the Fr/Fa boundary a facies change that indicates a eustatic fall, sometimes resulting in a hiatus (e.g. Geldsetzer et al., 1993). In contrast to this is an unpublished section in Central Asia (at Zindankak, northern Altai Mts., Uzbekistan) which was mentioned in Schindler (1990). There the U'KW consists of 73 cm of dark carbonate sediments, intercalated in light-grey massive carbonate platform deposits. The KW Horizon yields neritic brachiopods as well as pelagic conodonts, entomozoan ostracodes and homoctenids. This facies change does not necessarily indicate a deepening, but in no case does it mean a shallowing.

Of special interest are those sections in which the immediate pre- or post-event sequences show particular features that may give indications of the event causes. Thus, e.g., in a section described by Wang et al. (1991), the black shaly mudstones of the *Pa. linguiformis* Zone superpose limestone breccias and debris that are about 6 m in thickness. This sedimentological feature, in combination with a strong negative excursion of $\partial^{13}C$ and $\partial^{18}O$, and a moderate enrichment of Ir, is interpreted by some of the authors as evidence for oceanic impact(s) near South China, "although the projectile(s) might have been much smaller than the K/T impactor(s)".

In the cratonic area of South Morocco as well as in parts of the Montagne Noire, the deposition of the dark, KW-like facies began already prior to the U'KW and continued into the Early Famennian.

Duration of the Event

The time interval of mass extinction in conodonts was calculated by Sandberg et al. (1988a, b). This was based on the sequence of the U'KW at Schmidt Quarry (Ense area, Germany). There the Frasnian *Pa. linguiformis* fauna is separated from the overlying Famennian *Pa. triangularis* fauna by a 5 cm shale lacking a conodont record. The time span of this layer has been calculated as about 12,500 years. Thus the conodont mass-extinction event lasted 12,500 years at a maximum or only days at a minimum. With respect to other fossil groups the extinction happened stepwise between the base and top of the entire KW Horizon. Schindler (1990) calculated that the time span was on the order of a few 100,000 years. To the present author these calculations appear to be realistic.

Published Assumptions on the Causation

In the literature of the last two decades we find a plethora of suggested causes for either the Kellwasser Event or the disappearance of bioherrnal reefs. As long as a causal link or correlation between these two events is not entirely verified, both cases may be considered separately. With respect to the disappearance of reefs, suggested causes include: drowning (Playford, 1980; Johnson et al., 1986; Fuchs, 1987); emergence (Krebs, 1974); sea-level changes (House, 1975; Eder and Franke, 1982; Sandberg et al., 1988a, b); invasion of cool water masses into equatorial areas (Copper, 1977); eutrophization due to increased plankton production (Kasig and Wilder, 1983); increased sedimentation and organic production rates (Dreesen et al., 1985); and anoxic conditions in connection with the KW Event (Walliser, 1984, 1986; Geldsetzer et al., 1987).

As for causes of the anoxic KW sediments and the KW Event itself, the following assumptions have been made: result of an asteroid impact (McLaren, 1970; Geldsetzer et al., 1987; Goodfellow et al., 1988; Sandberg et al., 1988a, b; Wang et al., 1991; Claeys et al., 1992; McGhee, 1994 [in contrast to, e.g., McGhee, 1989; McGhee et al., 1984]; Oliver and Pedder, 1994); increase in temperature (Thomson and Newton, 1988); decrease in temperature (Wilde and Berry, 1984, 1986; Copper, 1986; Kalvoda, 1986); general sea-level fluctuation (Johnson et al., 1985; House, 1985; Oliver, 1980; Buggisch, 1991), especially regression (Johnson et al., 1985, 1986; Dreesen, 1987; Ji, 1988; Morrow and Geldsetzer, 1988; Sandberg et al., 1988a, b); rise of the anoxic water masses or/and oceanic overturn (Wilde and Berry, 1984, 1988; Walliser, 1986b; Walliser et al., 1988; Schindler, 1990). Of course, some of the authors cited prefer a combination of factors, but they either emphasize the mentioned cause or take it as the ultimate cause. Thus, e.g., Geldsetzer as well as McGhee (refer. cited) developed reasonable models for the occurrence of the extinction linked together with the formation of anoxic KW sediments. These models could work without an asteroid impact, but nevertheless they pay tribute to a popular hypothesis by proposing an impact as the ultimate cause, although compelling evidence is lacking.

Discussion

Searching for both ultimate and proximate causes, as well as for the processes that connect the two, we have to consider at least the following items:

(1) The KW Event happened in the course of a biocrisis that was characterized by a diversity reduction in many fossil groups, and which lasted several million years. (In order to avoid confusion, it should be called "Late Frasnian Crisis" rather than "Kellwasser Crisis").

(2) It is not yet unquestionably proven that the Fr/Fa Extinction Event is restricted to the time immediately before the *Pa. linguiformis/Pa. triangularis* conodont boundary, i.e. to the U'KW Horizon. There are indications and/or assumptions that some later smaller extinction pulses affected brachiopods, homoctenids and probably also stromatopors and corals, which had survived the main pulse at the Fr/Fa boundary.

(3) The KW Extinction Event is stepped, though short-lasting, with a duration on the order of 10^5 to $3 \cdot 10^5$ a. A first strong extinction pulse occurred at the base of the KW Horizon, i.e. within the *Pa. linguiformis* Zone, whereas the main extinction pulse coincided with the close of the zone.

(4) The KW Bio-Event coincides with the KW Litho-Event, i.e. with the occurrence of black-shale sediments in the upper part of the *Pa. linguiformis* Zone (U'KW Horizon).

(5) The observed signals for a regression are based on the increased percentage of the shallow-water conodont genus *Icriodus* and on sedimentological features such as hiatuses. Of course, the latter may result from post-event erosional forces, even under sea-water cover. According to the sea-level curve provided by Johnson et al. (1985), the time in discussion indeed shows a general turn from Frasnian transgressive to Famennian regressive periods. If this had begun already during the time of the U'KW Horizon, it could well have produced the observed features without claiming a rapid and strong eustatic fall due to glaciation in the southern hemisphere. We also should note that the proportionate increase in *Icriodus* starts not at the base but within the U'KW Horizon, continuing into the lowermost Famennian, where it even became stronger.

(6) In spite of exceptional occurrences, as e.g. in cratonic areas of NW Africa, older KW-like intercalations also occur regularly in the Frasnian, especially within the early parts of the Late *P. rhenana* Zone (L'KW Horizon), and as thin black-shale intercalations

between the two KW Horizons. It is a reasonable assumption that these black-shale litho-events can be attributed to equivalent causes and processes.

(7) It cannot be excluded that repeated impacts may have occurred during an interval of about 1 Ma or more. But it is most improbable that all of these would have been strong enough to produce the litho- and bio-events present in the interval between the L' and U'KW Horizons.

(8) The U'KW Horizon spans an interval of about 100 to 300 ka. As with other global bio-events that occur contemporaneously with profound facies changes, we have to ask whether an impact can trigger a complete change of the oceanic and sedimentary system for hundreds of thousands or, as in other cases, even millions of years. Of course, it cannot be excluded that an impact into an ocean with an extended anoxic layer could produce oceanic overturn and successive extinctions. But there remain still the unanswered questions of the duration and extent of such an overturn. Even if this mechanism – an asteroid impact induces oceanic overturn – cannot be excluded, Wilde and Berry (1986) convincingly showed that only relatively small changes in physical parameters are necessary to trigger such an overturn. Because it is even theoretically impossible to prove the non-existence of a non-existing impact, I prefer to presume geological causes for anoxic or oceanic overturn events, thereby also following the recognition that a less complicated and less spectacular solution must not necessarily be wrong.

(9) There is no real indication for an asteroid impact at the time of the KW Event. The observed suspect signals, i.e. the concentration of Ir and/or microtectites, are too weak and can be attributed to "normal" processes, and in most cases are younger than the KW Event.

Conclusions

The Kellwasser Event is one of the 7 Phanerozoic first order bio-events. It is combined with an anoxic geo-event. The enormous magnitude of the bio-event is due to the fact that it happened in the course of a preceding and still continuing, long-lasting biotic crisis. The associated geo-event was caused by flooding of anoxic waters over vast areas of the epeiric seas, due to a rise of the oceanic anoxic zone. Thus the first step of mass extinction took place at the time of flooding, i.e. at the base of the U'KW Horizon, within the *Pa. linguiformis* Zone. During the U'KW interval further extinctions occurred, as well as environmental changes such as a weak oxygenation of the bottom waters, possibly combined with a eustatic shallowing. In addition, there are indications of additional fluctuations, or lowering of the mean temperature in shallow marine waters. In most areas the main extinction pulse took place, when at the *Pa. linguiformis/ Pa. triangularis* boundary – with the exception of deeper parts of the basins – the anoxic to hypoxic settings were replaced by near-to-normal conditions. As an aftermath of the mass extinction event one or the other smaller extinction pulses followed within the lowermost parts of Early Famennian.

Post-KW-Event adaptive radiations. The time-span for the recovery from the KW Event was different in each of the affected groups. As already shown by Ziegler (1962) and Sandberg et al. (1988a, b), the new diversification in the conodont genus *Palmatolepis* started soon after the event and reached a first maximum during Late *triangularis* and Early *crepida* Zones. This was confirmed by the detailed study of Schülke (1995) in which he demonstrated that the diversification began with an intraspecific increase in variability and ultimately led to the origination of new taxa.

The radiation of ammonoids "occupied a considerable period of time spanning the *Raymondiceras* to *Paratorleyoceras* Genozones" (Becker, 1993b), reaching a maximum within the *Cheiloceras (Ch.) subpartitum* Zone. Becker (1993b) called this interval the Nehden Event, whereas Walliser (1985) used the term *Cheiloceras* Event for the first radiation event of *Cheiloceras* that occurred with the *subpartitum* Zone.

Mid-Famennian Event 1 (M'Fa 1 Event, Condroz-Event)

Condroz Event (Becker, 1990, 1993b). This bipartite event occurs at the base and top of the *Praemeroceras petterae* Zone *(Praemeroceras* Genozone; Late *Palmatolepis rhomboidea* Zone). A high percentage of Tornoceratidae and Cheiloceratidae became extinct at this event.

The lower extinction horizon coincides with the disappearance of the formerly widespread black *Cheiloceras* Shale with its pyritic fossils. This facies change resulted from the late Famennian regressive tendency recognized in many parts of the world, as, e.g., in Belgium (Dreesen, 1982), Poland (Szulzewski, 1986) and the Tafilalt region, Morocco (Hollard, 1960, 1967).

Mid-Famennian Event 2 (M'Fa 2 Event, Enkeberg Event)

Enkeberg Event (House, 1985). This event is called after the Enkeberg section (Rhenish Schiefergebirge), where House recognized a fall in diversity, the final extinction of *Cheiloceras* and the first appearance of Clymeniina. However, Becker (1993b) pointed out that the abrupt faunal turnover in this section is due to a lithological change. In the view of Becker the Enkeberg Event is rather a "gradual faunal turnover" which lasted a longer time. Becker's range chart of Nehdenian goniatites shows a 20 % (8 out of 39 species) species reduction at the base of the *Maeneceras biferum* Zone, whereas at the top of the zone 65 % of the species (34 out of 52, including the genus *Cheiloceras*) became extinct.

Becker attributes the first of the two steps to a transgressive pulse (Event 12 in Sandberg et al., 1989; Middle II e in Johnson and Sandberg, 1989; within Early *Palmatolepis marginifera* Zone), and the second step with the subsequent sea-level fall that continued as part of the overall Famennian regression.

The *annulata* Event (L'Fa Event)

Late Famennian Event; *annulata* Event (Walliser, 1981: unnamed in Fig. 1; Walliser, 1984a). This event is a typical example of a short-lived geo-event that is detectable worldwide though it did not cause extraordinary extinctions or originations but rather a blooming of apparently specially adapted taxa (often uncorrectly termed opportunists).

The *annulata* Event occurs in the middle part of the Famennian *(Platyclymenia annulata* Zone, *Prionoceras* Genozone, basal Upper Devonian IV Stufe; Late *Palmatolepis trachytera* Zone). Lithologically it is characterized by dark to black sediments, consisting of either shale or shale with concretional limestones. These sediments are intercalated in sequences of different facies, ranging from basinal shales through pelagic cephalopod limestones to marls containing cyrtospiriferids.

The event horizon yields the presumably epiplanktonic pelecypod *Buchiola* in abundance, together with *Guerichina*, orthocone cephalopods, entomozoan ostracods and sometimes thin-shelled pelecypods and brachiopods. Ammonoids, especially clymenids, occur in great abundance but low diversity. Obviously the change of conditions that caused the black-shale sedimentation also triggered a blooming of a few specialized taxa as well as a regional expansion of their occurrence.

The intercalation of black event sediments into a well oxygenated sequence represents only a very short time interval of not more than a few ten thousand years, probably even shorter. According to Becker (1992), even a second black-shale horizon can be present within the *Pl. annulata* Zone. This pattern and the worldwide synchronous occurrence of black sediments support the assumption that the event was caused by a short-term flooding of epeiric sea areas by oxygen-depleted waters due to a rise of the oceanic anoxic layer or to oceanic overturn according to the model of Wilde and Berry (1984, 1986).

Becker (1993b) mentions a Dasberg Event, characterizied by a radiation of Gonioclymeniina. Within the early *Clymenia* Stufe, at the base of the *acusticosta* Zone (Korn, 1981), Platycleminidae — but not all Platyclymeniina — are replaced by Gonioclymeniina. This radiation event coincides with a widespread recognizable transgression mentioned in Johnson et al. (1985) as the base of T-R-cycle II f, and in Sandberg et al. (1989) as event 16.

Hangenberg Event (D/C)

Devonian-Carboniferous Boundary Event; Terminal Devonian Event; Hangenberg Event (Walliser, 1980a, b: unnamed in Fig. 2; 1985).

With regard to conodont zonation, the stratigraphic position of this event is near the close of the Middle *Siphonodella praesulcata* Zone. This probably coincides with the boundary between the LE/LN spore Zones (*Retispora lepidophyta - Hymenozonotriletes explanatus* Zone/*Retispora lepidophyta - Verrucosisporites nitidus* Zone), and with the ammonoid zone boundary between *Wocklumeria sphaeroides* and *Cymaclymenia evoluta*.

The Hangenberg event is herein still called the D/C Event although the newly defined Devonian-Carboniferous boundary unfortunately has been fixed by the International Commission on Stratigraphy at a time level that is somewhat younger than the event, presumably about 0.3 to 0.8 Ma. At the boundary stratotype (La Serre, Montagne Noire/France) the boundary is now fixed within an oolitic sequence that even contains lithoclasts, i.e. within the worst possible facies. This choice was made due to the desire to fix the boundary at a level which can be recognized by the first occurrence of a new conodont species within an evolutionary lineage, in this case *Siphonodella sulcata*, descending from *S. praesulcata*. Of course, the event boundary can be recognized in many sections around the globe directly by the facies change and even

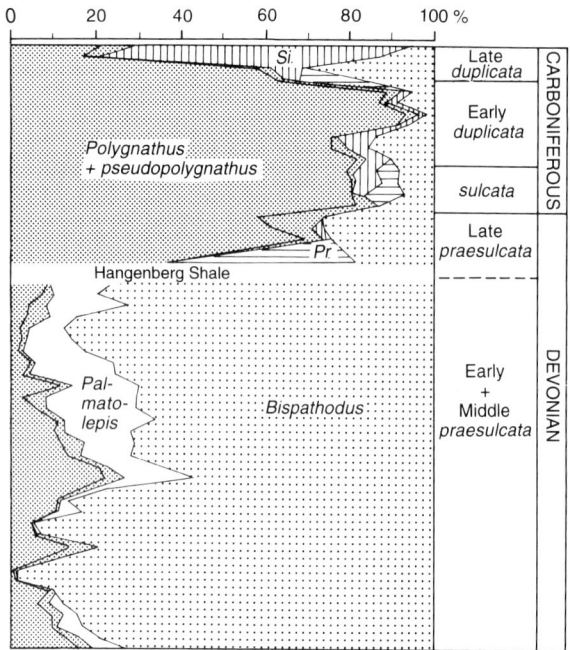

Fig. 5. The Hangenberg Event, indicated by a strong change in conodont biofacies. D/C boundary section at Puech de la Suque; after Girard (1994a). Pr.: *Protognathodus*, Si: *Siphonodella*

faunistically, as Girard (1994a) showed by her analysis of a boundary sequence at Puech de la Suque/Montagne Noire. There a spectacular change occurs about the turn from Middle to Late *praesulcata* Zones, interpreted by Girard as indication of a sea-level fall, preceded by a short transgressive pulse (see Fig. 5). These clear indications of the Hangenberg Event are confirmed by comparable results from sections in Germany (Clausen et al., 1989), Carnic Alps (Dreesen, 1992), South China (Ji et al., 1989) and the U.S. Mid-Continent (Over, 1992).

Although in the literature the Hangenberg Event is not included in the "big 5" Phanerozoic bio-events, it has to be emphasized that it is in fact one of the most severe bio-events in Phanerozoic history. It caused the drastic biotic turnover from the Middle to Late Palaeozoic faunal regime, and showed at least two different characteristic patterns. On the one hand, drastic mass extinction took place such as, e.g., with the clymenid ammonoids. On the other hand, in most of the other affected fossil groups the event extinguished or diminished those few taxa that had survived the preceding crisis. In some of the groups, e.g. trilobites, this crises had already begun in the late Frasnian, and culminated in the Kellwasser Event. The few taxa that survived this event continued without a successful recovery. In contrast to that, other groups such as the palmatolepid conodonts show a long-lasting period of diversification after the Kellwasser Event. However, during several zones prior to the D/C Event, *Palmatolepis* suffered a continual diversity reduction which was as serious as the disappearance of the last few taxa during and short after the event.

Also in goniatites a certain diversity reduction in both genus and species level took place in the course of late Famennian. By the Hangenberg Event, all still remaining taxa became extinct, with the exception of *Mimimitoceras* and the *Imitoceras* lineage, the latter giving rise to the Carboniferous ammonoid radiation. In contrast to the goniatites, the clymenids represented in the late Famennian a flourishing group of high diversity. Then, by the Hangenberg Event, they vanished nearly totally. Only one species, *Cymaclymenia striata*, survived for a short while after. In so far, the range chart of goniatitid and clymenid genera in House (1993a, Fig. 1) must be corrected: with the exception of "*Cymaclymenia*" and "*Mimimitoceras*, gen. nov. A & B", all other genera vanish at the base of the Hangenberg Shale. According to Becker (1993b), thus "the relative extinction rate" in genera "is about 85 percent. At species level the faunal disruption is even more strongly pronounced".

An overturn from Devonian to Carboniferous faunas, characterized in part by strong extinction, can also be recognized in groups other than conodonts and goniatites. Thus, in hemipelagic and pelagic ostracodes 50 % of species disappeared (Blumenstengel, 1993). This strong reduction in ostracode diversity is confirmed by Nemerovskaya et al. (1993). In trilobites the Phacopida became extinct, and within Proetida only a few Brachimetopidae and Proetacea survived. From this surviving stock a new and last radiation of trilobites started short by after the event (Brauckmann et al., 1993). Most probably the placoderm fishes also disappeared with the event.

In contrast to the pelagic and hemipelagic faunas, neritic shallow water environments appear to have been not strongly affected by the Hangenberg Event. This is, e.g., well documented for corals, ostraocodes, brachiopods, and shallow water conodonts in the Omolon region in NE-Russia by Simakov et al. (1983) and Shilo et al. (1984). Within shallow water bivalves from Central and Western Europe Amler (1993) confirms a gradual transition, rather than a sharp break at the event boundary.

In many sections within the pelagic facies realm, the D/C Event is characterized by an abrupt lithological change. An example is the section at the railway cut near Oberrödinghausen (Rhenish Schiefergebirge, Germany) which has served as a boundary reference section for many decades. There the late Famennian nodular cephalopod limestone is sharply separated from the succeeding Hangenberg Shale. At the base of the latter, a few dm of black shales occur. These yield the surviving *Cym. striata*, besides abundant *Guerichia venustiformis*. The clymenid species apparently did not survive into the overlying parts of the Hangenberg Shale, which contains a high amount of silt fraction in this section. In many other sections the Hangenberg shale is free of any silt fraction, weathering light grey. In addition, this type of shale also occurs in deeper parts of the basin where the Famennian is represented by shales. It might be relevant that this light grey Hangenberg shale occurs also in sections at many other regions outside Europe.

Significant lithological changes coincident with the Hangenberg Event and the base of the Hangenberg Limestone (*sulcata* Event in Kalvoda and Kukal, 1987) are also recognizable in shallow water environments. This is evident in Belgium (van Steenwinkel, 1993) and Moravia (Kalvoda and Kukal, 1987). The latter authors also emphasized similar sedimentologic pattern in South China and North America.

The pure Hangenberg Shale, developed in areas and positions without any input of silt fraction, is very smooth and typically shows evidence of ductile deformation which suggests that the load by following sedimentation exceeded the degree of stabilization within the shale. By such slidings a mixture of different aged sediments and fossils may occur. This is, e.g., the case at Drewer Quarry (Rhenish Schiefergebirge), where Korn 1991 erraneously reported the co-occurrence of Devonian and Carboniferous taxa within the Hangenberg Shale.

The described sequence, as well as its fossil content, mirrors the environmental changes in connection with the D/C Event. Within the late Famennian cephalopod limestone, signals are present which indicate that the long-lasting regressive tendency persisted within that well-oxygenated facies regime. In contrast to this, the basal black shales of the Hangenberg Shale represent an anoxic event, which was perhaps, but not necessarily, due to a short transgressive pulse. Analysing the causes of the event, we should consider that in shallow marine environments not strong perturbations occured, and that in many areas a maximal lowstand with condensed sedimentation or even with a hiatus in shallow water environments (e.g. in the Strunian facies; Van Steenwinkel, 1993)

was reached during the time of Hangenberg Shale. This sea-level lowstand probably is connected with a cooling episode in the southern hemisphere as documented in glacial deposits within the Parnaiba Basin (Loboziak et al., 1992). Moreover, the mass extinction in pelagic and hemipelagic biota took place at the base of the interval, i.e. directly coincident with the onset of black shale deposition. Bless et al. (1993) questioned a mass extinction at this level, and assumed rather a "gradual process of mass extinction" produced by the major regression associated with the deposition of the Hangenberg Shale. According to these authors the faunal overturn started instead at the base of the overlying Hangenberg Limestone. However, if we exclude those data of Korn (1981, 1991, 1993) which are based on disturbed sediments, then there is no basis for the mentioned assumptions of Bless et al. (1993). In contrast, the Hangenberg Event turns out to represent an extremely sudden mass extinction event. In this connection it should be emphasized that the term "faunal overturn" comprises the whole interval from extinctions to subsequent radiations, where the extinctions constitute the first and deciding step.

The described patterns can be explained by the following scenario. The moderate late Famennian regression, which continued into the middle "Hangenbergian" time, was interrupted by a relatively short anoxic episode. Thereby deeper parts of epeiric seas were flooded by oxygen depleted waters that formed due to a rise of the oceanic anoxic layer or due to an oceanic overturn. It is still questionable whether this anoxic event was caused by a transgressive pulse or by a change of oceanic parameters, e.g. due to a cooling episode in the southern hemisphere. Obviously, such sudden environmental turnovers following long-lasting, constant or only slightly fluctuating conditions are those main factors responsible for extinction events.

Carboniferous Global Events

With respect to global events, the Carboniferous still needs intensive investigations. Apart from the M'Tn and M'C Events described below, and a few diversification events within several fossil groups, the Carboniferous — and especially the Upper Carboniferous — appears to be a time in which the evolution of biota was not often punctuated by drastic events. However, there were a few major facies changes that can be recognized worldwide in shallow epicontinental basins. Thus, e.g., the Tournaisian alum shale is succeeded in many regions by a sequence of siliceous

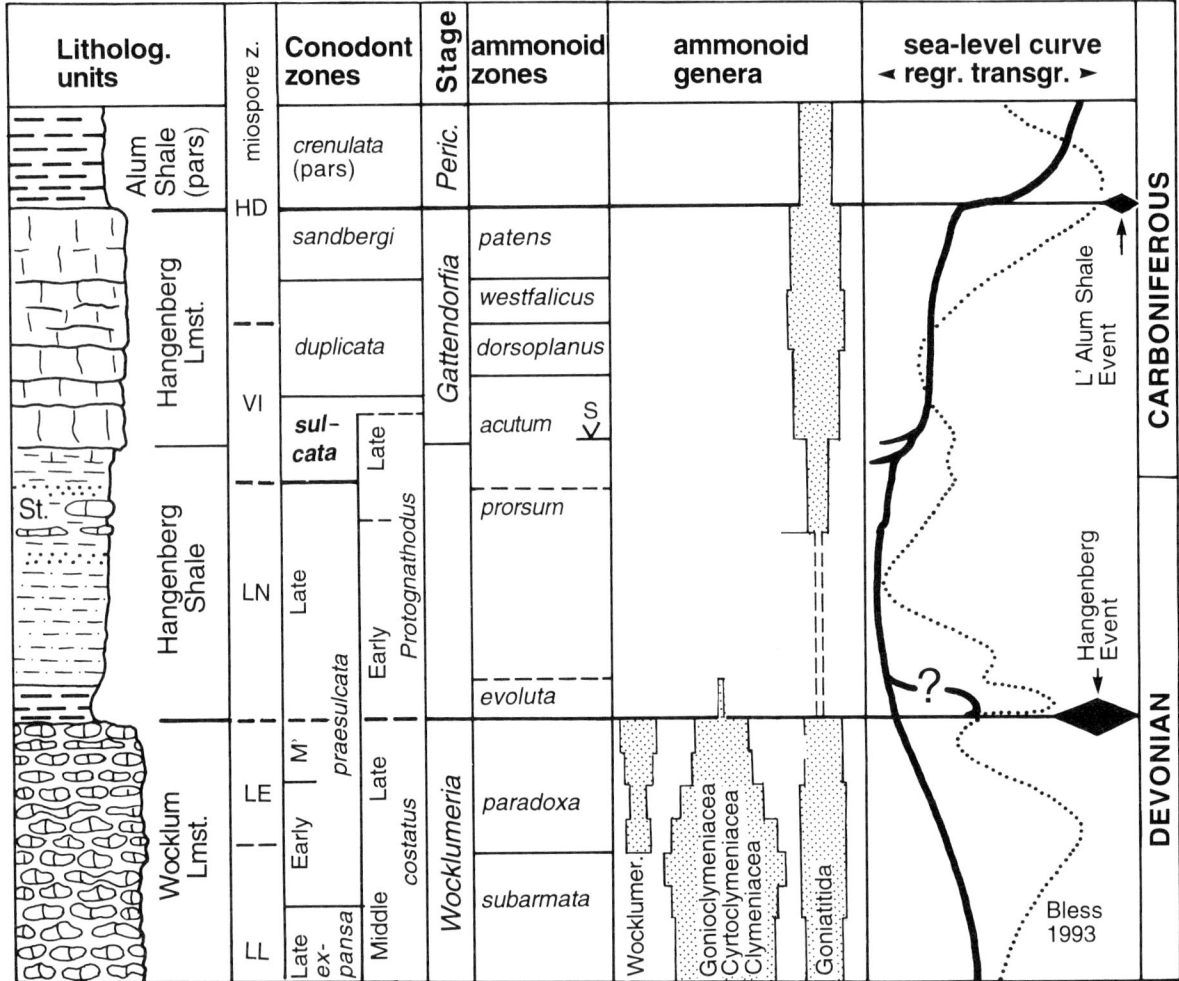

Fig. 6. Stratigraphic position of late Devonian and early Carboniferous events. Note that correlations between different stratigraphies are still uncertain with respect to the first occurrence of *Siphonodella sulcata* and the base of the Hangenberg Shale. s: first occurrence of *Gattendorfia subinvoluta*. Lithological units based on the Oberrödinghausen section (Rhenisch Schiefergebirge), but not to scale. St: Stockum Lmst.. Distribution and diversity of ammonoid genera based on House (1993) and Korn (1993). Range of miospore zones according to Bless (1993)

shales or bedded cherts. Within this sequence pelagic, cephalopod-bearing limestones are also intercalated (Erdbach or *Pericyclus* Limestone). This siliceous sequence is in turn overlain by black shales that pass upwards into dark, "normal" basinal shales.

Several extinction events in crinoids can be deduced from the range charts of Lane and Sevastopulo (1990). At the Osagean-Meramecian boundary, i.e. when the Keokuk Limestone was replaced by the Warsaw Shale, about 50 % of the North American genera disappeared. Because contemporaneously 25 % of the European genera vanished, the cited authors infered a global environmental cause for the extinctions. In addition, another strong extinction event occurred in crinoids near the Namurian base.

As a consequence of extinction events adaptive radiations typically occur more or less shortly afterwards. On account of this coupling they are not particularly named in this chapter. As a consequence of the high-order Hangenberg Extinction Event, extremely high diversification rates occurred within the lowermost Carboniferous near the bases of the *Siphonodella sulcata* and *S. acutum* Zones (see Fig. 5). There, apart from other groups, conodonts (especially *Protognatho-*

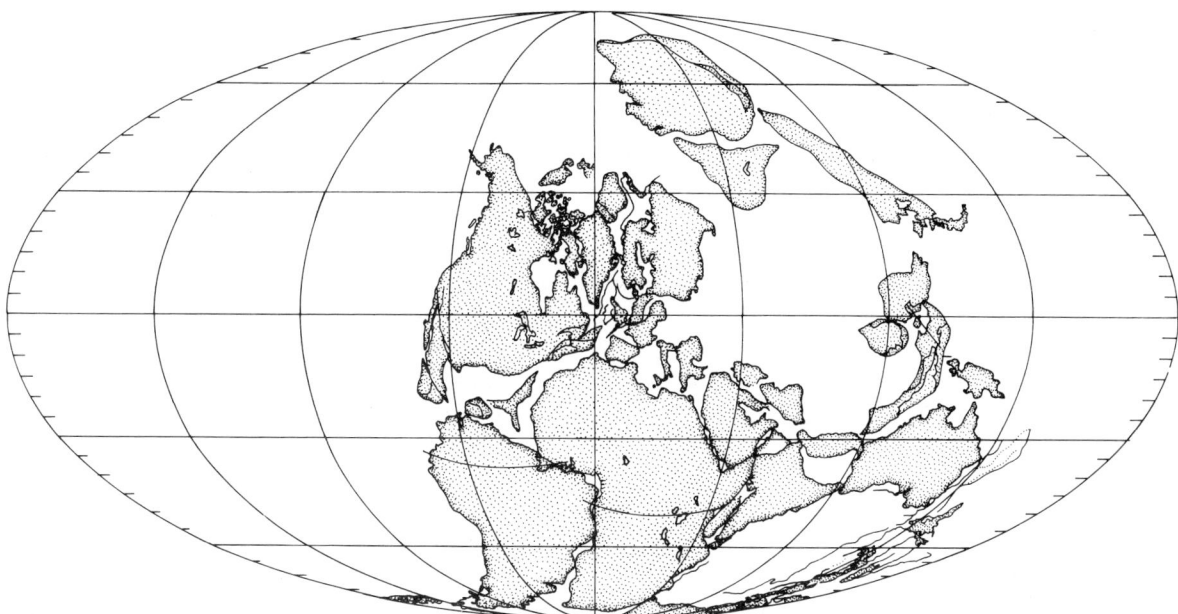

Fig. 7. Middle Carboniferous geographic setting (from Scotese and McKerrow, 1990)

dus, Siphonodella, and lateron *Gnathodus*), ammonoids, trilobites, ostracodes, and among the echinoderms especially the blastoids (Waters, 1990), rapidly increased in diversity.

Mid-Tournaisian Event (M'Tn)

This Event has been called Lower Alum Shale Event by Becker (1993a). It is one of those environmental changes that can be detected worldwide by the occurrence and spreading of anoxic facies. Because Becker (1993a) gave a comprehensive analysis of this event, here only a short summary and a comment is given. Becker agrees with the interpretation of Krebs (1969) that the base of the Lower Alum Shale represents a palaeogeographic turning point of first order, but he disagrees with Walliser (1984b), probably based on a misinterpretation of the text. Of course, I also agree in general with Krebs (1969) but expressed my still existing presumption that "the commencement of this profound change most probably coincides with the base of the Hangenberg Shale". With "commencement" I mean an introductory step or first pulse, leading finally, i.e. by the Lower Alum Shale, to a long-lasting sea-level highstand.

The transgression of the Lower Alum Shale coincides with the Lower/Middle Tournaisian and Lower/Upper Hastarian boundaries as well as with the base of the *Siphonodella crenulata* Zone. The recognition of the overturn with respect to ammonoid evolution gave rise to the boundary between the older *Gattendorfia*-Stufe (Early Carboniferous I, cu I, cd I) and the succeeding *Pericyclus*-Stufe (Early Carboniferous II, cu II, cd II). The transgression event caused a global disruption of pelagic biota. Thus the extinction among goniatites reached more than 50 % of genus-level taxa of the latest *Gattendorfia*-Stufe.

Mid-Carboniferous Event (M'C)

Mid-Carboniferous Event; Early Namurian Event (E'Na Event); Namurian E Event (House, 1993). This event can well be recognized in the evolution of conodont and ammonoid faunas. According to House (1993) "this break gives the largest number of extinct families for any 2 Ma unit in the history of Ammonoidea". This concerns the Nomismocerataceae and Neoglyphiocerataceae which became extinct, as well as the Pericyclataceae, Dimorphocerataceae and Goniatitaceae that lost a high percentage of their diversity. However, according to Nemirovskaya and Nigmadganov (1994), the last eumorphocerids and first homceratids appear to overlap as documented in one sample containing a conodont fauna intermediate to the pre- and post-event faunas.

With regard to conodonts, the genera *Gnathodus*

and *Lochriea*, which were dominant in early Carboniferous, became extinct. But already shortly before, two of the post-event dominant genera, *Declinognathodus* and *Idiognathoides* originated as descendences of *Gnathodus*. However, the radiation of these two new genera took place not earlier than a short while after the extinction event

Other marine invertebrates also lost drastically in diversity. According to Lane in Nemirovskaya and Nigmadganov (1994) 83 % of corals, 65 % of crinoids, 40-50 % of foraminifers, and about 15 % of brachiopods became extinct. Of course, as far as can be assumed from published data, this faunal overturn was not an abrupt change in all affected fossil groups but rather a relatively rapid transition.

The M'C Event closely coincides with a strong regression that is documented by unconformities in serveral cratonic areas (Ramsbottom, 1977). In addition, about that time the formerly widespread distributed Early Carboniferous shale and carbonate facies became influenced and soon widely replaced by terrigenous sediments. This change obviously is correlated with the final phase of orogeny in the Variscan Belt.

Conclusion

In the Devonian and Carboniferous numerous global events can be recognized. Their magnitude with respect to faunal changes range from nearly zero (e.g. *annulata* Event) to highest orders (e.g. Hangenberg and Kellwasser Events). In most cases the bio-events cover a certain time interval, though very short in comparison with the intervals between the events. In so far most of the bio-events turn out to be nothing else than quick faunal turnovers which are known since long time and gave then rise to the Devonian and Carboniferous subdivision into series and stages.

Most sensitive to environmental changes appear to be ammonoids and conodonts. However this might be an artefact based on the good knowledge of species ranges that is partly much better than in other faunal groups.

All Devonian and Carboniferous global bio-events can proximately be attributed to environmental changes, especially to sea-level changes and anoxic events. The latter are often connected with transgressions (e.g. Frasne Event). However, in cases where short-termed black-shale depositions occur (e.g. *annulata* Horizon, but probably also Kellwasser Horizons) it is assumed that they are caused by flooding of oxygen-depleted water masses over large areas, due to a rise of the oceanic oxygen-depleted zone, without any major eustatic movements. In such cases an oceanic overturn in the sense of Wilde and Berry (1984, 1986) has also to be considered.

The recognition that sea-level changes and anoxic events are the proximate causes for global bio-events does not exclude the importance of further parameters such as changes in climate, physical conditions and chemical composition of atmosphere and hydrosphere. All those parameters are important parts of the complex biosphere, and may amplify or even trigger the effect on the biota by eustatic changes and anoxic events.

Besides of their connection with anoxic events, global bio-events often occur with transgressions and sometimes with regressions. It is assumed that thereby particularly the turn itself is the extinction-triggering process.

The magnitude of extinctions during a global event does not only depend on the magnitude and speed of environmental changes. The extinction rate appears to be extraordinarily high if the event occurs after a long-lasting crisis in the biota. In addition, the systagenetic state of an affected faunal group, i.e. the degree of its diversification and relative specialization, is decisive for the magnitude of extinctions. Thus, the importance of a global bio-event depends on both environmental changes and evolutionary state.

Acknowledgements

The help of Gabriela Meyer in handling the text and Cornelia Kaubisch in preparing the figures is highly acknowledged.

References

Alberti, H., Groos-Uffenorde, H., Streel, M., Uffenorde, H. and Walliser, O.H., 1974. The stratigraphical significance of the *Protognathodus* fauna from Stockum (Devonian/Carboniferous boundary, Rhenish Schiefergebirge). Newsl. Stratigr. 3, 4, 263-276.

Alvarez, L.W., Alvarez, W., Asaro, F. and Michel, H.V., 1980. Extraterrestrial cause for the Cretaceous-Tertiary extinction. Science 208, 1095-1108, Washington.

Amler, M.R., 1993. Shallow marine bivalves at the Devonian/Carboniferous boundary from the Velbert Anticline (Rheinisches Schiefergebirge). Annales Soc. géol. Belgique 115, 1992, 405-423.

Becker, R.T., 1993a. Analysis of ammonoid palaeogeography in relation to the global Hangenberg (terminal Devonian) and Lower Alum Shale (Middle Tournaisian) Events. Annales Soc. géol. Belgique 115, 459-473.

Becker, R.Th., 1993b. Anoxia, eustatic changes, and Upper Devonian to lowermost Carboniferous global ammonoid diversity. In: House, M.R. (ed.), The Ammonoidea: Environment, Ecology, and Evolutionary Change. Systematics Ass. Spec. Vol. 47, 115-163, Clarendon Press, Oxford.

Becker, R.T., House, M.R. and Ashouri, A.-R., 1988. Potential stratotype section for the Frasnian/ Famennian boundary at El Atrous, Tafilalt, Morocco. Doc. subm. to the Devonian Subcomm., IUGS, Rennes, August 1988, 6 p., Rennes.

Bensaid, M., Bultynck, P., Sartenaer, P., Walliser, O.H. and Ziegler, W., 1985. The Givetian-Frasnian Boundary in pre-Sahara Morocco. Cour. Forsch.-Inst. Senckenberg 75, 287-300, Frankfurt/Main.

Bigey, F.P. and Curie, M., 1987. Devonian Bryozoa and global events: A review. 2. Internat. Symp. Devonian Syst., Abstr., p. 35, Calgary.

Bless, M.J.M., 1993. Comparison between Eustatic T-R Cycles around the Devonian-Carboniferous boundary and the distribution of the ostracode taxon *Pseudoleperditia* gr. *venulosa*. Ann. Soc. Géol. Belgique 115, 1992, 475-487.

Bless, M.J.M., Becker, R.T., Higgs, K., Paproth, E. and Streel, M., 1993. Eustatic cycles around the Devonian-Carboniferous boundary and the sedimentary and fossil record in Sauerland (Federal Republic of Germany). Ann. Soc. géol. Belgique 115, 1993, 689-702.

Blumenstengel, H., 1993. Ostracods from the Devonian-Carboniferous boundary beds in Thuringia (Germany). Ann. Soc. Géol. Belgique 113, 1993, 483-489.

Brauckmann, C., Chlupac, I. and Feist, R., 1993. Trilobites at the Devonian-Carboniferous boundary. Ann. Soc. Géol. Belgique 115, 1992, 507-518.

Buggisch, W., 1972. Zur Geologie und Geochemie der Kellwasserkalke und ihrer begleitenden Sedimente (Unteres Oberdevon). Abh. hess. L.-Amt Bodenforsch. 62, 1-68, Wiesbaden.

Buggisch, W., 1991. The global Frasnian-Famennian "Kellwasser"-Event. Geologische Rundschau 80, 49-72.

Caputo, M.V. and Crowell, J.C., 1985. Migration of glacial centers across Gondwana during Paleozoic Era. Geol. Soc. Am. Bull. 96, 1020-1036, Boulder.

Carls, P., Gandl, J., Groos-Uffenorde, H., Jahnke, H. and Walliser, O.H., 1972. Neue Daten zur Grenze Unter-/Mittel-Devon. Newsl. Stratigr. 2, 115-147.

Chlupác, I., 1980. Klonk at Suchomasty. Stratotype of the Silurian/Devonian boundary. In: Schönlaub, H.P. (ed.), Second European Conodont Symposium (ECOS II). Abh. Geol. B.A. 35, 177-180, Wien.

Chlupác, I., 1982. Preliminary submission for Lower-Middle Devonian boundary stratotype in the Barrandian area. Cour. Forsch.-Inst. Senckenberg 55, 85-96, Frankfurt/Main.

Chlupác, I., Lukes, P. and Zikmundová, J., 1979. The Lower/Middle Devonian boundary beds in the Barrandian area, Czechoslovakia. Geologica et Palaeontologica 13, 125-156, Marburg.

Clausen, C.D., Leuteritz, K. and Ziegler, W. (With a contribution of Korn, D.), 1989. Ausgewählte Profile an der Devon/Karbon-Grenze im Sauerland (Rheinisches Schiefergebirge). Fortschr. Geol. Rheinld. u. Westf. 35, 161-226, Krefeld.

Claeys, P., Casier, J.-G. and Margolis, S.V., 1992. Microtectites and mass extinctions: evidence for a Late Devonian asteroid impact. Science 257, 1102-1104.

Cockbain, A.E., 1988. The distribution of stromatoporoids in the Frasnian and Famennian. 5. Internat. Symp. Fossil Cnidaria V, Abstr., p. 32, Brisbane.

Copper, P., 1977. Paleolatitudes in the Devonian of Brazil and the Frasnian-Famennian mass extinction. Palaeogeogr., Palaeoclimatol., Palaeoecol. 21, 165-207, Amsterdam.

Copper, P., 1986. Frasnian/Famennian mass extinction and cold-water oceans. Geology 14, 835-839, Boulder.

Denckmann, A., 1895. Zur Stratigraphie des Oberdevon im Kellerwalde und in einigen benachbarten Devon-Gebieten. Jb. kgl. preuß. geol. L.-Anst. u. Bergakad. für 1894, 15, 8-64, Berlin.

Dreesen, R., 1982. Storm generated oolithic ironstones of the Famennian (Fa 1b - Fa 2a) in the Vesdre and Dinant Synclinoria (Upper Devonian, Belgium). Ann. Soc. géol. Belgique 105, 105-129.

Dreesen, R., 1984. Stratigraphic correlation of Famennian oolithic ironstones in the Havelange (Dinant Basin) and Verviers boreholes (Vesdre Massif) (Upper Devonian, Belgium). Bull. Soc. Belge Géol. 93, 197-211.

Dreesen, R.J.M., 1987. Event-Stratigraphy of the Belgian Famennian (Uppermost Devonian, Ardennes Shelf). In: Vogel, A., Miller, H. and Greiling, R. (eds.), The Rhenish Massif, 22-36, Vieweg, Braunschweig-Wiesbaden.

Dreesen, R.M.J., 1992. Conodont biofacies analysis

of the D/C boundary beds in the Carnic Alps. Jb. Geol. B.-A. 1992, 49-56; Wien.

Dreesen, R., Kasig, W., Paproth, E. and Wilder, H., 1985. Recent investigations within the Devonian and Carboniferous north and south of the Stavelot-Venn Massif. N. Jb. Geol. Paläont. Abh. 171, 237-264, Stuttgart.

Ebert, J., 1993. Global Events im Grenzbereich Mittel-/Ober-Devon. Göttinger Arb. Geol. Paläont. 59, 106 pp., Göttingen.

Eder, W. and Franke, W., 1982. Death of Devonian reefs. N. Jb. Geol. Paläont. Abh. 163, 241-243, Stuttgart.

Eder, W., Engel, W., Franke, W., Langenstrassen, F., Walliser, O.H. and Witten, W., 1977. Überblick über die paläogeographische Entwicklung des östlichen Rheinischen Schiefergebirges. Exk.-Führer Geotagung '77, I, Exk. A, 2-11, Göttingen.

Farsan, N.M., 1986. Frasnian mass extinction - a single catastrophic event or cumulative? Lecture Notes Earth Sci. 8, 189-197, Springer, Berlin Heidelberg New York.

Farsan, N.M., 1994. Tentaculiten: Ontogenese, Systematik, Phylogenese, Biostratonomie und Morphologie. Abh. senckenberg. naturf. Ges. 547, 128 pp., Frankfurt/Main.

Feist, R. and Schindler, E., 1994. Trilobites during the Frasnian Kellwasser Crisis in European Late Devonian cephalopod limestones. Cour. Forsch.-Inst. Senckenberg 169, 195-223, Frankfurt/Main.

Fuchs, A., 1987. Conodont biostratigraphy of the Elbingerode Reef Complex, Harz Mountains. Acta Geol. Polonica 37, 33-50, Warszawa.

Geldsetzer, H.H.J., Goodfellow, W.D., McLaren, D.J. and Orchard, M.J., 1987. Sulfur-isotope anomaly associated with the Frasnian-Famennian extinction, Medicine Lake, Alberta, Canada. Geology 15, 393-396, Boulder.

Geldsetzer, H.H.J., Goodfellow, W.D. and McLaren, D.J., 1993. The Frasnian-Famennian extinction event in a stable cratonic shelf setting: Trout River, Northwest Territories, Canada. Palaeogeography, Palaeoclimatology, Palaeoecology 104, 81-95.

Girard, C., 1994a. Conodont Biofacies and Event stratigraphy across the D/C Boundary in the stratotype area (Montagne Noire, France). Courier Forsch.-Inst. Senckenberg 168, 299-309, Frankfurt/Main.

Girard, C., 1994b. Les communautés de conodontes et les crises Kellwasser et Hangenberg de la fin du Dévonien en Montagne Noire (Sud de la France). Analyse faunistique et géochemique. Thèse d'Univ. de Montpellier II.

Goodfellow, W.D., Geldsetzer, H.H.J., McLaren, D.J., Orchard, M.J. and Klapper, G., 1988. The Frasnian-Famennian exctinction: Current results and possible causes. Can. Soc. Petrol. Geol. Mem. 14 (III), 9-21, Calgary.

Groos-Uffenorde, H. and Schindler, E., 1990. The effect of global events on entomozoacean Ostracoda. In: Whatley, R. and Maybury, C. (eds.), Ostracoda and Global Events. 101-112, Chapman and Hall, London.

Halstead, L.B., 1988. Extinction and survival of the jawless vertebrates, the Agnatha. In: Larwood, G.P. (ed.), Extinction and Survival in the Fossil Record. Systematics Assoc. Special Vol. 34, 257-267, Clarendon Press, Oxford.

Haude, R., 1972. Bau und Funktion der *Scyphocrinites*-Lobolithen. Lethaia 5, 95-125.

Hladil, J., 1992. Are there turbidites in the Silurian/Devonian Boundary Stratotype? (Klonk near Suchomasty, Barrandian, Czechoslovakia). Facies 26, 35-54, Erlangen.

Hladil, J., Ceichan, P. and Berousek, P., 1992. Rebuilding of the shallow water dwellers: otomari-Kacak and Kellwasser events. global Bioevents, Abstr., Göttingen 1992: 50-51, Göttingen.

Hollard, H., 1960. Une phase tectonique infra-famennienne dans le Tafilalt et le Maïder (Maroc présaharien). C. R. Acad. Sci., Paris 250, 7, 1303-1305.

Hollard, H., 1967. Le Dévonien du Maroc et du Sahara nord-occidental. Int. Sympos. Devonian Syst., Alberta Soc. Petrol. geol. I, 203-244.

Hou, H.-F., 1987. Devonian events of South China. 2. Internat. Symp. Devon. Syst., Abstr., p. 116, Calgary.

Hou, H.-F., Ji, Q. and Wang, J., 1988. Preliminary report on Frasnian-Famennian events in South China. Can. Soc. Petrol. Geol. Mem. 14 (III), 63-69, Calgary.

House, M.R., 1975. Faunas and time in the marine Devonian. Proc. Yorkshire Geol. Soc. 40 (4/27), 459-490, Hull.

House, M.R., 1985. Correlation of mid-Palaeozoic ammonoid evolutionary events with global sedimentary perturbations. Nature 313, 17-22, London.

House, M.R., 1993a. Earliest Carboniferous goniatite recovery after the Hangenberg Event. Ann. Soc. géol. Belgique 115, 1992, 559-579.

House, M.R., 1993b. Fluctuations in ammonoid evolution and possible environmental controls. In: House, M.R. (ed.), The Ammonoidea: Environment, Ecology, and Evolutionary Change. Systematics Assoc. Spec. Vol. 47, 13-34, Clarendon Press, Oxford.

House, M.R., Becker, R.T., Feist, R. and Klapper, G., 1988. Stratotype proposal for the Frasnian/Famennian boundary in the Montagne Noire. Doc. subm. to the Subcomm., IUGS, Rennes, August 1988, 14 p., Rennes.

Jahnke, H. and Shi, Yan (with a contribution by Haude, R.), 1989. The Silurian-Devonian boundary strata and the Early Devonian of the Shidian-Baoshan area (W. Yunnan, China). Cour. Forsch.-Inst. Senckenberg 110, 137-193, Frankfurt/Main.

Ji, Q., 1988. A preliminary report on the Frasnian-Famennian extinction event in South China. Cour. Forsch.-Inst. Senckenberg 102, 243, Frankfurt/Main.

Ji, Q., et al., 1989. The Dapoushang section: an excellent section for the Devonian-Carboniferous boundary stratotype in China. 165 pp. Science Press, Beijing.

Joachimski, M.M. and Buggisch, W., 1993. Anoxic events in the late Frasnian. Causes of the Frasnian-Famennian faunal crisis? Geology 21, 675-678.

Johnson, J.G. 1971. A quantitative approach to faunal province analysis. Am. J. Sci. 270, 257-280, New Haven.

Johnson, J.G., 1974. Extinction of perched faunas. Geology 2, 479-482, Boulder.

Johnson, J.G. and Boucot, A.J., 1973. Devonian brachiopods. In: Hallam, A. (ed.), Atlas of Palaeobiogeography, 89-96, Elsevier, Amsterdam, London, New York.

Johnson, J.G. and Murphy, M.A., 1984. Time-rock model for Siluro-Devonian continental shelf, western United States. Geol. Soc. Amer. Bull. 95, 1349-1359.

Johnson, J.G. and Sandberg, C.A., 1989. Devonian eustatic events in the western United States and their biostratigraphic responses. Canadian Soc. Petrol. Geologists, Mem. 14 (III), 171-178.

Johnson, J.G., Klapper, G. and Sandberg, C.A., 1985. Devonian eustatic fluctuations in Euramerica. Geol. Soc. Am. Bull. 96, 567-587, Boulder.

Johnson, J.G., Klapper, G. and Sandberg, C.A., 1986. Late Devonian eustatic cycles around margin of Old Red Continent. Annal. Soc. géol. Belg. 109, 141-147, Liège.

Kalvoda, J., 1986. Upper Frasnian and Lower Tournaisian events and evolution of calcareous Foraminifera - close links to climatic changes. Lecture Notes Earth Sci. 8, 225-236, Springer, Berlin Heidelberg New York.

Kalvoda, J. and Kukal, Z., 1987. Devonian-Carboniferous Boundary in the Moravian Karst at Lesni Lom Quarry, Brno-Lisen, Czechoslovakia. Cour. Forsch.-Inst. Senckenberg 98, 95-117.

Kasig, W. and Wilder, H., 1983. The sedimentary development of the Western Rheinisches Schiefergebirge and the Ardennes (Germany/Belgium). In: Martin, H. and Eder, F.W. (eds.), Intracontinental fold belts, 185-209, Springer, Berlin Heidelberg New York.

Kayser, E., 1873. Studien aus dem Gebiete des rheinischen Devon. IV. Über die Fauna des Nierenkalks vom Enkeberge und der Schiefer von Nehden bei Brilon, und über die Gliederung des Oberdevon im Rheinischen Schiefergebirge. Z. dt. geol. Ges. 25, 602-674, Berlin.

Klapper, G., Ziegler, W. and Mashkova, T.V., 1978. Conodonts and correlation of Lower/Middle Devonian boundary beds in the Barrandian area of Czechoslovakia. Geologica et Palaeontologica 12, 103-116.

Korn, D., 1981. Ein neues, Ammonoideen-führendes Profil an der Devon-Karbon-Grenze im Sauerland (Rhein. Schiefergebirge). N. Jb. Geol. Paläont. Mh., 1981, 513-526.

Korn, D., 1991. Threedimensionally preserved clymeniids from the Hangenberg Black Shale of Drewer (Cephalopoda, Ammonoidea; Devonian-Carboniferous boundary; Rhenish Massif). N. Jb. Geol. Paläont. Mh., 1991, 553-563.

Korn, D., 1993. The ammonoid faunal change near the Devonian-Carboniferous boundary. Ann. Soc. géol. Belgique 115, 1992, 581-583.

Krebs, W., 1969. Über Schwarzschiefer und bituminöse Kalke im mitteleuropäischen Variscikum. Erdöl und Kohle, Erdgas, Petroch. 27, 2-6, 62-67.

Krebs, W., 1974. Devonian carbonate complexes of Central Europe. Soc. Econ. Paleont. Mineral., Spec. Publ. 18, 155-208, Tulsa.

Lane, N.G. and Sevastopulo, G.D., 1990. Biogeography of Lower Carboniferous crinoids. In: McKerrow, and Scotese, (eds.), Palaeozoic Palaeogeography and Biogeography. Geolog. Soc., Mem. 12, 333-338, London.

Lardeux, H., 1969. Les tentaculites d'Europe occidentale et d'Afrique du Nord. Cahiers de Paléontologie, Edit. Centre Nat. Rech. Sci, 238 p., Paris.

Lethiers, F. and Feist, R., 1991. La crise des ostracodes benthiques au passage Frasnien-Famennien de Coumiac (Montagne Noire, France méridionale). C.R. Acad. Sci. Paris 312, Série II, 1057-1063.

Loboziak, S., Streel, M., Caputo, M.V. and De Melo, J.H.G., 1992. Middle Devonian to Lower Carboniferous miospore stratigraphy in the central Parraiba Basin (Brazil). Ann. Soc. géol. Belgique 115, 215-226.

Marchant, T.R., 1987. Calcareous Foraminifera from the Frasnian of Western Canada. 2. Internat. Symp. Devonian Syst., Abstr. p. 155, Calgary.

Matyja, H. and Narkiewicz, M., 1992. Conodont Biofacies Succession near the Frasnian/Famennian Boundary – Some Polish Examples. Cour. Forsch.-Inst. Senckenberg 154, 125-147, Frankfurt/Main.

McGhee, G.R., Jr., 1981. The Frasnian-Famennian extinctions: A search for extraterrestrial causes. Field Mus. Nat. Hist. 52, 3-5, Chicago.

McGhee, G.R., Jr., 1989. The Frasnian-Famennian extinction event. In: Donovan, S.K. (ed.), Mass extinctions: Processes and evidence. Columbia University Press, Chapter 7, 133-151, New York.

McGhee, G.R., Jr., 1994. Comets, Asteroids, and the Late Devonian Mass Extinction. Palaios 9, 513-515.

McGhee, G.R., Jr., Gilmore, J.S., Orth, C.J. and Olsen, E., 1984. No geochemical evidence for an asteroidal impact at late Devonian mass extinction horizon. Nature 308, 629-631.

McLaren, D.J., 1970. Presidental address: Time, life, and boundaries. J. Paleont. 44, 801-815, Tulsa.

Morrow, D.W. and Geldsetzer, H.H.J., 1988. Devonian of the Eastern Canadian Cordillera. Can. Soc. Petrol. Geol. Mem. 14 (I), 85-121, Calgary.

Morzadec, P., 1992. Evolution des Asteropyginae (Trilobita) et variations eustatiques au Dévonien. Lethaia 25, 85-96.

Nemirovskaya, T. and Nigmadganov, I., 1994. The Mid-Carboniferous Conodont Event. Cour. Forsch.-Inst. Senckenberg 168, 319-333; Frankfurt/Main.

Nemirovskaya, T.I., Chermnykh, V.A., Kononova, L.I. and Pazukhin, V.A., 1993. Conodonts of the Devonian-Carboniferous boundary section, Kozhim, Polar Urals, Russia. Ann. Soc. géol. Belgique 115, 1992, 629-647.

Niklas, K.J., Phillipps, T.L. and Carozzi, A.V., 1976. Morphology and paleoecology of Protosalvinia from the Upper Devonian (Famennian) of the Middle Amazon Basin of Brazil. Palaeontographica, B, 155, 1-30, Stuttgart.

Oliver, W.A., Jr., 1980. Corals in the Malvinokaffric Realm. Münsterische Forschungen zur Geologie u. Paläontologie 52, 13-27.

Oliver, W.A. and Pedder, A.E.H., 1994. Crises in the Devonian history of the rugose corals. Palaeobiology 20, 178-190.

Over, D.J., 1992. Conodonts and the Devonian-Carboniferous boundary in the upper Woodford shales, Arbuckle Mountains, South-central Oklahoma. J. Palaeont. 66, 293-311.

Paeckelmann, W., 1924. Das Devon und Carbon der Umgebung von Balve in Westf. Jb. Preuß. Geol. L.-Anst. XIV, 57-97. Berlin.

Pedder, A.E.H., 1982. The rugose coral record across the Frasnian/Famennian boundary. Geol. Soc. Am. Spec. Pap. 190, 485-489, Boulder.

Playford, P.E., 1980. Devonian "Great Barrier Reef" of Canning Basin, Western Australia. Am. Assoc. Petrol. Geol. Bull. 64, 814-840, Tulsa.

Playford, P.E., McLaren, D.J., Orth, C.J., Gilmore, J.S. and Goodfellow, W.D., 1984. Iridium anomaly in the Upper Devonian of the Canning Basin, Western Australia. Science 226, 437-439, Washington.

Ramsbottom, W.H.C., 1977. Major cycles of transgression and regression (mesothems) in the Namurian. Proc. Yorks geol. Soc. 41, 261-291.

Riegel, W., 1975a. Palynological sequence from Lower Emsian to Givetian of the Eifel region. Comm. internat. Microflore paléozoique, Newsl. 10, 6.

Riegel, W., 1975b. Die dispersen Sporen der Ems-, Eifel- und Givet-Stufe der Eifel (Rheinisches Schiefergebirge) und ihre stratigraphische und paläofloristische Bedeutung. Habil.-Thesis, 282 pp, Göttingen (unpubl.).

Roemer, F.A., 1850. Beiträge zur geologischen Kenntnis des nordwestlichen Harzgebirges. 1. Abt. Palaeontographica 3 (1. Lfg.), 1-67, Cassel.

Ross, C.A. and Ross, J.R.P., 1987. Late Paleozoic sea levels and depositional sequences. Cushman Found. Foram. Research, Spec. Publ. 24, 137-149, Washington.

Sandberg, C.W., Poole, F.G. and Johnson, J.G., 1989. Upper Devonian of western United States. Canadian Soc. Petrol. Geologists, Mem. 14 (III), 183-220.

Sandberg, C.A., Ziegler, W. and Dreesen, R., 1987a. Abrupt conodont biofacies changes redate and delimit the Frasnian (Late Devonian) extinction event in Euramerica (Abstract). Terra Cognita 7, 209-210, Strasbourg.

Sandberg, C.A., Ziegler, W. and Dreesen, R., 1987b. Frasnian/Famennian boundary and stratotype. Doc. subm. to the Devonian Subcomm., IUGS, Calgary, August 1987, 5 p., Calgary.

Sandberg, C.A., Ziegler, W. and Dreesen, R., 1988a. Late Frasnian mass extinction: Associated sea-level changes reflected by conodont faunas and biofacies. Cour. Forsch.-Inst. Senckenberg 102, 253-254, Frankfurt/M.

Sandberg, C.A., Ziegler, W., Dreesen, R. and Butler, J.L., 1988b. Late Frasnian mass extinction: Conodont event stratigraphy, global changes, and possible causes. Cour. Forsch.-Inst. Senckenberg 102, 263-307, Frankfurt/Main.

Schindler, E., 1990. Die Kellwasser-Krise (hohe Frasne-Stufe, Ober-Devon). Göttinger Arb. Geol. Paläont. 46, IV+115 p., Göttingen.

Schülke, I., 1995. Evolutive Prozesse bei *Palmatolepis* in der frühen Famenne-Stufe (Conodonten, Ober-Devon). Göttinger Arb. Geol. Paläont. 67, 108 p., Göttingen.

Scotese, C.R. and McKerrow, W.S., 1990. Revised World maps and introduction. In: McKerrow, W.S. and Scotese, C.R. (eds.), Palaeozoic Palaeogeography and Biogeography. Geol. Soc. London, Mem. 12, 1-21.

Shilo, N.A., Bouckaert, J., Afanasjeva, G.A., Bless, M.J.M., Conil, R., Erlanger, O.A., Gagiev, M.H., Lazarev, S.S., Onoprienko, Y.I., Poty, E., Razina, T.A., Simakov, K.V., Smirnova, L.V., Streel, M. and Swennen, R., 1984. Sedimentological and palaeontological atlas of the late Famennian and Tournaisian deposits in the Omolon region (NE-USSR). Ann. Soc. géol. Belgique 107, 137-247.

Simakov, K.V., Bless, M.J.M., Bouckaert, J., Conil, R., Gagiev, M.H., Kolesov, Y.V., Onoprienko, Y.I., Poty, E., Razina, T.P., Shilo, N.A., Smirnova, L.V., Streel, M. and Swennen, R., 1983. Upper Famennian and Tournaisian deposits of the Omolon region (NE-USSR). Ann. Soc. géol. Belgique 106, 335-399.

Scrutton, C.T., 1988. Patterns of extinction and Survival in Palaeozoic corals. In: Larwood, G.P. (ed.), Extinction and survival in the fossil record. Systematics Assoc. Special Vol. 34, 65-88, Clarendon Press, Oxford.

Sorauf, J.E. and Pedder, A.E.H., 1986. Late Devonian rugose corals and the Frasnian-Famennian crisis. Can. J. Earth Sci. 23, 1265-1287, Saskatoon.

Stearn, C.W., 1987. Effect of the Frasnian-Famennian extinction event on the stromatoporoids. Geology 15, 677-679, Boulder.

Steenwinkel, M. van, 1993. The Devonian-Carboniferous boundary: Comparison between the Dinant Synclinorium and the northern border of the Rhenish Slate Mountains. Ann. Soc. géol. Belgique 115, 665-681.

Streel, M. and Vanguestaine, M., 1988. Palynomorph distribution in the 'Extinction Layer' near the Frasnian/Famennian boundary in the shelf facies in Belgium. Doc. subm. to the Devonian Subcomm., IUGS, Rennes, August 1988, 6 p., Rennes.

Streel, M., Vanguestaine, M., Dreesen, R. and Thorez, J., 1987. Palynology (acritarchs and miospores) of the "barren black shale" near the Frasnian/Famennian boundary level at Hony (Belgium). Doc. subm. to the Devonian Subcomm., IUGS, Calgary, August 1987, 6 p., Calgary.

Szulzewski, M., 1986. Late Devonian events in Poland. Ann. Soc. géol. Belgique 109, 263-265.

Talent, J.A. and Yolkin, E.A., 1987. Transgression-Regression Patterns for the Devonian of Australia and Southern West Siberia. Cour. Forsch.-Inst. Senckenberg 92, 253-249, Frankfurt/Main.

Tappan, H., 1981. Extinction or survival and diversification: Patterns of selectivity during Paleozoic crises (abs.). Geol., climatol., Biol. Implic., Lunar Planet. Inst. Contr. 449, 54.

Tappan, H., 1982. Extinction or survival: Selectivity and causes of Phanerozoic crises. Geol. Soc. Am. Spec. Pap. 190, 265-276, Boulder.

Thompson, J.B. and Newton, C.R., 1988. Late Devonian mass extinction: Episodic climatic cooling or warming? Can. Soc. Petrol. Geol. Mem. 14 (III), 29-34, Calgary.

Vachard, D., 1994. Foraminifers et Moravamminids du Givétien et du Frasnien du domaine Ligerien (Massif Armoricain, France). Palaeontographica A 231, 1-92.

Walliser, O.H., 1980. The geosynclinal development of the Variscides with special regard to the Rhenohercynian Zone. In: Closs, H., Gehlen, K. von, Illies, H., Kuntz, E., Neumann, J. and Seibold, E. (eds.), Mobile Earth. Internat. Geodyn. Proj., Final Rep. FR Germany, 185-195, Boppard.

Walliser, O.H., 1981. The geosynclinal development of the Rheinische Schiefergebirge. Geol. Mijnb. 60, 89-96, s'Gravenhage.

Walliser, O.H., 1983. Statement to the boundaries of the Devonian System, its Series and Stages. Document submitted to the Subcommission of Devonian Stratigraphy, 4 pp., Montpellier.

Walliser, O.H., 1984a. Geologic processes and global events. Terra cognita 4, 17-20.

Walliser, O.H., 1984b. Pleading for a Natural D/C-Boundary. Cour. Forsch.-Inst. Senckenberg 67, 241-246, Frankfurt/M.

Walliser, O.H, 1985. Natural boundaries and Commission boundaries in the Devonian. Cour. Forsch.-Inst. Senckenberg 75, 401-408.

Walliser, O.H., 1986a. The IGCP Project 216 "Global biological events in Earth history". Lecture Notes Earth Sci. 8, 1-4, Springer, Berlin Heidelberg New York.

Walliser, O.H., 1986b. Towards a more critical approach to bio-events. In: Walliser, O.H. (ed.), Global Bio-Events, Lecture Notes in Earth Sciences 8, 5-16, Springer, Berlin Heidelberg New York.

Walliser, O.H. and Michels, D., 1983. Der Ursprung des Rheinischen Schelfes im Devon. N. Jb. Geol. Paläont. Abh. 166, 3-18.

Walliser, O.H., Lottmann, J. and Schindler, E., 1988. Global events in the Devonian of the Kellerwald and Harz Mountains. Cour. Forsch.-Inst. Senckenberg 102, 190-193, Frankfurt/Main.

Wang, K. and Bai, S., 1988. Faunal changes and events near the Frasnian-Famennian boundary of South China. Can. Soc. Petrol. Geol. Mem. 14 (III), 71-78, Calgary.

Wang, K., Orth, Ch.J., Attrep, M.J., Chatterton, B.D.E., Hou, H. and Geldsetzer, H.H.J., 1991. Geochemical evidence for a catastrophic biotic event at the Frasnian/Famennian boundary in south China. Geology 19, 776-779.

Waters, J.A., 1990. The palaeobiogeography of the Blastoidea (Echinodermata). In: McKerrow, W.S. and Scotese, C.R. (eds.) Palaeozoic Palaeogeography and Biogeography. Geol. Soc., Mem. 12, 339-352, London.

Wedekind, R., 1913. Die Goniatitenkalke des unteren Oberdevon vom Martenberg bei Adorf. Sitz.-Ber. naturforsch. Freunde Berlin 1913 (1), 23-77, Berlin.

Wilde, P. and Berry, W.B.N., 1984. Destabilization of the oceanic density structure and its significance to marine "extinction" events. Palaeogeogr., Palaeoclimatol., Paleoecol. 48, 143-162, Amsterdam.

Wilde, P. and Berry, W.B.N., 1986. The role of oceanographic factors in the generation of global bio-events. Lecture Notes Earth Sci. 8, 75-91, Springer, Berlin Heidelberg New York.

Yolkin, E.A., Weddige, K., Izokh, N.G. and Erina, M.V., 1994. New Emsian conodont zonation (Lower Devonian). Cour. Forsch.-Inst. Senckenberg 168, 139-157, Frankfurt/Main.

Yolkin, E.A., Apekina, L.S., Erina, M.V., Izokh, N.G., Kim, A.I., Talent, J.A., Walliser, O.H., Weddige, K., Werner, R. and Ziegler, W., 1989. Polygnathid lineages across the Pragian-Emsian Boundary, Zinzilban Gorge, Zerafshan, USSR. Cour. Forsch.-Inst. Senckenberg 110, 111-121, Frankfurt/Main.

Ziegler, W., 1962. Taxonomie und Phylogenie oberdevonischer Conodonten und ihre stratigraphische Bedeutung. Abh. hess. L.-Amt für Bodenforsch. 38, 166 pp., Wiesbaden.

Author´s address:

Otto H. Walliser, Institut und Museum für Geologie und Paläontologie, Goldschmidt-Strasse 3, D - 37077 Göttingen, Germany

Permian Global Bio-Events

Douglas H. ERWIN

Abstract. Permian global bio-events have been largely overshadowed by the end-Permian mass extinction, the highest-order bio-event of the Phanerozoic, which seems to have affected much of the Upper Permian. Regional marine bio-events are known from Lower Permian, but lack of attention and correlation difficulties have made it difficult to determine whether these events are truly global in nature. At least two global tetrapods events are known from taxonomic studies during the Lower Permian but have not been tied to specific sections. The end-Permian Event is associated with numerous geologic, climatic and geochemical perturbations, several of which may have contributed to the extensive extinctions. Sea-level change and associated habitat destruction, climatic instability and other factors were particularly important.

Contents

Introduction	251
Lower Permian Events	254
End-Permian Mass Extinction	254
Marine Extinctions	254
The Terrestrial Record	257
Causes of the Mass Extinction	258
Conclusion	261

Introduction

The end-Permian mass extinction brought a close to the Paleozoic, and with it the marine community types which had dominated the oceans since the Ordovician. One of the most extensive evolutionary radiations since the Ordovician occurred following the extinction, establishing the major marine community types which continue to dominate modern oceans. On land, tetrapods and insects experienced dramatic extinctions at the close of the Permian. Changes in floras occurred over a longer time span during the Permian, as the continents entered a phase of warmer climate following the Permo-Carboniferous glaciation. This brought an end to the Paleophytic Flora and the development of the Mesophytic Flora. Although anecdotal evidence suggests a variety of regional bio-events have affected various groups prior to the mass extinction, including gastropods, brachiopods, some echinoderm groups, and perhaps reef ecosystems, it remains unclear whether any of these events represents truly global bio-events. This uncertainty reflects both the emphasis on terminal Permian events as well as the continuing difficulties of global correlation during the Permian. More information is available about early to mid-Permian tetrapod bio-events but inter-regional correlation problems and taxonomic inconsistencies plague analyses of these episodes. Because of these problems, this chapter will only briefly consider events prior to the end-Permian mass extinction and will concentrate on the extinction itself.

The Permian Period, particularly the Late Permian, experienced far more extensive geologic change than most similar intervals of geologic time. The forma-

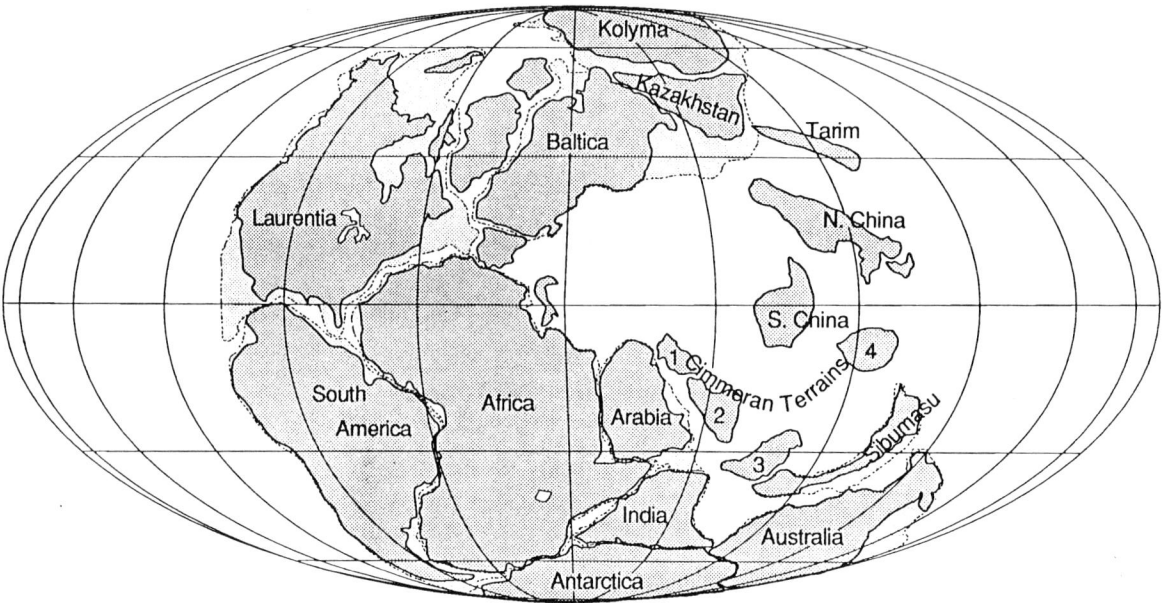

Fig. 1. Paleogeographic reconstruction for the Late Permian (Guadalupian). The position of the Chinese and Cimmeran terrains is highly uncertain

tion of the supercontinent of Pangea during the Early Permian was undoubtedly the most significant of these geologically, and had considerable impact on climate patterns, biogeography, and other processes. Laurasia and Gondwana collided with a rotational, northward movement during the Late Carboniferous and Early Permian. Soon after the collision a variety of small microplates (now part of south-east Asia and China) rifted off the northern margin of Gondwana and moved north. The collision of these micro-continents with Laurasia near the close of the Triassic re-established Pangea just as the opening of the Atlantic began its final break-up. However the movement of these microplates makes reconstructing the configuration of this region during the Permian very difficult. Thus the reconstruction shown in Fig. 1, while relatively reliable for the major continental units of Pangea, is largely conjectural in the proto-Tethyan region. Ideally biogeographic data could resolve many of these problems, but much of the paleobotanical data showing close floristic affinity between different regions, for example, is contradicted by paleomagnetic evidence that the regions were independent tectonic units during the same time. In part these difficulties may reflect the subtle difference between tectonic regions and biogeographic regions, but they also reflect patterns of endemism and vicariance. All in all, biogeographic data provides few consistent patterns to aid in deciphering the relationships between these tectonic units.

The movement of Gondwana across the South Pole during the Carboniferous and Permian led to the Permo-Carboniferous glaciation continued until near the end of the Lower Permian (Crowell, 1983, 1995; but see Dickins, 1983, 1985). The characteristic cyclothems of the Carboniferous mid-latitudes persisted through the Lower Permian. As the northward drift of Pangea moved Siberia near the North Pole a final, bi-polar pulse of continental ice developed (discussed by Stanley, 1988, who incorrectly believed these deposits were of latest Permian age). By the end of the Early Permian, global climates had begun to ameliorate, although tracking this global warming is complicated by the simultaneous northward drift of Pangea into equatorial latitudes. This warming trend continued through the Permian and into the Triassic. The formation of Pangea was associated with orogenic episodes throughout the Hercynian, Appalachian and Uralian regions. The resulting mountain chains created regional climatic shifts which further exacerbate global climate patterns.

Other geologic events include the end of the Kiaman superchron near the beginning of the Late Permian, widespread pyroclastic volcanism in South China and the formation of the Siberian flood basalts, the largest flood basalt province of the Phanerozoic, at the Permo-Triassic boundary. A series of geochemical anomalies signify profound changes in ocean chemistry near the close of the period. These are described in more detail elsewhere in this volume (Holser and Magaritz) and include the rapid oxidation of a large volume of organic carbon (recognized by a shift in $\partial^{13}C$ from +3 to -1), a shift in $\partial^{34}S$ and one of the largest shifts in strontium ratios in the Phanerozoic. These isotopic changes occur during an apparent rapid marine regression at the close of the period which has been estimated at 210 m (Forney, 1975) to 280 m (Holser and Magaritz, 1987) although these are almost certainly over-estimates (see below). These geologic, geochemical and climatic events of the Permian are summarized in Fig. 2.

As noted above, complicating any discussion of global bio-events is the lack of a well-established global correlation scheme for the Permian. The extensive facies-specificity of many groups and the unique histories of many basins has made correlations between the various regional stratigraphic stages

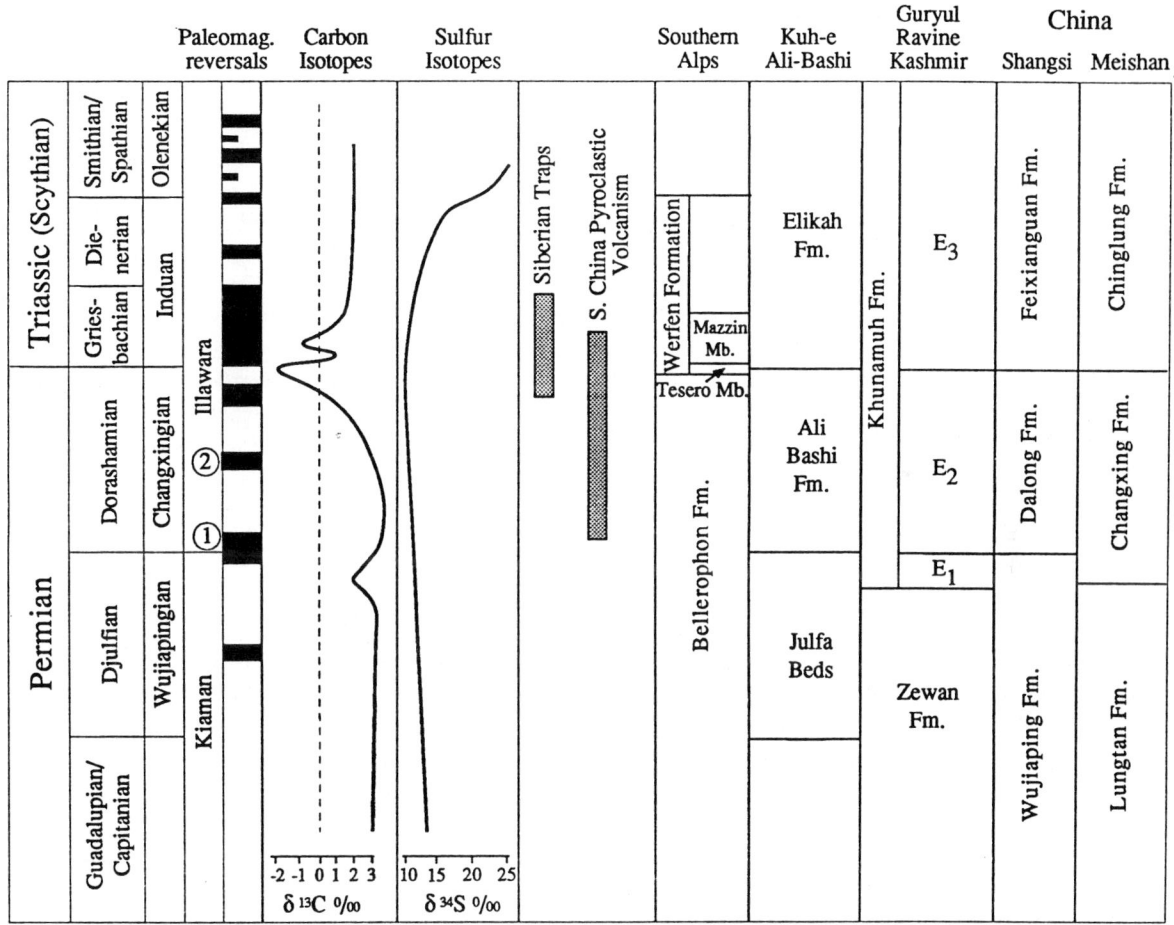

Fig. 2. Major geologic, geochemical and climatic events of the Late Permian, based on a variety of sources which are more completely detailed in Erwin (1993), and correlations for the major Permo/Triassic boundary sections, based largely on Sweet et al. (1992). From left to right, this chart shows the late Permian stages, the paleomagnetic reversal scheme (with normal polarity in black), including the Kiaman/Illawara Superchron boundary near the Wujiapingian/Changxingian boundary, the shifts in $\partial^{13}C$ and $\partial^{34}S$, volcanism near the boundary, and the proposed boundary correlations. Note that the continuity of sections remains uncertain. (1) denotes a reversal in the *Neogondollela orientalis* zone; (2) denotes a reversal at the base of the *N. changxingensis* conodont zone. Paleomagnetic information from Heller et al. (1988), Steiner et al. (1989) and Molostovskiy (1992)

difficult, hampering agreement on global correlations and global stages. Currently geologists working on the Permian are faced with a number of competing schemes. This continues to be an active area of research and recent advances, particularly in conodont correlation, suggest that resolution of these difficulties may soon be at hand. In this chapter I will adopt the scheme shown in Fig. 2. Not all Permian workers will agree with it, nor with the correlations shown, but it seems to me to be reasonable and at least moderately defensible. Increased refinement of global correlations, both marine and terrestrial, is likely to reveal additional, smaller global bio-events prior to the end-Permian mass extinction.

Lower Permian Events

Characterization of Lower Permian marine bio-events has received less attention than the end-Permian event. However, scattered research and anecdotal evidence suggest that several important bioevents may have occurred. For example, Ingavat-Helmcke and Helmcke (1986) documented two extinction horizons among Permian fusulinaceans in Thailand. The first event occurs near the end of the Sakmarian and coincides with both the onset of the Uralian orogeny and the end of the Permo-Carboniferous glaciation. It involves the disappearance of a variety of Arctic-Tethyan foraminifera and the development of an Eastern Tethyan fauna. A second bio-event occurred at the close of the Murgabian (*Neoschwagerina* Zone) with the disappearance of a number of fusulinids. In my own work in the southwestern United States I believe I have identified a fairly significant change in faunal assemblages between the middle and upper Wolfcampian, particularly among the gastropod assemblages. Considerable work remains to characterize this event and to determine whether it is of more than regional extent.

A number of authors have identified apparent bio-events near the close of the Permian, but it remains unclear whether or not these form part of the end-Permian extinction or not. For example, Schäfer and Fois-Erickson (1986) describe the decimation of the diverse Tethyan bryozoan faunas at the close of the Kazanian (Guadalupian) and during the early Djulfian. Similar patterns have been described for trilobites, some articulate brachiopods and tabulate and rugose corals (Erwin, 1994; but more recent data from South China indicates that rugose corals persisted to the P/Tr boundary: Ezaki, 1994). As discussed in Erwin (1993), such data are often plagued by preservational and correlation problems. Thus, it is difficult to determine whether such patterns indicate an end-Guadalupian event independent of the end-Permian extinction or the earliest phase of the end-Permian event perhaps masked in some degree by loss of habitat area and other factors. Detailed sampling and correlation through the Djulfian and Dorashamian will be required to resolve these questions.

Turning to land, terrestrial floras undergo a long-term conversion throughout the Permian as the Paleophytic flora gives way to the Mesophytic flora (Knoll, 1984). However, this occurs in response to the global warming during the period and no discrete bio-events have been identified. Such is not the case with tetrapods. Taxonomic compilations by Benton suggest increased tetrapod extinctions during the Artinskian (46% of amphibians; 70% of reptiles) and Ufimian (55% and 58%, respectively) (Maxwell, 1992; see also Benton, 1987, 1988). The net result of these extinctions was to reduce the overall proportion of amphibians in the tetrapod fauna and increase the proportion of 'reptiles'. Furthermore, pelycosaurs dominated the Early Permian reptilian faunas but declined during the Late Permian as therapsids underwent a spectacular radiation. No field studies have investigated either of these episodes to determine whether the extinctions occurred as discrete, global bio-events or involved a longer decline through the stage.

End-Permian Mass Extinction

Since the end-Permian mass extinction has been the topic of a number of comprehensive reviews (Erwin, 1990, 1993, 1994; Holser and Magaritz, 1987; Maxwell, 1989; Sweet et al., 1992; Teichert, 1990; Yang and Yin, 1987), which should be consulted for additional information, this chapter will be restricted to a briefer overview of this bio-event. Erwin (1993, 1994) serves as the basis for the following discussion.

Marine Extinctions

Boundary Sections. Permo-Triassic boundary sections were exhastively reviewed by Teichert (1990) and Erwin (1993); localities of the most important sections are shown on Fig. 3. A new perspective on the Pakistan, Alpine and Chinese sections is provided by Wignall and Hallam (1992, 1993). Boundary sections play a key role in understanding all bio-events, for they document the rate and nature of the extinction. Yet reliance upon the

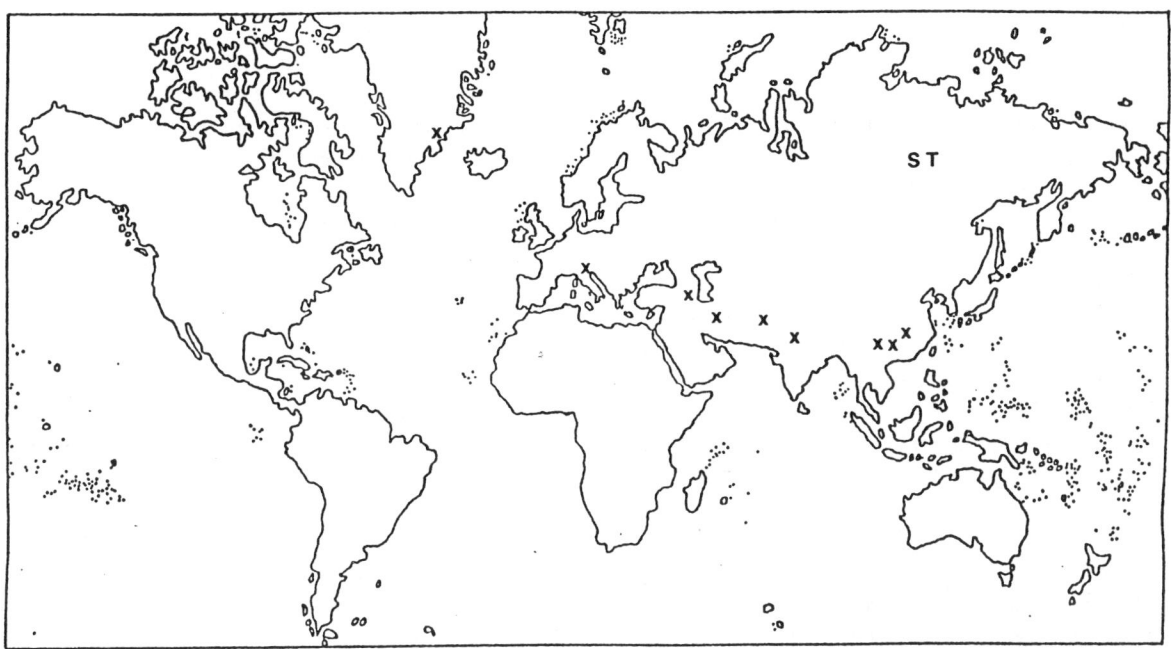

Fig. 3. Location of important Permo-Triassic marine boundary sections. The location of the Siberian flood basalts is noted by ST

evidence from boundary sections, particularly when such sections are few in number or concentrated in a single region, can unduly influence impressions of a bio-event. Many such sections, with broad geographic distribution, must be employed to analyze patterns of disappearance.

New biostratigraphic results, summarized by Sweet et al. (1992) are of particular interest. The traditional view has been that the Upper Permian includes, in order, the Guadalupian, Djulfian and Dorashamian Stages, with the Changxingian Stage largely equivalent to the Dorashamian. The ammonoid *Otoceras woodwardi* denotes the basal Triassic. There has been considerable debate over whether the upper portion of the Changxingian (*Neogondolella deflecta - N. changxingensis* conodont zone) is equivalent to the Upper Dorashamian (*Pseudotirolites* Ammonoid Zone). Sweet (1992) employed graphic correlation to develop a very different scheme, in which *Otoceras woodwardi* is correlative with *Neogondolella changxingensis*, and thus either the Changxing Formation is Griesbachian rather than latest Permian, or *Otoceras woodwardi* is a latest Permian species. Although Sweet's conclusions are not entirely new, they have caused considerable consternation. If correct, either the extinction is diachronous, earlier on Pangea than in China, or the extinction actually occurred in what has been considered the earliest Triassic. Graphic correlation is an imprecise tool however, since it ignores taphonomic problems, facies control and other confounding factors. Correlation difficulties limit paleontologists' ability to determine the rate of extinction, since they make it difficult to establish whether the apparent boundary is in fact isochronous.

Unfortunately analysis of this boundary has long been plagued by an apparent major marine regression at or near the boundary. The estimated magnitude of the regression ranges from 210 Ma (Forney, 1975) to 280 Ma (Holser and Magaritz, 1987). Correction for isostatic adjustment reduces these values by about 30%, but such estimates still assume that the modern hypsometric curve applies to the Permian. The evidentiary basis for a regression of this magnitude is not great and better estimates of the magnitude of the regression are needed, preferably in the context of a sequence stratigraphic study. Wignall and Hallam (1992, 1993) have provided such studies and their results discount claims for a pronounced regression near the boundary. In fact, they suggest that the sequence boundary occurs in the uppermost Changxingian and seas were advancing at the boundary itself.

While there is room for considerable work on the sequence stratigraphy of this interval (for example, through application of backstripping to better constrain sea-level changes) Wignall and Hallam deserve considerable credit for pioneering in the application of these techniques to the boundary interval.

The regression has led to an emphasis on the presence or absence of continuous boundary sections, in ignorance of the fact that a l l sections contain gaps of varying lengths. The relevant issue is the duration and distribution of these gaps relative to the required level of stratigraphic resolution (Sadler, 1981). Teichert (1990) concluded that the general lack of continuous sedimentation across the Permo/Triassic boundary had fostered the impression among many paleontologists that the extinction was catastrophic. Teichert's examination of the relevant boundary sections (excluding the southern Alps) led him to conclude that the extinction began during the Djulfian and continued through the Changxingian and Griesbachian. Based on available global correlations and absolute dates, he estimated the duration of the extinction at six to ten million years. Developments in age dating, biostratigraphy, isotopic analysis and other areas since Teichert wrote his contribution suggest that reexamination of his conclusions is timely.

Teichert ignored one of the most promising regions for studying the Permo-Triassic boundary, the southern Alps. Although this was the region where the first wholly marine Permo-Triassic boundary was identified, Assereto and colleagues (1973) concluded that an unconformity encompassed much of the latest Permian. By 1986 work by a number of European geologists had demonstrated that the sections were far more complete than previously acknowledged (papers in Cassinis, 1988). The gradual shift in carbon isotopes (Holser et al., 1989; Holser and Schönlaub, 1991) in contrast to the rather abrupt shifts found in South China (Baud et al., 1989) suggests the Italian-Austrian sections may be less condensed (or more continuous, in traditional terms) than those in South China. Traditionally the Permo/Triassic boundary was placed at the contact between the underlying Bellerophon Formation and the overlying Werfen Formation in the Southern Alps. The boundary actually appears to occur within the Tesero Horizon, a thin oolitic unit at the base of the Werfen. Curiously there is a pronounced increase in fungal spores and other debris in the upper Bellerophon and in the Tesero Horizon, a pattern repeated in sections in Greenland, Israel and elsewhere in proto-Tethys (Visscher and Brugman, 1988; Eshet, 1992).

What of the rate of the marine extinction? Recent uranium-lead dates for zircons from the ash comprising the boundary clay at the top of the Changxing Formation in South China yield a date of 251 ± 3.4 Ma (Claoue-Long et al., 1991); previous estimates put the boundary at 248-245 Ma. This new date effectively shortens the apparent duration of the extinction, a result consistent with recent biostratigraphic results. Global compilations at both the familial and generic levels suggest that the extinction began near the close of the Guadalupian and accelerated through the Djulfian, reaching a peak with a rapid pulse near the close of the Dorashamian (Teichert, 1990; Erwin, 1993, 1994). On a global scale this pattern still appears to be valid, and in part reflects the disappearance of a large number of marine basins during the early Late Permian.

However, sections in South China, and to a lesser extent those in Italy and a few other localities, display considerable evidence for rapid regional extinction near the boundary itself. For example, some 95% of the species of foraminifera, ammonoids, bivalves and articulate brachiopods found in the Changxing Formation disappear at the boundary (Yin et al., 1984). Xu (1991) employed a variant of graphic correlation to correlate 40 well-studied sections across South China. Of the 218 invertebrate species included in the analysis, only 18 pass into the Lower Triassic; most disappear in beds immediately below the boundary. Other studies of the Chinese sections have reached similar conclusions (see discussion in Erwin, 1993, 1994). Although Xu concluded that the extinction was essentially catastrophic, correlation and taxonomic problems compel one to view Xu's conclusions with caution. Thus the weight of recent evidence suggests the extinction was far more rapid than suggested by Teichert (1990) and Erwin (1990): probably 1-2 million years, and perhaps an even shorter interval.

Extinction Patterns. The end-Permian extinction eliminated about 49% of marine families and from 78 to 84% of marine genera from the Guadalupian through the Changxingian (Sepkoski, 1984, 1989, 1992; Erwin, 1993). Raup's (1979) application of rarefaction analysis suggested that as many as 96% of all marine species may have become extinct, but this value is almost certainly an overestimate since Raup was forced to ignore variations in clade extinction patterns and the differences in the species/family and genus/family ratios. Nonetheless, an overall extinction rate of durably skeletonized marine species of 85-90% seems reasonable. The greatest effect of the extinction was felt among inshore taxa, such as the fusulinid foraminifera, and among sessile, epifaunal filter feeders. Among the gastropods, which overall suffered among the lowest percentage extinction of any major group, the epifaunal, suspension feeders were eliminated (Erwin, in prep.). Sepkoski (1984) characterized this pattern as the decimation of the Paleozoic evolutionary fauna, which suffered a 79% familial extinction relative to the 27% family-level extinction among the Modern Fauna.

There are a variety of reef types in upper Permian rocks, and calcisponge/algal reefs persist into the Changxingian in South China and the Greek island of Skyros but disappear below the boundary (Flügel and Reinhardt, 1989). The fauna on the South China reefs was quite diverse and shows no evidence for gradual extinction.

Marine taxa can be divided into four groups based upon their response to the extinction event (see Teichert [1990] for a slightly different analysis): A number of groups continued across the boundary in seeming ignorance of the event, including conodonts, non-fusulinid foraminifera, bivalves, nautiloid cephalopods and bellerophontid gastropods. Another group appears to have been declining well before the end of the period, perhaps due to the drying out of a number of marine basins during the long-term regression which began in the early Guadalupian. This group includes blastoids, camerate crinoids and tabulate corals. The third groups also appears to have begun declining early but survived into the Dorashamian/ Dzulfian with reduced diversity. These include the last few trilobites, bryozoa, rugose corals and perhaps crinoids. The fourth group includes the fusulinid foraminifera, articulate brachiopods and ammonoid cephalopods and several groups of gastropods. All were important components of latest Permian marine communities but suffered dramatic declines in diversity during the final phase of the mass extinction. The final group includes a number of clades which were of reduced importance during the Late Permian, suffered varying degrees of extinction, but then rebounded strongly in the Triassic. This assemblage includes a number of other gastropods, the surviving echinoids, sponges, and rhynchonellid, spiriferid and terebratulid brachiopods.

The quality of the extinction data is not as high as for a number of other extinction episodes. In part this is due to the magnitude of the apparent hiatus in many sections during the latest Permian. Good analyses of extinction patterns have been carried out in China, but these are occasionally plagued by taxonomic inconsistencies and inflation. My impression is that the extinction may have begun earlier outside South China (i.e. on Pangea), particularly among the fusulinids, gastropods, bivalves and cephalopods. This conclusion must be tempered by the acknowledged difficulties in correlation, sampling and the Signor-Lipps effect.

The Terrestrial Record

The terrestrial record is crucial to determining whether the mechanism of the extinction was limited to the marine realm or was more general, affecting the terrestrial biota as well. Until recently the evidence has been equivocal, but there is growing evidence that tetrapods and insects suffered considerable extinction during the Late Permian and there is additional evidence for a disturbance among plant assemblages. Unfortunately the paucity of the fossil record of these groups, coupled with the difficulty in correlating between marine and terrestrial sections, makes it difficult to determine either the rate of these extinctions or whether the terrestrial events were exactly contemporaneous with those in the oceans.

T e t r a p o d s . The disappearance of 27 (75%) families of tetrapods during the Late Permian was one of a series of vertebrate extinction events during the Permian and Triassic. This breaks down as a 67% drop in amphibian diversity and 78% extinction among 'reptiles'. In the earliest Triassic about 42% of the tetrapod families disappeared (Maxwell, 1992; King, 1991; Benton, 1987, 1988; Olson, 1986). The therapsids suffered the most during the Late Permian/ Early Triassic, in part because labarinthodont amphibians and early therapsids had largely disappeared during extinctions earlier in the Permian. Although the data remains a bit fuzzy, the extinction appears to have been both taxonomically and geographically widespread. There is clearly considerable latitude for more intensive sampling of productive vertebrate regions. For example, detailed sampling of the relevant intervals in the Karoo Basin of South Africa should produce a far better picture of extinction and survival patterns in this region.

P l a n t s . The end of the Permo-Carboniferous glaciation and the subsequent global warming led to the decline of the Paleophytic Flora of broad-leaved pteridosperms, cordiates and pecopterid ferns and the development of the Mesophytic Flora, an assemblage of more mesic plants dominated by conifers, ginkgoes, cycads, cycadeoids and new varieties of pteridophytes and pteridosperms. Knoll (1984) demonstrated that this replacement process occurs at different times on different continents from the Early Permian to the Early Triassic. The replacement appears to begin in low latitudes and spread toward the poles (Frederiksen, 1972; DiMichele and Aronson, 1992). The Permo-Triassic boundary is marked by a change in the pollen (Traverse, 1988) and a spike in fungal spores (described earlier) which is suggestive of a short-term disturbance in terrestrial ecosystems.

I n s e c t s . Perhaps the best indication of the extent of the terrestrial events is found among the insects. Twenty-seven of the 36 described insect orders have been described from the Permian. Of these, 8 orders disappeared during the Late Permian and a further 10 suffered considerable contraction. Only one or two insect orders have become extinct since the Permian (Labandeira and Sepkoski, 1993). Thus virtually all of the major insect extinctions occurred near the Permo-Triassic boundary.

Causes of the Mass Extinction

Perhaps the most troubling aspect of the end-Permian extinction event is that most of the suggested causes of the extinction can be eliminated for one reason or another, suggesting that paleontologists have considerably more to learn about this episode. It seems quite likely that a multitude of contributory causes must be invoked, some of them associated with the late Permian marine regression. In evaluating the possible causes of the extinction several points must be kept in mind. First, the proposed mechanism must be consistent with the available geologic, climatic and geochemical data (Fig. 2). Second, the proposed mechanism must act on the appropriate time scale. For example, simply relating the extinction to the formation of Pangea is insufficient since Pangea formed tens of millions of years before the extinction and persisted for tens of millions of years after the post-extinction recovery was well advanced. Third, since we now have good evidence that both marine and terrestrial ecosystems were affected by this event, although the marine realm does appear to have been more heavily impacted, the proposed mechanism must act in both areas. Finally, the magnitude of the action must be sufficient to plausibly account for the magnitude of the extinction. The difficulties associated with eliminating some 90% of all marine species are all too infrequently addressed. Past discussions of this extinction, in fact of most bio-events, attempt to focus on a single cause. Complex historical events rarely have a single cause (although the K/T extinction appears to provide a salutary counter example). The concentration of geologic events during the latest Permian immediately renders suspect, it seems to me, a simple explanation for this event.

Rather than exhaustively review the plethora of proposed causes, which have been extensively covered in other recent publications (Erwin, 1990, 1993, 1994; Maxwell, 1989), I will emphasize some more recent suggestions, including extra-terrestrial impact, volcanism, and several global warming/anoxia hypotheses. Finally, I will offer my own ideas on the subject. It should come as no surprise to discover that I believe there were a multitude of causes rather than a single catastrophic cause.

The realization that anomalously high concentrations of platinum group elements, principally iridium, offered a means of identifying the impact of extra-terrestrial objects (Alvarez et al., 1980), followed by Raup and Sepkoski's (1984, 1986) periodicity hypothesis initiated detailed geochemical investigations of many boundary intervals. Initial studies of P/Tr boundary sections in South China (Meishan and Wachapo Mountain; Asaro et al., 1982) and Armenia (Alekseev et al., 1983) failed to find anomalous concentrations of iridium. Later work in South China reported iridium levels of 5-8 pbb (Sun et al., 1984; Xu et al., 1989). However re-analyses of the same section failed to verify these results (Clark et al., 1986; Orth, 1989; Orth et al., 1990) although microspherules have been recovered (Xu et al., 1985, 1989; but see Yin et al., 1989, 1992). Such spherules may indicate impact of a stony meteorite (which would have a different elemental abundance pattern, one little different from the earth) but there is as yet no good evidence to support such a conclusion.

The Siberian flood basalts erupted over a period of less than a million years essentially at the Permo-Triassic boundary (within the limits of analytical precision; Campbell et al., 1992), although biostratigraphic evidence suggests that at least some of the flood basalt erupted during the Early Triassic. The latest Permian was also a time of widespread pyroclastic volcanism in South China. Sections of the Changxing Formation and correlative formations contain numerous bentonites and the P/Tr boundary itself is marked by a clay layer of volcanic origin. This widespread volcanism has generated a number of recent suggestions linking either the flood basalts or the pyroclastic volcanism to the mass extinction through climate (cooling produced by sulfate aerosols and a dust cloud or warming via release of CO_2).

The geochemistry and texture of the boundary clays in South China and the presence of abundant microspherules demonstrate the volcanic origin of the layer (Yin et al., 1984, 1989, 1992) with a likely volume of >1000 km^3 (Zhou and Kyte, 1988). Yin and his colleagues have argued that these massive pyroclastic eruptions produced a global dust cloud, eliminating photosynthesis in much the same way as has been proposed for the K/T impact, and triggering the mass extinction. However, not only were there a series of large pyroclastic eruptions in South China during the latest Permian, but they are no larger than numerous other, well-dated and well-studied events which had no discernible biologic effects (Erwin and Vogel, 1992). Massive pyroclastic eruptions would certainly cause considerable short-term global cooling and a drop in primary productivity. However, the effects appear to be too transient. When we examined 4 pyroclastic eruptions over the past 80 Ma of equivalent magnitude to the end-Permian event we could identify neither global nor regional effects in either the marine or terrestrial vertebrate record.

The Siberian traps are the largest flood basalt of the Phanerozoic, covering 2.5 x 10^6 km^2 to a depth of up to 3000 m. Recent isotopic analyses indicate the main phase of the eruption began very close to the boundary (Campbell et al., 1992). Campbell and colleagues suggested that the end-Permian mass extinction was triggered by the injection of large amounts of sulfur dioxide during the eruption of the traps which induced global cooling, acid rain and regression, the latter from expansion of a polar ice cap. The difficulty with this hypothesis lies in the connection drawn between the eruption and the extinction. As noted above, there is no evidence for a polar ice cap in the latest Permian, not for global cooling during this interval. Indeed there is considerable evidence for continued global warming across the boundary. Furthermore, sulfate particles undergo increasing aggregation as the volume of sulfates increases, which accelerates the rate at which such particles settle out of the atmosphere (Pinto et al., 1989). Thus while the eruption of the Siberian traps may have caused a brief environmental disaster, there is little evidence to support the contention that it was the major cause of the extinction.

The final set of extinction scenarios I will discuss were triggered by the discovery of the $\partial^{13}C$ shift as one approaches the Permo-Triassic boundary. Each of these hypotheses links this carbon oxidation event to a reduction in atmospheric oxygen, causing either global or marine anoxia and to an increase in atmospheric carbon dioxide leading to global warming. The details of these proposals differ considerably however. Holser and Schönlaub (1991) discuss the oxidation of organic carbon on the continental shelves as they were exposed by the marine regression. Gruszczynski et al. (1989, 1990), Hoffman et al. (1990), and Malkowski et al. (1989) suggest that greatly increased sequestration of organic carbon within a stratified ocean (in the form of sapropel-like deposits) followed by rapid marine overturn and oxidation of the previously sequestered carbon was responsible for the observed isotopic shift and the extinction. Their data suggests a far larger shift in $\partial^{13}C$ than any of the other groups (from +8 to -3) and also requires the oxidation of about ten times as much carbon as was sequestered. The authors do not identify the source of this additional carbon. The final model (Hallam, 1989; Wignall and Hallam, 1992, 1993) suggests that the spread of anoxic marine waters during the earliest Triassic transgression was actually responsible for the extinction, and claims the presence of finely laminated sediments at the boundary, the depauperate nature of earliest Triassic faunas and an apparent Cerium anomaly as evidence in favor of this scenario. It is not clear how this scenario explains the widespread terrestrial extinctions.

The difficulty with these scenarios lies in the presumed relationship between the shift in carbon isotopes and the presumed drop in atmospheric oxygen and increase in atmospheric carbon dioxide. It would appear that a variety of mechanisms are also possible causes of the shift in carbon isotopes. For example, regressions will also trigger release of gas hydrates, primarily methane. Gas hydrates are one of the major carbon reservoirs (<10,000 gigatons of carbon) today and consist of methane locked into an icy lattice (Kvenvolden, 1988). Gas hydrates are ubiquitous under permafrost and along the outer continental shelf where temperature/pressure values allow their accumulation in the sediment (generally below 300-500 m water depth). The methane (and smaller amounts of other gases) is produced during methanogenesis with an average $\partial^{13}C$ of -60 ‰. Marine regressions remove part of the overlying water column leading to the destabilization of the gas hydrates and the release of the methane to the atmosphere (Nisbet, 1990; Paull et al., 1991). The methane oxidizes first to carbon monoxide, then eventually to carbon dioxide. The input of carbon with such a high negative $\partial^{13}C$ value will produce a marked shift in atmospheric carbon values then oceanic organic carbon and carbonate.

Equations introduced by Spitzy and Degens (1985) permit one to determine the volume of carbon from a reservoir of a particular $\partial^{13}C$ value required to shift global $\partial^{13}C$ values by a given amount. Since we know $\partial^{13}C$ shifted from +3 to -1 near the Permo/Triassic boundary, the volume of carbon dioxide from a volcanic source (-5 $\partial^{13}C$), erosion of marine carbonates (-20 to -25 $\partial^{13}C$) or methane (-60 $\partial^{13}C$) required can be easily determined and compared with the likely output rates or reservoir sizes (these results are discussed in more detail elsewhere [Erwin, 1993]). The results reveal: (1) only gas hydrates or erosion of marine carbonates could have caused a shift of the magnitude seen; volcanic carbon dioxide is too light. (2) Release of a fairly small fraction of the current gas hydrate reservoir would be sufficient to shift $\partial^{13}C$ by the observed amount; calculating the volume of this reservoir in the Late Permian is difficult without better knowledge of the hypsometry and latitudinal temperature gradients, but the reservoir was certainly larger than today. (3) The change in atmospheric oxygen and carbon dioxide would be less than 2%, insufficient to induce either significant global warming or anoxia. (4) Paradoxically, the fact that the shift in $\partial^{13}C$ was only 4‰ suggests that current estimates for the magnitude of the regression are gross overestimates (consistent with the results of Wignall and Hallam, 1993). If sea-level had declined by 210 to 280 m as some have suggested (Holser and Magaritz, 1987) the bulk of the gas hydrate reservoir would have been released and the shift in $\partial^{13}C$ would have been larger than observed. The exposure of the continents may be as complete as Holser and Magaritz (1987) calculated, but deriving the vertical drop in sea-level requires knowledge of Permian hypsometry. The lack of a larger drop in $\partial^{13}C$ suggests that Permian hypsometry differed greatly from modern hypsometry and Later Permian epeiric seas included a large amount of land area between 0 and -50 meters (based on Guadalupian sea-levels. Thus while a combination of gas hydrate release and oxidation of eroded marine carbonates during the regression was probably the cause of the $\partial^{13}C$ shift, this does not appear to be related to the extinction itself.

Most recent explanations of the end-Permian bio-event focus on the various climatic and biologic effects of the extensive marine regression at the end of the period. In the past the regression and fluctuations of other climatic changes have been related to the formation of Pangea. As noted previously, Pangea formed tens of millions of years before the extinction and persisted into the Jurassic. Clearly Pangea itself was not directly involved with the mass extinction (Erwin, 1990) although the climatic consequences of the regression (increased albedo, seasonality, etc [Robinson, 1973]) may have been exacerbated by the supercontinent. Although the evidence in favor of a rapid extinction has been growing over the past few years, it is best developed in South China, and it remains unclear to what extent this reflects a global pattern.

As I have discussed elsewhere (Erwin, 1994) I believe the causes of the end-Permian extinction lie in a number of co-occurring causes operating in three phases. The extinction appears to have begun in the early Djulfian with the loss of many marine basins and the destruction of considerable habitat area. This regression increased exposure of Pangea, exacerbating climatic instability. Toward the end of the Permian, volcanic eruptions and perhaps an increase in atmospheric carbon dioxide further increased the climatic instability and environmental degradation. Finally, immediately prior to the boundary, the regression ended and a transgression began. It seems likely that these waters may have been dysaerobic and been a contributing factor to the extinctions. Much of the terrestrial extinction may have occurred at this time as near-shore terrestrial environments were overwhelmed by the transgression.

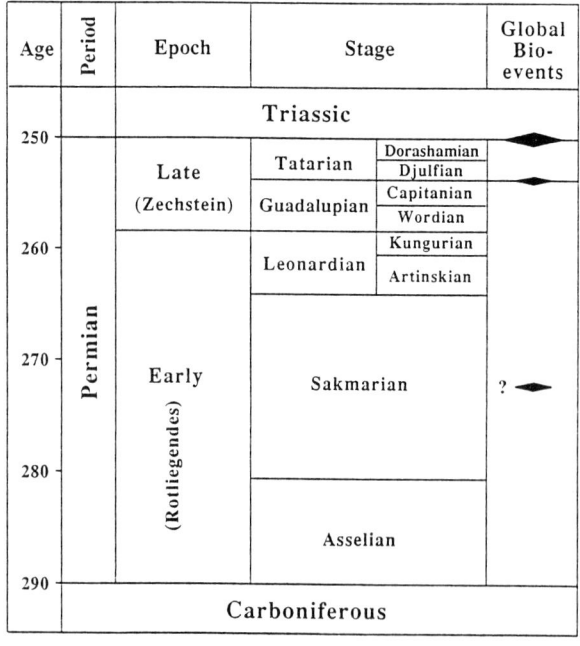

Fig. 4. Global bio-events in the Permian

Conclusion

It should by now be apparent that any discussion of the nature of the massive end-Permian extinction is plagued by considerable uncertainty. Significant advances made in understanding the events of the latest Permian over the past decade, principally through enhanced biostratigraphy, new isotopic analysis, the detailed, high-resolution biostratigraphic work on sections in South China and the resolution of the extent of the Late Permian extinctions in insects and tetrapods. Yet many questions remain, some of which have been noted in this chapter. Resolution of many of these issues requires enhanced global biostratigraphic correlation, but recent advances suggest that this is near at hand. Chief among the other outstanding issues are:

1) What was the magnitude of the latest Permian marine regression and subsequent transgression, and how closely does the shift in $\partial^{13}C$ track the regression?
2) Was the pace of extinction equally rapid on Pangea and South China, or was it more rapid in South China?
3) Did substantial latitudinal or other biogeographic effects play a role in the extinction?
4) A related issue focuses on the Lazarus taxa which disappeared near the end of the Permian only to re-emerge in the Middle Triassic. Where were they, and what ecologic, biogeographic or other attributes led to their survival?
5) Finally, higher quality biostratigraphic and taxonomic analysis (both terrestrial and marine) would allow more accurate plotting of the nature of the extinction.

In the past resolution of such issues has appeared unlikely, if not impossible. The resurgence of interest in the extinction over the past decade has demonstrated that there is much more we can learn about this event, which in turn may illuminate such issues as the role of mass extinctions in the history of life.

References

Alekseev, A.S., Barsokova, L.D., Koesov, G.M., Nazarov, M.A. and Grigoryan, A.G., 1983. Permian-Triassic boundary event: geochemical investigations of the Transcaucasia section. Abstracts 14th Lunar and Planetary Conf., pp. 3-4.

Alvarez, L.W., Alvarez, W., Asaro, F. and Mitchel, H.V., 1980. Extra-terrestrial cause for the Cretaceous-Tertiary extinction. Science 208, 1094-1108.

Asaro, F., Alvarez, L.W., Alvarez, W. and Mitchel, H.V., 1982. Geochemical anomalies near the Eocene/Oligocene and Permian/Triassic boundaries. In: Silver, L.T. and Schultz, P.H. (eds.), Geological Implications of Impacts of Large Asteroids and Comets on the Earth. Special Paper 190, Geological Society of America, 517-528.

Assereto, R., Bosellini, A., Fantini Sestini, N. and Sweet, W.C., 1973. The Permian-Triassic boundary in the southern Alps (Italy). In: Logan, A. and Hills, L.V. (eds.), The Permian and Triassic Systems and Their Mutual Boundary. Memoir 2, Canadian Society of Petroleum Geologists, 176-199.

Baud, A., Magaritz, M. and Holser, W.T., 1989. Permian-Triassic of the Tethys: carbon isotope studies. Geologische Rundschau 78, 649-677.

Benton, M., 1987. Mass extinctions among families of non-marine tetrapods: the data. Memoir. Soc. Geol. France No. 150, 21-32.

Benton, M.J., 1988. Mass extinctions in the fossil record of reptiles: paraphyly, patchiness and periodicity (?). In: Larwood, G.P. (ed.), Extinction and Survival in the Fossil Record. pp. 269-294. Oxford University Press, Oxford.

Campbell, I.H., Czmanske, G.K., Fedorenko, V.A., Hill, R.I. and Stepanov, V., 1992. Synchronism of the Siberian traps and the Permian-Triassic boundary. Science 258, 1760-1763.

Cassinis, G., (ed.), 1988. Proceedings of the Field Conference on: Permian and Permian-Triassic boundary in the South-Alpine segment of the Western Tethys, and additional regional reports. Memorie della Società Geologica Italiana 34, 1-366.

Claoue-Long, J.C., Zhang, Z.C., Ma, G.G. and Du, S.H., 1991. The age of the Permian-Triassic boundary. Earth and Planetary Science Letters 105, 182-190.

Clark, D.J., Wang, C.-Y., Orth, C.J. and Gilmore, J.S., 1986. Conodont survival and low iridium abundances across the Permian-Triassic boundary in South China. Science 233, 984-986.

Crowell, J.C., 1983. Ice ages recorded on Gondwanan continents. Geological Society of South Africa Transactions 86, 237-262.

Crowell, J.C., 1995. The ending of the Late Paleozoic Ice Age during the Permian Period. In: Scholle, P. (ed.), The Permian of the Northern Continents. Springer, Berlin Heidelberg New York.

Dickins, J.M., 1983. Permian to Triassic changes in life. Mem. Australasian Paleontols. 1, 297-303.

Dickins, J.M., 1985. Late Paleozoic glaciation. Journal of Geology and Geophysics 9, 163-169.

DiMichele, W.A. and Aronson, R.B., 1992. The Pennsylvanian-Permian vegetational transition: a terrestrial analogue to the onshore-offshore hypothesis. Evolution 46, 807-824.

Erwin, D.H., 1990. The end-Permian mass extinction. Annual Review of Ecology and Systematics 21, 69-91.

Erwin, D.H., 1993. The Great Dying: Life and Death in the Permian. 352 p. Columbia University Press, New York.

Erwin, D.H., 1994. The Permo-Triassic extinction. Nature 367, 231-236.

Erwin, D.H. and Vogel, T.A., 1992. Testing for causal relationships between large pyroclastic volcanic eruptions and mass extinctions. Geophysical Research Letters 19, 893-896.

Eshet, T., 1992. The palynological succession and palynological events in the Permo-Triassic boundary interval in Israel. In: Sweet, W.C., Yang, Z.Y., Dickins, J.M. and Yin, H.F. (eds.), Permo-Triassic Boundary Events in the Eastern Tethys. pp. 134-145. Cambridge University Press, Cambridge.

Ezaki, Y., 1994. Patterns and paleoenvironmental implications of end-Permian extinction of Rugosa in South China. Palaeogeography, Palaeoclimatology, Palaeoecology 107, 165-187.

Flügel, E. and Reinhardt, J., 1989. Uppermost Permian reefs in Skyros (Greece) and Sichuan (China): implications for the Late Permian extinction event. Palaios 4, 502-518.

Forney, G.G., 1975. Permo-Triassic sea level change. Journal of Geology 83, 773-779.

Frederiksen, N.O., 1972. The rise of the Mesophytic flora. Geoscience and Man 4, 17-28.

Gruszczynski, M., Halas, S., Hoffman, A. and Malkowski, K., 1989. A brachiopod calcite record of the oceanic carbon and oxygen isotope shifts at the Permian/Triassic transition. Nature 337, 64-68.

Gruszczynski, M., Hoffman, A., Malkowski, K., Zawidzka, K., Halas, S. and Zeng, Y., 1990. Carbon isotope drop across the Permian-Triassic boundary in SE Sichuan, China. N. Jb. Geol. Paläont. Mh 10, 600-606.

Hallam, A., 1989. The case for sea-level change as a dominant causal factor in mass extinction of marine invertebrates. Phil. Trans. Roy. Soc. B 235, 437-455.

Hallam, A., 1991. Why was there a delayed radiation after the end-Paleozoic extinctions? Historical Biology 5, 257-262.

Heller, F., Lowrie, W., Li, H.M. and Wang, J., 1988. Magnetostratigraphy of the Permo-Triassic boundary section at Shangsi (Guangyuan, Sichuan Province, China). Earth and Planetary Science Letters 88, 348-356.

Hoffman, A., Gruszczynski, M. and Malkowski, K., 1990. Oceanic $\partial^{13}C$ values as indicators of atmospheric oxygen depletion. Modern Geology 14, 211-221.

Holser, W.T. and Magaritz, M., 1987. Events near the Permian-Triassic boundary. Modern Geology 11, 155-180.

Holser, W.T. and Schönlaub, H.P., 1991. The Permian-Triassic boundary in the Carnic Alps of Austria (Gartnerkofel Region). Abhandlungen der Geologischen Bundesanstalt 45, 1-232.

Holser, W.T., Schönlaub, H.-P., Attrep, M., Jr., Boeckelmann, K., Klein, P. et al., 1989. A unique geochemical record at the Permian/Triassic boundary. Nature 337, 39-44.

Ingavat-Helmcke, R. and Helmcke, D., 1986. Permian fusulinacean faunas of Thailand - event controlled evolution. In: Walliser, O.H. (ed.), Global Bio-Events. pp. 240-248. Springer, Berlin Heidelberg New York.

King, G.M., 1991. Terrestrial tetrapods and the end-Permian event: a comparison of analyses. Historical Biology 5, 239-255.

Knoll, A.H., 1984. Patterns of extinction in the fossil record of vascular plants. In: Nitecki, M.H. (ed.), Extinctions. pp. 23-68. University of Chicago Press, Chicago.

Kvenvolden, K., 1988. Methane hydrate - a major reservoir of carbon in the shallow geosphere. Chemical Geology 71, 41-51.

Labandeira, C. and Sepkoski, J.J., Jr., 1993. Insect diversity in the fossil record. Science 261, 310-315.

Malkowski, K., Gruszczynski, M., Hoffman, A. and Halas, S., 1989. Oceanic stable isotope composition and a scenario for the Permo-Triassic crisis. Historical Biology 2, 289-309.

Maxwell, W.D., 1989. The end Permian mass extinction. In: Donovan, S.K. (ed.), Mass Extinctions: Processes and Evidence. pp. 152-173. Belhaven Press, London.

Maxwell, W.D., 1992. Permian and Early Triassic extinction of non-marine tetrapods. Palaeontology 35, 571-584.

Molostovskiy, E.A., 1992. Paleomagnetic stratigraphy of the Permian System. International

Geology Review 34, 1001-1007.

Nisbet, E.G., 1990. The end of the ice age. Canadian Journal of Earth Sciences 27, 148-157.

Olson, E.C., 1986. Problems of Permo-Triassic terrestrial vertebrate extinctions. Historical Biology 2, 17-35.

Orth, C.J., 1989. Geochemistry of the bio-event horizons. In: Donovan, S.K. (ed.), Mass Extinctions: Processes and Evidence. pp. 37-72. Belhaven Press, London.

Orth, C.J., Attrep, M., Jr., and Quintana, L.R., 1990. Iridium abundance patterns across bio-event horizons in the fossil record. In: Sharpton, V.L. and Ward, P.D. (eds.), Global Catastrophes in Earth History. pp. 45-60. Geological Society of America Special Paper 247, Boulder, CO: Geological Society of America.

Paull, C.K., Ussler, W., III, and Dillon, W.P., 1991. Is the extent of glaciation limited by marine gashydrates? Geophysical Research Letters 18, 432-434.

Pinto, J.P., Turco, R.P. and Toon, O.B., 1989. Self-limiting physical and chemical effects in volcanic eruption clouds. Journal of Geophysical Research 94, 11165-11174.

Raup, D.M., 1979. Size of the Permo-Triassic bottleneck and its evolutionary implications. Science 206, 217-218.

Raup, D.M. and Sepkoski, J.J., Jr., 1984. Periodicity of extinctions in the geologic past. Proceedings of the National Academy of Sciences, USA 81, 801-805.

Raup, D.M. and Sepkoski, J.J., Jr., 1986. Periodic extinction of families and genera. Science 231, 833-836.

Robinson, P.L., 1973. Palaeoclimatology and continental drift. In: Darlington, D.H. and Runcorn, S.K. (eds.), Implications of Continental Drift to the Earth Sciences. pp. 451-476. Academic Press, London.

Sadler, P., 1981. Sediment accumulation rates and the completeness of stratigraphic sections. Journal of Geology 89, 569-584.

Schäfer, P. and Fois-Erickson, E., 1986. Triassic bryozoa and the evolutionary crisis of Paleozoic stenolaemata. In: Walliser, O.H. (ed.), Global Bioevents. pp. 251-255. Springer, Berlin Heidelberg New York.

Sepkoski, J.J., Jr., 1984. A kinetic model of Phanerozoic taxonomic diversity. III. Post-Paleozoic families and mass extinctions. Paleobiology 10, 246-267.

Sepkoski, J.J., Jr., 1989. Periodicity in extinction and the problem of catastrophism in the history of life. Journal of the Geological Society of London 146, 7-19.

Sepkoski, J.J., Jr., 1992. A compendium of fossil marine animal families, 2d ed.; Milwaukee Public Museum Contributions in Biology and Geology 83, 1-155.

Signor, P.W., III, and Lipps, J.H., 1982. Sampling bias, gradual extinction patterns, and catastrophes in the fossil record. In: Silver, L.T. and Schultz, P.H. (eds.), Geological Implications of Impacts of Large Asteroids and Comets on Earth. Geol. Soc. Amer. Sp. Pap. 190, 291-296.

Spitzy, A. and Degens, E.T., 1985. Modeling stable isotopic fluctuations through geologic time. Mitteilungen Geologisch-Paläontologisches Institut Universität Hamburg 59, 155-166.

Stanley, S.M., 1988. Paleozoic mass extinctions: shared patterns suggest global cooling as a common cause. American Journal of Science 288, 334-352.

Steiner, M., Ogg, J., Zhang, Z. and Sun, S., 1989. The late Permian/Early Triassic magnetic polarity time scale and plate motions of South China. Journal of Geophysical Research 94, 7343-7363.

Sun, Y.Y., Chai, Z.F., Ma, S.L., Mao, W.Y., Xu, D.Y., Zhang, Q.W., Yang, Z.Z., Sheng, J.Z., Chen, C.Z., Rui, L., Liang, X.L. and Hi, J.W., 1984. The discovery of iridium anomaly in the Permian-Triassic boundary clay in Chagxing, Zhijing, China and its significance. In: Tu, G. (ed.), Developments in Geosciences. pp. 235-245. Beijing, Science Press.

Sweet, W.C., 1992. A conodont-based high-resolution biostratigraphy for the Permo-Triassic boundary interval. In: Sweet, W.C., Yang, Z.Y., Dickins, J.M. and Yin, H.F. (eds.), Permo-Triassic Boundary Events in the Eastern Tethys. pp. 120-133. Cambridge University Press, Cambridge.

Sweet, W.C., Yang, Z.Y., Dickins, J.M. and Yin, H.F., eds., 1992. Permo-Triassic Boundary Events in the Eastern Tethys. Cambridge University Press, Cambridge.

Teichert, C., 1990. The Permian-Triassic boundary revisited. In: Kauffman, E.G. and Walliser, O.H. (eds.), Extinction Events in Earth History. pp. 199-238. Springer, Berlin Heidelberg New York.

Traverse, A., 1988. Plant evolution dances to a different beat. Historical Biology 1, 277-302.

Visscher, H. and Brugman, W.A., 1988. The Permian-Triassic boundary in the Southern Alps: a palynological approach. Memorie della Societa Geologica Italiana 34, 121-128.

Wignall, P.B. and Hallam, A., 1992. Anoxia as a cause of the Permian/Triassic extinction: facies evidence from northern Italy and the western United States. Palaeogeography, Palaeoclimatology, Paleoclimatology 48, 143-162.

Wignall, P.B. and Hallam, A., 1993. Greisbachian (earliest Triassic) paleoenvironmental changes in the Salt Range, Pakistan and southeast China and their bearing on the Permo-Triassic extinction. Palaeogeography, Palaeoclimatology, Paleoclimatology 102, 215-237.

Xu, D.Y., Ma, S.-L., Chai, Z.-F., Mao, X.-Y., Sun, Y.-Y. et al., 1985. Abundance variation of iridium and trace elements at the Permian/Triassic boundary at Shagsi in China. Nature 314, 154-156.

Xu, D.Y., Zhang, Q.W., Sun, Y.Y., Yan, Z., Chai, Z.F. and He, J.W., 1989. Astrogeological Events in China. Van Nostrand Reinhold, New York.

Xu, G., 1991. Stratigraphical time-correlation and mass extinction event near Permian-Triassic boundary in South China. Journal of China University of Geosciences 2, 36-46.

Xu, G. and Grant, R.E., 1994. Brachiopods near the Permian-Triassic boundary in South China. Smithsonian Contributions to Paleobiology 76, 1-68.

Yang, Z.Y. and Yin, H.F., 1987. Achievements in the study of Permo-Triassic events in South China. Advances in Science of China, Earth Sciences 2, 23-43.

Yin, Y.S., Chai, Z., Ma, S., Mao, Z., Xu, D. et al., 1984. The discovery of iridium anomaly in the Permian-Triassic boundary clay in Changxing, Zhejiang, China and its significance. pp. 235-245. In: Developments in Geosciences: Contributions to 27th Annual International Geological Congress, Moscow Academica Sinica, Beijing.

Yin, H.F., Xu, G.R. and Ding, M.H., 1984. Palaeozoic-Mesozoic alternation of marine biota in South China. Scientific Papers on Geology for International Exchange - Prepared for the 27th International Geological Congress, 195-207.

Yin, H.F., Huang, S., Zhang, K.X., Yang, F.Q., Ding, M.H., Ziamei, B.I. and Suzian, Z., 1989. Volcanism at the Permian-Triassic boundary in South China and its effects on mass extinction. Acta Geologica Sinica 2, 417-431.

Yin, H.F., Huang, S., Zhang, K.X., Hansen, H.J., Yang, F.Q., Ding, M.H. and Die, X.M., 1992. The effects of volcanism on the Permo-Triassic mass extinction in South China. In: Sweet, W.C., Yang, Z.Y., Dickins, J.M. and Yin, H.F. (eds.), Permo-Triassic Boundary Events in the Eastern Tethys. pp. 146-157. Cambridge University Press, Cambridge.

Zhou, L. and Kyte, F.T., 1988. The Permian-Triassic boundary event: a geochemical study of three Chinese sections. Earth and Planetary Sciences 90, 411-421.

Manuscript received March 1993
Revision received July 1994

Author's address:

Douglas H. Erwin, Department of Paleobiology, National Museum of Natural History, NHB-121, Smithsonian Institution, Washington D.C. 20560, U.S.A.

Major Bio-Events in the Triassic and Jurassic

Anthony HALLAM

Abstract: The term bio-event has usually been understood in terms of mass extinctions, and this is the sense in which it is used here. Numerous bio-events could be recognised if attention were restricted to the ammonites, the fossil group with the highest rate of turnover in time, but to qualify as major events it is necessary to show that other groups were affected also. Because of this requirement only a few major events are recognisable.

For the Triassic there are four events, two of which affected both marine and terrestrial organisms. Early in the period, the Smithian-Spathian boundary marks a severe extinction phase for ammonites, conodonts and bivalves. The mid-Carnian event is important for a number of marine invertebrate groups, especially in the Alpine region, but only for ammonites is its global significance fully established. The late Carnian event affected some marine organisms but more particularly it has been claimed to be associated with a profound turnover in terrestrial tetrapod faunas, but this claim remains controversial. The end-Triassic event is one of the five biggest for marine organisms in the whole Phanerozoic, and there were also notable extinctions among terrestrial tetrapods and plants.

Within the Jurassic only two major events are discernible, both of which are most clearly manifested in Europe, with neither being of established global significance. The early Toarcian event profoundly affected all marine benthos and nektobenthos. The Tithonian event is less striking and clear-cut in time, but is important for some marine groups.

With regard to possible causal factors, sea-level change, on either regional or global scale, seems to be implicated in most if not all of the events, because of loss of habitat area in epicontinental seas due either to regression or to the spread of anoxic bottom waters during the succeeding transgression. Anoxia is most obvious for the Toarcian event but was probably important also for the Triassic-Jurassic boundary event. Climatic change may have been important for the late Carnian event, but there is no satisfactory evidence supporting either volcanism or bolide impact as causal factors.

Contents

Introduction	265
Triassic	268
Smithian-Spathian Boundary	268
Mid-Carnian	269
Late Carnian	269
End-Triassic	270
Jurassic	271
Major Faunal Groups	271
Jurassic Events	275
Possible Causes of the Extinction Events	276
Sea-Level Change	276
Climate	278
Volcanism	278
Bolide Impact	278

Introduction

By the beginning of the Triassic the supercontinent Pangaea had been assembled, comprising a northerly (Laurasia) and southerly (Gondwanaland) component

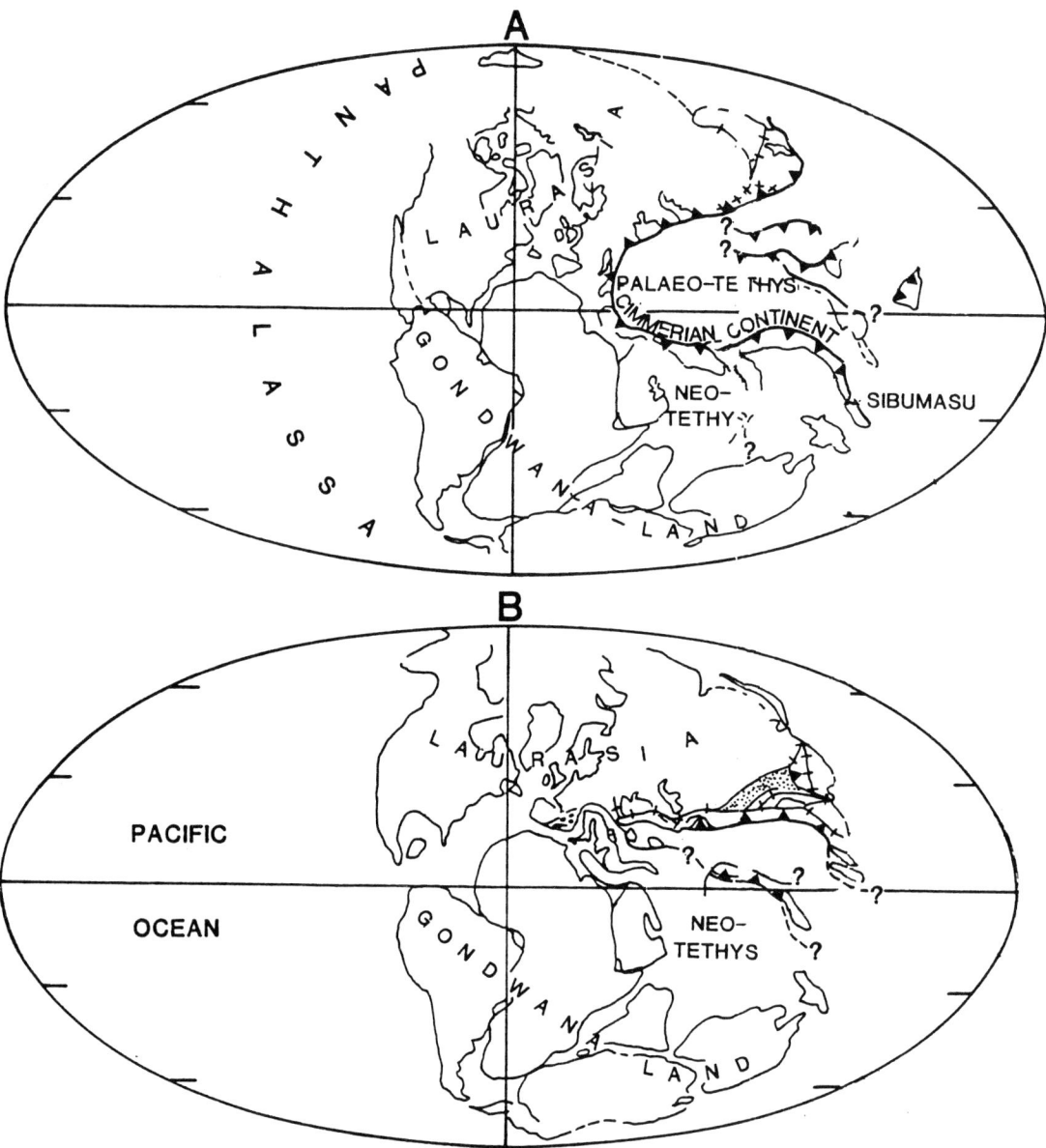

Fig. 1. Sketch diagrams to illustrate the collision of the Cimmerian continent with Eurasia in the late Jurassic, with concomitant destruction of Palaeo-Tethys. A, Early Triassic; B, Late Jurassic. Simplified from Sengör et al. (1988)

separated by the Palaeotethys, which contained a number of large islands in its easterly part (Fig. 1A). In late Triassic time rifting began in the region of what is now eastern Northern America and northwestern Africa, leading eventually to seafloor spreading and the opening of the central sector of the Atlantic Ocean, from mid-Jurassic times onwards. Further east, an elongate Cimmerian continent split off from Gondwanaland, with the opening of Neotethys, and eventually collided with eastern Laurasia in the late Jurassic, leading to the disappearance of Palaeotethys (Fig. 1B). The Indian Ocean also began to be created as the eastern part of Gondwanaland began disintegrating. The climate of Triassic-Jurassic times was substantially more equable than today, in the sense that there were no polar ice caps and a much lower latitudinal temperature gradient, but seasonal extremes of temperature on the large continental masses must have been

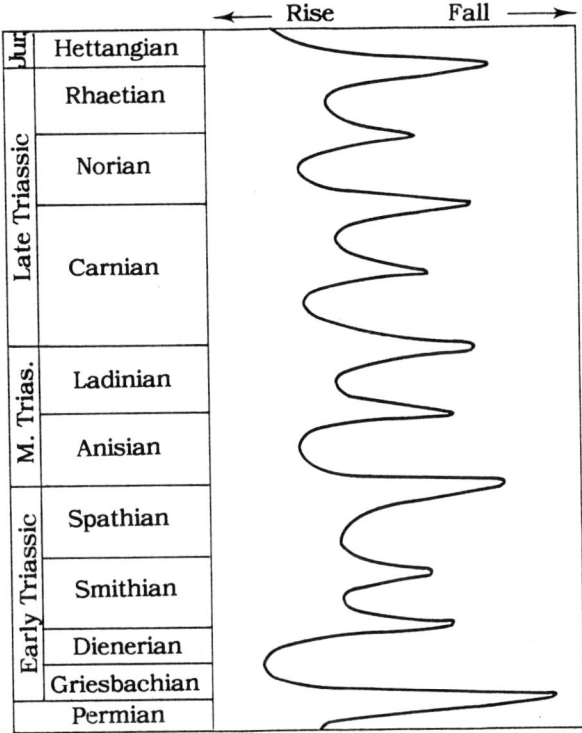

Fig. 2. List of Triassic stages, with Embry's (1988) sea-level curve, based on excellent data from Arctic Canada. This is thought to give a better indication of Triassic eustasy than the curve of Haq et al. (1987)

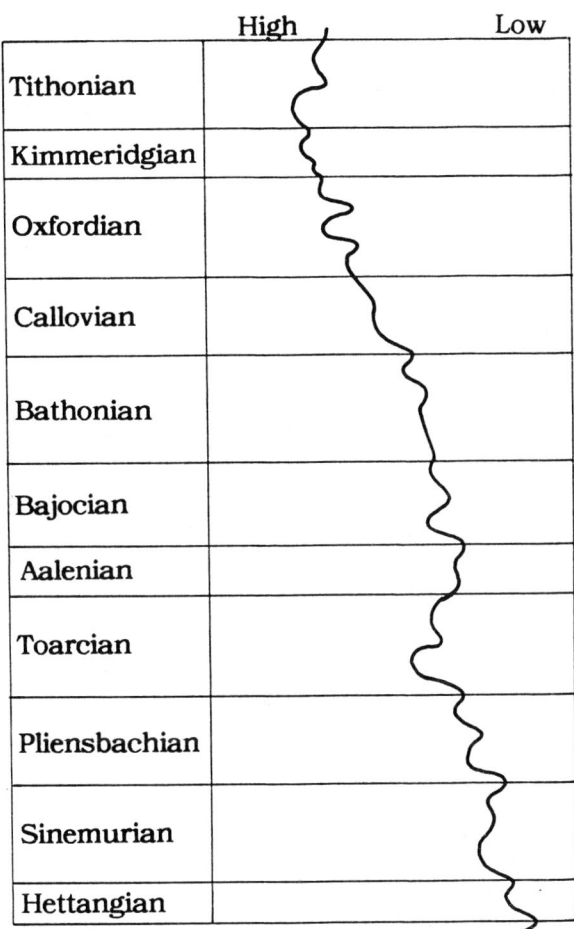

Fig. 3. List of Jurassic stages, with Hallam's (1988) sea-level curve, modified at the Callovian-Oxfordian boundary from the work of Norris (1993)

considerable, and continental interiors in low latitudes were very arid (Crowley and North, 1991; Frakes, 1979; Frakes et al., 1992; Hallam, 1985, 1993). Global sea level stood very low at the end of the Palaeozoic. There was a major rise at the beginning of the Triassic, leading to widespread flooding of continental margins. Thereafter the general trend was of rise to a Triassic maximum early in the latter part of the period, followed by a fall to a new minimum at the end. The general trend during the Jurassic was of rise of sea level from a Hettangian low to a Kimmeridgian-early Tithonian maximum. A number of shorter-phase cycles can be recognised superimposed on these general trends (Hallam, 1992 and Figs. 2 and 3).

The term bio-events, as used in recent literature, refers to major biotic turnover in geologically brief episodes of time. It has usually related to mass extinctions but could also involve migrations and radiations. These latter two phenomena could follow on extinction, being the consequence of the cessation of pre-emptive competition after niche vacation (Hallam, 1990b). Generally extinctions are more clear-cut and short-term than radiations, and migrations are of necessity of no more than regional significance. Therefore the main focus in this chapter will be on extinction events.

The next important matter to address is whether the extinction events are global or merely regional in extent. While it is obvious that the major episodes of mass extinction are global there are only a few of these. There are many other lesser events, however, that may qualify as mass extinctions for some groups, if they relate to episodes of high biotic turnover, but which may be only regional phenomena. On the other hand, if a fossil group finally disappears from the stratigraphic record, it may reflect a genuinely global event even if the evidence is not clear in all parts of the world, because otherwise it might have been expected to return from regional refugia when environmental conditions ameliorated.

In attempting to draw up a list of events for the Triassic and Jurassic we must face up to the fact of lack of sufficient evidence in many parts of the world. Compared, for instance, to the Cretaceous and much of the Palaeozoic, epicontinental seas were relatively restricted for long periods of time, and good marine sections in more than one continent are not abundant. For the sort of detailed studies now required, there are not many well researched marine sections across the world, and even fewer well dated non-marine sections. For the Jurassic, there is a heavy reliance on European, and for the Triassic on European and North American data.

For such reasons it cannot yet be conclusively established in some cases how many events are genuinely global. They include only one of the "big five" Phanerozoic mass extinctions, at the end of the Triassic. Only two others are included in Sepkoski's (1986) list of 29 Phanerozoic marine extinction events, based on a compilation of generic data; these are in the Pliensbachian and Tithonian. A third possible event, in the Carnian, also marked by a peak in generic extinctions, is dismissed by Sepkoski as a probable artifact, due perhaps both to high turnover and backward smearing of the extinction record from the end-Triassic event.

At the other extreme, numerous "bio-events" can be discerned among the ammonites, whose high rate of turnover compared with other fossil groups provides the basis for early Mesozoic stratigraphy. Ammonite species and genera define zones and stages often in a very clear-cut way, which involved complete extinction of the older taxa. Because of provinciality, however, many if not most such extinctions were only of regional extent, and ammonite extinctions were for the most part not reflected in extinctions of other groups of marine organisms. Therefore, if the ammonite extinctions were the result of environmental perturbations, such perturbations could not have been of great significance for the marine biota as a whole. In this article only major bio-events, affecting more than just ammonites, will be considered.

Treatment of the Triassic and Jurassic will be somewhat different. Knowledge is still sparse on the stratigraphic distribution of organisms in the Triassic, whereas there is a much fuller knowledge for the Jurassic, at least in Europe. Therefore information for the Triassic is relatively more tentative and provisional, and attention is directed immediately to the most likely events. For the Jurassic, there are enough data to allow initially a discussion of a number of important fossil groups, utilising the evaluation of experts, before paying attention to events. In the final part of the chapter possible causes are briefly discussed.

Triassic

There has been dispute about whether the Rhaetian should be recognised as a separate stage; it is here treated as the youngest substage of the Norian.

Ammonites and conodonts are the best studied groups from a stratigraphic point of view (Tozer, 1981a, b; Clark, 1987). Because of their evident stenotopy ammonite turnover should reflect any notable event. According to E.T. Tozer (pers. comm.) four big events are discernible:

1) Smithian-Spathian boundary: This was marked by a nearly total extinction followed by a major radiation.

2) Lower-Upper Carnian: This marked a fairly large turnover; the earliest Upper Carnian faunas are not very well known but they are much different from the Ladinian.

3) Middle-Upper Norian: A large extinction comparable to 2), followed by radiation of mostly small forms, for example heteromorphs. The very large leiostracans crossed this boundary and persisted probably to the end of the period.

4) Norian-Hettangian (=Triassic-Jurassic boundary): A nearly complete extinction took place, with only phylloceratids continuing into the Jurassic.

Ammonite non-events include the Ladinian-Carnian and Carnian-Norian boundaries.

Smithian-Spathian Boundary

Besides the major extinction event noted above for the ammonites, there was evidently a major event also affecting the conodonts. The biggest Triassic conodont diversity was in the Smithian, with a big drop to almost nothing at the Smithian-Spathian boundary. There was indeed a species diversity reduction at each of the stage boundaries of the Lower Triassic (Griesbachian, Dienerian, Smithian, Spathian). After this subperiod conodont diversity declined, with final extinction at the end of the Triassic (Clark, 1987). Relevant information is available for one other group, the Bivalvia. *Claraia* and *Eumorphotis* are the two most important Lower Triassic bivalve genera. *Claraia* is abundant in the Griesbachian and attained its acme in the Dienerian. Many species vanished at the top of the Dienerian or in the Lower Smithian. The genus is missing from the Spathian, except for the doubtful occurrence of one species in small quantities. *Eumorphotis* reached its acme in the Dienerian and Lower Smithian; nearly half the species disappear at the Smithian-Spathian boundary (Yin, 1990).

With regard to non-marine biota, Benton (1986) records a major extinction peak for tetrapods at the Lower-Middle Triassic boundary. The lack of stratigraphic precision for non-marine strata is such that this hitherto unremarked event could well correspond to the slightly older Smithian-Spathian boundary event in the marine realm. According to Balme (1970) the available palynological evidence in Europe suggests that the end of the Lower Triassic marks a more critical floral change than the beginning of the Mesozoic.

Mid-Carnian

As noted above, the Lower-Upper Carnian boundary marks the time of an important extinction among the ammonites. Though the stratigraphic data are no more precise than mid-Carnian there is evidence of an important phase of extinction among several other marine groups; further research may enable this to be pinned down more precisely to the Lower-Upper Carnian boundary.

Thus, among Triassic crinoids the Encrinidae are the most conspicuous elements of Anisian to Lower Carnian faunas. They reached a peak of diversity in the early Carnian but disappeared completely by the mid-Carnian (Johnson and Simms, 1989; Simms, 1991). Similar patterns of an abrupt decline in diversity following an early Carnian peak have also been documented among echinoids (Smith, 1990) and scallops (Johnson and Simms, 1989). However, while it is true that the bivalve species of the Lower Carnian Cassian Formation are generally very different from those of the Upper Norian Kössen Formation in the same region (the Alps), Newton et al. (1987) have found that a large number of bivalve species in the Lower Norian of Oregon are also present in the Cassian Formation, suggesting that their extinction was of regional not global significance.

Other groups are more poorly documented. Bryozoans appear to show a pattern similar to that of crinoids and the other groups cited in that, from a peak in the early Carnian, the group declined from 22 to 13 species in the late Carnian (Schäfer and Fois, 1987). However, gastropods appear to have experienced a steady increase in extinction rate from Ladinian to Norian times without any significant peak in the Carnian (Erwin, 1990). There is clearly a need for more precise data and for more groups, at the level of substage rather than stage. With regard to reef taxa, there is some evidence for a lack of synchroneity in change up the stratigraphic succession. One of the dominant reef builders, *Tubiphytes*, underwent a major decline in the mid-Carnian whereas calcisponges, bryozoans, and rhodophyte and chlorophyte algae declined in the late Carnian, roughly synchronous with the diversification of foraminifera, tabulozoans, spongimorphs and corals (Stanley, 1988, Fig. 3). Stanley considers that this major mid- to late Carnian turnover in reef biota was global in extent. In detailed work on the Newark Supergroup of eastern North America, Cornet and Olsen (1990) identified an episode of elevated palynofloral turnover near or at the end of the mid-Carnian.

Late Carnian

Benton (1986) was the first to point out the lack of synchroneity between the marine and non-marine extinction events in the Carnian, with a mid-Carnian event for the marine invertebrates and a late Carnian event for terrestrial vertebrates. It should be noted, however, that, besides the more equivocal record of reef biota, four out of six families of marine tetrapods became extinct at the end of the Carnian (Benton, 1988). There has been some dispute about the relative importance of the late Carnian and end-Triassic turnover in terrestrial vertebrates, but Benton's most recent review (1991) makes an attempt to establish that the earlier event was much the more significant. According to him, nine diverse families of tetrapods (thirteen, including families with only one species) died out during the late Carnian, compared to six in the latest Triassic. At species level the late Carnian event far exceeded the end-Triassic event in magnitude. Many of the tetrapod families disappearing were relatively diverse, averaging four species each, while the six families that disappeared at the end of the Triassic had a mean terminal diversity of 1.3 species each. The post-Carnian diversity reduction was almost certainly the result of depauperate faunas, not poor preservation or inadequate collecting. Terrestrial plants also show a peak in extinctions at the Carnian-Norian boundary (Boulter et al., 1988). Olsen and Sues (1986) also record a late Carnian extinction rate peak for pollen and spores in the Newark Supergroup.

Benton's views have been challenged by other terrestrial vertebrate researchers. Weems (1992) recognises no extinction event at the end of the Carnian, and nor do Hunt and Lucas (1992). Hunt and Lucas point out that the claim of an extinction event at this time is based on areas where there is a major facies change at the Carnian-Norian boundary, as in Germany, or a significant unconformity that spans the early Norian, as in Argentina. The Chinle Group of the western United States exhibits, however, a relatively continuous depositional and fossil record across the boundary, with no major facies change; no significant extinction event is recorded. A similar pattern is true in India, though the data were somewhat poorer.

End-Triassic

The mass extinction in the marine realm at the end of the Triassic is generally recognised as being one of the five biggest in the Phanerozoic (Hallam, 1990a). In their analysis of extinction rate (families/Ma) Raup and Sepkoski (1982) place it fourth in importance behind the end-Permian, end-Ordovician and end-Cretaceous mass extinctions. Using a different extinction metric, percent extinction of genera per million years, the end-Triassic peak is higher than the end-Cretaceous (Raup and Sepkoski, 1988).

The global mass extinction event is seen most clearly with the two most abundant and diverse megafossil groups, the ammonites and bivalves. No fewer than six ammonite superfamilies became extinct either at or shortly before the boundary, after diversity had reached a high level in the Norian; possibly no more than one genus survived the boundary (Tozer, 1981a). The bivalves do not show any significant change at family level, but nearly half the genera disappeared, and in Europe, the only continent for which there is an adequate record, nearly all the species became extinct (Hallam, 1981).

For at least three groups, taxa dominant in the Palaeozoic that survived the end-Palaeozoic mass extinction finally succumbed at the end of the Triassic. These include the last of the conodonts and murchisoniacean gastropods; brachiopods underwent a severe reduction of athyridaceans, spiriferoids and dielasmatids. According to Batten (1973) the gastropod turnover was much more significant than that at the end of the Permian. Reef communities underwent a major episode of mass extinction that affected both the scleractinian corals and calcisponges (Stanley, 1988). On the other hand, the main extinction event among echinoids and crinoids had taken place in the Carnian and there is little notable change in the low-diversity faunas across the Triassic-Jurassic boundary (Smith, 1990, Simms, 1990).

With regard to marine micro-organisms, there is no satisfactory information on a global scale about the foraminifera, but it is clear that ostracods underwent a major mass extinction, with the virtual disappearance of the Palaeocopida and severe reductions in the Cytheracea and Bairdacea (Whatley, 1988). Dinoflagellates underwent a major episode of extinction (J.B. Riding, pers. comm.; see his text below) and coccoliths experienced an almost complete turnover in the Austrian Alps, the only region for which data are available; only one species survived (Bown and Lord, 1990).

The all-important question must be asked - was the extinction a catastrophic event at the very end of the period, or was it preceded by a gradual decline in diversity? Even where relevant data are abundant, as for the end-Cretaceous extinctions, this has proved to be a difficult question to answer, because of sampling problems. In consequence there is as yet no consensual agreement. In the case of the end-Triassic event a decisive answer is even less possible because of the severe sampling problems posed by the limited number of satisfactory sections, but there are a few pointers. In a comprehensive review of bivalve extinctions Hallam (1981) came to the conclusion that a final catastrophic event had been preceded by some diversity decline subsequent to the early Norian, though some of this latter phenomenon was perhaps attributable to the progressive reduction in volume of the appropriate facies. On the other hand, there is no hint of any temporal diversity reduction in the spectacular reef assemblages in the Austrian Alps until their sudden disappearance at the end of the Triassic (E. Flügel, pers. comm.). Dr. C.R. Newton has supplied the following detailed information about Norian Cordilleran terrane bivalves in North America.

For several terranes - Wallowa, Wrangellia, Alexander - there are now nice silicified bivalve faunas from both early Norian (mainly kerri) and late Norian intervals. Curiously enough, in each of these terranes the late Norian faunas are much less diverse than their early Norian counterparts, even though ecologically the silicified faunas seem congruent. In general, there is a reduction of within-habitat species richness from 35-37 species in the kerri zone faunas to 19-23 species in the late Norian assemblages. This amounts to an approximate 40% reduction in overall species richness, although the precise percentages vary somewhat between terranes. In no case have I found a late Norian silicified assemblage whose species richness exceeds that of ecologically congruent early Norian faunas from the same terrane. This consistent pattern suggests a within-Norian loss of species richness among bivalves. I am not certain whether this is a progressive loss throughout the intervening portions of the Norian, or whether the loss occurred catastrophically, or in steps. The reduction in diversity is clear, but many questions about patterns and mechanisms remain!

Turning to the terrestrial realm, eastern North America is the only region where there is detailed documentation of the pattern of tetrapod change across the Triassic-Jurassic boundary. On the basis of data from both skeletons and footprints, Olsen has claimed that at least ten families became extinct at the end of the Norian, the most striking event in the whole Carnian-Sinemurian sequence of the Newark Supergroup, probably lasting for several hundred thousand years (Olsen et al., 1987; Olsen et al., 1990). The contemporary extinction event in the palynoflora allows a reasonably precise correlation with the extinction event in the marine record. As already noted, Benton (1991) has played down the importance of the vertebrate extinction event, affecting dinosaurs and other groups, pointing out that a much more significant turnover took place between the Carnian and the Norian. Weems (1992) also denies the existence of a catastrophic extinction event among terrestrial vertebrates at this time.

It has been stated that the terrestrial plant record is less affected by mass extinction events than the marine invertebrate record (Knoll, 1984), specifically that of the Triassic-Jurassic boundary by Ash (1986) and Traverse (1988); neither megafossils nor palynomorphs show much change across the boundary. (Many of the changes involved seed ferns, with several families becoming extinct at or near the boundary). Close examination of certain classic sections suggests, however, that there might have been a significant extinction event. Thus in East Greenland it has been well established for many years that the "Rhaetian" *Lepidopteris* and Hettangian *Thaumatopteris* megafloral zones are quite distinct from each other. According to Pedersen and Lund (1980), there is a very low percentage of seven species common to the two zones, whereas the spores and pollen have many species in common. They suggested that some indistinguishable palynomorphs have been produced by different megafossil plants, which implies that palynologists are likely to underestimate the degree of floral turnover at a given boundary. It is therefore noteworthy that, although the change in pollen and spores through the late Triassic to early Jurassic interval in Europe is not especially marked, the biggest "species" turnover takes place exactly at the Triassic-Jurassic boundary (Visscher and Brugman, 1981; Traverse, 1988). Even more striking is the pronounced palynomorph turnover at this horizon in the Newark Supergroup (Olsen and Sues, 1986; Olsen et al., 1990).

Jurassic

Major Faunal Groups

As noted earlier, it is desirable firstly to consider the evidence of turnover for a number of important marine groups for which a substantial amount of information on stratigraphic distribution is available. For all groups information from Europe is the most precise, and for the microfossils there are few published data from anywhere else. There is very little information available on the stratigraphic distribution through the Jurassic of non-marine organisms; the tetrapod record being particularly poor.

A m m o n i t e s.   The high turnover exhibited by the ammonites can be interpreted if one wishes as a series of extinction and radiation events within this group. It has long been recognised in Europe, since the work of Klüpfel (1917), that there is a relationship between sedimentary cycles related to varying depth of sea and the evolution of ammonites. Shallowing episodes, corresponding elsewhere with marine regression, appear to correlate with extinction of taxa and deepening/transgressive episodes with radiation of the survivors, and/or immigration of new groups from other regions (Hallam, 1987). A good example from the Lower Jurassic is the extinction of the boreal family Amaltheidae at the end of the Pliensbachian, and the earliest Toarcian immigration from the Tethyan Realm and subsequent radiation of the Hildoceratidae and Dactylioceratidae (Hallam, 1991).

Two Jurassic ammonite specialists, whose research has been concentrated on the Middle Jurassic, have been consulted for their views on bio-events. Phylogenetic analysis of Middle Jurassic ammonites by Professor J.H. Callomon indicates a net increase in the average overall phyletic diversity from the Aalenian to the Callovian which appears to correlate with the average rate of eustatic sea-level rise during the Jurassic. He perceives, however, no correlation between higher order eustatic fluctuations as published by Haq et al. (1987) and Hallam (1988) and the turnover of ammonite genera and species on a global scale. The data are sufficiently extensive for one to be able to say that the absence of such a correlation is statistically probably significant.

Professor G.E.G. Westermann has supplied the following text.

The majority of well established Jurassic ammonite zones (i.e. standard zones) are separated by regional changes of the entire assemblage thus comprising low-level events. The dominant species do not belong to continuous clades, but to immigrant clades (Westermann, 1988). Abiotic factors presumably caused "instant" shifts in the relative abundance and/or geographic distribution of entire or partial faunas between the zones. In a number of cases, range overlap can be demonstrated, at least in single basins of provinces, so that displacive competition is indicated. With the onset of strong bioprovincialism in the Later Bajocian, however, such faunal turnovers or bio-events are much more frequent in the low-latitude Tethyan Realm than in the high-latitude Boreal Realm. The latter is of low diversity, usually with clades dominating long successions of zones (Callomon, 1985), and bio-events are recognised only at the onset and termination of clades (e.g. Kossmoceratidae and Cardioceratidae). In the Tethyan Realm, which is much larger and subdivided into many more provinces, the much more diverse faunas expand and contract extensively, resulting in more numerous bio-events as outlined above. This contrast is reflected in a simple statistic: Post-Bajocian subzonal index species of the Boreal Realm belong to only 28 genera; those of a similar number of Tethyan subzones to 49 genera (Westermann, 1988). Boreal clades also tend to be longer than Tethyan clades presumably resulting from lower competition.

Within the Middle Jurassic, the only cosmopolitan ammonite events recognised are (1) around the Aalenian/Bajocian stage boundary and (2) around the Lower/Upper Bajocian substage boundary. In the first event Phymatoceratidae and Graphoceratidae become extinct (the latter restricted to west-Tethyan and south-Asian areas); and Sonniniidae, Strigoceratidae, Oppeliidae and Lissoceratidae arise, the former three having evolved from Hammatoceratidae (which survive). The Sonniniidae arise in the low latitudes of the eastern Pacific and expand eastward through the Hispanic Corridor, the cratonic, proto-Central Atlantic seaway leading to western Tethys. At this time the eastern Pacific margin has several species in common with the Submediterranean Province, indicating that the Hispanic Corridor was open at this time (although the eustatic curve shows a lowstand). In the Boreal Realm, which at this time is essentially restricted to the northernmost Pacific and adjacent, mostly cratonic areas in the north, this event may be seen in the extinction of the endemic Hammotoceratidae, but no new taxa evolve here.

The mid-Bajocian event is apparently global. It is preceded by the extremely uniform fauna of the Humphriesianum Standard Zone in the pan-tropical Tethyan Realm, with similar species of stephanoceratids and sonniniids in the eastern Pacific, Mediterranean, northwest Europe, southwest Asia and even New Guinea. The Hispanic Corridor, still a cratonic seaway connecting eastern Pacific and western Tethys, was again active in faunal exchange (Westermann, 1981). But there is no record of coeval, high-latitude faunas. At the substage boundary the sonniniids became extinct, and the perisphinctaceans and spiroceratids appeared throughout the pan-tropical Tethyan Realm. Both are cryptogenic, but the former seem to have earlier Andean ancestors and immigrated into Tethys (*Praeleptosphinctes*). The consistent range overlap of stephanoceratids again indicates displacive competition. Along the eastern Pacific margin, however, a very distinctive fauna developed now, the Eurycephalitinae, later accompanied by the Neuqueniceratinae, also endemic. These faunas, extending to the Early Callovian, are so distinct that an East-Pacific Subrealm has been recognised (Westermann, 1981). By inference, the Hispanic Corridor was essentially closed during this time interval. Coeval with these low-latitude events, the Boreal Realm became clearly separated, with the appearance of the long-ranging Cardioceratidae, later accompanied by the Kossmoceratidae (Callomon, 1985). Significantly, the Boreal event is accompanied by expansion of the Arctic seas (largely cratonic) onto European platforms, including the reopening of the Norway-Greenland seas which had been closed since earliest mid-Jurassic times. It stands therefore to reason that both the basal Bajocian and the mid-Bajocian events are related to the opening and closing of cratonic seaways which influence the temperature, salinity, currents etc, of the major marine basins. Seaway activity was not synchronised and this is partly isostatically controlled. The influence of eustatic sea-level changes on the shelf seas may have been more important for single basins and provinces as discussed below. Other mid-Jurassic faunistic changes tend to be limited to single faunal provinces or even basins and these may more closely follow the usual model of transgression-related faunal expansion. Examples are the early Callovian dispersal of the Macrocephalitidae from the southwest Pacific area into the Mediterranean and European seas (Westermann and Callomon, 1988). But the beginning of the homogeneous perisphinctacean fauna in western Tethys and the eastern Pacific in mid-Callovian time is due to the reactivation of the Hispanic Corridor, which now opened by drifting to form the true Central Atlantic Ocean (Westermann, 1981).

This analysis, for about one third of Jurassic time, indicates how complex the relationship was between ammonite provinciality and evolutionary and migrationary events, and palaeogeographic changes. Nothing comparable has yet been recognised in other groups, except for bivalve migrations such as through the Hispanic Corridor (Hallam, 1987).

Unfortunately no such authoritative statements are available for other parts of the Jurassic, but it is important to address the question of ammonite extinctions in the Tithonian, since this has repeatedly been recognised as an extinction event by Raup and Sepkoski (1988). Sepkoski (1982) records that seven out of a total of eleven ammonite families became extinct in the Tithonian.

However, the literature recording more precise stratigraphic information reveals that only three of the seven, the Aspidoceratidae, Atoxioceratidae and Dorsoplanitidae, persisted to the end of the stage, and that among these only a minority of genera persisted. The duration of the Tithonian is poorly known but, whatever modern timescale is used, 5 Ma seems to be a reasonable minimum estimate. Thus what may appear superficially to be an important extinction event (7 families out of 11) turns out on closer analysis to be of no great significance compared with other extinction events in ammonite history. Indeed, the turnover of taxa at the Tithonian-Berriasian boundary is less marked than that at the Berriasian-Valanginian boundary both in Europe and South America (Hallam, 1986).

B i v a l v e s . The stage-by-stage diversity and turnover of bivalve genera and species in Europe has been studied by Hallam (1976). Diversity shows a sharp rise through the first three stages of the Jurassic, before falling in the Toarcian. Thereafter the diversity rises to a maximum in the Bathonian, after which it stays more or less stable, with no significant diminution. There are no notable extinction peaks among the genera and only one among the species, in the Pliensbachian. No comparable analysis is yet possible for other parts of the world, but some information is available from Chile. No comparably striking extinction peak is recognisable in the Pliensbachian, and distinctive genera important in the Sinemurian and Pliensbachian of the Western Cordilleras of the Americas, such as *Weyla* and *Lithiotis* (the correct name for "*Plicatostylus*") persisted into the mid-Toarcian. It has to be borne in mind, however, that the Upper Pliensbachian and Lower Toarcian is generally developed in facies poor in bivalves, so that any short-term extinction event might not be readily recognisable.

Because Hallam's (1976) study did not extend into the Cretaceous it was not possible to obtain data for extinctions in the youngest Jurassic stage, the Tithonian. It has subsequently become clear that there was a major Tithonian extinction at species level in Europe, but there is no indication of a contemporary extinction in the excellent and richly fossiliferous secions across the Jurassic-Cretaceous boundary in central Chile (Hallam, 1986; Hallam et al., 1986). Nor is a significant marine bio-event recognisable in Japan; many bivalve species survived from the Tithonian into the earliest Cretaceous (Hayami, 1989).

B r a c h i o p o d s . No one has attempted a detailed analysis of brachiopod turnover throughout the Jurassic. It is apparent, however, that there was a notable extinction event in the early Toarcian, with the disappearance of many familiar Lower and Middle Liassic taxa including the last of the spiriferids (Hallam, 1986). There was also a Tithonian extinction event among brachiopods in Europe (Sandy, 1988).

F o r a m i n i f e r a . The Liassic (or Lower Jurassic) forams in the Paris Basin exhibit a steady increase in species diversity from the Hettangian to the Pliensbachian, but with no notable change across stage boundaries. Many species disappeared after the Pliensbachian and the Toarcian marks an almost complete renewal of the fauna (Bizon and Oertli, 1961).

The stratigraphic distribution of Jurassic foram species in Great Britain is dealt with in Jenkins and Murray (1981). The Liassic picture is similar to that of northern France except that an episode of increased turnover is signalled at the Sinemurian-Pliensbachian boundary; this is thought to be true of the whole of north west Europe. However, consultation of the range chart (Fig. 6.1.1.) shows only two species going extinct at this boundary. The pronounced Liassic episode of extinction extends from the late Pliensbachian to the oldest (Tenuicostatum) Toarcian zone.

Turning to younger Jurassic faunas, there was a major influx of species in the earliest Bathonian. Of twenty-four Bathonian species (twelve in the Upper Bathonian) only seven persisted into the Callovian. There is little to no turnover of species at either the Callovian-Oxfordian or Oxfordian-Kimmeridgian boundaries, and diversity diminished considerably after the earliest Kimmeridgian, which is thought to be a consequence of unsuitable facies. In summary, most Jurassic foram species compare with the bivalves in being long-ranging, through several stages, and exhibiting only one significant extinction event within the period.

Ostracods. After the end-Triassic extinction event diversity was low in the Hettangian and increased in the next two stages (Whatley, 1988). Hettangian to Pliensbachian faunas were dominated by the suborder Metacopina, which went extinct immediately afterwards. The time of the drastic turnover can be located precisely at the Tenuicostatum-Falciferum Zone boundary, in the Lower Toarcian (Lord, 1974; Bate and Robinson, 1978). With one exception the mid- and late Jurassic were times of gradual rather than abrupt change (Bate and Robinson, 1978; Whatley, 1988). The Bathonian-Callovian boundary appears as a faunal break that few Aalenian-Bathonian genera and almost no species pass (Bate and Robinson, 1978). Whatley (1988) records an extinction peak in the Bathonian. This turnover is not reflected in other fossil groups apart from ammonites and warrants closer examination. In Great Britain the Upper Kimmeridgian to Portlandian faunas are much reduced, probably owing to facies restrictions, and an extinction peak as recorded by Whatley (1988) in the "Purbeckian", i.e. latest Jurassic to earliest Cretaceous.

Coccoliths. This group extends throughout the Jurassic, with an acme in the Upper Callovian and Lower Oxfordian, following a gradual diversity increase through the Liassic into the Middle Jurassic. No marked extinction interval is discernible. In particular, all nine Toarcian species continue right through the stage. Twelve new species enter the succession in the Upper Callovian-Lower Oxfordian, much the biggest increase concentrated in a limited time interval. Many species appear to continue into the Cretaceous but others disappear (Hamilton, 1982).

Dinoflagellates. This stratigraphically important group of microfossils has lacked a comprehensive modern coverage in the literature, and so Dr. Riding's supplied text is here presented in its entirety.

Jurassic Dinoflagellate Cysts*. Dinoflagellates are micro-organisms, the majority of which are unicellular, aquatic, marine and have a planktonic biflagellate stage in their life cycle. Most species are photosynthetic and form a significant element of the marine phytoplankton. Certain taxa form a resistant resting cyst which is the only potentially fossilisable stage of the dinoflagellate life cycle (see, for example, Evitt, 1985; Taylor, 1987). Physiological and cytological studies show that the dinoflagellates are a primitive eucaryotic group with probably Precambrian origins (Evitt, 1985). The fossil record begins with a stratigraphically and geographically isolated occurrence in the Late Silurian (Sarjeant, 1978). Then follows a gap of some 200 Ma before the continuous Late Triassic to Quaternary record begins. The dinoflagellate cyst record diversified rapidly throughout the Jurassic (Bujak and Williams, 1979).

Several factors affect the utility of dinoflagellate cysts in the assessment of global bio-events. Studies of modern dinoflagellates and their cysts reveal that only some 10% of modern species produce preservable (sporopollenin) resting cysts, and these co-exist with the remaining majority of non cyst-producers. Numbers of cyst species, therefore, are not related to the numbers of motile species (Evitt, 1981; 1985). Moreover, absolute numbers of cysts does not reflect the total dinoflagellate biomass. The proportions of cyst-producers to non cyst-producers may have varied greatly in the geological past; in the modern floras, this ratio also varies with latitude. Any statements regarding oceanic productivity and/or levels of photosynthetic activity based on either diversity or counts of fossil dinoflagellate cysts are, therefore, at best partially accurate or at worst profoundly spurious. Furthermore the record is known to be selective, best exemplified by *Ceratium* Schrank 1793; the many extant marine species of this distinctive genus do not produce resistant resting cysts (Evitt, 1985). Sporopollenin cysts clearly related to *Ceratium*, however, are present in sedimentary rocks of Late Jurassic to Cretaceous age (Wall and Evitt, 1975). The relationship between the cyst record and motile dinoflagellate populations is therefore difficult to quantify.

Norian/Rhaetian (Late Triassic) to Sinemurian (Early Jurassic) dinoflagellate cyst associations are normally of relatively low (<10) species diversity (Helby, Morgan and Partridge, 1987; Riding and Thomas, 1992; Woollam and Riding, 1983). A more diverse (c. 20 species) suite of Norian dinoflagellate cysts, however, was reported from Arctic Canada by Bujak and Fisher (1976). This association, the *Sverdrupiella* complex, is most diverse in northern hemisphere high latitudes and has a circum-Pacific distribution pattern (Helby, Wiggins and Wilson, 1987). In Britain, six of the eight recorded Rhaetian dinoflagellate cyst taxa have range tops at the Triassic/Jurassic boundary; *Beaumontella langii* (Wall, 1965) Below 1987 and *Dapcodinium priscum* Evitt 1961 are the two species which span this junction (Riding and Thomas, 1992). The record of cyst-producing dinoflagellates in Europe, therefore, shows a marked increase in extinctions at this system boundary paralleling many other fossil groups (e.g. Colbert, 1965). Dinoflagellate cyst diversity increased markedly during the late Pliensbachian and Toarcian. In the late Pliensbachian, a suite of forms including *Luehndea spinosa* Morgenroth 1970, *Mancodinium semitabulatum* Morgenroth 1970 and *Nannoceratopsis gracilis* Alberti 1961 emerged. This event is co-incident with Supercycle UAB-4 of Haq et al. (1987). A distinctive and diverse complex of small dinoflagellate cysts (the Parvocysta suite of Riding, 1984) is a numerically significant component of European Toarcian and earliest Aalenian associations.

The early Toarcian anoxic event, represented by the Jet Rock of Yorkshire and the Posidonienschiefer of Germany, severely affected the dinoflagellate cyst record (Loh et al., 1986; Wille, 1982; Wille and Gocht, 1979). In Germany, dinoflagellate cyst diversity was profoundly reduced in these bituminous strata, a situation mirrored by the benthos and nektobenthos. However, Wille (1982) discovered that not all dinoflagellate cyst species reacted in the same way to these anoxic conditions. The dinoflagellate cyst is, unlike the motile cellulosic thecate stage, a non-motile benthonic resting stage (Dale, 1983). It follows that, if the bottom waters are anoxic, successful excystment of the protoplasm will be precluded unless resuspended into oxygenated waters. The partially opened excystment apertures of cysts recovered from the Posidonienschiefer of south-west Germany support this argument as these forms appear to have been killed when attempting to excyst (Wille, 1982; Wille and Gocht, 1979). Wille (1982) contended that only those cysts which formed in shallow, non-anoxic waters had the potential to excyst and complete their life cycles. Loh et al. (1986) reported a dinoflagellate cyst "blackout" in the Posidonienschiefer of north-west Germany and suggested that, whatever the causal factors, they were more intense than in south-west Germany as Wille (1982) recorded a low diversity assemblage throughout these strata. It seems likely, therefore, that upon the onset of anoxic bottom conditions, dinoflagellate cysts could not complete their life cycles except those living in shallow, neritic areas. These included species such as *M. semitabulatum* and *N. gracilis*, which survived this facies (Wille, 1982, Fig. 2).

The Aalenian and Bajocian were important times in fossil dinoflagellate evolution as the gonyaulacacean lineage rapidly diversified. These forms experimented with excystment apertures (Riding, 1987). Dinoflagellate cyst diversity increased rapidly and at a relatively steady rate throughout the Mid Jurassic and into the Late Jurassic (Bujak and Williams, 1979, Fig. 6); with no major species turnover events (Woollam and Riding, 1983, Fig. 8). Following a fall in the late Bathonian, sea level rose throughout the Callovian to Kimmeridgian (Haq et al., 1987). The increase in shelf area may be the principal reason for the rise in Callovian to Kimmeridgian dinoflagellate cyst diversity (Wall et al., 1977). There is an apparent marked decrease in diversity during the Portlandian. The phenomenon is gradual and the trend reverses at the Jurassic-Cretaceous boundary (Bujak and Williams, 1979). This downward trend, and the parallel molluscan faunal crisis, may be related to eustatic fall at this time (Haq et al., 1987).

Although difficult to relate to productivity etc., the dinoflagellate cyst record in itself may contain valuable palaeoecological information (see, for example, Riding et al., 1985; 1991) and requires detailed comparison to other floral and faunal data. Dinoflagellate cysts are, therefore, more useful in palaeoecological studies than Evitt (1981, 1985) envisaged. Nevertheless, the selectivity and incompleteness of the dinoflagellate fossil record must always be considered during such studies.

J.B. Riding, Biostratigraphy & Sedimentology Group, British Geological Survey, Nottingham

* Published with the approval of the Director, British Geological Survey.

Jurassic Events

As has been made clear already, numerous marine bio-events can be recognised if attention is focussed on the exceptionally high rate of ammonite turnover, but unless they are indicated in the record of other groups they must be dismissed as minor events. Attention is here confined to what are evidently major events affecting the marine biota as a whole, and it should be apparent from what has been written above that these are very few.

1) E a r l y   T o a r c i a n. This event was first recognised by Hallam (1961). It evidently corresponds to the Pliensbachian extinction peak of Sepkoski (1986) and Raup and Sepkoski (1988) but careful examination strongly indicates that the main extinction in Europe corresponds closely to the widespread development of early Toarcian (Falciferum Zone) black shales. All groups of benthos were severely affected but this is not the case for planktonic groups such as the coccolithophorids and certain types of crinoid. Whereas the event in Europe was the most striking one within the whole period it is far from clear that it was global in extent. The evidence from South America suggests otherwise, and indeed many species occurring in that region already in the Hettangian to Pliens-

bachian migrated to Europe in the late Toarcian, presumably to occupy niches vacated as a consequence of the benthic extinction event (Hallam, 1986, 1987).

2) T i t h o n i a n. Raup and Sepkoski's Tithonian extinction peak appears to be a reality at least for Europe, as shown by bivalves and other benthic groups, but no benthic turnover is recognisable across the Tithonian-Berriasian boundary in South America (Hallam, 1986) nor at the stratigraphically equivalent horizon in the boreal realm, in northern Siberia (V. Zakharov, pers. comm.) and nowhere across the world is the change in ammonite faunas especially striking, in fact it is rather less than at a number of other Mesozoic stage boundaries. With regard to the non-marine record, about which much less is known, there was apparently no significant change in dinosaur faunas across the Jurassic-Cretaceous boundary in the classic sections of the North American Western Interior (S.G. Lucas, pers. comm.).

Possible Causes of the Extinction Events

Table 1 presents a list of the major bio-events, involving substantial extinctions, recognised for the Triassic and Jurassic, together with their main characteristics. Locations of the principal regions where the events have been recognised, or searched for, are given in Figs. 4 and 5. Further research may well allow the recognition of more events, but they are likely to be of relatively minor importance. Not enough is known yet about the duration of the major events, but there is reasonably good evidence that at least the end-Triassic and early Toarcian events had catastrophic components, that is, some if not most of the extinctions were accomplished in a very brief time span, perhaps no more than a few millennia.

The cause or causes of all mass extinction events is still the subject of active debate and no final answers are yet possible on the Triassic and Jurassic events. Nevertheless the number of possibilities for phenomena which affected extensive regions if not the whole world, is not limitless. In fact, judging from recent literature, there are only four serious contenders, sea-level change, with associated anoxia, climate, volcanism and bolide impact.

Sea-Level Change

Two types of sea-level change, which may be global (eustatic) or regional, have been invoked to account for mass extinctions of marine benthos and nektobenthos. Both rapid fall and rise, the latter associated with the spread of anoxic bottom waters, would have caused severe restriction of habitat area of epicontinental seas (Hallam, 1989, 1992) and restriction of habitat area is acknowledged to be the prime cause of increased extinction rate at the present as well as in the past (Eldredge, 1991).

The most clear-cut association between extinction and facies is that in the early Toarcian, marked by an extensive spread of black shales in Europe, deposited at a time of rapid global sea-level rise (Hallam, 1987). Carbon isotope data also establish this as an oceanic anoxic event comparable to that at the Cenomanian-Turonian boundary (Jenkyns, 1988). This phenomenon could well be implicated also in the end-Triassic extinction event, because the earliest Hettangian deposits laid down at a time of rapid sea-level rise are in dysaerobic or anaerobic facies over extensive regions (Hallam, 1981, 1990a). This transgressive event was

Table 1. Major bio-events in the Triassic and Jurassic.

Event	Main characteristics
Tithonian	An important event among marine invertebrates in Europe but not in South America. Evidence lacking from rest of world
Early Toarcian	Profoundly affected marine benthos in Europe, but not established yet as a major extinction event elsewhere in the world
End-Triassic	The most important extinction event within the two periods, drastically affecting most marine life. Also some terrestrial vertebrates and plants. Obviously global and probably catastrophic to some extent.
Late Carnian	Predominantly a terrestrial event, affecting plants and vertebrates though controversy persists about the latter. Also some marine groups, e.g. reef biota
Mid-Carnian	Extinction of some marine groups including crinoids, echinoids, scallops, bryozoans and some reef organisms. May be only regional in extent and confined to Europe.
Smithian-Spathian boundary	Extinction of ammonites, conodonts and bivalves

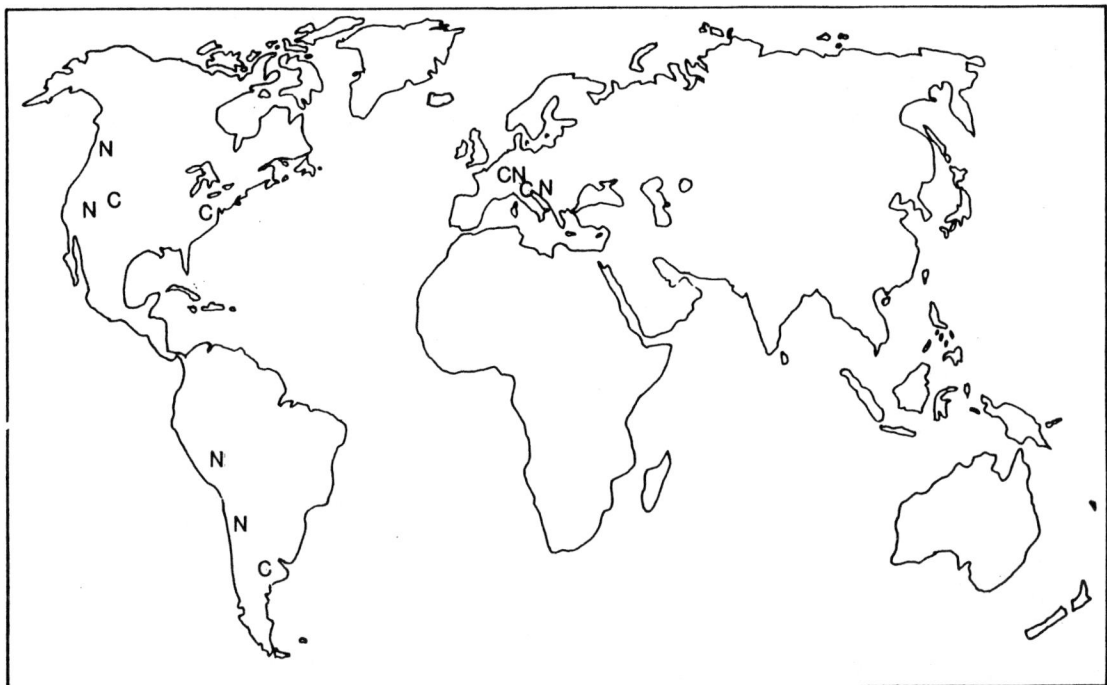

Fig. 4. Location of the principal regions where Triassic extinction events have been recognised or searched for.

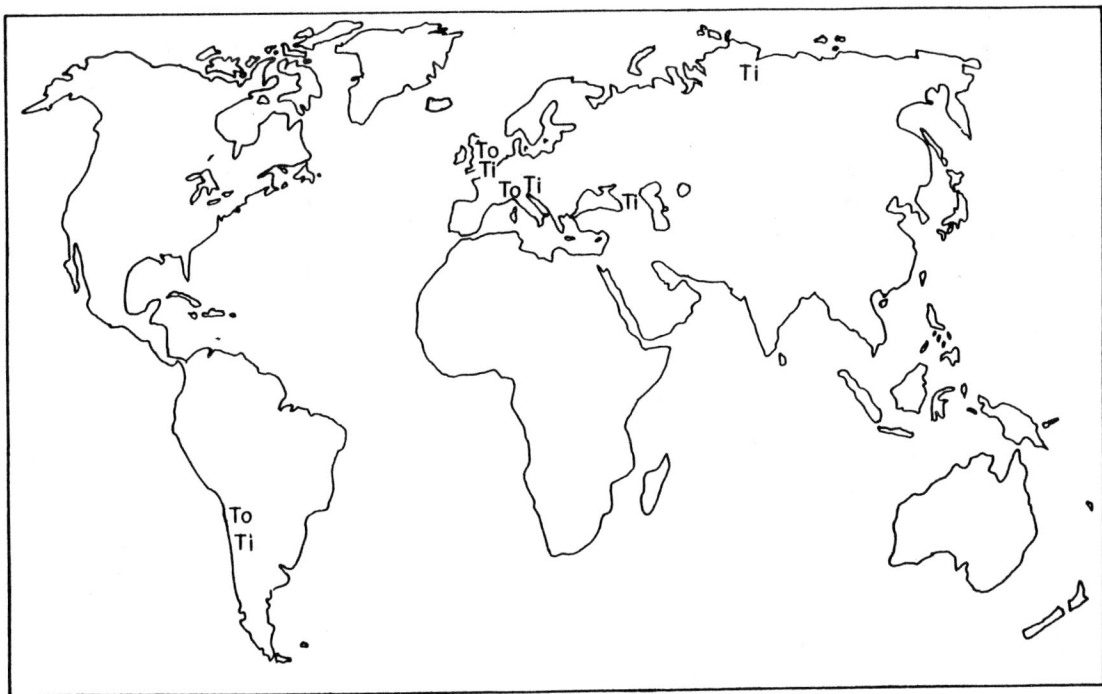

Fig. 5. Location of the principal regions where Jurassic extinction events have been recognised or searched for.

immediately preceded, however, by a rapid sea-level fall, and so regression could have been involved as well. The Triassic-Jurassic boundary regression-transgression couplet could possibly relate to a major tensional event causing rifting and volcanicity in the southern North Atlantic region (Hallam, 1990a; Cathles and Hallam, 1991).

The Tithonian extinction event in Europe corresponds geographically with an episode of regional regression. The lack of evidence either of regression or mass extinction across the Jurassic-Cretaceous boundary in South America strongly supports the interpretation of a causal relationship between regression and extinction (Hallam, 1986).

Regression may also be implicated in the two marine extinction events within the Triassic. The extinctions which led to such a contrast between the rich Carnian and Norian faunas of the Alps could relate to the regional regression marked by the non-marine Raibl Formation; in western North America, however, many Carnian species persist into the Norian. The Exxon eustatic curve for the Triassic (Haq et al., 1987) depicts two episodes of sea-level fall, early in the mid-Carnian and late Carnian. Embry's (1988) Triassic eustatic curve (Fig. 2), based on excellent sections in northern Canada, indicates a major fall in the mid-Carnian and an even greater one at the end of the stage. With regard to the Smithian-Spathian boundary extinction event, which has not been recognised before and demands much further research, the Exxon curve depicts little change across the stage boundary, but the Lower Triassic is one of the least satisfactory parts of their whole curve. Embry's curve indicates a sea-level fall in the late Smithian but greater falls at the Dienerian-Smithian and Spathian-Anisian boundaries, where no extinction events have been recognised.

Climate

Disappearance of the Alpine reef ecosystem at the end of the Triassic has been attributed to a fall of temperature, but this interpretation receives no support from oxygen-isotope analysis of a classic section across the Triassic-Jurassic boundary, which indicates no significant change (Hallam and Goodfellow, 1990). A better case for climatic change as a cause of extinctions can be made for the late Carnian event affecting terrestrial organisms. Simms and Ruffell (1989, 1990) have documented good evidence for a Carnian humid phase over extensive regions of the northern hemisphere, being succeeded by more arid conditions in the Norian.

One must bear in mind also that a major sea-level fall could have had significant consequences for continental climate, with an increase in seasonal range of temperature, and increased aridity of continental interiors (Hallam, 1989). This could well have a bearing on the end-Triassic extinctions among terrestrial groups.

Volcanism

That most of the Deccan Traps of India appear to have been erupted within the magnetic zone embracing the Cretaceous-Tertiary boundary has provoked interest in a possible relation between large-scale flood basalt eruptions and mass extinctions, perhaps connected with the emission of aerosols into the atmosphere and stratosphere (Rampino et al., 1988). The most obvious candidate for the end-Triassic extinctions would appear to be the Karoo volcanics of southern Africa, but these are evidently too young. Numerous radiometric dates indicate a Pliensbachian-Toarcian peak (Fitch and Miller, 1984) and palynological dating indicates an onset of volcanicity no earlier than late Sinemurian (Aldiss et al., 1984). However, the initiation of a significant episode of basaltic volcanicity in eastern North America, and probably elsewhere on the North Atlantic margins, can now be precisely dated both palynologically and radiometrically as coinciding almost exactly with the beginning of the Jurassic, but slightly postdating the Triassic-Jurassic boundary (Olsen et al., 1987). Using a correlation based on Milankovitch cyclicity, Olsen et al. (1990) suggested that the volcanicity began about 60,000 years after the boundary. It would therefore appear that the North Atlantic volcanic province is also too young to be implicated directly in the extinctions, but there may be an indirect connection through tensional tectonic activity, as already noted in the discussion of sea-level change.

Bolide Impact

Because the large Manicouagan crater in Quebec has been dated as late Triassic, with an error embracing the Triassic-Jurassic boundary, a claim has been made that an impact event might have been responsible for the end-Triassic extinctions (Olsen et al., 1987). The most obvious place to look for supporting evidence such as iridium anomalies and shocked quartz is in the Newark Supergroup of eastern North America, the most proximal well-studied sequence, where the Triassic-Jurassic boundary is quite well constrained by palynological data. No such evidence has yet been found despite prolonged search for shocked quartz by Mark Anders, and iridium analyses by Frank Asaro. New U-Pb zircon dating of the Manicouagan impact event, apparently more reliable that any previous effort, gives a date of 214 ± 1 Ma (Hodych and Dunning, 1992). This is appreciably older than a recent date for the Triassic-Jurassic boundary, using the same method, of 202 ± 1 Ma (Dunning and Hodych, 1990).

A Russian geologist claimed to have found shocked quartz at the Triassic-Jurassic boundary in the classic Kendelbach section of Austria. This claim has not been borne out by further research on this section, and on another Triassic-Jurassic boundary section in England. In neither case have iridium anomalies or shocked quartz been found (Hallam, 1990; Orth et al., 1990). Shocked quartz grains have been reported from a section at Corfino in northern Italy by Bice et al. (1992) but the dating as Triassic-Jurassic boundary is insecure in the absence of ammonites, being based on the disappearance of the bivalve *Rhaetavicula contorta*. In the better dated and more complete Kendelbach section this species disappears well below the boundary. Furthermore it is necessary, to support the association of extinction with a bolide impact, to demonstrate not merely the presence of shocked minerals at the critical boundary, but their absence from subjacent and suprajacent strata. If they are found at more than one horizon it would be more plausible to invoke a background effect in quartz grain sedimentation related to impacts at different times, or a non-impact origin. In fact no fewer than three horizons with shocked quartz grains are reported by Bice et al., within a few metres of what they claim to be the Triassic-Jurassic boundary. The authors admit that the Manicouagan impact is an unlikely candidate, because the most modern dating suggests an age approximately 12 Ma before the boundary, and favour instead the comet shower hypothesis of Hut et al. (1987), which seems weak on both evidential and theoretical grounds (Hallam, 1989).

A final criticism of bolide impact at the end of the Triassic is that according to Benton (1991) and Weems (1992) there is no catastrophic extinction among the terrestrial vertebrates to be accounted for, a situation quite unlike that at the end of the Cretaceous. This suggests that the manifestly significant extinction at this time is essentially a phenomenon of the marine realm, implying that the explanation involves a complex of oceanographic factors.

The occurrence of either an iridium anomaly or increased concentrations of cosmic dust at certain horizons in the European Jurassic has led to the speculation that bolide impacts could have been the cause. Thus a significant iridium enrichment was found in a ferruginous crust at or near the Lower-Middle Jurassic boundary in the Southern Alps (Rocchia et al., 1986) and a concentration of iron-nickel spherules of cosmic origin at the Middle-Upper Jurassic boundary in southern Poland (Brochwicz-Lewinski et al., 1984). In neither case do these mark mass extinction horizons, and the same is true for a phosphatic limestone layer at the Jurassic-Cretaceous boundary in northern Siberia, where an anomalously high concentration of iridium has also been found (Zakharov et al., 1992). The sedimentary facies in all cases suggests condensation, and the alternative explanation considered by the first two sets of authors, and strongly promoted by Zakharov and his colleagues, is likely to be the correct one. Strongly reduced sedimentation rate, as in the deep ocean today, leads to a concentration of cosmic background fallout, including platinum-group metal-bearing minerals. In particular, the topmost Callovian is a horizon of widespread condensation in Europe, the consequence of a rapid rise of sea level (Norris, 1993).

It has become evident in recent years that many, if not most, of the small platinum-group-element spikes found in the geological record can be accounted for satisfactorily by purely Earth-bound events, including not only reduction of sedimentation rate but element redistribution in diagenesis, at and across redox boundaries (Colodner et al., 1992). As Jansa (1993) has pointed out, furthermore, there is no single indicator that can provide sufficient proof of an impact event.

Acknowledgements

I am indebted to Mike Benton, John Callomon, Cathy Newton, Jim Riding, Tim Tozer and Gerd Westermann for providing material that has helped in the production of this article.

References

Aldiss, D.T., Benson, J.M. and Rundle, C.C., 1984. Early Jurassic pillow lavas and palynomorphs in the Karoo of eastern Botswana. Nature 310, 302-304.

Ash, S., 1986. Fossil plants and the Triassic-Jurassic boundary. In: Padian, K. (ed.), The beginning of the age of dinosaurs. pp. 21-29. Cambridge Univ. Press, Cambridge.

Balme, B.E., 1970. Palynology of Permian and Triassic strata in the Salt Range and Surghar Range, West Pakistan. Dept. Geol. Univ. Kansas Spec.

Publ. 4, 305-451.

Bate, R.H. and Robinson, E. (eds.), 1978. A stratigraphic index of British Ostracoda. Seel House Press, Liverpool.

Batten, R.L., 1973. The vicissitudes of the gastropods during the interval of Guadaloupian-Ladinian time. In: Logan, A. and Hills, L.V. (eds.), The Permian and Triassic systems and their mutual boundary. Canad. Soc. Petrol. Geol. Mem. 2, 596-607.

Benton, M.J., 1986. More than one event in the Late Triassic extinction. Nature 321, 857-859.

Benton, M.J., 1988. Mass extinctions and the fossil record of reptiles: paraphyly, patchiness and periodicity. In: Larwood, G.P. (ed.), Extinction and survival in the fossil record. pp. 269-294. Clarendon Press, Oxford.

Benton, M.J., 1991. What really happened in the Late Triassic? Histor. Biol. 5, 263-278.

Bice, D.M., Newton, C.R., McCauley, S., Reiners, P.W. and McRoberts, C.A., 1992. Shocked quartz at the Triassic-Jurassic boundary in Italy. Science 255, 443-446.

Bizon, G. and Oertli, H., 1961. Contribution à l'étude micropaléontologique (Foraminifères, Ostracodes) du Lias du bassin de Paris. Mem. Bur. Rech. Géol. Min. 4, 107-119.

Boulter, M.C., Spicer, R.A. and Thomas, B.A., 1988. In: Larwood, G.P. (ed.), Patterns of plant extinction from some palaeobotanical evidence. pp. 1-36. Clarendon Press, Oxford.

Bown, P.R. and Lord, A.R., 1990. The occurrence of calcareous nannofossils in the Triassic/Jurassic boundary interval. Cahiers Univ. Cathol. Lyon Ser. Sci. 3, 127-136.

Brochwicz-Lewinski, W., Gasiewicz, A., Suffczynski, S., Szatkowski, K. and Zbik, M., 1984. Lacunes et condensations à la limite Jurassique moyen-supérieur dans le Sud de la Pologne: manifestation d'un phénomène mondial? C.R. Acad. Sci. Paris 299, Ser. II, 1359-1362.

Bujak, J.P. and Fisher, M.J., 1976. Dinoflagellate cysts from the Upper Triassic of arctic Canada. Micropaleont. 22, 44-70.

Bujak, J.P. and Williams, G.L., 1979. Dinoflagellate cyst diversity through time. Mar. Micropaleont. 4, 1-12.

Callomon, J.H., 1985. The evolution of the Jurassic ammonite family Cardioceratidae. Spec. Pap. Palaeont. 33, 49-90.

Cathles, L.M. and Hallam, A., 1991. Stress-induced changes in plate density, Vail sequences, epeirogeny, and short-lived global sea level fluctuations. Tectonics 10, 659-671.

Clark, D.L., 1987. Conodonts: the final fifty million years. In: Aldridge, R.J. (ed.), Palaeobiology of conodonts. pp. 165-174. Horwood, Chichester.

Colbert, E.H., 1965. The age of reptiles. Norton, New York.

Colodner, D.C., Boyle, E.A., Edmond, J.M. and Thomson, J., 1992. Post-depositional mobility of platinum, iridium and rhenium in marine sediments. Nature 358, 402-404.

Cornet, B. and Olsen, P.E., 1990. Early to Middle Carnian (Triassic) flora and fauna of the Richmond and Taylorsville basins, Virginia and Maryland, USA. Virginia Museum of Natural History Guidebook 1, 1-83.

Crowley, T.J. and North, G.R., 1991. Paleoclimatology. Oxford Univ. Press, Oxford.

Dale, B., 1983. Dinoflagellate resting cysts: "benthic plankton". In: Fryzell, G.A. (ed.), Survival strategies of the algae. pp. 69-136. Cambridge Univ. Press, Cambridge.

Dunning, G.R. and Hodych, J.P., 1990. U/Pb zircon and baddeleyite ages for the Palisades and Gettysburg sills of the northeastern United States. Implications for the age of the Triassic/Jurassic boundary. Geology 18, 795-798.

Eldredge, N., 1991. The miner's canary: unravelling the mysteries of extinction. Prentice Hall Press, New York.

Embry, A.F., 1988. Triassic sea-level changes: evidence from the Canadian Arctic archipelago. Soc. Econ. Paleont. Mineral. Spec. Publ. 42, 249-260.

Erwin, D.H., 1990. Carboniferous-Triassic gastropod diversity patterns and the Permo-Triassic mass extinction. Paleobiol. 16, 187-203.

Evitt, W.R., 1981. The difference it makes that dinoflagellates did it differently. Internat. Commission for Palynology Newsletter 4 (1), 6-7.

Evitt, W.R., 1985. Sporopollenin dinoflagellate cysts. Their morphology and interpretation. Am. Ass. Stratig. Palynol. Foundation, Dallas.

Fitch, F.J. and Miller, J.A., 1984. Dating Karoo igneous rocks by the conventional K-Ar and $^{40}Ar/^{39}Ar$ age spectrum methods. Geol. Soc. S. Afr. Spec. Publ. 13, 247-266.

Frakes, L.A., 1979. Climates through geologic time. Elsevier, Amsterdam.

Frakes, L.A., Francis, J.E. and Syktus, J.I., 1992. Climate modes of the Phanerozoic. Cambridge Univ. Press.

Hallam, A., 1961. Cyclothems, transgressions and

faunal change in the Lias of north west Europe. Trans. Edinburgh Geol. Soc. 18, 132-174.

Hallam, A., 1976. Stratigraphic distribution and ecology of European Jurassic bivalves. Lethaia 9, 245-259.

Hallam, A., 1981. The end-Triassic bivalve extinction event. Palaeogeog., Palaeoclimatol., Palaeoecol. 35, 1-44.

Hallam, A., 1985. A review of Mesozoic climates. J. geol. Soc. Lond. 142, 433-445.

Hallam, A. 1986. The Pliensbachian and Tithonian extinction events. Nature 319, 765-768.

Hallam, A., 1987. Radiations and extinctions in relation to environmental change in the marine Lower Jurassic of north west Europe. Paleobiol. 13, 152-168.

Hallam, A., 1988. A re-evaluation of Jurassic eustasy in the light of new data and the revised Exxon curve. Soc. Econ. Paleont. Mineral. Spec Publ. 42, 261-273.

Hallam, A., 1989. The case for sea-level change as a dominant causal factor in mass extinction of marine invertebrates. Phil. Trans. Roy. Soc. B 325, 437-455.

Hallam, A., 1990a. The end-Triassic mass extinction event. Geol. Soc. Am. Spec. Paper 247, 577-583.

Hallam, A., 1990b. Biotic and abiotic factors in the evolution of early Mesozoic marine molluscs. In: Ross, R.M. and Allmon, W.D. (eds.), Causes of evolution - a paleontological perspective. pp. 249-260. Chicago Univ. Press, Chicago.

Hallam, A., 1992. Phanerozoic sea-level changes. Columbia Univ. Press, New York.

Hallam, A., 1993. Jurassic climates as inferred from the sedimentary and fossil record. Phil. Trans. Roy. Soc. London B 341, 287-296.

Hallam, A. and Goodfellow, W.D., 1990. Facies and geochemical evidence bearing on the end-Triassic disappearance of the Alpine reef ecosystem. Histor. Biol. 4, 131-138.

Hallam, A., Biro Bagoczy, L. and Perez, E., 1986. Facies analysis of the Lo Valdes Formation (Tithonian-Hauterivian) of the High Cordillera of central Chile, and the palaeogeographic evolution of the Andean Basin. Geol. Mag. 123, 425-435.

Hamilton, G.B., 1982. Triassic and Jurassic calcareous nannofossils. In: Lord, A.R. (ed.), A stratigraphic index of calcareous nannofossils. pp. 17-39. Horwood, Chichester.

Haq, B.U., Hardenbol, J. and Vail, P.R., 1987. The chronology of fluctuating sea level since the Triassic. Science 235, 1156-1167.

Hayami, I., 1989. Outlook on the post-Paleozoic historical biogeography of pectinids in the Western Pacific Region. In: Ohba, H. et al. (eds.), Current aspects of biogeography in West Pacific and East Asian regions. Nature and Culture 1, 3-25. Univ. Museum Tokyo.

Helby, R., Morgan, R. and Partridge, A.D., 1987. A palynological zonation of the Australian Mesozoic. Assoc. Australas. Palaeont. Mem. 4, 1094.

Helby, R., Wiggins, V.D. and Wilson, G.J., 1987. The circum-Pacific occurrence of the Late Triassic dinoflagellate *Sverdrupiella*. Aust. J. Earth Sci. 34, 151-152.

Hodych, J.P. and Dunning, G.R., 1992. Did the Manicouagan impact trigger end-of-Triassic mass extinction? Geology 20, 51-54.

Hunt, A.P. and Lucas, S.G., 1992. No tetrapod extinction event at the Carnian-Norian boundary (Late Triassic): evidence from the western United States and India. Abstracts 29th Int. geol. Congr., Kyoto, 66.

Hut, P., Alvarez, W., Elder, W.P., Hansen, T.A., Kauffman, E.G., Keller, G., Shoemaker, E.M. and Weissman, P.-R., 1987. Comet showers as a cause of mass extinctions. Nature 329, 118-126.

Jansa, L.F., 1993. Cometary impacts into ocean: their recognition and the threshold constraint for biological extinctions. Palaeogeogr., Palaeoclimatol., Palaeoecol. 104, 271-286.

Jenkins, D.G. and Murray, J.W. (eds.), 1981. Stratigraphic atlas of fossil Foraminifera. Horwood, Chichester.

Jenkyns, H.C., 1988. The early Toarcian (Jurassic) anoxic event: stratigraphic, sedimentary, and geochemical evidence. Am. J. Sci. 288, 101-151.

Johnson, A.L.A. and Simms, M.J., 1989. The timing and cause of Late Triassic marine invertebrate extinctions: evidence from scallops and crinoids. In: Donovan, S.K. (ed.), Mass extinctions, processes and evidence. pp. 174-194. Belhaven, London.

Klüpfel, W., 1917. Über die Sedimente der Flachsee im Lothringer Jura. Geol. Rundschau 7, 97-109.

Knoll, A.H., 1984. Patterns of extinction in the fossil record of vascular plants. In: Nitecki, M.H. (ed.), Extinctions. pp. 21-68. Chicago Univ. Press, Chicago.

Loh, H., Maul, B., Prauss, M. and Riegel, W., 1986. Primary production, maceral formation and carbonate species in the Posidonia Shale of NW Germany. Mitt Geol. Paläont. Inst. Univ. Hamburg 60, 397-421.

Lord, A.R., 1974. Ostracods from the Domerian and Toarcian of England. Palaeont. 17, 599-622.

Newton, C.R., Whalen, M.T., Thompson, J.B., Prins, N. and Dellalla, D., 1987. Systematics and paleoecology of Norian (Late Triassic) bivalves from a tropical island arc: Wallowa terrane, Oregon. Paleont. Soc. Mem. 22, 1-83.

Norris, M.S., 1993. A facies analysis of Middle-Upper Jurassic boundary beds in Europe. Unpubl. PhD thesis, University of Birmingham.

Olsen, P.E. and Sues, H.-D., 1986. Correlation of continental Late Triassic and Early Jurassic sediments and patterns of the Triassic-Jurassic tetrapod transition. In: Padian, K. (ed.), The beginning of the age of dinosaurs. pp. 321-351. Cambridge Univ. Press, Cambridge.

Olsen, P.E., Shubin, N.H. and Anders, M.H., 1987. New Early Jurassic tetrapod assemblages constrain Triassic-Jurassic tetrapod extinction event. Science 237, 1025-1029.

Olsen, P.E., Fowell, S.J. and Cornet, B., 1990. The Triassic/Jurassic boundary in continental rocks of eastern North America; a progress report. Geol. Soc. Am. Spec. Paper 247, 585-594.

Orth, C.J., Attrep, M. and Quintana, L.R., 1990. Iridium abundance patterns across bio-event horizons in the fossil record. Geol. Soc. Am. Spec. Paper 247, 45-60.

Pedersen, K.R. and Lund, J.J., 1980. Palynology of the plant-bearing Rhaetian to Hettangian Kap Stewart Formation, Scoresby Sund, East Greenland. Rev. Palaeobot. Palynol. 32, 1-69.

Raup, D.M. and Sepkoski, J.J., 1982. Mass extinctions in the marine fossil record. Science 215, 1501-1503.

Raup, D.M. and Sepkoski, J.J., 1988. Testing the periodicity of extinction. Science 241, 94-96.

Riding, J.B., 1984. A palynological investigation of the Toarcian to early Aalenian strata from the Blea Wyke area, Ravenscar, North Yorkshire. Proc. Yorks. Geol. Soc. 45, 109-122.

Riding, J.B., 1987. Dinoflagellate cyst stratigraphy of the Nettleton Bottom Borehole (Jurassic: Hettangian to Kimmeridgian), Lincolnshire, England. Proc. Yorks. Geol. Soc. 46, 231-266.

Riding, J.B. and Thomas, J.E., 1992. Dinoflagellate cysts of the Jurassic System. In: Powell, A.J. (ed.), The stratigraphic distribution of dinoflagellate cysts. Chapman and Hall, London.

Riding, J.B., Penn, I.E. and Woollam, R., 1985. Dinoflagellate cysts from the type area of the Bathonian stage (Middle Jurassic; south west England). Rev. Palaeobot. Palynol. 45, 149-169.

Riding, J.B., Walton, W. and Shaw, D., 1991. Toarcian to Bathonian (Jurassic) palynology of the Inner Hebrides, north west Scotland. Palynol. 15.

Rocchia, R., Boclet, D., Bonté, P., Castellarin, A., Jéhanno, C., 1986. An iridium anomaly in the Middle-Lower Jurassic of the Venetian region, northern Italy. J. geophys. Res. 91, B13, E259-262.

Sandy, M.R., 1988. Tithonian Brachiopods. Mem. Soc. geol. France N.S. 154, 71-74.

Sarjeant, W.A.S., 1978. *Arpylorus antiquus* Calandra, Emdn., a dinoflagellate cyst from the Upper Silurian. Palynol. 2, 167-179.

Schäfer, P. and Fois, E., 1987. Systematics and evolution of Triassic Bryozoa. Geologica et Palaeontologica 21, 173-225.

Sengör, A.M.C., Altiner, D., Cin, A., Ustaömer, T. and Hsü, K.J., 1988. Origin and assembly of the Tethyside orogenic collage at the expense of Gondwanaland. In: Audley-Charles, M.G. and Hallam, A. (eds.), Gondwana and Tethys. pp. 119-181. Oxford Univ. Press, Oxford.

Sepkoski, J.J., 1982. A compendium of fossil marine families. Milwaukee Public Museum Contr. Bio. Geol. 51, 1-125.

Sepkoski, J.J., 1986. Phanerozoic overview of mass extinction. In: Raup, D.M. and Jablonski, D. (eds.), Patterns and processes in the history of life. pp. 277-295. Springer, Berlin Heidelberg New York.

Simms, M.J., 1991. The radiation of post-Palaeozoic echinoderms. In: Taylor, P.D. and Larwood, G.P. (eds.), Major evolutionary radiation. pp. 288-304. Clarendon Press, Oxford.

Simms, M.J. and Ruffell, A.H., 1989. Synchroneity of climatic change and extinctions in the late Triassic. Geology 17, 265-268.

Simms, M.J. and Ruffell, A.H., 1990. Climatic and biotic change in the late Triassic. J. Geol. Soc. Lond. 147, 321-327.

Smith, A.B., 1990. Echinoid evolution from the Triassic to Lower Liassic. Cahiers Univ. Cathol. Lyon, sér. Sci. 3, 79-117.

Stanley, G.D., 1988. The history of early Mesozoic reef communities, a three step process. Palaios 3, 170-183.

Taylor, F.R.J., (ed.), 1987. The biology of dinoflagellates. Botan. Monographs vol. 27. Blackwell Scientif. Publ., Oxford.

Tozer, E.T., 1981a. Triassic Ammonoidea: classification, evolution and relationship with Permian and Jurassic forms. In: House, M.R. and Senior, J.R.

(eds.), The Ammonoidea. pp. 66-100. Acad. Press, London.

Tozer, E.T., 1981b. Triassic Ammonoidea: geographic and stratigraphic distribution. In: House, M.R. and Senior, J.R. (eds.), The Ammonoidea. pp. 397-431. Acad. Press, London.

Traverse, A., 1988. Plant evolution dances to a different beat; plant and animal evolutionary mechanisms compared. Histor. Biol. 1, 277-302.

Visscher, H. and Brugman, W.A., 1981. Ranges of selected palynomorphs in the Alpine Triassic of Europe. Rev. Palaeobot. Palynol. 34, 115-128.

Wall, D., 1965. Microplankton, pollen and spores from the Lower Jurassic in Britain. Micropaleont. 11, 151-190.

Wall, D. and Evitt, W.R., 1975. A comparison of the modern genus *Ceratium* Schrank, 1793, with certain Cretaceous marine dinoflagellates. Micropaleont. 2, 14-44.

Wall, D., Dale, B., Lohman, G.P. and Smith, W.K., 1977. The environmental and climatic distribution of dinoflagellate cysts in modern marine sediments from regions in the North and South Atlantic Oceans and adjacent seas. Mar. Micropaleont. 2, 121-200.

Weems, R.E., 1992. The "terminal Triassic catastrophic extinction event" in perspective: a review of Carboniferous through Early Jurassic terrestrial vertebrate extinction patterns. Palaeogeog., Palaeoclimatol., Palaeoecol. 94, 1-29.

Westermann, G.E.G., 1981. Ammonite biochronology and paleogeography of the circum-Pacific Middle Jurassic. In: House, M.R. and Senior, J.R. (eds.), The Ammonoidea. Acad. Press, London.

Westermann, G.E.G., 1988. Duration of Jurassic stages based on averaged and scaled subzones. Recent Researches in Geology 12, 90-100.

Westermann, G.E.G. and Callomon, J.H., 1988. The Macrocephalitinae and associated Bathonian and early Callovian (Jurassic) ammonoids of the Sula Islands and New Guinea. Palaeontographica A 203, 1-90.

Whatley, R.C., 1988. Patterns and rates of evolution among Mesozoic Ostracoda. In: Hanai, T., Ikeya, N. and Ishizaki, K. (eds.), Evolutionary biology of Ostracoda. pp. 1021-1040. Kodansha, Tokyo.

Wille, W., 1982. Evolution and ecology of Upper Liassic dinoflagellates from SW Germany. N. Jb. Geol. Paläont. Abh. 164, 74-132.

Wille, W. and Gocht, H., 1979. Dinoflagellaten aus dem Lias Südwestdeutschlands. N. Jb. Geol. Paläont. Abh. 158, 221-258.

Woollam, R., and Riding, J.B., 1983. Dinoflagellate cyst zonation of the English Jurassic. Inst. Geol. Sci. Rep. 83/2, 1-44.

Yin Hongfu, 1990. Paleogeographic distribution and stratigraphical range of the Lower Triassic *Claraia, Pseudoclaraia* and *Eumorphotis* (Bivalvia). J. China Univ. Geosciences 1, 98-110.

Zakharov, V.A., Lapukhov, A.S. and Shenfil, O.V., 1992. Are the iridium anomalies always evidence of impact character of Earth biosphere turnovers? 5th Internat. Conf. on Global Bioevents Göttingen 1992. Abstract Vol., p. 125. Göttingen.

Manuscript received October 1992

Author´s address:

Anthony Hallam, School of Earth Sciences, University of Birmingham,
Birmingham B15 2TT, U.K.

Cretaceous Bio-Events

Erle G. KAUFFMAN and Malcolm B. HART

Abstract. Bio-events occur at local, regional, and global scales, reflecting short-term, extraordinary, environmental changes. They can be classified as Diversification Bio-Events (punctuated evolution, population blooms, colonization and immigration bio-events), or Diversity Reduction Bio-Events (mass mortality, ecosystem shock, extinction and emigration bio-events). High-resolution (cm-dm scale) stratigraphic, geochemical, and paleobiological analyses demonstrate that many regional and most global bio-events are complex, multicausal phenomena. They may consist of two or more, closely spaced levels ("steps") of biological response to rapid environmental changes - perturbations and their feedback loops - in ocean-climate systems. This is especially true for regional to global mass extinctions, which also tend to be ecologically graded, affecting more tropical, more stenotopic taxa/ecosystems first and most profoundly, and more poleward and/or eurytopic biotas later, and to a lesser degree. Comparisons of the stratigraphic expression of local, regional, and global bio-events are presented for the Cenomanian-Turonian (Ce-Tu: middle Cretaceous) Greenhorn Cyclothem at the Pueblo, Colorado reference section, where seven regional and one global bio-event (the C-T mass extinction) intervals are well defined. Regional and global bio-events are documented for the Americas, Europe, North Africa, and India. Global bio-event intervals include: the Jurassic-Cretaceous mass extinction interval; the Early Aptian Selli Bio-Event; the Late Aptian mass extinction interval; the Middle-Late Albian substage boundary bio-events; the Albian-Cenomanian stage boundary bio-events; the Cenomanian-Turonian boundary mass extinction interval; the Turonian-Coniacian stage boundary bio-events; the Coniacian/Santonian stage boundary bio-events; the Santonian/Campanian stage boundary bio-events; the 68 Ma (Middle Maastrichtian) extinction interval; and the Cretaceous-Tertiary boundary mass extinction interval.

Contents

Introduction	285
Comparing Local, Regional, and Global Cretaceous Bio-Events: a Case History from the Pueblo Area, Colorado, U.S.A.	288
Regional and Global Cretaceous Bio-Events of the Americas	295
Regional and Global Cretaceous Bio-Events in Northern and Southern Europe	300
Regional and Global Cretaceous Bio-Events in North Africa	303
Regional and Global Cretaceous Bio-Eevents in India (and Indian Ocean)	303
Conclusions	304

Introduction

Biological events, or bio-events, comprise short-term (days to <100 ka) evolutionary, ecologic, and/or biogeographic responses to rapid environmental changes of both terrestrial and extraterrestrial origin. They represent ecostratigraphic units (sensu Kauffman, 1986, 1988a; Kauffman et al., 1991; Boucot, 1986;

Sageman et al., 1995, in press), and are important components of high-resolution event/cycle chronostratigraphy within and between depositional basins. In some cases, they occur on a global scale. Sageman et al. (1995, in press) have classified bio-events into: (a) Diversification Bio-Events, in which there is an abrupt increase in diversity, population size, and/or biomass (e.g. widespread colonization events, population bursts or epiboles, immigration events sourced in species-rich areas, and rapid evolutionary events); and (b) Diversity Reduction Bio-Events, in which population size, biomass, and/or diversity rapidly decline due to, e.g., ecosystem shock, mass mortalities, abrupt restriction of community or biogeographic range, regional emigration events among species-rich communities, and regional to global extinction events. Kauffman (1986, 1988a), Kauffman et al. (1991), and Sageman et al. (1995, in press), provide detailed discussions of various types of bio-events.

Diversification Bio-Events reflect: (a) short-term improvement in environmental conditions for life, and especially favorable increases in temperature, nutrients, light availability, water or atmospheric chemistry (especially dissolved oxygen), and habitat quality, heterogeneity, and stability; (b) rapid expansion and diversification of favorable ecospace leading to widespread colonization of newly available niches, removal of biological, chemical, and/or physical constraints on population size, species occurrence and diversity, etc.; and (c) rapid evolution of favorable new adaptive traits or more complex species interactions in ecosystems, leading to niche expansion or partitioning, occupation of new niches, and rapid evolutionary radiations, especially following regional to global mass extinctions.

Diversity Reduction Bio-Events may be caused by: (a) unpredictable environmental perturbations brought about by abrupt climatic and/or oceanographic changes, or by short-term, large-scale sedimentologic, tectonic, volcanic, or extraterrestrial events; (b) by the breaching of narrow ecological thresholds (e.g. critical temperature barriers) due to large-scale environmental changes over short to long time frames; (c) or by major evolutionary and ecological changes that have a pervasive, short-term effect on large and complex ecosystems (e.g. removal of keystone species, important trophic resources, or critical habitats; the evolution of new competitors, predators, or disease strains that broadly affect an ecosystem).

Most bio-events are autocyclic, local to regional in scale (Sageman et al., 1995, in press), and affect, at best, large biomes within a single continent or ocean basin (i.e. a single depositional basin in the geological record), a number appear to be allocyclically regulated, and interregional to global in scale through Phanerozoic time. The latter are of special interest to us because they infer the existence of very rapid, global-scale changes in environmental parameters of the Earth, especially within the ocean-climate system. These are difficult to conceive from the perspective of human history, but similar environmental fluctuations may be inherent in the rates and patterns of modern global change, and the expanding biodiversity crisis.

The study of interregional and global bio-events has had a long history, largely focused on mass extinctions leading to abrupt changes in the global biota. Much of this research, especially in recent years, has concentrated on the Cretaceous Period, which encompasses four global mass extinctions (Jurassic-Cretaceous, Late Aptian, Cenomanian-Turonian, Maastrichtian-Danian bio-events: Sepkoski, 1993), and which spans one of the most dynamic periods of environmental change in Earth history.

During the Cretaceous, global ecosystems were subjected to: (a) Major tectonic changes including accelerated spreading and a major rearrangement of Earth's plates, increasing the rates and magnitude of volcanism, eustatic sea-level fluctuations, continental tectonism, and paleogeographic change (Larson, 1991a,b); the Cretaceous was one of the most volcanically active periods in Earth history (Kauffman and Caldwell, 1993); (b) the rise and wane of the Pacific superplume (Larson, 1991a,b) with its regional effects on plate movements, volcanism, outgassing, and aquatic and atmospheric chemistry (Fig. 1); (c) rapid tectonoeustatic changes in sea level at several scales (Haq et al., 1987), including the highest stand (300 m above present: Pitman, 1978; McDonough and Cross, 1991) since the Ordovician; (d) a greenhouse climate interval related to broad-scale global warming which peaked in the basal Turonian (Jenkyns et al., 1994), atmospheric CO_2 levels from 2X to nearly 4X modern levels (Berner, 1994), no long-term polar ice, an

Fig. 1 (opposite page). Integrated analysis of Mesozoic and Cenozoic magnetostratigraphy (Harland et al., 1990), ocean crust production (Larson, 1991a), sea surface paleotemperatures (Savin, 1977; Arthur et al., 1985), sea-level changes (Haq et al., 1988), black shale deposition (Jenkyns, 1980), and world oil resources (Irving et al., 1974; Tissot, 1979). Modified from Larson (1991a) and Larson et al. (1993)

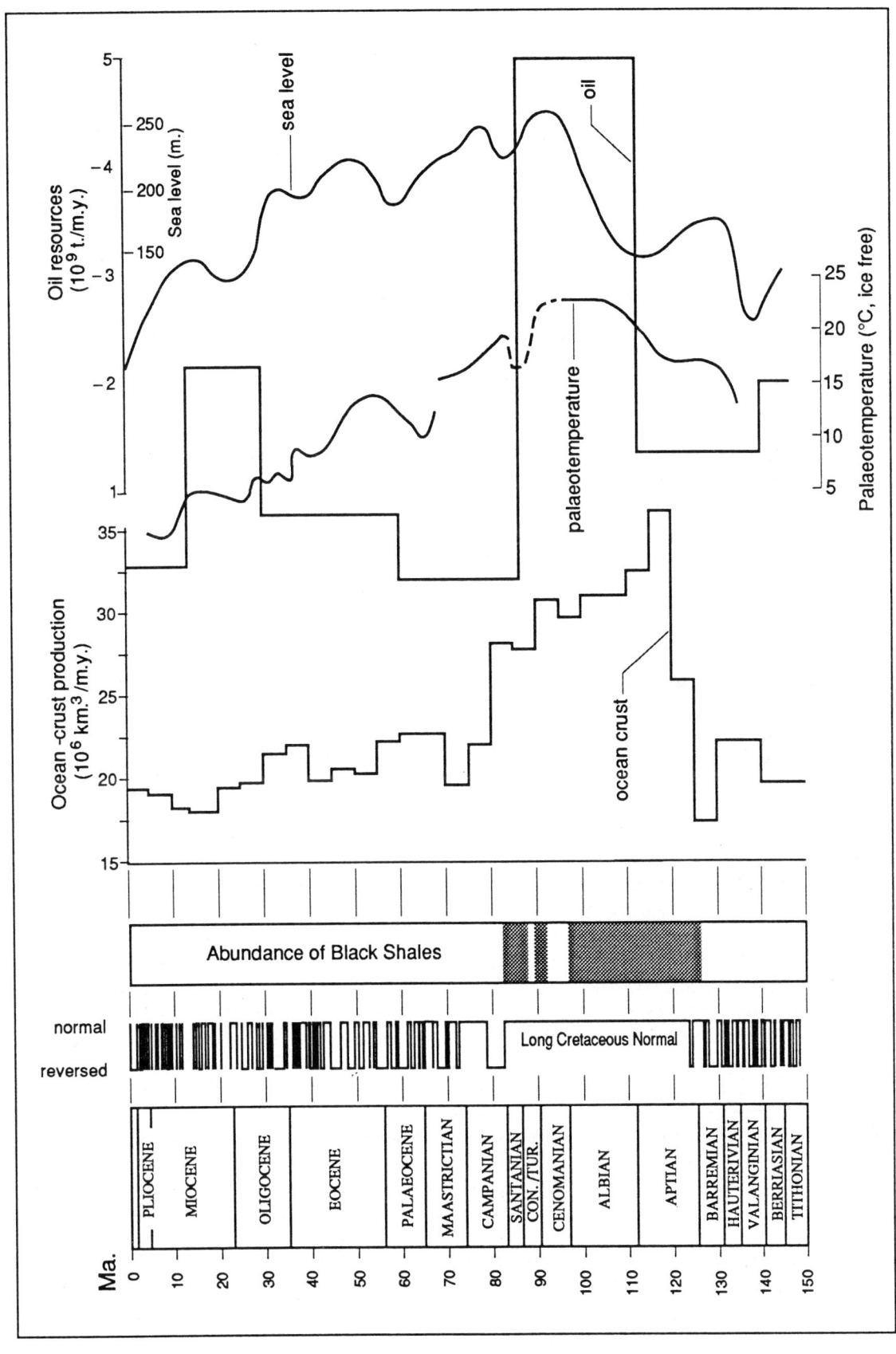

expanded tropical climate zone (episodically including a core supertropical climate zone - Supertethys) which largely lay north of the paleoequator (Kauffman and Johnson, 1988; Johnson, 1993), a series of large-scale climate changes associated with sea-level fluctuations and especially cooling during the Campanian-Maastrichtian regression, and episodically enhanced effects of Milankovitch climate cyclicity on sedimentation, watermass characteristics, and ecosystem structure; (e) major changes in the global biota associated with, for example, the great diversification of angiosperms and early mammals, a major radiation among calcareous marine plankton, and a major change in the structure of tropical reef-associated communities (from coral-algal-dominated to rudistid bivalve-dominated), and a series of four mass extinctions, all leading to significant changes in global ecosystem structure; and (f) a significant number of extraterrestrial comet and meteorite impacts, including the largest yet recorded in Phanerozoic history - the Chicxulub impact at or very near to the Cretaceous-Tertiary boundary.

Despite these large-scale environmental changes, Jurassic and Cretaceous biotas had broadly adapted in their evolution to pervasive warm, relatively equable, background greenhouse environments, and in general may have had narrower adaptive ranges, and more specialized niche parameters than modern biotas adapted to a climatically more dynamic icehouse world. Cretaceous biotas were probably more perturbation- and extinction-prone than their modern counterparts, enhancing the development of interregional to global bio-events. The study of Cretaceous bio-events may therefore shed much light on the nature of biological responses to a Greenhouse world, and its demise. These observations, in turn, may lead to predictive modeling that can be applied to modern global change as we rapidly shift from an icehouse to a greenhouse world as a result of global warming.

In this chapter we present an example contrasting local, regional, and global bio-events from the North American reference section of the Late Albian - Middle Cenomanian Greenhorn Cyclothem near Pueblo, Colorado (see also Sageman et al., 1995 in press). We then develop a history of regional bio-events for the Cretaceous of Eurasia and the Americas, and identify those which are shared between these well-studied areas and which therefore represent interregional to global bio-events.

Comparing Local, Regional, and Global Cretaceous Bio-Events: a Case History from the Pueblo Area, Colorado, U.S.A.

High-resolution event and cycle stratigraphy (Kauffman, 1988a; Kauffman et al., 1991) involves cm-dm scale examination of fresh outcrop and core in search of physically, chemically, and/or biologically defined surfaces, or thin stratigraphic intervals, which reflect short-term, regionally pervasive, stratigraphic phenomena. Examples are: magnetic reversals; pervasive volcanic ash falls (bentonites); regional storm and mass flow deposits; prominent geochemical excursions in elemental and stable isotope values; bio-events such as mass mortalities, extinction levels, colonization surfaces, and population blooms; high-frequency sequence stratigraphic or parasequence boundaries and flooding surfaces; and bedding cycle boundaries and events associated with Milankovitch climate cycle deposits. These events and surfaces, tested for their regional extent through linear and graphic correlation methods (Edwards, 1984, 1989), especially those integrating event, cycle, and biozone boundaries with radiometric ages into a single graphic matrix (Kauffman, 1988a), form the basis for a highly refined chronostratigraphy within and between depositional basins. This chronostratigraphy attains, for example, average levels of refinement between 14 and 100 ka per event-bounded interval for various Cretaceous sequences in the Western Interior Basin of North America (Kauffman, 1988a; 1995 in press; Kauffman et al., 1991). As an integral part of this chronology, biological events of all types are commonly documented and tested against the regional composite chronostratigraphy (especially against bentonites and biozone boundaries) to determine their regional extent, their characteristic expression, and their probable causes. The discovery rate of local and regional bio-events using this stratigraphic approach has been high; the chronostratigraphic refinement of the system, as a whole, further allows identification of regional and global bio-events responding to allocyclic environmental processes, and their differentiation from local events reflecting autocyclic processes.

Of special interest in this type of high-resolution analysis of regional to global bio-events is the discovery that many mass extinctions, previously considered to be single short-term ecological crises or catastrophes (e.g. the Cretaceous-Tertiary boundary mass extinction; "catastrophe" sensu Schindewolf, 1962; Alvarez et al., 1980), are composed, in the observed data sets, of a series of individual short-term bio-events ("steps" of Kauffman, 1988b) clustered within a <1-3 Ma interval of time. For many mass extinctions and other large-scale bio-events, these intervals are also characterized by extraordinary geochemical and/or sedimentologic fluctuations, defining frequent large-scale disruptions of ocean/climate systems.

Figure 2 compares the expression of local, regional, and global bio-events (as tested against radiometry, biostratigraphy, and event/cycle chronostratigraphy) for the Cenomanian - Middle Turonian portion of the Greenhorn Cyclothem, a second-order depositional sequence in the Western Interior Basin, from the standard North American reference section at Pueblo, Colorado (modified and expanded from Kauffman, 1995 in press; and Sageman et al., 1995 in press). The types of diversification or reduction bio-events recognized in this section are coded by letters and explained in the key at the bottom of the figure. The magnitude of these bio-events is differentiated by the line widths, with local bio-events (not consistently found outside of this and nearby sections) represented by the thinnest lines, regional to basinwide bio-events by intermediate-width lines, and global bio-events represented by the heaviest lines. Where similar bio-events are extremely abundant and stratigraphically closely spaced (e.g. a sequence of mass mortality beds), the interval is enclosed by a box. Groups of temporally clustered regional to global bio-events reflecting a general response to large-scale environmental changes or perturbations are labeled with a letter (A-H, subsequently discussed), in the third column from the right of Fig. 2.

Several general trends are obvious in this data set: (1) the global bio-event (e.g. the Cenomanian-Turonian mass extinction) is actually composed of a series of closely spaced, ecologically selective bio-events, or extinction steps, which show up in both the observed data (Kauffman, 1988b; Fig. 2), and in statistically predicted data (Harries, Kauffman and Elder, in manuscript), and which do not represent a single catastrophe; (2) the number of bio-events for this stratigraphic interval diminishes from the local, through the regional, to the global scale; (3) bio-events are temporally clustered within certain stratigraphic intervals which represent major short-term changes or perturbations in ocean-climate systems; bio-events are more scattered during background environmental conditions; (4) some regional bio-events reflect intervals, or clusters, of local bio-events of the same character (e.g. closely spaced *Entolium* mass mortality surfaces in the Hartland Shale, and *Collignoniceras woollgari* mass mortality surfaces in the Fairport Member). Whereas individual bio-event surfaces in these clusters have limited lateral distribution, the same kind of bio-events occur in clusters, in the same narrow time interval, across the basin; (5) the taxonomic expression of the same bio-event may vary from basinal to shoreward facies, but the general character of the species diversity and abundance data remains consistent across these bio-event horizons/intervals in all facies (Sageman et al., 1995 in press).

Briefly, the characteristics of the regional North American and global bio-events, or groups of closely related bio-events, as expressed at the standard reference section of the Greenhorn Cyclothem at Pueblo, CO, are as follows (the letters below are keyed to those in the IDRB and RBE columns on Fig. 2).

The Thatcher Limestone Bio-Events

These comprise a series of abrupt colonization and evolutionary diversification bio-events among subtropical to warm temperate molluscs and foraminifera, associated with very rapid immigration of a subtropical water mass into the southern half of the Western Interior Seaway (Kauffman, 1988a). At Pueblo, Colorado, this event is represented by the 15-40 cm-thick Thatcher Limestone Member of the Graneros Shale, and a 0.5 - 1 m interval of moderately calcareous dark clay shale below it (Fig. 2A) (Kauffman, 1985), collectively representing a 3rd-order maximum flooding event. This interval spans 50-100 ka at the base of the Middle Cenomanian *Calycoceras tarrantense* biozone (spanning 95.8-96 Ma; Kauffman et al., 1993). Molluscan diversity increases from 4-5 to over 60 species in this interval, including small caprinid rudists; immigrant taxa are largely related to

Fig. 2 (next two pages). A typical example of the distribution and density of local (thin lines), regional (moderately thick lines) and global-scale bio-events (very thick lines), as documented through high-resolution (cm-dm scale) stratigraphic analysis of the Cenomanian-Turonian Greenhorn Cyclothem, a second-order eustatic sea level cycle and depositional sequence, near Pueblo, Colorado, U.S.A. See key to symbols at bottom of figure. Note abundance of bio-events and their clustering around times of major environmental (facies) changes or perturbations, especially around the Cenomanian-Turonian mass extinction interval. Event and cycle stratification dominates many Cretaceous sequences in this region. Note also that the mass extinction interval (heavy lines and boxes) is composed of a number of discrete regional and global extinction bio-events distributed through >1 Ma of time, and not a single event. The diagram is extensively discussed in text

290

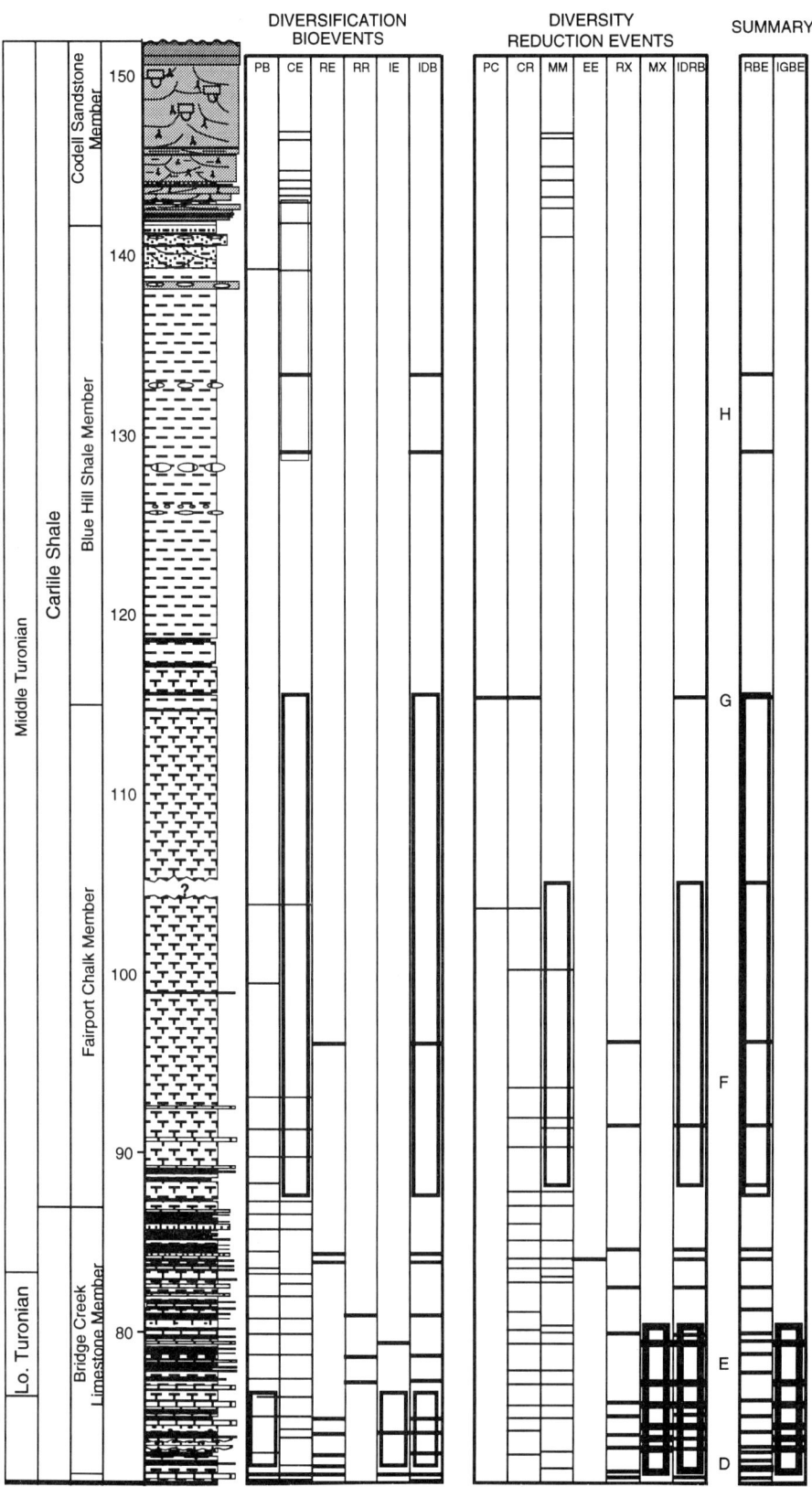

CENOMANIAN-TURONIAN BIOEVENTS, GREENHORN CYCLOTHEM, PUEBLO, CO, USA

Key to line width
- Local bioevent
- Regional bioevent
- Global bioevent
- Regional group of bioevents
- Global group of bioevents

Key to symbols

PB	Population/biomass blooms
CE	Colonization events
RE	Rapid evolutionary events
RR	Rapid radiations
IE	Immigration events
IDB	Integrated diversification bioevents
PC	Population collapse
CR	Community reduction events
MM	Mass mortalities
EE	Emigration events
RX	Regional extinction levels
MX	Mass extinction levels
RBE	Regional bioevents
IGBE	Interregional to global bioevents
IDRB	Integrated diversity reduction bioevents

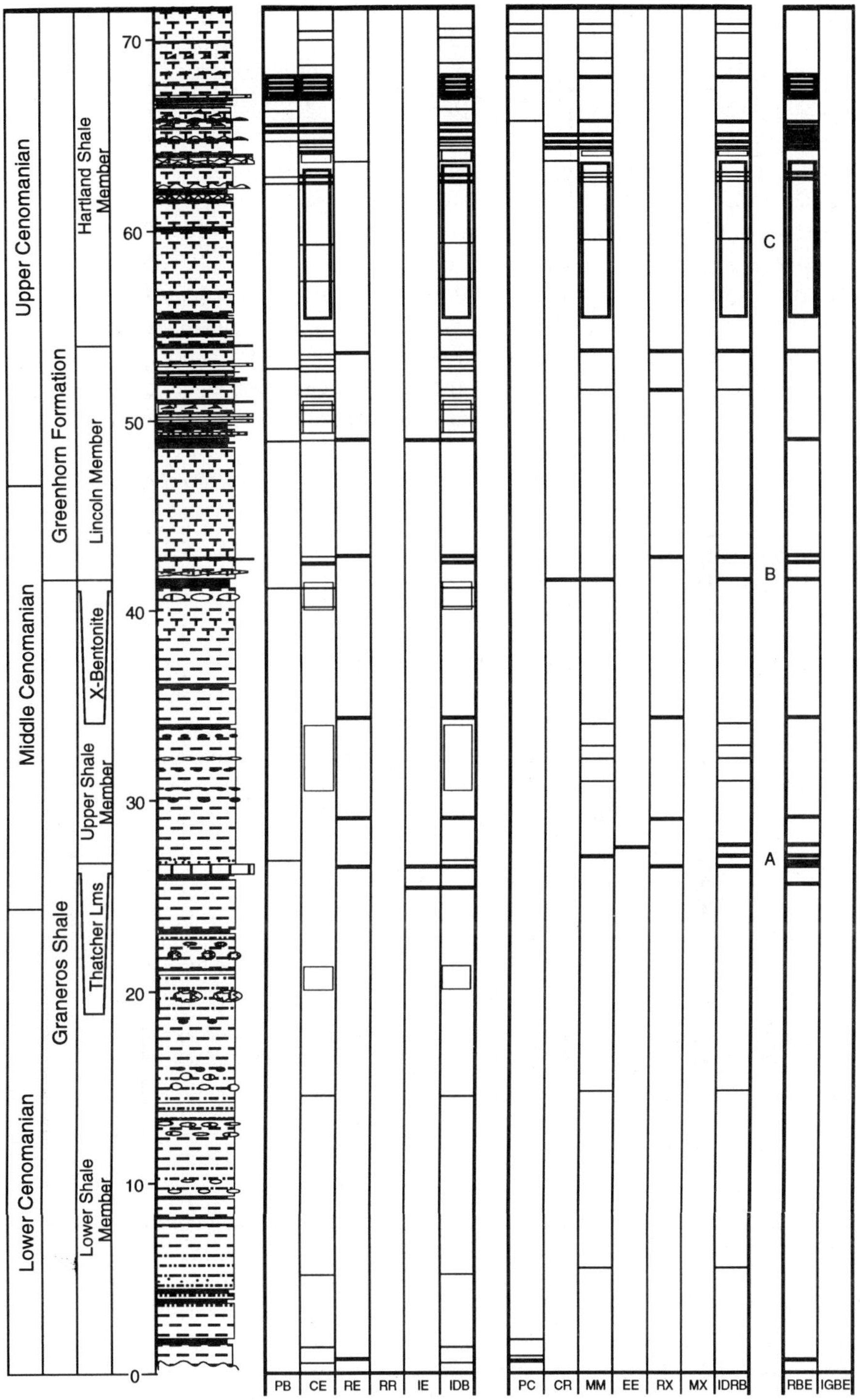

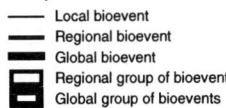

the Woodbine fauna of Texas (Stephenson, 1952) and are further associated with a flood of planktic and calcareous benthic foraminifer species (Gupta and Eicher, personal communication, 1993), including species of *Hedbergella, Heterohelix, Gumbelitria*, and *Globigerinelloides*. This diversification bio-event was rapidly followed, within another 50-100 ka, by an equally abrupt set of emigration bio-events among warm temperate biotas, by population and community decline, and by regional extinction bio-events (Kauffman, 1988a) as the water mass retreated to the Gulf Coast. Molluscan diversity declined to <10 mild temperate species, mainly ammonites and low-oxygen tolerant inoceramid and ostreid bivalves, and virtually all calcareous foraminifers disappeared from the Western Interior Basin (Eicher and Diner, 1985).

The X-bentonite Bio-Events

These are represented by a series of coeval regional community diversification and reduction bio-events in the middle part of the Middle Cenomanian, spanning less than 100 ka (94.95-95.05 Ma; base of *Cunningtoniceras amphibolum* biozone), and recognizable to varying degrees from southern Alberta, Canada, to north Texas. At Pueblo (Fig. 2A), the upper 1 m or less of the Graneros Shale marks an abrupt change from non-calcareous to moderately calcareous shale with limestone concretions and lenses. This interval is associated with rapid immigration of a warm southern water mass. Molluscan diversity changes abruptly from <5 to >15 species, mainly with the addition of southern warm water immigrants, and the abrupt reappearance of planktic foraminifers (Eicher and Diner, 1985; *Hedbergella* and *Heterohelix* spp.). This brief set of diversification bio-events was abruptly terminated by mass mortalities and limited species-level extinctions associated with emplacement of the 30-60 cm-thick "X bentonite" ash fall near the peak of a third-order relative sea level lowstand. Limited planktic and benthic mass mortality surfaces directly underlie the ash, where a number of typical Middle Cenomanian molluscs became extinct, and all benthic foraminifers abruptly disappeared for the next 700 ka. This ash fall is directly overlain by planktic foraminifer population bursts, a widespread oyster colonization event (*Ostrea beloiti* beds), and rapid evolution of a few new late Middle Cenomanian molluscan taxa (*Plesiacanthoceras wyomingense* biozone). These diversification bioevents are possibly associated with nutrient increase in the water column as a result of the ash fall, accelerated immigration of warmer southern waters into the basin, and a relatively abrupt increase in productivity as evidenced by the rapid change from dark, somewhat calcareous clay shales to hemipelagic calcareous shales and shaley chalks (Scott, 1969; Kauffman, 1985).

The Hartland Shale Oxygen Depletion Bio-Events (Fig. 2A)

Increased rates of immigration of a southern warm temperate water mass into the Western Interior Basin late in the Greenhorn transgression (middle Late Cenomanian *Metoicoceras mosbyense* biozone) produced an episodically stratified water column, regional oxygen depletion events (RODEs: the lower and upper Hartland Oxygen Depletion Events of Sageman, 1991), a thick series of laminated, organic-rich shales, and generally deleterious benthic life conditions for about 500 ka. But this generally dysoxic interval was also associated regionally with dynamic, short-term changes in water chemistry and circulation, and short intervals of benthic environments favorable for marine life. These windows of opportunity gave rise to a very closely spaced (1-10 cm) sequence of <1 to 3 year-long colonization events (based on cohort analysis of benthic bivalves), followed by mass mortality surfaces, among benthic, low-oxygen adapted molluscs (e.g. *Entolium, Inoceramus, Phelopteria*, small oysters), spread over 10-15 m of section. These are interspersed with longer-term colonization events among benthic bivalves associated with sedimentary bypass, condensation, and mass flow events producing calcarenite beds, 1-5 cm thick, within this sequence. The upper part of this sequence is further characterized by primary biological response in benthic community structure to depositional cycles reflecting Milankovitch climate cycles, similar to those described by Kauffman for younger Cretaceous strata (1988a). Whereas it is difficult to regionally trace single, closely spaced, colonization and mass mortality surfaces in this dynamic environmental interval, the RODE intervals themselves, each spanning about 200 ka and containing the same types of bio-events, contain a unique set of bio-events that can be regionally traced from Texas to Montana (Sageman, 1991).

The OAE-II Bio-Events (Figs. 2A,B)

In the latest Cenomanian, initiation of Oceanic Anoxic Event II was represented in the Western Interior Basin by an abrupt increase in the rate of sea level rise, continued rapid immigration of southern warm temperate to subtropical water masses (Kauffman, 1984), abrupt positive shifts in $\partial^{13}C$ and $\partial^{18}O$ values, initiation of a series of trace element advection and/or enrichment events, and dynamic changes in the burial rates of organic carbon (Kauffman, 1988b, 1995 in press; Pratt et al., 1993). At the Pueblo, Colorado, reference section this interval is about 250-300 ka long and spans the *Sciponoceras gracile* to basal *Neocardioceras juddii* biozones. A series of stratigraphically closely spaced, regional diversification events, population blooms, benthic colonization events, and waves of immigration of subtropical to warm-temperate marine molluscan lineages are associated with the rapid northward advance of southern warm water masses. Molluscan diversity increases within 2 m from around 25 species in the uppermost Hartland Shale levels to over 260 species in the main part of the *S. gracile* biozone (lower Bridge Creek Limestone and equivalents), about 100-150 ka later. After a long interval without benthic foraminifers, or with rare agglutinated species, an abrupt, widespread, apparently coeval colonization event among diverse benthic foraminifers occurs at the initiation of this interval (Benthonic Zone of Eicher and Diner, 1985). These diversification bio-events were soon countered by a regional series of molluscan extinctions (Elder, 1987, 1989; Kauffman, 1988b), responding to deteriorating conditions in the lower water column and benthic zone, maximum oxygen depletion, and the first trace element enrichment events including Iridium in the Ce-Tu boundary sequence (Orth et al., 1993). These mark initiation of the Cenomanian-Turonian mass extinction interval about 520-540 ka below the Ce-Tu boundary (Kauffman, 1995 in press).

The Cenomanian-Turonian (C-T) Mass Extinction Bio-Events

This widely recognized global complex of bio-events is represented in high-resolution stratigraphic data at Pueblo, Colorado, by a closely spaced, ecologically graded series of extinction events, or steps, in the observed data spanning nearly 1 Ma of latest Cenomanian and earliest Turonian time in the Western Interior Basin (Fig. 2B) (Elder, 1985, 1987, 1989; Kauffman, 1988b, 1995 in press). This mass extinction was unusual in that it occurred near the maximum flooding interval of the Greenhorn Cyclothem (Kauffman, 1984), a second-order, tectonoeustatic, depositional sequence. Many of the individual extinction events in this sequence are intercontinental to global in nature. Similar records are found in Europe (e.g. papers in Jarvis et al., 1988) and South America (Koutsoukos and Hart, 1990a,b; Villamil et al., 1995 in press). The Cenomanian-Turonian mass extinction interval, which actually initiated in the Caribbean Province with the late Early to middle Late Cenomanian demise of tropical reef ecosystems (Johnson and Kauffman, 1990), started later (latest Cenomanian, 93.9 Ma) in the more temperate Western Interior Basin, as seen at the Pueblo, Colorado section, and proceeded through a series of five Late Cenomanian and two earliest Turonian steps of extinction in the observed data (e.g. at Pueblo; Elder, 1985, 1987; Kauffman, 1988b, 1995 in press). Initial extinction levels in the *S. gracile* biozone, during a general time of taxonomic diversification, are characterized by the abrupt global loss of keeled planktic foraminifers (the *Rotalipora* extinction) and a few important subtropical molluscan lineages. Four additional Late Cenomanian extinction steps, each spanning only thousands of years, involve the progressive loss of warm to mild temperate and/or cosmopolitan molluscs and calcareous plankton. The final large extinction event occurred at the Ce-Tu boundary, and smaller Early Turonian extinction events mainly involve cosmopolitan and temperate molluscan lineages (Kauffman, 1988b, 1995 in press; Harries and Kauffman, 1990). The mass extinction interval in the Western Interior Basin mainly involves regional extinction events among endemic North American molluscs, and global events among more cosmopolitan molluscs and calcareous plankton (data in Elder, 1985, 1987, 1989; Harries and Kauffman, 1990; Kauffman, 1988b, 1995 in press). The entire mass extinction is enveloped in a stratigraphic interval characterized by extraordinary variations in oceanic chemistry and temperature, and spans two to three major global iridium spikes (Orth et al., 1993) as well as preliminary physical evidence (microsphere concentrations) for an impact shower. Extinction steps are characteristically associated with these major geochemical perturbations, whether earth-bound or extraterrestrial in origin, representing ocean-climate perturbations which progressively exceeded the adaptive ranges of tropical to temperate Cenomanian marine taxa. Kauffman (1995 in press) describes this extinction "event", and its causes, in considerable detail. There are few diversification bio-events and little origination, mai nly among crisis progenitor bivalve lineages, during the Ce-Tu mass extinction interval (Kauffman and Harries, 1995 in press). Instead, successive extinction bio-events during the last 520 ka of the Cenomanian resulted in a progressive loss of marine diversity and change in community structure from more equitable, diverse, level bottom communities to more inequitable, low-diversity communities.

The Fairport Regional Oxygen Depletion Bio-Events

These are associated with a 900 ka-long sequence of well-laminated, organic-rich chalky shales of the early Middle Turonian Fairport Chalky Shale Member (*Collignoniceras woollgari* biozone), Carlile Shale, in the Western Interior Basin (Fig. 2B). These mark the early highstand systems tract of the Greenhorn Cyclothem/depositional sequence. These facies reflect the development of highly stratified southern and northern water masses in the basin (Kauffman, 1984, 1988a) following the highest sea level stand of the Cretaceous (300 m above present stand). Both the benthic zone and large portions of the water column were episodically affected by dysoxic to anoxic conditions which, however, were frequently punctuated by brief storm or current-generated oxygenation events. These conditions produced an interval characterized by closely spaced, alternating, short-term colonization and mass mortality events, mainly among juvenile and young adult populations of ammonites (especially *Collignoniceras woollgari*) and benthic bivalves (e.g. *Inoceramus, Pseudoperna*). As in the Hartland Regional Oxygen Depletion Events, individual colonization or mass mortality bio-event surfaces of the Fairport sequence are difficult to trace regionally, but the 100-300 ka-long interval itself, characterized by hundreds of such surfaces, can be recognized from Montana into New Mexico. Retreat of southern water masses with falling sea level caused a moderate lowering of water temperatures and rapid waves of emigration of southern warm-water elements in the biota during this time. A series of regional extinction levels among molluscs and benthic foraminifers further characterize this interval (Kauffman, 1988a; 1995 in press).

Middle Turonian Downlap Bio-Events

This mid-Middle Turonian downlap surface lies at the contact between the Fairport Chalk and Blue Hill Shale members (90.7 Ma), Carlile Shale, at Pueblo, Colorado, and elsewhere in the Western Interior Basin. It marks the junction of the *Collignoniceras woollgari* and *Prionocyclus hyatti* biozones, and is associated with an abrupt acceleration in the rate of eustatic sea level fall. This was characterized by a rapid shift from carbonate-dominated to clay-dominated sedimentation with downlap of siliciclastic wedges throughout the basin, and an increase in both oxygen and turbidity in the water column. An abrupt regional depletion of both benthic and calcareous planktic populations, a significant drop in species diversity, moderate species-level extinction, and abrupt restriction of low-oxygen adapted, *Inoceramus*-dominated, benthic communities resulted within a very short time interval throughout the basin. These changes comprise a set of closely related, regional diversity reduction bio-events. The final Turonian emigration of warm-temperate molluscs toward the Gulf Coast was also coincident with this surface, which probably represents less than a thousand years in duration.

The Blue Hill Shale Colonization Bio-Events

During mid-regression (mid-highstand systems tract) of the second-order Greenhorn marine cycle/depositional sequence, eventually leading to the development of a sequence boundary near the Middle-Late Turonian boundary, the Western Interior Seaway rapidly became shallower and broadly oxygenated above the 90.7 Ma downlap surface. This reflected an increase in the rate of sea level fall and breakdown of water mass stratification for the next 800 ka. Pervasive bioturbation characterized benthic sediments for the first time in the marine cycle, and times of fine sediment bypass and development of coarser-grained lag deposits produced numerous surfaces favorable to regional colonization and diversification bio-events among diverse benthic molluscs in the middle and upper Blue Hill Shale Member, Carlile Shale. These concretion levels, and their sparse but diverse, temperate, benthic molluscan faunas, have been traced as chronostratigraphic surfaces over a five-state area in the Western Interior Cretaceous Basin. For each, colonization seems to have lasted only a few years at most, based on molluscan growth characteristics and size-cohort analyses of shells.

The data presented here from Pueblo, Colorado, demonstrate the diverse characteristics of regional bio-events, at different scales, in the field, and the complex nature of global bio-events like the Cenomanian-Turonian mass extinction interval. These data also show the advantages of high-resolution event/cycle chronostratigraphic analyses in focusing attention on event stratification, including diverse bio-events, in the rock record. Whereas the general purpose of lithostratigraphy and biostratigraphy is to seek natural divisions of the stratigraphic record through evaluation of lithologic and taxonomic similarities within rock units (e.g. formations, members, biozones), event and cycle chronostratigraphy seek to document the dynamic history of the Earth and its biosphere by focusing on short-term, regional, biological, geochemical, and physical events in the stratigraphic record, irrespective of broader similarities or differences among biostratigraphic or lithostratigraphic units. In the North American Cretaceous, this research suggests that event and cycle deposition actually dominate the stratigraphic record, even in seemingly uniform basinal facies of a single lithogenetic unit. This is predictable considering the relatively stable, equable nature of Cretaceous greenhouse environments which, like the biotas that adapted to them, may have been highly sensitive to even small-scale perturbations in ocean-climate systems.

Regional and Global Cretaceous Bio-Events of the Americas

There is a major contrast in the documentation of pre-Albian and younger regional to global bio-events in the Americas. The pre-Albian paleobiologic history of this hemisphere has not been documented with sufficiently high-resolution data, and in enough areas, to interpret the timing and patterns of major bio-events or to correlate them precisely on a global scale. In contrast, Albian and younger strata and biotas have been more extensively studied than in most other parts of the Cretaceous world; local, regional, and global bio-events have been documented in great detail for most the middle and Late Cretaceous (examples in Kauffman, 1986, 1988a, Kauffman et al., 1991, and references therein), and tied to a high-resolution chronostratigraphic, sequence stratigraphic, and biostratigraphic system with global correlation potential (e.g. see Kauffman et al., 1993, and references therein). Figure 3 shows the relative Cretaceous sea level curve for North America (from Kauffman and Caldwell, 1993), the distribution of regional and oceanic dysoxic or anoxic events, and the distribution of important regional and global bio-events defined herein. These bio-events are briefly described below, from oldest to youngest, with key references to the data that support their age, distribution and magnitude.

The Jurassic-Cretaceous (J-K) Boundary Mass Extinction Bio-Event (141 Ma)

Sepkoski (1993) has most recently determined that the J-K boundary represents a second-order mass extinction boundary (about 7% familial extinction), and is thus potentially a global bio-event. The terrestrial record of this extinction is still poorly documented in the Americas. Upchurch and Wolfe (1993) did not distinguish latest Jurassic from Neocomian megafloral assemblages where preserved in the Kootenay and Nikanassin megafloras of western Canada. But some marine data document the importance of this event. Bakker (1993) noted that the long-necked Late Jurassic Plesiosaurs (Muraenosaurs and Cryptocleidids) suffered near-total extinction at the Jurassic-Cretaceous boundary. In northern Canada, Jeletzky (1984) extensively discussed the problems of defining the Jurassic-Cretaceous boundary based on molluscs, noting that (p. 177), "none of the previously proposed variants of the delimitation of the Jurassic and Cretaceous Systems is based on a well defined biochronological event that permits an easy recognition of the Jurassic-Cretaceous boundary on a global scale." He chooses the definition of the J-K boundary favored by most specialists: the base of the Berriasian Stage is the base of the *Grandis-Jacobi* Zone as developed in southeastern France, which overlies unfossiliferous limestones of presumed latest Tithonian age. Nevertheless, Jeletzky (1984, Fig. 5) showed only about a 21 percent survival among ammonite genera across the J-K boundary in various North and South American localities, but high survival levels among lineages of the bivalve *Buchia*, and even among several species of the ammonite *Berriasella*. Ascoli et al. (1984) documented a depletion of 18 of 22 ostracode species, and 28 of 35 foraminifer species in three steps at the end of the Kimmeridgian, mid-Tithonian, and at the end of the Tithonian, without significant origination, on the Scotian Shelf. 30-35% of this extinction took place at the J-K boundary. These steps all had coeval Western European equivalents and thus interregional to global expression. The J-K extinction bio-event in North America therefore appears to be well defined, but not well documented with high-resolution stratigraphic data for diverse organisms.

The Late Aptian Mass Extinction Bio-Event (112-114 Ma).

Sepkoski (1993) noted that the Aptian-Albian boundary interval contained a global second-order mass extinction in familial and generic data. The interval is especially poorly studied in the Americas, and data in support of this bio-event are sparse. In the terrestrial realm, Upchurch and Wolfe (1993) noted no major vegetational changes across this boundary interval, nor are there significant changes in fresh water molluscs (Kauffman and Hanley, in manuscript). Johnson and Kauffman (1990) noted a small Late Aptian extinction peak among primitive tropical rudistid bivalve lineages (and by proxy, probably other taxa within reef ecosystems) utilizing per-taxon extinction rate analyses. This Caribbean-wide extinction event follows the normal pattern of demise of tropical reef ecosystems, occurring some time before the major portion of the mass extinction at the end of the Aptian. Young's (1974) ammonite data from the Gulf Coastal Plain shows a significant break in zonal species, but not genera or lineages, across the Aptian-Albian boundary, with dominant lineages of *Kasanskyella* and *Hypacanthoplites* spanning the boundary, whereas *Dufreynoya* and less common ammonite genera disappear. The Aptian mass extinction occurs near the end of OAE Ia, but precise causal mechanisms have not been investigated through high-resolution geochemical, stratigraphic, and paleobiological analyses.

The Middle-Late Albian Boundary Bio-Event (103 Ma)

There is a significant regional bio-event at the Middle-Upper Albian boundary throughout Europe and parts of India (subsequently discussed). Although these strata are poorly represented in North America (northern interior and Gulf Coast only), and have not been well documented in South America, some evidence exists that this even may also affect the North American record. Based on the work of Jeletzky (summarized in 1968, 1970), there appears to be a significant, short-lived extinction among Middle Albian ammonite genera and species in the Northern Interior Subprovince of Canada (partially summarized in Kauffman et al., 1993), with the loss of important species groups belonging to *Gastroplites, Stelckiceras*, and *Pseudopulchellia*. This is followed by a thick stratigraphic interval almost barren of ammonites and other molluscs which may reflect the regional effects of the onset of OAE Ic (Jenkyns, 1980) and its first incursion into epicontinental seas in the early Late Albian. This same event seems to be poorly defined in existing data from the warm temperate Gulf Coast subprovince, however. In this region, only *Oxytropidoceras* disappears among major zonal ammonites (as well as several less common genera), whereas species of such dominant, biostratigraphically useful genera as *Metengonoceras* and *Manuaniceras* continue well up into the Late Albian. Much more high-resolution data are needed to confirm this as a regional bio-event, especially among other faunal groups and from terrestrial ecosystems.

The Albian-Cenomanian Boundary Bio-Events (97.2-98.3 Ma)

This traditional division between the Early and Late Cretaceous has been widely considered to represent a major, abrupt change in at least marine biotas. Yet this boundary has been remarkably difficult to define in North America because of difficulties in correlation to the European stage sequence (reviewed in Cobban and Kennedy, 1989). The traditional boundary in the North American Western Interior Basin has been taken at the top of the endemic *Neogastroplites* ammonite range zones (top of the *N. maclearni* biozone 97.2 Ma; Kauffman et al., 1993, Fig. 6) and on the Gulf Coast at the top of the *Drakeoceras gabrielense - D. drakei* biozone, marking the end of a relatively short-term (500 ka) regional extinction interval characterized by the loss of such common ammonite genera as *Drakeoceras, Mortoniceras, Pervinquieria, Eopachydiscus, Boesites, Adkinsites*, and *Venezoliceras*, as well as typical and numerically dominant Late Albian bivalve lineages among *Texigryphaea, Neithea*, Trigoniidae, Cardiidae, Inoceramidae (*Birostrina* and the *Inoceramus comancheanus* plexus), certain turritellid gastropod lineages, and many groups of echinoids. The Gulf Coast record seems to support a major extinction interval, overlain by sparsely fossiliferous Early Cenomanian strata with different biotas, at the designated Albian-Cenomanian boundary. But in the Western Interior Basin, Cobban and Kennedy (1989) have recently argued that species of *Metengonoceras* associated with at least the highest three *Neogastroplites* biozones (see Kauffman et al., 1993), are closely related to earliest Cenomanian forms on the Gulf Coast (though not conspecific). If

Fig. 3 (opposite page). A relative sea level curve for the Western Interior Basin of North America (left: from Kauffman and Caldwell, 1993) compared to the Haq et al. (1987) global cycle chart (middle; onlap-offlap curve only), regional oxygen depletion events (RODE: Kauffman, 1984), global oceanic anoxic events (OAE), and both regional and global bio-events expressed in the North American Cretaceous (right two columns). See text for detailed discussion

RELATIVE SEA LEVEL HISTORY, OXYGEN DEPLETION EVENTS, AND REGIONAL TO GLOBAL BIOEVENTS, CRETACEOUS OF NORTH AMERICA

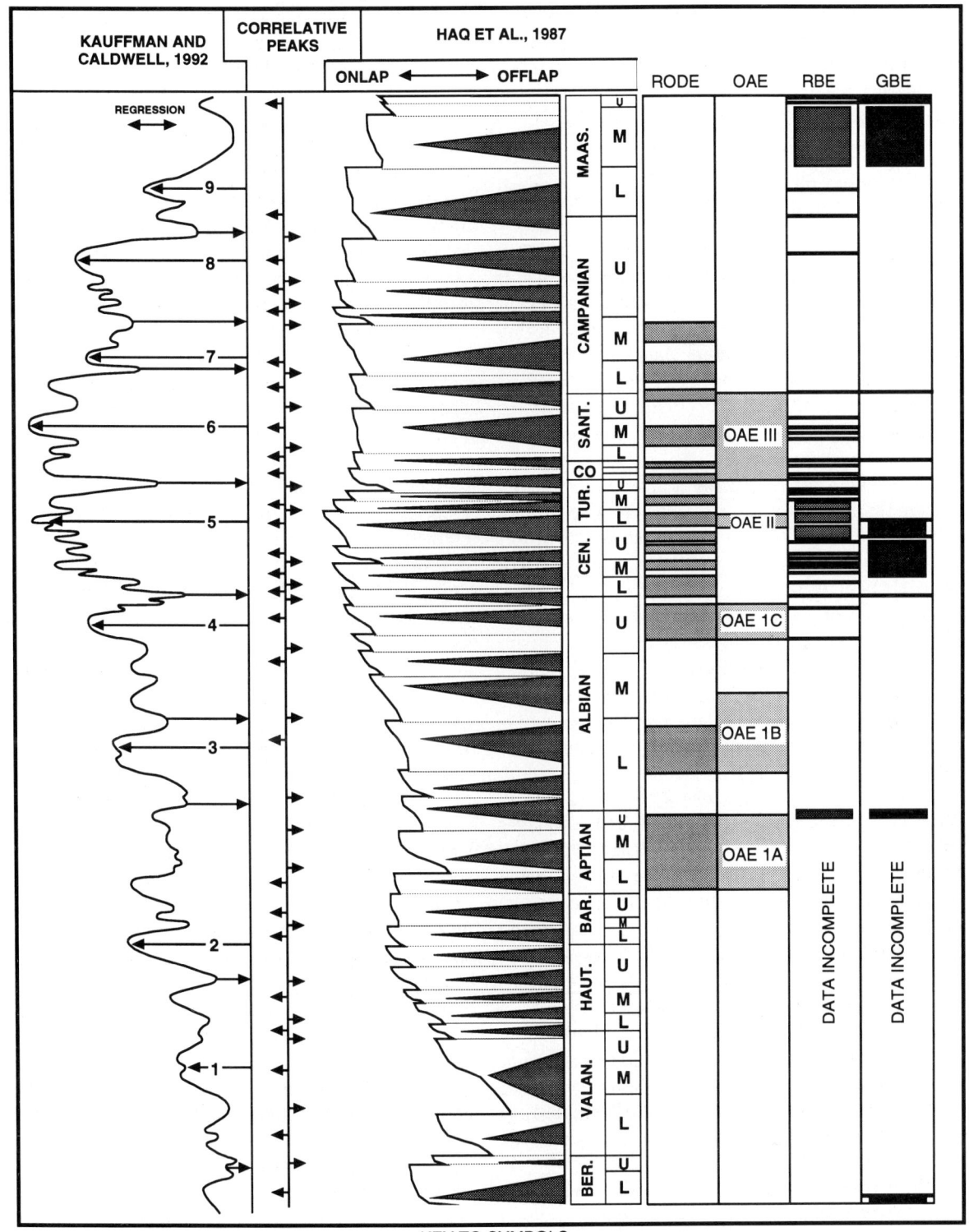

KEY TO SYMBOLS

GBE Global Bioevents
RBE Regional Bioevents
OAE Oceanic Anoxic Events
RODE Regional Oxygen Depletion Events

this is the case, and there is still considerable debate, there would be no evidence of a regional bio-event at the Albian-Cenomanian boundary in the Western Interior Basin of North America, with most known lineages passing through this level. The two suggested boundaries are illustrated and briefly discussed in Kauffman et al. (1993).

The Cenomanian-Turonian (Ce-Tu) Boundary Mass Extinction Bio-Events (93.4 Ma)

This is one of the best-studied mass extinctions in the Americas. In North America, Eicher and Diner (1985, 1989), Leckie (1985), Caldwell et al. (1993), and Watkins et al. (1993), have documented patterns of extinction among calcareous plankton, and Elder (1985, 1987, 1989), Kauffman (1988b, 1995 in press), Johnson and Kauffman (1990), and Harries and Kauffman (1990) have documented molluscan extinction patterns.

In South America, Tomas Villamil and Claudia Arango (in manuscript) have recently compiled a high-resolution foraminiferal and molluscan record of this mass extinction interval in Colombia, tied to detailed trace element (Orth et al., 1993), stable isotope, organic carbon, and event/cycle/sequence stratigraphic analyses. In the Cretaceous sedimentary basins of the Brazilian Atlantic margin (Campos, Santos, Sergipe-Alagoas basins, etc.), microfaunal changes globally associated with the Late Cenomanian extinction events have been documented by Koutsoukos and Hart (1990a,b), and by Koutsoukos et al. (1990). Whereas the genus *Rotalipora* is extremely rare in these basins, the majority of other faunal changes associated with this stratigraphic level have been documented. Ce-Tu boundary research is now being extended into Venezuela, where similar patterns are being found, and more extensively into Brazil, by E.G. Kauffman, C.C. Johnson, and T. Villamil of the University of Colorado, and a group headed by L.M. Pratt at Indiana University, in conjunction with E. Koutsoukos and other PETROBRAS scientists.

The Cenomanian-Turonian boundary extinction occurs during peak greenhouse conditions, with CO_2 levels 2-3 times present levels (Berner, 1994), and just prior to the highest temperature peak (Jenkyns et al., 1994) and eustatic sea level rise of the Mesozoic (300 m above present stand: Pitman, 1978; Haq et al., 1987; as documented in the Western Interior of North America by McDonough and Cross, 1991). It also occurs during a time of extraordinary global fluctuations in geochemical parameters defining temperature, water chemistry, and carbon cycling, as represented in the stable isotopic data for North America by Arthur et al. (1985), Pratt (1985), Pratt et al. (1993), Kyser et al. (1993), Orth (1989) and Orth et al. (1993), among others, and for South America (Colombia) by Orth et al. (1993), and Villamil et al. (1994 in press). These data have been summarized by Kauffman (1988b; 1995, in press) as they relate to Cenomanian-Turonian Mass Extinction events.

These data all show that the Cenomanian-Turonian boundary mass extinction in the Americas was not a single bio-event, but rather a 1-1.4 Ma-long complex of short-term biological responses ("steps" of Kauffman, 1988b) to an interval of major ocean-climate perturbations recorded in the geochemical, sedimentological and paleobiological record. Based on preliminary identification of microspheres at two and possibly three stratigraphic levels in Colombia (Villamil and Arango, in manuscript), and tentative identification of shocked quartz grains from the latest Cenomanian of central Colorado reported by Rampino et al. (1993), some of these perturbations, and related extinction steps, may be caused by multiple extra-terrestrial perturbations (presumably a comet storm, sensu Hut et al., 1987), especially oceanic impacts and their dynamic environmental feedback processes (Kauffman, 1988b; 1995 in press).

The extinction steps that collectively represent the Cenomanian-Turonian (Ce-Tu) Mass Extinction Bio-event (sensu Kauffman, 1988b; 1995 in press), as known from large observed data sets throughout North America and northern South America, are summarized below (age determinations are taken from the calculated time scale of Kauffman et al., 1993, based on new radiometric ages of Obradovich, 1993): (a) A series of latest Early to middle Late Cenomanian extinction events among rudistid bivalves (especially primitive lineages among the Diceratidae, Requieni-idae, Monopleuridae, and Caprinidae) and other tropical taxa resulted in collapse of the Caribbean Province reef ecosystems by 500 ka-1 Ma below the Ce-Tu boundary (Johnson and Kauffman, 1990; Johnson, 1993). Data for this extinction event, drawn mainly from Mexican localities, is still imprecisely dated, however. This event may be intercontinental or even global, for similar age rudistid extinctions have been documented by J. Philip (1991, personal communication) in the Mediterranean region prior to the Ce-Tu boundary. (b) In the subtropical to temperate climate zones, the mass extinction initiated with the abrupt loss of a few warm temperate to subtropical ammonite and bivalve taxa, about 520-540 ka below the Ce-Tu boundary (Elder, 1985, 1987, 1989), associated with major positive shifts in stable isotope values (Arthur et al., 1985; Pratt, 1985; Pratt et al., 1993), and the first regional to global trace element enrichment peak, including an iridium spike (Orth, 1989, Orth et al., 1993); this comprises extinction step MX1a of Kauffman (1988b, 1995 in press). (c) Global extinction of keeled planktonic foraminifers, the *Rotalipora* extinction, 490 ka below the Ce-Tu boundary, is associated with possible microspheres in Colombia (Villamil and Arango, in manuscript), the first major positive peak in stable isotope fluctuations (Pratt, 1985), and initiation of a broad interval of moderate Iridium enrichment in North America (Orth et al., 1993). (d) A second step of subtropical to warm temperate molluscan extinction (Elder, 1985, 1987, 1989; MX1b of Kauffman, 1988b, 1995 in press) occurs about 390 ka below the Ce-Tu boundary during one of the most geochemically dynamic intervals of the Late Cenomanian, with strong negative excursions in $\partial^{13}C$ and $\partial^{18}O$ and two major trace element spikes, including relatively low-level Ir enrichment associated with the extinction. (e) One of the largest, probably global, extinction levels among subtropical to warm temperate molluscs (Elder, 1985, 1987, 1989; MX2 of Kauffman, 1988b, 1995 in press), depleting known diversity by 55% in North America, occurs 340-350 ka below the Ce-Tu boundary. This is at the base of the *Neocardioceras juddi* biozone (Elder, 1987, 1989: Kauffman, 1988b, 1995 in press; Kauffman et al., 1993), and is associated with the peak positive spike in the global $\partial^{13}C$ excursion, rapid positive and negative fluctuations in $\partial^{18}O$ values, and a major trace element enrichment horizon, including a significant iridium spike. Possible microspheres have been found at this level in Colombia (Villamil and Arango, in manuscript). (f) A second major (regional) extinction (MX3 of Kauffman, 1988b, 1995 in press), among more warm temperate and cosmopolitan molluscs and benthic foraminifers occurs about 320 ka below the Ce-Tu boundary near a peak positive excursion of $\partial^{13}C$ and a trace element enrichment spike. (g) A regional, moderate level (30-35%) loss of planktic foraminifers and temperate molluscan species (Elder, 1985, 1987, 1989; MX4 of Kauffman, 1988b, 1995 in press) occurs about 225 ka below the Ce-Tu boundary, associated with a major positive spike in the global $\partial^{13}C$ excursion. (h) At the Ce-Tu boundary, in a zone of rapid stable isotope fluctuations, the last positive peaks in the global $\partial^{13}C$ and $\partial^{18}O$ isotopic excursions, abrupt C_{org} reduction, a U-Th trace element enrichment horizon, and near peak global sea level rise, a final major molluscan extinction occurs (70% of surviving species; data from Elder, 1985, 1987, 1989). This is associated with a major depletion in planktic foraminifers (Eicher and Diner, 1985) and a marked nannoplankton extinction event (Watkins et al., 1993) (MX5 of Kauffman, 1988b, 1995 in press). (i) Abrupt extinction steps continue into the basal Turonian in North America, during the survival and early recovery intervals of the Ce-Tu mass extinction interval (Harries and Kauffman, 1990), and are considered part of the Ce/Tu "bio-event." The first of these (MX6 of Kauffman, 1988b; 1995 in press) involved the abrupt loss of a few temperate molluscan taxa 200-230 ka above the Ce-Tu boundary during a time of continued short-term stable isotope fluctuations and trace element enrichment events. (j) The final important extinction event (MX7 of Kauffman, 1988b; 1995 in press) involves abrupt loss of prevalent nannoplankton species (Watkins et al., 1993) shortly followed by temperate molluscan species (about 40-45% of existing diversity: Harries and Kauffman, 1990). These extinctions were associated with strong negative $\partial^{13}C$ and C_{org} excursions, and a positive $\partial^{18}O$ excursion. Important short-term regional origination events among molluscs occur between MX3 and MX4, between the nannoplankton extinction and MX5, and between the nannoplankton extinction and MX-7. Whereas these stepwise extinction events are predicated on observed, high-resolution range zone data and are thus subject to criticism because of the Signor-Lipps effect (1982), recent evaluation of these data using the Marshall (1991) statistical method for plotting predicted ranges (Harries and Kauffman, in manuscript) shows that many of the steps of extinction continue to show up in predicted species range data, whereas others become more graded in aspect; there is no evidence in predicted range data for a single mass extinction bio-event.

The Turonian-Coniacian Faunal Turnover (88.7 Ma)

In North America, as also in Europe, there is a large-scale, short-term marine faunal turnover within a few ka around the Turonian-Coniacian boundary (88.7 Ma). This occurs in the lower part of the Niobrara Cyclothem in North America, near the first of four 3rd-order maximum flooding intervals (T7A of Kauffman, 1985) which produce thick sequences of limestone-shale/marl bedding cycles (Milankovitch cycles) in basinal facies. Environmental conditions were very similar to those at the Cenomanian-

Turonian mass extinction boundary, including a significant positive $\partial^{13}C$ excursion, and abrupt loss of over 55 percent of both endemic and cosmopolitan molluscan taxa in the Americas, especially among ammonites and bivalves. Neither benthic nor planktic foraminifers reflect this extinction event, however. The bio-event is marked by loss of cosmopolitan inoceramid bivalves of the *Mytiloides straitoconcentricus, M. incertus-M. fiegei, M. dresdensis*, and *M. mytiloidiformis* lineages as well as the pervasive ammonite genus *Prionocyclus*. These inoceramids are largely replaced, normally within a meter of rock (+/- 100 ka), by typical Coniacian *Cremnoceramus (Cr? waltersdorfensis, Cr. rotundatus, Cr. erectus; Cr. deformis* lineage) and less common *Inoceramus* s.s. species, together with the dominant ammonite genera *Forresteria, Peroniceras,* and *Gauthericeras*, most of which are also cosmopolitan forms.

The Coniacian-Santonian Faunal Turnover (86.6 Ma)

Another marked, short-term replacement of dominant North American and cosmopolitan molluscs marks a 1-2 m interval (+/- 200 ka) at the Coniacian-Santonian boundary. Among ammonites, Coniacian *Peroniceras, Scaphites, Phlycticrioceras,* and small *Baculites* spp. disappear and are replaced by *Clioscaphites* and larger species of *Baculites*. Among Inoceramidae, dominant lineages of the *Magadiceramus subquadratus* group, and surviving *Mytiloides, Cremnoceramus,* and *Volviceramus* are abruptly replaced by new genera (*Sphenoceramus, Cladoceramus, Cordiceramus*), which rapidly radiate during Santonian time with rising global sea level. This turnover takes place under generally equable environmental conditions associated with continued global tectonoeustatic sea level highstand, during the second maximum flooding interval (T7B of Kauffman, 1985) of the Niobrara Cyclothem in North America. No detailed geochemical data are available yet to determine possible causes for this bio-event.

The Santonian-Campanian Faunal Turnover (83.5 Ma)

The final sharp faunal turnover at Cretaceous stage boundaries in North America occurs at the Santonian-Campanian boundary, where the Cladoceramid and most Sphenoceramid bivalves disappear, and are replaced by radiations among typical Campanian *Platyceramus* and *Endocostea*. Similarly, the *Clioscaphites* and *Desmoscaphites* ammonite plexes disappear and are replaced by the multinodose *Scaphites* (*S. leei, S. hippocrepis* lineages). Detailed data on other groups is not yet available. This stage boundary is less well defined by faunal turnover than those preceeding it, however, mainly because of the continuity of other bivalve lineages among *Sphenoceramus, Platyceramus, Cordiceramus,* and *Endocostea*. This bio-event occurs at or very near to a 3rd-order sequence boundary (R7C of Kauffman, 1985) during a Coniacian-Early Campanian period of generally elevated sea level. The faunal replacement marks initiation of a regional oxygen reduction event as evidenced by abrupt increase in organic carbon and pyrite values, and increased preservation of lamination in overlying basal Campanian strata.

The 68 Ma Extinction Bio-Event

Kauffman (1988b, Fig. 3) argued that the Cretaceous-Tertiary mass extinction was stepwise in nature and actually began between 67.5 and 68 Ma with the abrupt extinction of rudistid bivalve-dominated reef ecosystems (Johnson and Kauffman, 1990), the loss of most inoceramid bivalve lineages (MacLeod, 1994), and a great reduction in the diversity of shallow, warm water, level bottom communities. New radiometric ages (Obradovich, 1993) suggest now that this event took place between 68 and 68.5 Ma. Since this proposal, Johnson and Kauffman (1995 in press) have confirmed with detailed stratigraphic work that Caribbean Province rudistid reef ecosystems collapsed suddenly within the concurrent planktic range zones of *Gansserina gansseri* and *Quadrum trifidum*, 68-68.5 Ma ago. Swinburne (1991) documented a similar Middle Maastrichtian demise of rudistid reefs in the Mediterranean Region, but did not give a precise date. Ward et al. (1991) and MacLeod (1994) have most recently documented that the main Maastrichtian lineages of Inoceramidae (excepting *Tenuipteria*) also became extinct within the Middle Maastrichtian (about 67-69 Ma). Johnson (1992, and personal communication, 1994) has shown that the North American angiosperms showed an abrupt Middle Maastrichtian extinction (40-60% of standing diversity) at about 68 Ma. Collectively, these data suggest that a major extinction interval around 68 Ma ago characterized the Middle Maastrichtian, and may have been global in aspect in the marine realm. Proof of this as a global bio-event will depend upon more precise correlation between marine basins, and to continental facies. Causal mechanisms for this event, or series of closely spaced events, are still under investigation; whereas there is a general global cooling and eustatic fall under way at this time, there are no indications that rates of change were beyond background levels, and there are no independent detailed geochemical data available to suggest significant perturbation in the ocean-climate system.

The Cretaceous-Tertiary Mass Extinction Bio-Event

Perhaps the most widely accepted of all Cretaceous bio-events, and that which is most catastrophic in nature, is the terminal Cretaceous mass extinction associated with evidence for impact of a major bolide (probably a comet) on Earth 65.5 Ma ago — the Chicxulub, Mexico, impact event (Alvarez et al., 1992; Hildebrand et al., 1991; Quezada-Mumeton et al., 1992). Abrupt trace element and stable isotope excursions are associated with this impact. A major extinction among calcareous plankton, dominant Mesozoic Mollusca (e.g. rudistids, inoceramids, ammonites, belemnites, diverse oysters, actaeonellid and nerineid gastropods, etc.), echinoids, corals, marine and terrestrial reptiles, and plants at or very near to the Cretaceous-Tertiary boundary impact is recorded in high-resolution observed data presented by many authors (e.g. see papers in Birkelund and Bromley, 1979; Christensen and Birkelund, 1979; Silver and Schultz, 1982; Lamolda, Kauffman and Walliser, 1988; Sharpton and Ward, 1990; and Kauffman and Walliser, 1990, among many contributions). Alvarez et al. (1980, 1982), have suggested that the extinction was catastrophic and wholly related to the K-T boundary impact. Globally, the last specimens of many dominant Cretaceous groups do occur exactly at the K-T boundary (Alvarez et al., 1984), and the last dinosaurs occur within a half-meter of the boundary impact level. This defines the true extinction point of diverse lineages.

However, many authors have pointed out that important, abrupt extinctions and mass killings, some of global extent, also occurred before and even after the K-T boundary event (e.g. Kauffman, 1988b, and references cited above for the 68 Ma event). For many groups, the major loss of biomass, and of many species and genera, preceded the K-T boundary event. These observations do not detract from the importance of the impact-related mass extinction event at the end of the Cretaceous, but only add to its complexity and suggest, as with all other well studied extinction intervals, that it was not a perfect bio-event, i.e., a single global catastrophe. Nor do they suggest that other extinction events clustered around the K-T boundary were not impact-related. McHone and Dietz (1991) recorded 11 impact craters temporally clustered around the K-T boundary, and two Iridium spikes have been recorded at the Brazos River in Texas (Hansen et al., 1993 and references therein), suggesting the possibility of a comet storm (sensu Hut et al., 1987) during this time. This proposal needs to be carefully tested with high-resolution geochemical and stratigraphic data spanning at least 5 Ma across the K-T boundary.

In addition to the 68 Ma bio-event referenced above, characterized by the early demise of rudistid reef ecosystems, shallow warm water molluscs and echinoderms, typical lineages of inoceramid bivalves, and significant numbers of angiosperms, Kauffman (1988b) proposed at least two additional, short-term extinction steps prior to the K-T boundary. One occurred about 66.5 Ma, marking a moderate, short-term demise in warm water molluscs, and another within a few 100 ka of the K-T boundary expressed among foraminifers (e.g. see Keller, 1989a,b) and nannoplankton (see Perch-Nielsen et al., 1982). New high-resolution stratigraphic data suggests that the first of these two steps is probably younger, closer to 66 Ma, within the *A. mayaroensis* biozone, and may be regional rather than global, involving accelerated loss of ammonites in western Europe (e.g. see Ward et al., 1991) and shelf-associated marine molluscs in the Texas-Mexico region (e.g. see Hansen et al., 1987, 1993), among other things. The younger of these pre-K-T events appears global for the marine plankton, and may also involve the abrupt loss of some angiosperm taxa (Nichols et al., 1990) in North America, additional molluscan taxa through Texas and Mexico (Kauffman and Hansen, in manuscript), and possibly most of the final dinosaur lineages. But again, precise correlation of the timing of these disparate latest Maastrichtian extinction events remains difficult.

Regional and Global Cretaceous Bio-Events in Northern and Southern Europe

During the Cretaceous, Northern and Southern Europe were very different regions in terms of regional sedimentary environments and faunal provinces. Northern Europe, comprising northern France, the United Kingdom, Belgium, the Netherlands, northern Germany, Denmark and Poland, comprised a wide shelf area that during the Late Cretaceous sea level highstand was characterized by chalk facies. The Early Cretaceous strata of this area are characteristic of either fresh-water and terrestrial, or shallow marine environments. In Southern Europe (Spain, southeastern France, southern Switzerland, Italy and Greece) the Cretaceous succession is typified by pelagic limestones (often dominated by Milankovitch cyclicity), pelagic mudstones and dark, organic-rich mudstones rimming carbonate platforms.

In Northern Europe, marine strata of Early Cretaceous age are characterized by blue-gray mudstones, the Speeton Clay Formation and equivalents in northern Germany and the United Kingdom, and the Gault Clay Formation and equivalents in the United Kingdom and northern France. The faunal changes in these successions have been described by Fletcher (1973), papers in Casey and Rawson (1973), Hart et al. (1989), and a wide range of other United Kingdom, German and Polish paleontologists. Many of the changes are local immigration events, especially those that are characterized by the diversification of ammonite, belemnite and bivalve faunas. None of these "bio-events" in northern Europe, apart from that in the Late Cenomanian, is coincident with OAEs, as the environments were too shallow.

Within the Lower Cretaceous strata of southern France (Fig. 4) there are a number of anoxic events, characterized by dark, well laminated, organic rich shales and carbonates representing pelagic or hemipelagic environments. These have variously been described as:

Age	OAE	S.E. France	Central Italy
Late Albian	1c	Breistroffer	Amadeus
Middle Albian			
Early Albian	1b	Paquier	Urbino/Nerone
Late Aptian		Jakob	"113"
Early Aptian	1a	Goguel	Selli

The first of these events is probably Early Aptian in age. In Central Italy, this has become known as the Selli Event (see Coccioni et al., 1989, 1992). This event is stratigraphically situated just above magnetic polarity event MO in the earliest Aptian, within the upper part of the *G. blowi* zone. The Selli Event equates with the Groguel Event of the Vocation Trough (southeastern France) and may be co-eval with the Fischschiefer Event in northern Germany.

Within this earliest Aptian interval there is a level of significant faunal and floral turnover (Mutterlose, 1992a, Fig. 8). This is a correlative world-wide event (Erba and Mutterlose, 1992; Mutterlose, 1992b; Mutterlose, 1990), and has been described as the nannoconid "crisis", with taxa such as *N. abundans* and *N. borealis* becoming extinct. Both belemnites and ammonites suffer a major change at this level, with new taxonomic groups appearing above the Selli Event. The long-ranging planktonic foraminifer, *Hedbergella*, also appears at this level.

Although the faunal and floral changes are identifiable world-wide, the Selli Event and its associated black shales (Breheret, 1985a,b; Moullade, 1966) are best seen in the Umbria Marche Apennines of Central Italy. Here, within the lower levels of the marne à Fucoidi (or Scisti a Fucoidi), there is a series of dark, organic-rich clays interbedded with thin pelagic limestones. Thurow (personal communication, 1994) has shown that there is a major excursion $\partial^{13}C$ isotopic levels, attaining +4 ‰ (PDB) in this interval; this is comparable to those recorded from the Late Cenomanian anoxic event (OAE II). Within the pelagic facies of Central Italy and southeastern France (Vocontian Trough) anoxic events (and associated faunal changes) are also recorded within the Late Aptian, Early Albian and Late Albian, representing OAEs 1a, 1b, and 1c (see Jenkyns, 1980). Whereas the '113' and Jakob Events appear to be synchronous, they are not associated with any major extinction events. Within the earliest Albian the Urbino and Paquier Events are not synchronous, although the 'black shale rhythms' above and below are, in general, found over the same interval. Breheret et al. (1989), have shown that the Paquier Event in the earliest Albian is characterized by dark, organic-rich shales with a characteristic planktonic foraminiferal fauna. Below the 'event' the fauna is dominated by *T. roberti*, *T. bejaouensis* and *H. trochoidea* while the assemblage found in the sediments overlying the event contains *T. roberti*, *H. trochoidea*, *H. rischi*, and other, large hedbergellid foraminifers. Within the strata immediately adjacent to, and within, the Paquier Event only small, primitive hedbergellids are found. Breheret et al. (1989) attributed this to a major expansion of the oxygen minimum zone within the water column, restricting the planktonic foraminifera to only the surface waters. If this interpretation is correct it would be an exact parallel with that described by Jarvis et al. (1988) for the Late Cenomanian bio-event.

The Late Cenomanian Event (Thomel Event in southeastern France = Bonarelli Event) is well known throughout Europe and has been described by so many authors that it is almost impossible to list all of them in any one publication. Jarvis et al. (1988) have documented many of these and a recent paper by Jeans et

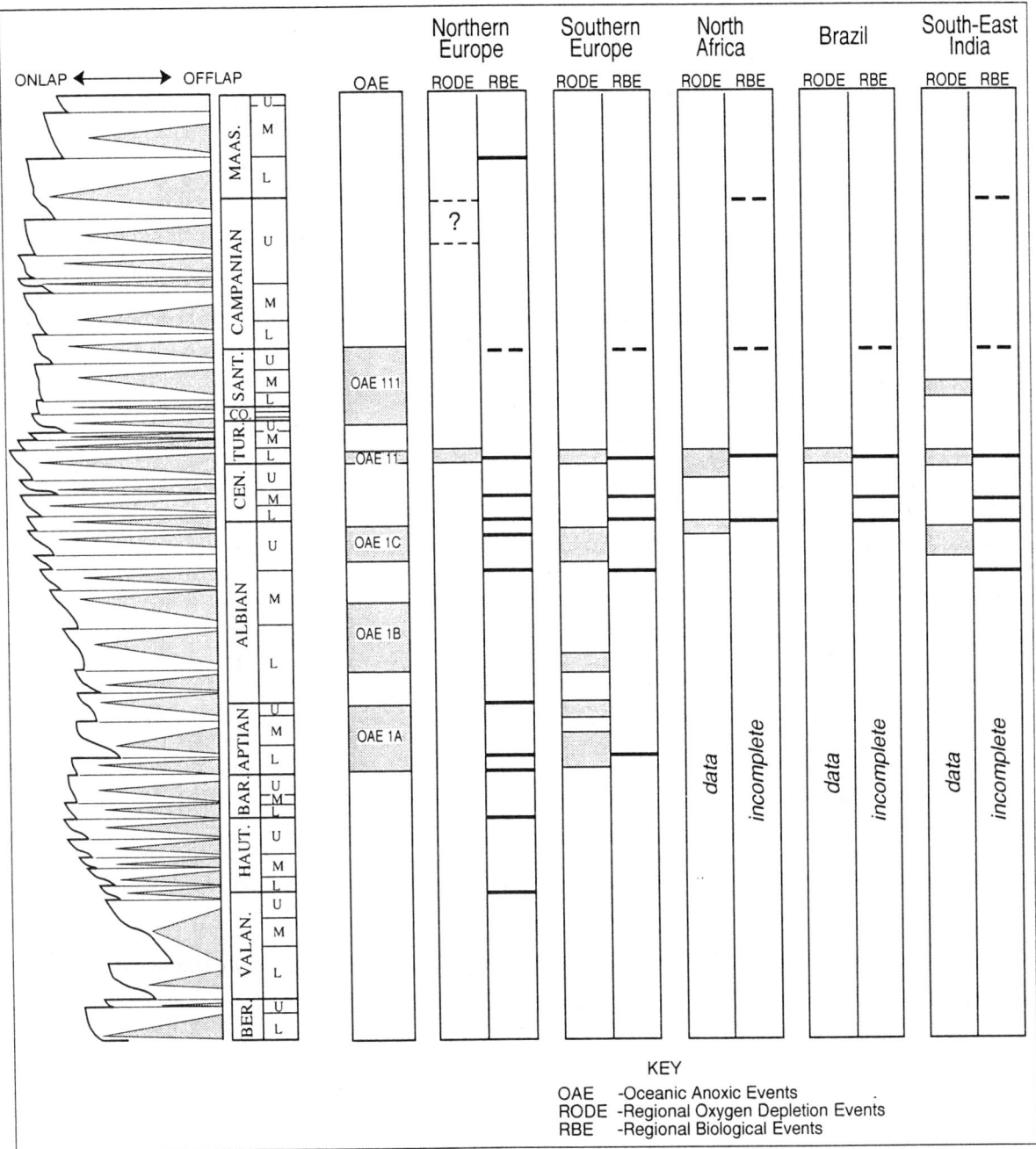

Fig. 4. The global cycle chart of Haq et al. (1987), as shown in Fig. 2, together with the global Oceanic Anoxic Events (OAE), regional oxygen depletion events, and regional bio-events in northern Europe, Southern Europe, North Africa, Brazil, and southeastern India

al. (1991) has added later contributions. In general, these data suggest that it shares many characteristics with the same interval in North America (see previous discussion). OAE III, spanning the Coniacian - Santonian, is not known on-shore in northwestern Europe and while there are well-known faunal changes at this level (Mortimore and Pomerol, 1991; Hart et al., 1989) they are not dramatic. For off-shore southwestern England (Hart and Ball, 1986; Hart and Duane, 1989) claystones of this age are known from the Goban Spur, together with an interval of poor benthic faunas and badly etched planktonic foraminifera. Hart and Ball (1986) also recorded dark mudstones of Late Campanian age, again with reduced benthic faunas. There are no other records of Campanian 'events' in northwestern Europe, although Nyong and Olsson (1984) and Olsson and Nyong (1984) have described the effect of an oxygen minimum zone of this age on the New Jersey coastal plain (U.S.A.).

The Late Maastrichtian (K/T) extinction event is well known from Denmark (Stevens Klint, Nye Klev), Austria, Spain and Italy (Gubbio). All of these localities are well documented in the international literature.

Regional and Global Cretaceous Bio-Events in North Africa

The Cretaceous successions of Libya and Tunisia are well known, with those of Egypt less studied. Although Barr (1968) has described the foraminifera of the Libyan successions in some detail, it is the successions in Tunisia that have attracted the most attention. In the Kalaat Senan Region of Central Tunisia, Robaszynski et al. (1993), and Hardenbol et al. (1993), have described the sequence stratigraphy, macropaleontology and micropaleontology of the Late Cenomanian Bahloul Formation. This is an expanded succession of black laminite and black marly limestones. Many of these limestones contain calcispheres (called calcispherite by Robaszynski et al., 1993). These calcified dinoflagellates are particularly characteristic of the Late Cenomanian extinction event and are regarded as a disaster taxon by Hart (1991). Robaszynski et al. (1993) also describe some dark mudstones in the uppermost Albian (again associated with calcisphere-rich limestones). It is unclear as to whether or not this is evidence of OAE Ic. In the higher levels of the Cretaceous, the El Kef section is now regarded as a reference point for all work on the K/T boundary. Keller (1988) reviews the extinction history of calcareous plankton in this area.

Regional and Global Cretaceous Bio-Events in India (and Indian Ocean)

The Cretaceous Basins of Eastern Indian (Manhanadi, Krishna-Godavani and Cauvery) are now becoming better known as oil exploration by the Oil & Natural Gas Commission (ONGC) develops. One of the most studied basins is the Cauvery Basin in southeastern India (Tamil Nadu). In this basin there is a well-known on-shore succession, together with a very complete Cretaceous succession known from ONGC boreholes. The Lower and mid-Cretaceous (Banerji and Sastri, 1979; Phansalkar and Kumar Mary, 1983; Sundaram and Rao, 1986; Ramasamy and Banerji, 1991; Kale and Phansalkar, 1992a,b) is characterized by a dark clay/mudstone succession that is generally known as the Utatur Formation. Venkatachalapathy et al. (1994) have shown that this extends from the *T. roberti* zone of the Aptian to the *M. sagali* zone of the Middle Turonian. There are a number of faunal changes within this succession but no real evidence of OAE Ia, Ib or Ic. Off-shore, and in the deeper parts of the basin, Govindan (1993) has recorded dark, organic-rich clays that are probably co-eval with OAE Ic. The pattern of the organic content recorded by Govindan is almost identical to that detected on the Brazilian Continental Margin (Koutsoukos et al., 1991) over the same interval. Recent work by Tewari and Hart (in preparation) has shown a number of distinct foraminiferal changes in the latest Albian that are probably the local equivalent of those associated with OAE Ic. Govindan (1982, 1993) has also described the presence of the Late Cenomanian event and this has been confirmed by Venkatachalapathy et al. (1994) and Tewari and Hart (in manuscript). The succession in the Ariyalur - Pondocherry Depression of the Cauvery Basin (Govindan, 1993, Fig. 1), deposited during Late Turonian to Maastrichtian time, is typified by shallow-water facies (including reefal limestone) and the end of the Cretaceous is represented by a continental succession. In the deeper parts of the basin, Govindan (1993) records the presence of OAE III.

In the Indian Ocean the Late Cenomanian oceanic anoxic event (OAE II) has been described from the ODP boreholes adjacent to the Exmouth Plateau. A thin black mudstone is present in the uppermost Cenomanian (von Rad et al., 1992; Wonders, 1992). There is no record of the Coniacian - Santonian event although Tewari (personal communication) has detected some changes in foraminiferal faunas of this age in Sites 761, 762, and 763.

Conclusions

Cretaceous bio-events can be classified as diversification bio-events (e.g. population blooms, rapid evolutionary radiations, widespread colonization and immigration events), or as diversity reduction bio-events (population decline, community shock, mass mortalities, regional and mass extinctions, and emigration events). Bio-events can be defined at local, regional, and global scales. Regional and global bio-events are most strongly influenced by allocyclic environmental phenomena in Earth history. Most global-scale bio-events are expressed as mass extinctions, and as biotic turnovers associated with certain stage or substage boundaries. The majority of large-scale regional to global bio-events in the Cretaceous are also associated with major perturbations of the ocean-climate system including extraterrestrial impacts, oceanic anoxic events, rapid changes in rates of sea level rise or fall, rapid temperature excursions, and extraordinary fluctuations in ocean and/or atmospheric chemistry.

High-resolution stratigraphic analysis of regional to global-scale bio-events demonstrates that most, if not all, do not occur as single event surfaces, representing catastrophes. Rather they are represented by a complex of stratigraphically tightly clustered, short lived bio-events in the observed data, defining stepwise patterns of change. These sequential bio-events are characteristically ecologically graded, initiating with and most profoundly affecting more tropical and structurally complex ecosystems, and progressively affecting more tolerant, less complex temperate to polar ecosystems through time. This reflects sequential breaching of a series of progressively more stable ecological thresholds from pole to equator, and through time. Individual bio-events within these complexes may be extraordinarily developed, however, defining widespread catastrophes which are essentially isochronous worldwide, for example that associated with the giant Chicxulub bolide impact at or very close to the Cretaceous-Tertiary boundary.

Stepwise patterns of bio-events associated with mass extinctions or other large-scale biotic crises are closely correlative, in many cases, with short-term, large-scale fluctuations/perturbations in the ocean-climate system, suggesting cause and effect. These environmental perturbations, and their negative feedback loops, are most commonly reflected in both geochemical and sedimentological data (e.g. trace element spikes, stable isotope excursions, rapid shifts in carbonate or organic carbon values, microtektite and shocked quartz horizons, widespread dissolution surfaces, etc.).

An analysis of regional Cretaceous bio-events reflected in detailed stratigraphic data from North and South America, the Caribbean Province, northern and southern Europe, and India is herein presented, with brief syntheses and references to the data sets. Those Cretaceous bio-events that appear to be interregional or global in nature are as follows: The Jurassic-Cretaceous, Late Aptian, Cenomanian-Turonian, and Cretaceous-Tertiary mass extinctions, and significant biotic turnovers at the Lower-Middle Cenomanian boundary, in the middle part of the Middle Cenomanian, and at the Santonian-Campanian stage boundary.

If we can understand the complexities of global and regional bio-events through very high-resolution stratigraphic analyses, integrating physical, geochemical, and paleobiological data bases, we can better define their causes and their effects on Earth and life history. By integrating these detailed studies with a highly resolved system of chronology and regional correlation, blending biostratigraphy, chronostratigraphy, and geochronology, we can more confidently separate out those components of bio-event complexes that are regional from those that are global in scale. But the assumption that complex bio-events like mass extinctions represent single, catastrophic global phenomena of the same age everywhere is not supported by our current knowledge of the well studied mass extinction events, including the end-Ordovician, Frasnian-Famennian (Late Devonian), Permo-Triassic, Cenomanian-Turonian, Cretaceous-Tertiary, and Eocene-Oligocene biodiversity crises. All have a complex stratigraphic history, indicating multicausal events which may, in part, be globally isochronous.

The great variability in the adaptive ranges and life strategies of living organisms, and of whole ecosystems, and the resiliency of the global biota that has ensured its survival and continued diversification across all Phanerozoic mass extinctions, collectively argue for biologically complex, temporally expanded response to even the most severe terrestrial or extraterrestrial perturbations through time.

References

Alvarez, L.W., Alvarez, W., Asaro, F. and Michel, H.V., 1980. Extra-terrestrial cause for the Cretaceous-Tertiary mass extinction. Science 208, 1095-1108.

Alvarez, W., Alvarez, L., Asaro, F. and Michel, H.V., 1982. Current status of the impact theory for the terminal Cretaceous extinction. In: Silver, L.T. and Schultz, P.H. (eds.), Geological Implications

Of Impacts Of Large Asteroids And Comets On Earth. Geological Society of America, Special Paper 190, 305-315.

Alvarez, W., Kauffman, E.G., Surlyk, F., Alvarez, L., Asaro, F. and Michel, H.V., 1984. The impact theory of mass extinctions and the marine and vertebrate fossil record across the Cretaceous-Tertiary boundary. Science 223, 1135-1141.

Alvarez, W., Smit, J., Lowrie, W., Asaro, F., Margolis, S.V., Claeys, P., Kastner, M. and Hildebrand, A.R., 1992. Proximal impact deposits at the Cretaceous-Tertiary boundary in the Gulf of Mexico: A restudy of DSDP leg 77, sites 536 and 540. Geology 20, 697-700.

Arthur, M.A., Dean, W.E., Pollastro, R.M., Claypool, G.E. and Scholle, P.A., 1985. Comparative geochemical and mineralogical studies of two cyclic transgressive pelagic limestone units, Cretaceous Western Interior Basin, U.S. In: Pratt, L.M., Kauffman, E.G. and Zelt, F.B. (eds.), Fine-grained deposits and biofacies of the Cretaceous Western Interior Seaway: Evidence of Cyclic Sedimentary Processes, Society of Economic Paleontologists and Mineralogists, 2nd Annual Midyear Meeting, Field Trip Guidebook No. 4, 16-27.

Arthur, M.A., Dean, W.A. and Schlanger, S.O., 1985. Variations in the global carbon cycle during the Cretaceous related to climate, volcanism and changes in atmospheric CO_2. Natural variations Archaean to Present. American Geophysical Union Monography 32, 504-529.

Ascoli, P., Poag, C.W. and Remane, J., 1984. Microfossil zonation across the Jurassic-Cretaceous boundary on the Atlantic margin of North America. In: Westerman, G.E.G. (ed.), Jurassic-Cretaceous Biochronology and Paleogeography of North America. Geological Association of Canada Special Paper 27, 31-48.

Bakker, R.T., 1993. Plesiosaur extinction cycles - events that mark the beginning, middle and end of the Cretaceous. In: Caldwell, W.G.E. and Kauffman, E.G. (eds.), Evolution of the Western Interior Basin. Geological Association of Canada Special Paper 39, 641-664.

Banerji, R.K. and Sastri, V.V., 1979. Quantification of foraminiferal biofacies and reconstruction of palaeobiogeography of the Cauvery Basin. Journal of the Geological Society of India 20, 571-586.

Barr, F.T., 1968. Upper Cretaceous stratigraphy of Jabal Al Akhdar, Northern Cyrenaica. Geology and Archaeology of Northern Cyrenaica, Libya. Petroleum Exploration Society of Libya, 131-142.

Berner, R.A., 1994. GEOCARB II: A revised model for atmospheric CO_2 over Phanerozoic time. American Journal of Science 294, 56-91.

Birkelund, T. and Bromley, R.G. (eds.), 1979. Cretaceous Tertiary Boundary Events. I. The Maastrichtian and Danian of Denmark. 210 pp. University of Copenhagen.

Boucot, A.J., 1986. Ecostratigraphic criteria for evaluating the magnitude, character and duration of bioevents. In: Walliser, O.H. (ed.), Global Bioevents. Lecture Notes in Earth Sciences 8, 25-45. Springer, Berlin Heidelberg New York.

Breheret, J.G., 1985a. Indices d'un événement anoxique étendu à la Tethys alpine à l'Aptien inférieur (événement Paquier). C.R. Acad. Sc. Paris Ser. II, 300, no. 8, 355-358.

Breheret, J.G., 1985b. Sédimentologie et diagenèse de la matière organique contenue dans le niveau Paquier, couche répérée de 'Albien inférieur vocoutien. C.R. Acad. Sc. Paris, Ser. II, 301, no. 15, 1151-1156.

Breheret, J.G., Caron, M. and Delamette, M., 1989. Niveau riche en matière organique dans l'Albien Vocoutien; quelques caractères du paléoenvironnement; essai d'interpretation génétique. Docum. Bur. Rech. géol. min., no. 110, 141-191.

Caldwell, W.G.E., Diner, R., Eicher, D.L., Fowler, S.P., North, B.R., Stelck, C.R. and von Holdt, W.L., 1993. Foraminiferal biostratigraphy of Cretaceous marine cyclothems. In: Caldwell, W.G.E. and Kauffman, E.G. (eds.), Evolution of the Western Interior Basin. Geological Association of Canada Special Paper 39, 477-520.

Casey, R. and Rawson, P.F., 1973. The Boreal Lower Cretaceous. Geological Journal Special Issue No. 5. 448 pp. Seel House Press, Liverpool.

Christensen, W.K. and Birkelund, T. (eds.), 1979. Cretaceous-Tertiary Boundary Events Symposium. II. Proceedings. 250 pp. University of Copenhagen.

Cobban, W.A. and Kennedy, W.J., 1989. The ammonite *Metengonoceras hyatt*, 1903, from the Mowry Shale (Cretaceous) of Montana and Wyoming. United States Geological Survey Bulletin 1787-L, L1-L11.

Coccioni, R., Franchi, R., Nesci, O., Wezel, C.F., Battistini, F. and Pallechi, P., 1989. Stratigraphy and mineralogy of the Selli Level (Early Aptian) at the base of the Marne a Fucoidi in the Umbrian-Marchean Appenines (Italy). In: Wiedmann, J. (ed.), Cretaceous of the Western Tethys. Proceedings 3rd Intern. Symp. Tübingen. p. 563-584.

Schweizerbart'sche Verlagsbuchhandlung.

Coccioni, R., Erba, E. and Premoli Silva, I., 1992. Barremian-Aptian calcareous plankton biostratigraphy from the Gorgo Cerbara section (Marche, central Italy) and implications for plankton evolution. Creaceous Research 13, 517-538.

Edwards, L.E., 1984. Insights on why graphic correlation (Shaw's method) works. Journal of Geology 92, 583-597.

Edwards, L.E., 1989. Supplemented graphic correlation: A powerful tool for paleontologists and non-paleontologists. Palaios 4, 127-143.

Eicher, D.L. and Diner, S.R., 1985. Foraminifera as indicators of water mass in the Cretaceous Greenhorn Sea, Western Interior. In: Pratt, L.M., Kauffman, E.G. and Zelt, F.B. (eds.), Fine-grained deposits and biofacies of the Cretaceous Western Interior Seaway: Evidence of Cyclic Sedimentary Processes, Society of Economic Paleontologists and Mineralogists, 2nd Annual Midyear Meeting, Field Trip Guidebook No. 4, 60-71.

Eicher, D.L. and Diner, S.R., 1989. Origin of the Cretaceous Bridge Creek cycles in the Western Interior, United States. Palaeogeography, Palaeoclimatology, and Palaeoecology 74, 127-146.

Elder, W.P., 1985. Biotic patterns across the Cenomanian-Turonian extinction boundary near Pueblo, Colorado. In: Pratt, L.M., Kauffman, E.G. and Zelt, F.B. (eds.), Fine-grained deposits and biofacies of the Cretaceous Western Interior Seaway: Evidence of Cyclic Sedimentary Processes, Society of Economic Paleontologists and Mineralogists, 2nd Annual Midyear Meeting, Field Trip Guidebook No. 4, 157-169.

Elder, W.P., 1987. Cenomanian-Turonian (Cretaceous) stage boundary extinctions in the Western Interior of the United States. Unpubl. Ph.D. Thesis, University of Colorado, 690 pp.

Elder, W.P., 1989. Molluscan extinction patterns across the Cenomanian-Turonian stage boundary in the Western Interior of the United States. Paleobiology 15, 299-320.

Erba, E. and Mutterlose, J., 1992. The floral and faunal turnover in the Early Aptian (Early Cretaceous). Abstract Volume, 5th Intern. Conf. on Global Bio-events, 31-32, Göttingen.

Fletcher, B.N., 1973. The distribution of Lower Cretaceous (Berriasian-Barremian) foraminifera in the Speeton Clay of Yorkshire, England. In: Casey, R. and Rawson, P.F. (eds.), The Boreal Lower Cretaceous. Geological Journal Special Issue No. 5, 161-168. Seel House Press, Liverpool.

Govindan, A., 1982. Imprint of global "Cretaceous Anoxic Events" in East Coast basins of India and their implications. Bulletin of the Oil and Natural Gas Commission 19, 257-270.

Govindan, A., 1993. Cretaceous anoxic events, sea level changes and microfauna in Cauvery Basin, India. In: Biswas, S.K. et al. (eds.), Proceedings of the Second Seminar on Petroliferous Basins of India, vol. 1, 161-176.

Hansen, T.A., Farrand, R.B., Montgomery, H.A., Billman, H.G. and Blechschmidt, G., 1987. Sedimentology and extinction patterns across the Cretaceous-Tertiary boundary interval in east Texas. Cretaceous Research 8, 229-252.

Hansen, T.A., Upshaw, Banks, III, Kauffman, E.G. and Gose, W., 1993. Patterns of molluscan extinction and recovery across the Cretaceous-Tertiary boundary in east Texas; report on new outcrops. Cretaceous Research 14, 685-706.

Haq, B.V., Hardenbol, J. and Vail, P.R., 1987. Chronology of fluctuating sea levels since the Triassic. Science 235, 1156-1167.

Haq, B.V., Hardenbol, J. and Vail, P.R., 1988. Mesozoic and Cenozoic chronostratigraphy and cycles of sea-level change. In: Wilgus, C.K. et al. (eds.), Sea level changes: An integrated approach. Society of Economic Paleontologists and Mineralogists Special Publication 42, 71-108.

Hardenbol, J., Caron, M., Amedro, F., Dupuis, C. and Robaszynski, F., 1993. The Cenomanian-Turonian boundary in central Tunisia in the context of a sequence-stratigraphic interpretation. Cretaceous Research 14, 449-454.

Harland, W.B., Armstrong, R.L., Cox, A.V., Craig, L.E., Smith, A.G. and Smith, D.G., 1990. A geologic time scale 1989. 263 pp. Cambridge University Press, Cambridge.

Harries, P.J. and Kauffman, E.G., 1990. Patterns of survival and recovery following the Cenomanian-Turonian (Late Cretaceous) mass extinction in the Western Interior Basin, United States. In: Kauffman, E.G. and Walliser, O.H. (eds.), Extinction Events in Earth History. Lecture Notes in Earth Sciences 30, 277-298. Springer, Berlin Heidelberg New York.

Hart, M.B., 1991. The Late Cenomanian calcisphere global bioevent. Proceedings of the Ussher Society 7, 413-417.

Hart, M.B. and Ball, K.C., 1986. Late Cretaceous anoxic events, sea-level changes and the evolution of the planktonic foraminifera. In: Summerhayes, C.P. and Shackleton, N.J. (eds.), North Atlantic

Palaeoceanography. Geolgcial Society Special Publication 21, 67-78.

Hart, M.B. and Duane, A.M., 1989. Late Cretaceous development of the Atlantic Continental Margin off S.W. England. Proceedings of the Ussher Society 7, 168-171.

Hart, M.B., Bailey, H.W., Crittenden, S., Fletcher, B.N., Price, R.J. and Swiecicki, A., 1989. Cretaceous. In: Jenkins, D.G. and Murray, J.W. (eds.), Stratigraphical Atlas of Fossil Foraminifera. British Micropalaeontological Society Series. p. 273-371. Ellis Horwood, Chichester.

Hildebrand, A.R., Penfield, G.T., Kring, D.A., Pilkington, M., Camargo, Z.A., Jacobsen, S.B. and Boynton, W.V., 1991. Chicxulub Crater: A possible Cretaceous/Tertiary boundary impact crater on the Yucatan Peninsula, Mexico. Geology 19, 867-871.

Hut, P., Alvarez, W., Elder, W.P., Hansen, T.A., Kauffman, E.G., Keller, G., Shoemaker, E.M. and Weissman, P.R., 1987. Comet showers as a possible cause of stepwise extinctions. Nature 329, 118-126.

Irving, E., North, F.K. and Couillard, R., 1974. Oil, climate and tectonics. Canadian Journal of Earth Sciences 11, 1-17.

Jarvis, I., Carson, G.A., Cooper, M.K.E., Hart, M.B., Leary, P.N., Tocher, B.A., Horne, D. and Rosenfeld, A., 1988. Microfossil assemblages and the Cenomanian-Turonian (Late Cretaceous) oceanic anoxic event. Cretaceous Research 9, 3-103.

Jeans, C.V., Long, D., Hall, M.A., Bland, D.J. and Cornford, C., 1991. The geochemistry of the Plenus Marls at Dover, England: evidence of fluctuating oceanographic conditions and of glacial control during the development of the Cenomanian-Turonian d^{13}C anomaly. Geological Magazine 128, 603-632.

Jeletzky, J.A., 1968. Macrofossil zones of the marine Cretaceous of the Western Interior of Canada and their correlation with the zones and stages of Europe and the Western Interior of the United States. Geological Survey of Canada, Paper 67-72, 66 p.

Jeletzky, J.A., 1970. Cretaceous macrofaunas. In: Geology and Economic Minerals of Canada. Geological Survey of Canada, Economic Geology Report 1, 5th edition, p. 649-662.

Jeletzky, J.A., 1984. Jurassic-Cretaceous boundary beds of western and Arctic Canada and the problem of the Tithonian - Berriasian stages in the Boreal Realm. In: Westerman, G.E.G. (ed.), Jurassic-Cretaceous Biochronology and Paleogeography of North America. Geological Association of Canada Special Paper 27, 175-256.

Jenkyns, H.C., 1980. Cretaceous anoxic events: From continents to oceans. Journal of the Geological Society, London, 137, 171-188.

Jenkyns, H.C., Gale, A.S. and Corfield, R., 1994. Carbon- and oxygen-isotope stratigraphy of the English Chalk and Italian Scaglia and its palaeoclimatic significance. Geological Magazine 131 (1), 1-34.

Johnson, C.C., 1993. Cretaceous biogeography of the Caribbean region. Unpubl. Ph.D. Thesis, University of Colorado, Boulder, CO, 651 pp.

Johnson, C.C. and Kauffman, E.G., 1990. Originations, radiations and extinctions of Cretaceous rudistid bivalve species in the Caribbean Province. In: Kauffman, E.G. and Walliser, O.H. (eds.), Extinction Events in Earth History. p. 305-324. Springer-Verlag, Berlin, Heidelberg, New York.

Johnson, C.C. and Kauffman, E.G., 1995 in press. Maastrichtian extinction patterns of Caribbean Province rudistids. In: MacLeod, N. and Keller, G. (eds.), The Cretaceous-Tertiary Mass Extinction: Biotic and Environmental Events. 38 MS pp. W.W. Norton & Co.

Johnson, K.R., 1992. Foliar physiognomy of Maastrichtian leaf floras from the northern Great Plains: Implications for paleoclimate. Society for Sedimentology, SEPM 1992 Theme Meeting, Fort Collins, CO, Abstracts, p. 36.

Kale, V.S. and Phansalkar, V.G., 1992a. Calcareous nannofossils from the Utatur Group, Trichinopoly District, Tamil Nadu, India. Journal of the Palaeontological Society of India 37, 85-102.

Kale, V.S. and Phansalkar, V.G., 1992b. Nannofossil biostratigraphy of the Utatur Group, Trichinopoly District, South India. Memoire di Scienze Geologiche, Padova 43, 89-107

Kauffman, E.G., 1984. Paleobiogeography and evolutionary response dynamic in the Cretaceous Western Interior Seaway of North America. In: Westermann, G.E.G., Jurassic-Cretaceous Biochronology and Paleogeography of North America. Geological Association of Canada Special Paper 27, 273-306.

Kauffman, E.G., 1985. Cretaceous evolution of the Western Interior Basin of the United States. In: Pratt, L.M., Kauffman, E.G. and Zelt, F.B. (eds.), Fine-grained Deposits and Biofacies of the Cretaceous Western Interior Seaway: Evidence of Cyclic Sedimentary Processes. Society of Economic

Paleontologists and Mineralogists, 2nd Annual Midyear Meeting, Golden, CO, Field Trip Guidebook No. 4, IV-XIII.

Kauffman, E.G., 1986. High-resolution event stratigraphy: Regional and global bioevents. In: Walliser, O.H. (ed.), Global Bioevents. Lecture Notes in Earch History 8, 279-335.

Kauffman, E.G., 1988a. Concepts and methods of high-resolution event stratigraphy. Annual Review of Earth and Planetary Science 16, 605-654.

Kauffman, E.G., 1988b. The dynamics of marine stepwise mass extinction. In: Lamolda, M.A., Kauffman, E.G. and Walliser, O.H., (eds.): Paleontology and Evolution: Extinction Events. Revista Espanola de Paleontologia, numero Extraordinario, p. 57-71.

Kauffman, E.G., 1995 in press. Global change leading to biodiversity crisis in a greenhouse world: The Cenomanian-Turonian (Cretaceous) mass extinction. In: Stanley, S.M., Knoll, A.H. and Kennett, J. (eds.), The Effects Of Past Global Change On Life. 49 MS pp. Washington, D.C., National Academy Press.

Kauffman, E.G. and Caldwell, W.G.E., 1993. The Western Interior Basin in space and time. In: Caldwell, W.G.E. and Kauffman, E.G. (eds.), Evolution Of The Western Interior Basin. Geological Association of Canada, Special Paper 39, 1-30.

Kauffman, E.G. and Harries, P.J., 1995 in press. The importance of crisis progenitors in recovery from mass extinction. In: Hart, M.B. (ed.), Geological Association of London Special Volume, 48 MS pp.

Kauffman, E.G. and Johnson, C.C., 1988. The morphological and ecological evolution of middle and Upper Cretaceous reef-building rudistids. Palaios 3, 194-216.

Kauffman, E.G. and Walliser, O.H. (eds.), 1990. Extinction Events in Earth History. Lecture Notes in Earth Sciences 30, 432 pp. Springer, Berlin Heidelberg New York.

Kauffman, E.G., Elder, W.P. and Sageman, B.B., 1991. High-resolution correlation: A new tool in chronostratigraphy. In: Einsele, G., Ricken, W. and Seilacher, A. (eds.), Cycles and Events in Stratigraphy. p. 795-819. Springer, Berlin Heidelberg New York.

Kauffman, E.G., Sageman, B.B., Kirkland, J.I., Elder, W.P., Harries, P.J. and Villamil, T., 1993. Molluscan biostratigraphy of the Western Interior Cretaceous Basin, North America. In: Caldwell, W.G.E. and Kauffman, E.G. (eds.), Evolution of The Western Interior Basin. Geological Association of Canada, Special Paper 39, 397-434.

Keller, G., 1988. Extinction, survivorship and evolution of planktic foraminifera across the Cretaceous/Tertiary boundary at El kef, Tunisia. Marine Micropaleontology 13, 239-263.

Keller, G., 1989a. Extended period of extinctions across the Cretaceous/Tertiary boundary in planktonic foraminifera of continental shelf sections: implications for impact and volcanism theories. Geological Society of America Bulletin 101, 1408-1419.

Keller, G., 1989b. Extended Cretaceous-Tertiary boundary extinctions and delayed population change in planktonic foraminiferal faunas from Brazos River, Texas. Paleoceanography 4, 287-332.

Koutsoukos, E.A.M. and Hart, M.B., 1990a. Cretaceous foraminiferal morphogroup distribution patterns, palaeocommunities and trophic structures: a case study from the Sergipe Basin, Brazil. Transactions of the Royal Society of Edinburgh, Earth Sciences 81, 221-246.

Koutsoukos, E.A.M. and Hart, M.B., 1990b. Radiolarians and Diatoms from the mid-Cretaceous successions of the Sergipe Basin, Northeastern Brazil: palaeogeographic assessment. Journal of Micropalaeontology 9, 45-64.

Koutsoukos, E.A.M., Leary, P.N. and Hart, M.B., 1990. Latest Cenomanian-Earliest Turonian low-oxygen tolerant benthonic Foraminifera: a case study from the Sergipe Basin (N.E. Brazil) and the Western Anglo-Paris Basin (Southern England). Palaeogeography, Palaeoclimatology, Palaeoecology 77, 145-177.

Koutsoukos, E.A.M., Mello, M.R., de Azambuja Filho, N.C., Hart, M.B. and Maxwell, J.R., 1991. The Upper Aptian-Albian succession of the Sergipe Basin, Brazil: an integrated palaeoenvironmental assessment. Bulletin of the American Association of Petroleum Geologists 75, 479-498.

Kyser, T.K., Caldwell, W.G.E., Whittaker, S.G. and Cadrin, A.J., 1993. Paleoenvironment and geochemistry of the northern portion of the Western Interior Seaway during Late Cretaceous time. In: Caldwell, W.G.E. and Kauffman, E.G. (eds.), Evolution Of The Western Interior Basin. Geological Association of Canada, Special Paper 39, 355-387.

Lamolda, M.A., Kauffman, E.G. and Walliser, O.H. (eds.), 1988. Palaeontology and Evolution: Extinction Events. Revista Espanola de Paleontologia,

numero Extraordinario, 155 pp.
Larson, R.L., 1991a. Latest pulse of the Earth: evidence for a mid-Cretaceous superplume. Geology 19, 547-550.
Larson, R.L., 1991b. Geological consequences of superplumes. Geology 19, 963-966.
Larson, R.L., Fischer, A.G., Erba, E. and Premoli Silva, I., 1993. Summary of workshop results. In: Larson, R.L. et al., Apticore-Albicore: A workshop on global events and rhythms of the mid-Cretaceous, Perugia, Ocotber 1992, 56 pp.
Leckie, R.M., 1985. Foraminifera of the Cenomanian-Turonian boundary interval, Greenhorn Formation, Rock Canyon Anticline, Pueblo, Colorado. In: Pratt, L.M., Kauffman, E.G. and Zelt, F.B. (eds.), Fine-grained deposits and biofacies of the Cretaceous Western Interior Seaway: Evidence of Cyclic Sedimentary Processes. Society of Economic Paleontologists and Mineralogists, 2nd Annual Midyear Meeting, Field Trip Guidebook No. 4, 139-150.
MacLeod, K.G., 1994. Bioturbation, Inoceramid extinction, and mid-Maastrichtian ecological change. Geology 22, n. 2, 139-142.
Marshall, C.R., 1991. Estimation of taxonomic ranges from the fossil record. In: Gilinsky, N.L. and Signor, P.W. (eds.), Analytical Paleobiology. The Paleontological Society, Short Courses in Paleontology 4, 19-38.
McDonough, K.J. and Cross, T., 1991. Late Cretaceous sea level from a paleoshoreline. Journal of Geophysical Research 96, 6591-6607.
McHone, J.F. and Dietz, R.S., 1991. Multiple impact craters and astroblemes: Earth's record. Geological Society of America, Annual Meeting, San Diego, CA, Abstract Volume, p. A 183.
Mortimore, R.N. and Pomerol, B., 1991. Stratigraphy and eustatic implications of trace fossils in the Upper Cretaceous chalk of Northern Europe. Palaois 6, 216-231.
Moullade, M., 1966. Etude stratigraphique et micropaléontologique du Crétacé inférieur de la "Fosse vocontienne". Doc. Lab. Géol. Fac. Sc. Lyon 15 (1-2), 369 pp.
Mutterlose, J., 1990. A belemnite scale for the Lower Cretaceous. Cretaceous Research 11, 1-15.
Mutterlose, J., 1992a. Migration and evolution patterns of floras and faunas in marine Early Cretaceous sediments of NW Europe. Palaeogeography, Palaeoclimatology, Palaeoecology 94, 261-282.
Mutterlose, J., 1992b. Biostratigraphy and palaeobiogeography of Early Cretaceous calcareous nannofossils. Cretaceous Research 13, 167-189.
Nichols, D.J., Fleming, R.F. and Frederiksen, N.O., 1990. Palynological evidence of effects of the terminal Cretaceous event on terrestrial floras in western North America. In: Kauffman, E.G. and Walliser, O.H. (eds.), Extinction Events in Earth History. Lecture Notes in Earth Sciences 30, 351-364. Springer, Berlin Heidelberg New York.
Nyong, E.E. and Olsson, R.K., 1984. A palaeoslope model of Campanian to Lower Maastrichtian Foraminifera in the North American Basin and adjacent continental margin. Marine Micropaleontology 8, 437-477.
Obradovich, J., 1993. A Cretaceous Time Scale. In: Caldwell, W.G.E. and Kauffman, E.G. (eds.), Evolution of the Western Interior Basin. Geological Association of Canada, Special Paper 39, 379-396.
Olsson, R.K. and Nyong, E.E., 1984. A palaeoslope model for Campanian-Lower Maastrichtian Foraminifera of New Jersey and Delaware. Journal of Foraminiferal Research 14, 50-68.
Orth, C.J., 1989. Geochemistry of the bio-event horizons. In: Donovan, S.K. (ed.), Mass Extinctions: Processes and Evidence. pp. 37-72. Columbia University Press, New York.
Orth, C.J., Attrap, Jr., M., Quintana, L.R., Elder, W.P., Kauffman, E.G., Diner, R. and Villamil, T., 1993. Elemental abundance anomalies in the Late Cenomanian extinction interval: A search for the source(s). Earth and Planetary Science Letters 117, 189-204.
Perch-Nielsen, K., McKenzie, J. and He, Q., 1982. Biostratigraphy and isotope stratigraphy and the "catastrophic" extinction of calcareous nannoplankton at the Cretaceous/Tertiary boundary. In: Silver, L.T. and Schultz, H.P. (eds.), Geological implications of impacts of large asteroids and comets on the Earth. Geological Society of America Special Paper 190, 353-371.
Phansalkar, V.G. and Kumar Mary, K., 1983. Biostratigraphy of Utatur and Trichinopoly Groups of the Upper Cretaceous of Trichinopoly district Tamilnadu. Prof. Kelkar Memorial Volume, 183-195.
Pitman, W.C. III, 1978. Relation between eustasy and stratigraphic sequences of passive margins. Geological Society of America Bulletin 89, 1389-1403.
Pratt, L.M., 1985. Isotopic studies of organic matter and carbonate in rocks of the Greenhorn marine cycle. In Pratt, L.M., Kauffman, E.G. and Zelt, F.B. (eds.), Fine-grained deposits and biofacies of the Cretaceous Western Interior Seaway: Evidence

of Cyclic Sedimentary Processes. Society of Economic Paleontologists and Mineralogists, 2nd Annual Midyear Meeting, Field Trip Guidebook No. 4, 38-48.

Pratt, L.M., Arthur, M.A., Dean, W.E. and Scholle, P.A., 1993. Paleooceanographic cycles and events during the Late Cretaceous in the Western Interior Seaway of North America. In: Caldwell, W.G.E. and Kauffman, E.G. (eds.), Evolution of the Western Interior Basin. Geological Association of Canada, Special Paper 39, 333-354.

Quezada-Mumeton, J.M., Marin, L.E., Sharpton, V.L., Ryder, G. and Schuraytz, B.C., 1992. The Chicxulub impact structure: Shock deformation and target composition. Lunar and Planetary Science Conference, Abstracts 23, 1121-1122.

Rad, U. von, Haq, B.U. et al., 1992. Proceedings of the Ocean Drilling program, Scientific Results 122, College Station, TX (Ocean Drilling Program).

Ramasamy, S. and Banerji, R.K., 1991. Geology, petrography and systematic stratigraphy of pre-Ariyalur sequence in Tiruchirapalli District, Tamil Nadu, India. Journal of the Geological Society of India 37, 577-594.

Rampino, M.R., Strothers, R.B., O'Neil, B. and Haggerty, B., 1993. Asteroid impacts, mass extinction events, and flood basalt eruptions - an external driver. 1990 Society of Economic Paleontologists and Mineralogists Meeting, Stratigraphic Record of Global Change, Pennsylvania State University, Abstract Volume, p. 57-58.

Robaszynski, F., Hardenbol, J., Caron, M., Amedro, F., Dupuis, C. Gonzalez Donoso, J.-M., Linares, D. and Gartner, S., 1993. Sequence stratigraphy in a distal environment: the Cenomanian of the Kalaat Senan Region (Central Tunisia). Bull. Centres Rech. Explor.-Prod. Elf-Aquitaine 17, 395-433.

Sageman, B.B., 1991. High-resolution event stratigraphy, carbon geochemistry, and paleobiology of the Upper Cenomanian Hartland Shale Member (Cretaceous), Greenhorn Formation, Western Interior, U.S.A.. Unpubl. Ph.D. Thesis, University of Colorado, Boulder, CO, 532 pp.

Sageman, B.B., Kauffman, E.G., Harries, P.J. and Elder, W.P., 1995 in press. Cenomanian-Turonian bioevents and ecostratigraphy in the Western Interior Basin: Contrasting scales of local, regional, and global events. In: Brett, C. (ed.), Bioevents in Stratigraphy. 68 MS pp.

Savin, S.M., 1977. The history of the Earth's surface temperature during the past 100 million years. Annual Review of the Earth and Planetary Sciences 5, 319-355.

Schindewolf, O., 1962. Neokatastrophismus? Deutsche Geologische Gesellschaft, Zeitschrift Jahrg. 114, n. 2, 430-445.

Scott, G.R., 1969. General and engineering geology of the northern part of Pueblo, Colorado. United States Geological Survey Bulletin 1262, 131 pp.

Sepkoski, J.J., Jr., 1993. Ten years in the library: new data confirm paleontological patterns. Paleobiology 19, n. 1, 43-51.

Sharpton, V.L. and Ward, P.D. (eds.), 1990. Global Catastrophes in Earth History. Geological Society of America Special Paper 247, 631 pp.

Signor, P.W. and Lipps, J.H., 1982. Sampling bias, gradual extinction patterns, and catastrophes in the fossil record. In: Silver, L.T. and Schultz, P.H. (eds.), Geological Implications of Impacts of Large Asteroids And Comets On The Earth. Geological Society of America, Special Paper 190, 291-296.

Silver, L.T. and Schultz, P.H. (eds.), 1982. Geological Implications Of Impacts Of Large Asteroids And Comets On The Earth. Geological Society of America, Special Paper 190, 528 pp.

Stephenson, L.W., 1952. Larger invertebrate fossils of the Woodbine Formation (Cenomanian) of Texas. United States Geological Survey, Professional Paper 242, 211 pp.

Sundaram, R. and Rao, P.S., 1986. Lithostratigraphy of Cretaceous and Palaeocene rocks of Tiruchirapalli District, Tamil Nadu, South India. Records of the Geological Survey of India 115, 9-23.

Swinburne, N.H.M., 1991. Tethyan extinctions, sea-level changes and the Sr-isotope curve in the 10 M.a. preceeding the K/T boundary. EOS, Transactions of the American Geophysical Union 72, suppl., p. 267.

Tissot, B., 1979. Effects on prolific petroleum source rocks and major coal deposits caused by sea-level changes. Nature 277, 463-465.

Upchurch, G.R. and Wolfe, J.A., 1993. Cretaceous vegetation of the Western Interior and adjacent regions of North America. In: Caldwell, W.G.E. and Kauffman, E.G. (eds.), Evolution of the Western Interior Basin, Geological Association of Canada Special Paper 39, 243-282.

Venkatachalapathy, R., Chinnamani, M. and Ragothaman, V., 1994. Lower age limit for the mid-Cretaceous sediments of the Thiruchirapalli area. Abstract volume, 14th Indian Colloquium on

Micropalaeontology and Stratigraphy, p. 28.

Villamil, T., Arango, C. Orth, C.J. and Pratt, L.R., 1995 in press. High-resolution analysis of the Cenomanian-Turonian boundary in Colombia: Evidence for sea-level rise, condensation and upwelling. In: Pindell, J. and Drake, C. (eds.), Tectonic Evolution of the Northern South American Passive Margin. Geological Society of America Special Paper, 32 MS pp.

Ward, P., Kennedy, W.J., MacLeod, K.G. and Mount, J., 1991. End Cretaceous molluscan extinction patterns in Bay of Biscay K/T boundary sections: Two different patterns. Geology 19, 1181-1184.

Watkins, D.K., Bralower, T.J., Covington, J.M. and Fischer, C.G., 1993. Biostratigraphy and paleoecology of the Upper Cretaceous calcareous nannofossils in the Western Interior Basin, North America. In: Caldwell, W.G.E. and Kauffman, E.G. (eds.), Evolution of the Western Interior Basin. Geological Association of Canada Special Paper 39, 521-538.

Wonders, A.A.H., 1992. Cretaceous planktonic foraminiferal biostratigraphy, Leg. 122, Exmouth Plateau, Australia. In: von Rad, U., Haq, B.U. et al. (eds), Proceedings of the Ocean Drilling Program, Scientific Results 122, College Station, TX (Ocean Drilling Program), 587-600.

Young, K., 1974. Lower Albian and Aptian (Cretaceous) ammonites of Texas. Geoscience and Man 8, 175-228.

Manuscript received June/November 1994
Revision received January 1995

Authors' addresses:

Erle G. Kauffman, Department of Geological Sciences, University of Colorado, Boulder, CO 80309-0250, U.S.A.

Malcolm B. Hart, Department of Geological Sciences, University of Plymouth, Plymouth, PL4 8AA, U.K.

The Man-Made Global Disaster
An Epilogue to the Subject of Global Bio-Events

Otto H. WALLISER

Abstract. The recent global environmental crisis is inherent in the cultural evolution of mankind. To ensure the future of mankind we must actually apply the knowledge which is provided us by the brain, i.e. the organ to whose special evolution we "owe" the disaster. Doing so we have to learn from the past for the future.

Contents

Introduction
Is the Disaster Inevitable?

Introduction

Today we know that global crises and even catastrophes in Earth's history have recurred again and again. They are caused by environmental changes that in most instances can be derived from the episodic and/or periodic changes and fluctuations of our permanently altering planet. Hence they are "natural" to a large extent. The crises and catastrophes represent in part serious impacts on ecosystems, as they may cause the extinction of a more or less large number of the organisms that constitute these ecosystems. The niches created thereby in the ecosystems will be the site of attraction for new developments. Although almost a paradox, mass extinctions caused by such catastrophes actually lead to an acceleration of evolution.

Without global bio-events, the development in many groups of organisms and in ecosystems would have happened in a different way and probably would not have been yet at such an advanced stage as it is now. Perhaps *Homo sapiens* L. - or a comparable species - would not have evolved to the present state, i.e. no creature would exist capable of bringing the blue planet, called Earth, to the verge of ruin and of reflecting on how it could have happened.

Is the Disaster Inevitable?

If the impending final 'out' for mankind, caused by all the catastrophic impacts on our global ecosystem, is to be averted, the causes must be analyzed and at the same time strategies for the future must be developed. In analyzing the causes, it has to be asked whether or not the present menacing situation can be explained as the inevitable consequence of biological evolution and cultural development of man.

Before turning to this question, it should be noted that the cultural development of mankind is not subject to the rules of biological evolution. However, all "achievements" of culture and civilization, e.g. the differentiation of languages, agriculture, the progress in the fields of medicine and hygiene, as well as writing and computer, pebble tool and nuclear weapons, are the result and thus an integral component of our biological evolution.

Biological evolution and cultural development, although not identical, share similar mechanisms and processes. Thus innovations play an important role.

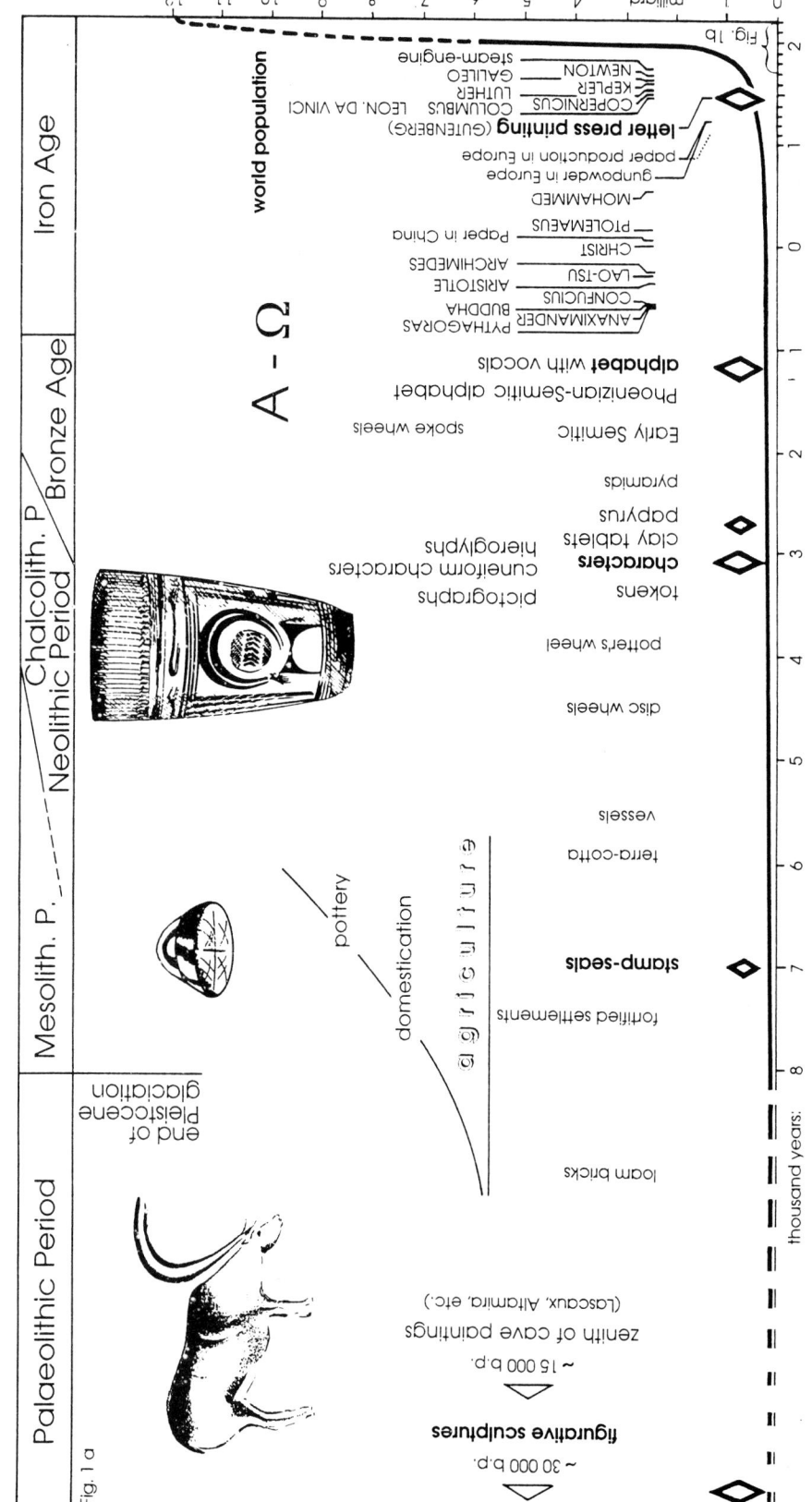

Fig. 1. The cultural development of mankind. Emphasis is given to those steps and innovations (rhombs) that improved the capacity of storing and distributing accumulated knowledge. Whereas the curves on the world population and on the destruction of tropical rain forests are based on known data, the other curves illustrate just the general trends

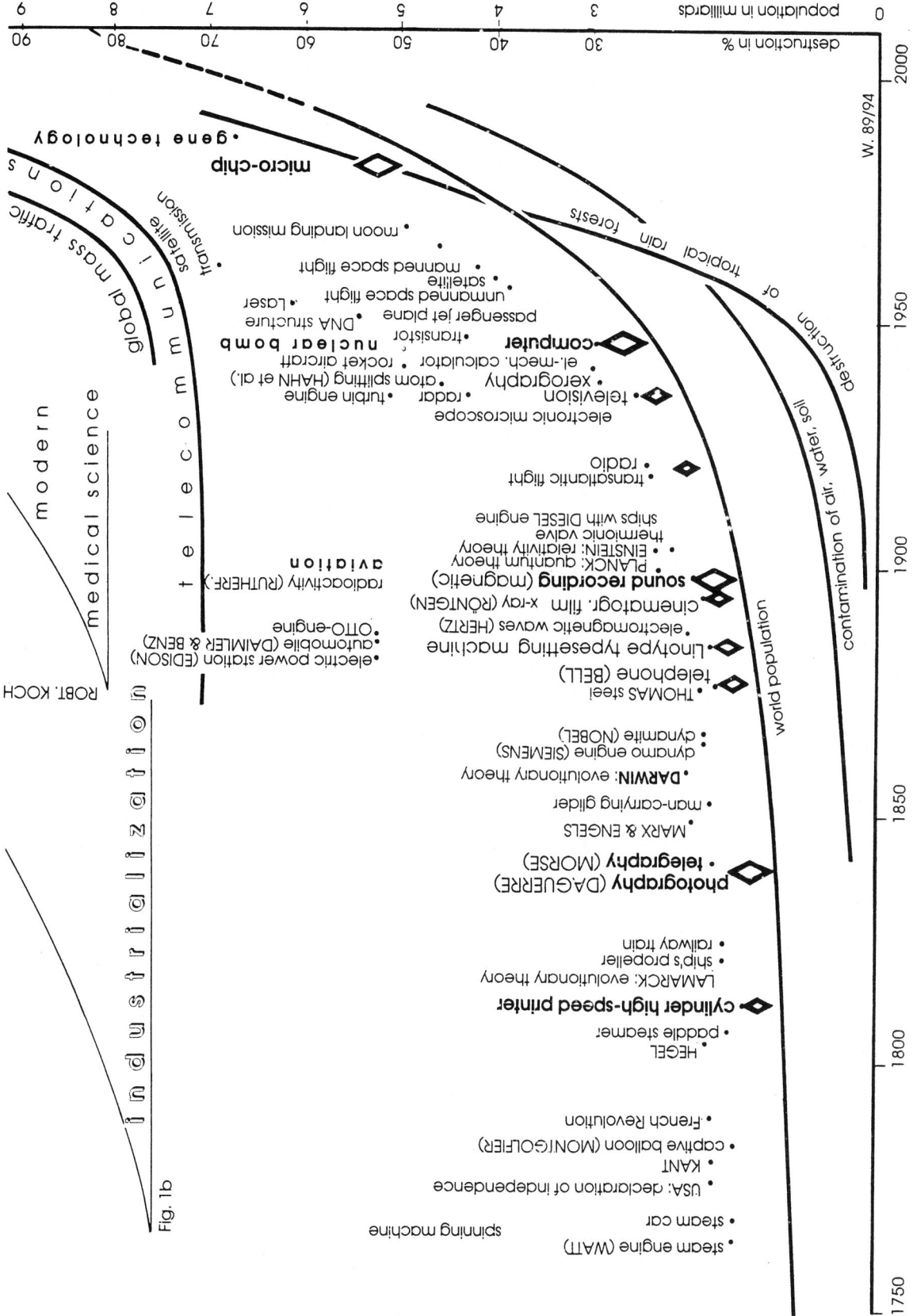

Fig. 1b

Biological innovations may lead to a radiative propagation and differentiation of the innovated new construction. Cultural and scientific innovations often produce an enormous step forward or even a radical change in the cultural development.

Important stages and innovations in the cultural development of modern *Homo sapiens* are summarized in Fig. 1. Herein, emphasis is given to the progress made in accumulating achieved knowledge, i.e. to storing and distributing knowledge. Accumulation of knowledge is the primary precondition for any scientific and technical, and thus cultural progress. In this context the term progress is applied neutrally without any judgement as to whether the resulting effects are positive or negative.

A priori it may be assumed, and it is actually confirmed by the course of development, that technical-scientific progress is more and more accelerated by the increasing accumulation and distribution of knowledge. In addition, feedback effects will increase the speed of development. This leads inevitably to an exponential course in the development curves of almost all the parameters connected with civilization.

The shift of these curves toward an exponential course corresponds in time approximately to the demographic curve. In all, this turning point lies in the present century, and for many parameters even only in its second half, such as e.g. the chemical contamination of air, water and soil, and the destruction of rain forests. This trend also distinctly indicates the complexity and close interaction of all parameters and mechanisms which have an effect on the cultural process.

The exponential course of the development curves by itself, however, represents not only a rapid, but also a gradual change in quantity and quality. There also occur abrupt punctuated developments creating completely new, hitherto unknown dimensions and thus new hazards. This is true, for example, for weapons, i.e. destruction techniques, whose development from pebble tool to the heaviest conventional bomb can be considered as a gradual increase in efficiency, although various innovative pushes are quite obvious (improving supplementation of the spear by an atlat; increase of the string tension by the construction of a crossbow; invention of gun powder). A new destructive dimension arose with the construction of the atomic and H-bombs which for the first time provided the possibility of exterminating mankind entirely whether by direct impact or radioactive contamination. The same applies to gene technology. For the first time in the almost 4 billion years of evolutionary history, directed manipulations in the genetic code of organisms can be made, i.e. new species can be constructed.

The cultural development up to the exponential acceleration and to the achievement of new dimensions bears an inevitability immanent in the development process. This also holds for the increasing risk potential for mankind which now is also intensifying toward the exponential.

It would be absolutely wrong to conclude from the preceding that the final disaster is inevitable. On the contrary, the mere fact that we recognize the dangers, their causes and many of the implied complex structures shows that we are mentally capable of developing the necessary strategies to ensure our future and to prevent the final disaster.

In order to elucidate the problem, some facts and postulates are given in shortened form:

Human beings are the only living creatures capable of destroying the global ecosystem and consequently their own basis of existence. Without human beings, the global ecosystem would flourish according to the rules of evolution. The only chance for mankind is to act and behave according to the ecological and evolutionary rules and requirements.

The exponential reproduction rate of our species and thus the impending human overpopulation of the Earth are considered to be the main danger. This fact is indeed so terrifying by itself that the true reasons for the present damages have been forgotten or are being deliberately hidden.

The damages and dangers involved are mostly sufficiently known: warming of the atmosphere due to the unrestrained output of "greenhouse gases", among others methane due to an enormous increase in the consumption of meat, the increase of CO_2 emitted by industry and traffic, and also for many other reasons. Even more dangerous is the destruction of the ozone layer which protects us and other organisms from deadly cosmic ultraviolet radiation. The pollution of the atmosphere by toxins such as the various gaseous trace elements, aerosols, acids and radioactive isotopes directly affects all organisms. Yet largely underestimated is the danger of the contamination of soils and waters, including the oceans. Not least to mention is the over-exploitation of all existing resources, among which the destructive exploitation of our global ecosystem involves the greatest danger: it is not only the climatic change caused by the destruction of tropical and boreal forests and the involved erosive destruction of some parts of the ecosystem, but also the catastrophic diminution of species diversity.

However, no comprehensive strategy for the future exists to this day which in consideration of the actually existing complex political, sociological and cultural circumstances has a chance of being rapidly realized. Hoping for such a strategy would be unrealistic and waiting for it would be irresponsible. It is essential to connect as many as possible of the already existing partial strategies and to put them into action.

A basic precondition for a successful realization of strategies for the future, however, is the recognition and the elimination of those mechanisms and modes of behaviour which have hitherto prevented the transformation of knowledge into necessary action.

Again and again so-called political and economic constraints are put forward in order to excuse non-activity. I will not discuss here the actual causes hiding behind such statements, as, among others, individual and collective egoism or maintenance of power. Instead I want to pose some selected questions. The answers to these important questions lead automatically to the strategies.

It is questionable, for example, whether the ideology of economic growth supported by the industrial countries can still be justified. The industrial countries account for about 1/4 of the world's total population, but they perform more than 3/4 of the world's production and consume 4/5 of the world resources. It may certainly be supposed that Earth would already have been plundered and contaminated long before had the underdeveloped countries reached the same level of consumption due to comparable productivity.

The situation in the energy sector is very similar. To assure their productivity, the industrial countries use 4/5 of the world's energy consumption. The largest part of this energy is won from non-renewable resources such as natural gas and petroleum or atomic power.

In fact, there should be no doubt that every human being has the same right to an acceptable standard of living. This however requires that the provoked waste of consumer goods, resources and energy is reduced to a reasonable extent.

Atomic power not only represents a high risk potential, but its use also requires that we impose on subsequent uncounted generations the task of keeping life-threatening atomic wastes in a safe place - waste that accumulated before their time and is not of the slightest use to them.

Further, we have to ask whether it is sensible to overproduce animal and plant food in the agricultural industries at the expense of the ecosystem and the consumer's health. Food overproduction requires monocultures leading to the degradation and contamination of the soil through the use of heavy agricultural machines and chemicals (fertilizers, herbicides, insecticides, and so forth) and often inducing disastrous erosion. Mass production of animals becomes possible only, on the one hand, by concentrated feed that in turn is produced in monocultures in the so-called Third World and sold at much too low prices, and on the other, by the use of chemical growth-promoting products.

Last but not least, strategies for the future require general consent. This consent should be possible through the knowledge that mankind cannot exist without a well-functioning global ecosystem. Thus the right to a functional global ecosystem is as much a primary human right as is the right of individual integrity and the right of freedom. It must be realized that the destruction of the natural basis of our existence is as criminal as is harm to a fellow being.

Another essential aspect is provided by the fact that many of the impacts on the biosphere made at present will show their effects only after a long time, after many generations. From Earth's history it is known that slow changes can lead to final catastrophes, often only after thousands or hundreds of thousands of years. The recovery of the global ecosystem will take as much time. Therefore we have to verify our present and future actions in respect to its compatibility with the future. The behaviour of every individual and every community must aim at assuring the basis for our future in a way that serves as a basis for the assurance of mankind's future. By this, we realize that the assurance of mankind's future is equivalent to the preservation of other creatures and their environments.

The New Ethics, i.e. a future-orientated pan-ecological human ethics, corresponds to the Categorical Imperative of KANT, but extended by the demand for responsibility for nature and the future. Great and revolutionary efforts are necessary to achieve KANT's extended Categorical Imperative and they must be based rationally and on the knowledge obtained from the analysis of the complex processes in the biosphere. By this, the understanding of the processes involved in global bio-events that occurred formerly in Earth's history play an important role. The situation now calls for quick action before it is too late once and for all.

Acknowledgement

Thanks are given to Gabriela Meyer for translating an earlier German version into the English language.

Author´s address:

Otto H. Walliser, Institut und Museum für Geologie und Paläontologie, Goldschmidt-Strasse 3, D - 37077 Göttingen, Germany

Global Event Stratigraphy

Chris BARNES, Anthony HALLAM, Dimitri KALJO, Erle G. KAUFFMAN, and Otto H. WALLISER

Abstract. A short characterization of 65 Phanerozoic global extinction events is given.

Introduction (by O.H. Walliser)

Stratigraphy is the description of strata, their content, and sequence. It aims (a) at the deciphering of processes and parameters that caused, formed, and changed the strata and their inorganic and organic content, and (b) at a detailed subdivision of the sequences.

Event stratigraphy is based on events, documented and perceptible in rock sequences. In order not to make the term "event" useless and impracticable, it should be restricted to relatively sudden changes, intercalations and so forth. Because sudden changes occur in many of the relevant parameters (e.g. lithology, sedimentology, ecology, chemistry, magnetism), many kinds of event stratigraphies exist which parallel each other. A combination of several of these event stratigraphies leads to a high resolution event stratigraphy (HIRES, Kauffman, 1988), and in a further combination with biostratigraphy to a highly integrated holostratigraphy (Walliser, 1986). Selected examples of these different stratigraphies are shown in Fig. 1.

Global Event stratigraphy, the theme of this chapter, differs from the above-mentioned and often regionally constrained event stratigraphies. It is restricted and based on those events which are globally traceable and of which the worldwide synchroneity is either proven or at least highly probable, based on available, although possibly yet incomplete data.

As mentioned elsewhere (e.g. Walliser, 1985), many of the original higher-rank stratigraphical boundaries – such as for Systems, Series and Stages – have been established because a faunal or/and lithological turnover has been recognized. However, the correlation of these boundaries from their type area to other regions was formerly subject to controversial discussions, mainly concerning the question of synchroneity.

Because in most cases world wide correlation is based on biostratigraphy, the question of the speed of spreading or migration of a taxon has been discussed repeatedly over many decades. But we know from recent observations, e.g. with the migration of *Crepidula formicata* along the coast of Great Britain, that even taxa with short-lived larva can spread around the globe in a period of not more than 1.000 a, i.e. about three hundredth (or even less) of a biostratigraphic zone. Important therefore is only the distribution of that environment to which a taxon is adapted. In this connection it should be mentioned that organisms have a restricted or worldwide distribution only in so far as their environments, i.e. the specific facies, are restricted or globally distributed without migratory barriers.

The main arguments against synchroneity have been (a) that a facies change might be diachronous with respect to different places, caused, e.g., by a slow

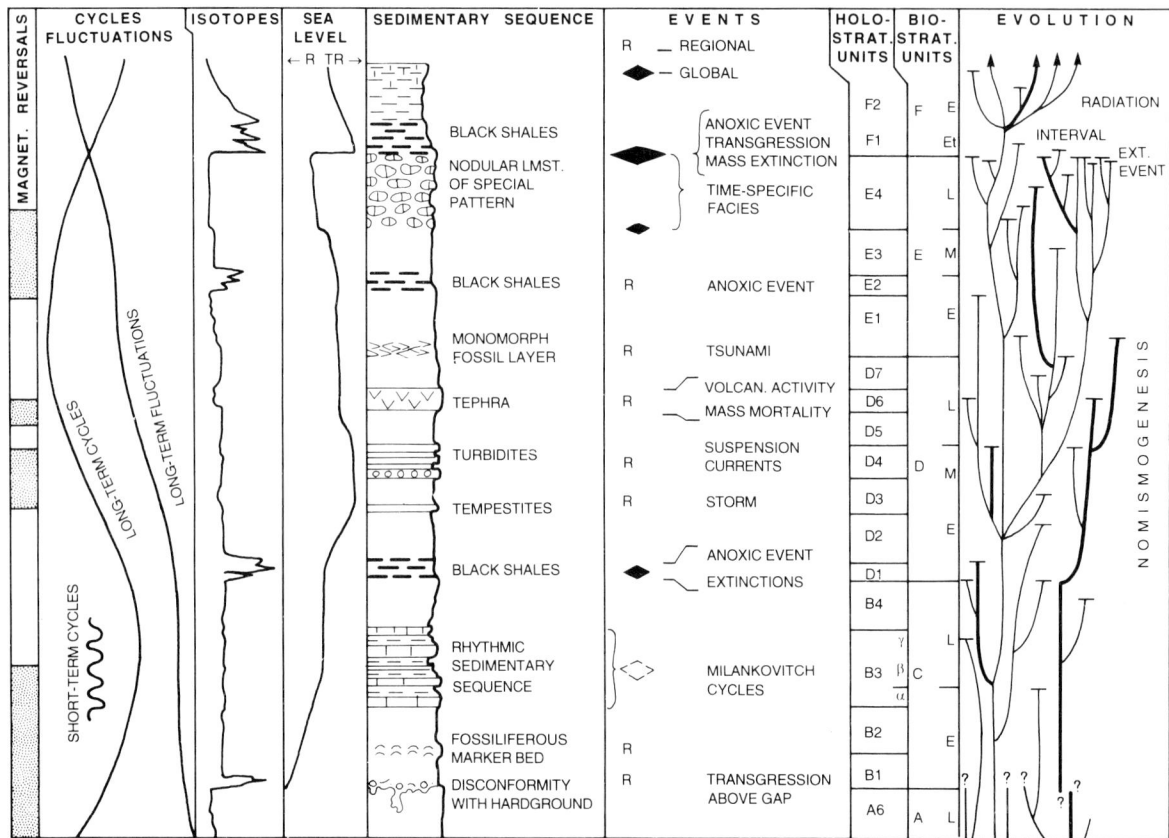

Fig. 1. Synoptic scheme showing selected parameters useful for event stratigraphy (Based on Kauffman [1988] and Walliser [1984]. Column biostratigraphic units: Et: earliest, E: early, M: middle, L: late. Column evolution: thick lines of species ranges = index species. Holostratigraphic units = high resolution stratigraphic units

transgression or regression, and (b) that a locally observed faunal turnover can simply be caused by a local facies change.

Without any doubt, the above-mentioned arguments are valid and verified by many detailed observations. But on the other hand, they do not necessarily exclude the possibility, that certain turnovers are worldwide and synchronous. It was a central task of IGCP Project 216 "Global Biological Events in Earth History" to investigate which of the known overturns are also traceable worldwide and globally synchronous, i.e. real global events.

The recognition of synchroneity became possible through the enormous progress in refining different stratigraphies during the last two decades. These investigations have benefited through the individual and cooperative efforts of members of both the International Commission on Stratigraphy and IGCP Project 216.

The majority of global bio-events are combined with litho-events, i.e. strong facies changes. These often can be easily recognized in the sections. Insofar, they are of great help for mapping geologists.

In the following, each Global Event is labeled by an abbreviation which indicates the stratigraphical position within the internationally accepted sequence of Stages, Series and Systems. Many of the listed events have received individual names, taken from a typical site or from an important fossil, including those which might have disappeared with the event or which appeared just after the event (e.g.: Late Eifelian Event = L'Eif, Kacak, Odershausen, *otomari*, *rouvillei* Event). As far as possible, these individual names are also quoted.

With the size of rhombs four different orders of Global Events are distinguished. As discussed in this volume (Walliser, 1995b), this is a very rough and relative classification.

The authorship of the following characterization is indicated at the beginning of each System.

QUATERNARY (by O.H. Walliser)

◆ T'Ho — Terminal Holocene Event; man-made disaster: Population burst of *Homo sapiens*. Drastic reduction of diversity; elimination of many biotopes; pollution of soil, water, air; increasing greenhouse effect; destruction of the ozone layer. All caused by *Homo sapiens* within a century (10^2a).

◆ B'Ho — Basal Holocene Event. Extinction of numerous genera of mainly large mammals.
Causes: Climatic change at the disappearance of the ice sheets, possibly in addition with human acitivity (overkill hypothesis).

TERTIARY (by O.H. Walliser)

◆ E'/L'Plio — Early/Late Pliocene Event. Extinction in western Atlantic Molluscs. Two phases at about 3 and 2.5 Ma.
Causes: Cooling at the beginning of Ice Age; sea-level drop.

◆ E'Ol — Early Oligocene Event. Extinctions in planktonic foraminifera and nannoplankton. Extinctions also in mammals (e.g. titanothers).
Causes: Extremely strong sea-level drop, in combination with a further cooling pulse (? deepening of the current strait between Antarctic and Australia).

◆ L'Eo — Late Eocene Events. 4 minor bio-events in a pulsative temporal pattern between M'/L'Eocene and Eo/Ol boundaries spanning about 3 Ma between 40 and 37 Ma. Extinctions in molluscs, planktonic foraminifera and nannoplankton of warm water affinities.
Causes: Episodes of climatic cooling. Towards the end of the Eocene Epoch the psychrosphere originated by the separation of Australia from Antarctica.

CRETACEOUS (by E.G. Kauffman)

◆ K/T — The Cretaceous-Tertiary boundary is marked by a major plankton extinction at the top of the *A. mayaroensis* and base of the Paleocene P0 plankton biozones. This level marks the final extinction of ammonites, belemnites, rudistid and inoceramid bivalves (*Tenuipteria*), nerineid and actaeonellid gastropods, marine reptiles, about 60+ percent of angiosperm species, many groups of echinoids, corals, other warm water molluscs, and primitive mammals. The last dinosaurs are found within 0.5 m of the boundary. This level may mark a slightly earlier extinction step, a few ka below the K/T boundary, which is also noted in palynomorphs and ocean plankton. A giant impact (the Chicxulub Crater) occurs at or very close to the K/T boundary (currently debated) and is recognized worldwide by an Ir and associated trace metal spike, shocked quartz and feldspar, microtektites and their alteration products, soot concentrations, and rare minerals (e.g. Fullerines). A regional tsunami deposit has been reported around the margins of the Caribbean Province at this level. The possibility of two closely spaced impacts is raised by a double Ir spike in a few Gulf Coast North American sites, by two possible boundary clays at scattered USA and Eurasian sites, and by stratigraphic separation of the shocked quartz-rich layer from the microtektite-rich layer in continental North America; this is still debated. Major $\partial^{13}C$ and $\partial^{18}O$ excursions mark the K/T boundary. According to the global cycle chart, the K/T boundary occurs just below the peak of global first-order Mesozoic sea-level fall, but on a minor third-order rise.

◆ L'K The Late Cretaceous or the 68 Ma Event: The first major mass killing and mass extincton events (steps) associated with the C/T mass extinction interval (65.5-68 Ma) occur globally at around 68 Ma ("Middle" Maastrichtian) The rudistid reef ecosystem collapses, with loss of many rudistid genera, species, and associated molluscan and coral taxa, during the *G. gansseri - Q. trifidum* concurrent range zone. Numerous warm water bivalves (e.g. large *Exogyra*), gastropods (e.g. among Neriniidae, Actaeonellidae) and echinoids vanish or become rare near this level; major lineages of Inoceramidae become extinct, as do many more specialized ammonites, 40-60 percent of North American angiosperm species, and numerous reptilian clades within 500 ka of the reef extinction. The 68 Ma event occurs with a 3rd-order maximum flooding peak on a generally falling global eustatic sea level trend, and continued but diminishing Cretaceous greenhouse conditions. Cretaceous diversity never recovers after this extinction.

◆ Sa/Ca The Sa/Ca boundary (83.5 Ma) is characterized by a moderate but short-term turnover in marine faunas, especially molluscs; the cosmopolitan inoceramid genera *Cladoceramus* and *Sphenoceramus* disappear along with ammonites like *Clioscaphites* and *Desmoschaphites* in North America, and are rapidly replaced by Campanian *Platyceramus*, *Endocostea*, and multinodose scaphitoid ammonites, respectively. The Sa/Ca Event is associated with a third-order maximum flooding event on a broad second-order sea-level fall. Initiation of the final oxygen depletion interval comprising OAE III occurs a short distance above the boundary.

◆ Co/Sa The Coniacian-Santonian boundary (86.6 Ma) is characterized by a very sharp, moderate scale marine faunal replacement, especially among molluscs. A regional rudistid reef extinction event occurs in the Caribbean Province. In North America, ammonite genera such as *Peroniceras*, *Scaphites* s.s., *Phlycticrioceras*, and small *Baculites* are replaced by *Clioscaphites* and large *Baculites*. Among bivalves, the genera *Magadiceramus*, surviving *Mytiloides*, *Cremnoceramus* and *Volviceramus* are abruptly replaced by new inoceramid genera (e.g. *Sphenoceramus, Cladoceramus, Cordiceramus*). The turnover takes place during a global, third-order maximum flooding interval and termination of the second regional oxygen depletion event comprising OAE III, which is recognized in many regions of North and South America, offshore England, and elsewhere by organic-rich dark shales.

◆ Tu/Co The Turonian-Coniacian boundary interval (88.7 Ma) is defined below by the highest ranges of Turonian *Prionocyclus germari, Scaphites corvensis, Mytiloides dresdensis, M. lusatiae*, and *M. problematicus* in Euramerica, and the base of the *Forresteria blancoi, Scaphites impendicostatus, S. frontierensis*, and *Cremnoceramus? rotundatus* biozone (USA). The Tu/Co boundary interval comprises an important interregional stepwise extinction level near a peak third-order sea-level highstand (middle Early Coniacian), global warming, and long-term greenhouse conditions. In North America, molluscan extinction involves 55% of Turonian standing diversity. A major extinction among Caribbean Province rudistids, and restriction of reef ecosystems, initiates the extinction interval in the middle Late Turonian. A marked foraminiferal turnover precedes the Tu/Co boundary by 200-300 ka. At the boundary, cremnoceramid-dominated inoceramid assemblages replace *Inoceramus-Mytiloides*-dominated assemblages, associated with a lineage- and generic-level change in ammonites. A modest positive $\partial^{13}C$ excursion spans the boundary, and in North America, there is a shift from organic-rich to organic-poor strata, marking the end of the first of four marine oxygen depletion events which comprise OAE III.

◆ Ce/Tu   The Cenomanian-Turonian boundary (93.4 Ma) is marked by the top of the *Neocardioceras juddi - Inoceramus pictus* biozone, and the base of *Watinoceras devonense - Nigericeras scotti - Mytiloides hattini* biozone in Europe, North and South America, and North Africa. This is the most dynamic middle Cretaceous interval, and spans a 1.4 Ma-long, second-order, stepwise mass extinction; 5-6 discrete extinction events lie below the boundary, initiating with collapse of paleotropical reef ecosystems first near the Lower-Middle Cenomanian boundary, and peaking again in the middle Late Cenomanian. The extinction proceeds through subtropical to temperate molluscan and plankton taxa during 400-450 ka of the latest Cenomanian. Notable among these steps are: the *Rotalipora* extinction level (MX1: PFX: 93.8 Ma) associated with an Ir spike and microspheres in the Americas, and with a calcisphere (disaster taxa) event in North Africa; the lower *Neocardioceras* zone molluscan extinction (MX2: 93.65 Ma); the basal *Neocardioceras juddi* biozone extinction event (MX3; 93.62 Ma); the top *N. juddi* molluscan extinction event (MX4; 93.55 Ma); and the Ce/Tu boundary extinction (MX-5; 93.4 Ma) characterized by widepread loss of typical Cenomanian ammonites, gastropods, and bivalves, including a change from *Inoceramus*-dominated to *Mytiloides*-dominated inoceramid bivalve assemblages. Three lesser extinction levels and four major origination levels lie within 900 ka-1 Ma above the Ce/Tu boundary. Collective extinction of foraminifer species is 48 percent, and nearly 70 percent for molluscan species in North America. The Ce/Tu extinction interval is associated with near-peak Mesozoic eustatic sea-level highstand (300 m above present stand in the Lower Turonian), by a global warming peak followed by an abrupt basal Turonian cooling, and by greenhouse climates reflecting elevated CO_2. The mass extinction is bracketed by a global positive $\partial^{13}C$ excursion with four regionally correlative subpeaks, with a major positive $\partial^{18}O$ excursion (two regionally correlative subpeaks), and with an interval of enhanced Corg preservation in marine strata worldwide, reflecting the main part of OAE II (=Bonarelli or Thomel Event) in the latest Cenomanian. A closely grouped set of trace element enrichment layers, especially among noble metals, including Iridium (2-3 major peaks), lies between 200 and 450 ka below the Ce/Tu boundary in North and South America, and western Europe; another Ir spike occurs above the boundary in Colombia. Microspheres are associated with Ir at two latest Cenomanian stratigraphic levels in North and South America; all Ir spikes are associated with a discrete extinction level, or step, but some extinction levels lack trace element enrichment. A Ce/Tu comet impact shower, and dynamic feedback loops in the ocean-climate system, is inferred but not yet proved by existing data.

◆ Al/Ce   The Albian-Cenomanian boundary interval (97.2-98.3 Ma), is marked by the top of the *Drakeoceras drakei - D. gabrielense* Biozone (USA-Mexico Gulf Coast) and characterized by a short-term (about 500 ka) global turnover in marine biotas, especially among warm temperate to subtropical molluscs and oceanic plankton, as well as regional calcisphere blooms (disaster taxa), e.g. in North Africa. In the Gulf and Atlantic Coast and Southern Interior Subprovinces of North America, this involves loss of typical Albian ammonite genera (*Drakeoceras, Mortoniceras, Pervinquieria, Eopachydiscus, Boesites, Adkinsites, Venezoliceras*) and bivalve lineages (*Texigryphaea*, most *Neithea*, "*Inoceramus*" *comancheanus* plexus, *Birostrina*). The Al/Ce boundary lies within or at the top of *Neogastroplites* biozones in the Northern Interior Subprovince. The Al/Ce boundary is associated with peak development followed by rapid breakup of OAE Ic (but continued regional oxygen depletion), and a very small 3rd-order sea-level lowstand associated with an otherwise rapidly increasing rate of second-order eustatic sea-level rise. It is further coeval with connection and mixing of southern and northern water masses in the Western Interior Basin of North America, intensive regional volcanism, and continued ocean tectonic activity associated with the rise of the Pacific Superplume and accelerated plate spreading.

M/L'Al — The Middle-Late Albian Boundary Bio-Event (103 Ma) is marked by the top of the *Stelckiceras liardense* biozone (north temperate North America), the top of the *Manuaniceras carbonarium* biozone and the base of the *M. powelli* biozone around the warm-temperate - subtropical North American Gulf Coast. The event is characterized in North America by short-term extinctions among ammonite genera and species groups, especially in north temperate faunas (e.g. *Gastroplites, Stelckiceras, Pseudopulchellia*), and warm temperate-subtropical *Oxytropidoceras s.s.* The M/L'Al event post-dates the termination of OAE Ib: regionally, dark shales with few fossils overlie this extinction, possibly marking the onset of OAE Ic, its spread into epicontinental seas, and widespread marine oxygen depletion. The event coincides with a very small third-order sea-level fall on a broad second-order eustatic rise. The event is recorded from North America, Europe, India and North Africa. There is a slight drop in ocean crust production at this boundary, although it remains overall at a high level for the Cretaceous.

L'Ap — The Late Aptian, second-order, mass extinction interval (112-114 Ma) spans the *Dufrenoyia rebeccae, D. justinae,* and *Kazanskyella spathi* biozones on the American Gulf Coastal Plain; it initiates with a short early-middle Late Aptian extinction interval among primitive rudistid lineages in the families Diceratidae, Requieniidae, Monopleuridae, and early Caprotinidae, associated with collapse of reef ecosystems in the Tethyan Realm, and moderate extinction among molluscs (especially bivalves, ammonites) in subtropical to warm temperate settings. The extinction is associated with later phases of OAE 1a, and a third-order sea-level rise leading to a late Late Aptian eustatic highstand, followed by a rapid second-order sea-level fall at the Aptian-Albian boundary. The L'Ap Event is coeval with accelerated rates of ocean crust production, outgassing, and plate spreading associated with emplacement of the Pacific Superplume, but to a lesser degree than during peak Middle Aptian spreading activity.

E'Ap — The Early Aptian Selli Bio-Events (116-118 Ma) occur just above magnetic polarity event M0, in the upper *G. blowi* biozone, and consist of a series of global plankton extinctions associated with a nannoconid crisis (disaster species), loss of *N. abundans* and *N. borealis,* and major extinctions among European ammonites and belemnites. This event is poorly defined in the Americas. It is associated with the first interregional to global OAE (the Selli Event, OAE 0), initiation of an accelerated plate spreading/ocean crust generation interval related to rapid rise of the Pacific Superplume, enhanced outgassing, and elevated rates of eustatic sea-level rise.

J/K — Jurassic-Cretaceous Boundary Bio-Event (141 Ma); a second-order mass extinction interval. Base of the *grandis - jacobi* Biozone in southeastern France, and global equivalents. Extinction of nearly all long-necked Plesiosaurs (Muraenosaurs, Cryptocleidids), 79 percent of North and South American ammonite genera, 80 to 82 percent of northeastern North American foraminifer and ostracode species, respectively. These are spread through three short-term extinction events, or steps, at the base, middle, and end of the Tithonian Stage, a span of 7-8 Ma. The J-K boundary event coincides a global third-order sea-level lowstand during early first-order eustatic rise.

JURASSIC AND TRIASSIC (by A. Hallam)

◆ **E'To** — Early Toarcian: Profoundly affected marine benthos in Europe, but not established yet as a major extinction event elsewhere in the world.

◆ **Tr/J** — End-Triassic: The most important extinction event within the two periods, drastically affecting most marine life. Also some terrestrial vertebrates and plants. Obviously global and probably catastrophic to some extent.

◆ **L'Ca** — Late Carnian: Predominantly a terrestrial event, affecting plants and vertebrates though controversy persists about the latter. Also some marine groups, e.g. reef biota

◆ **M'Ca** — Mid-Carnian: Extinction of some marine groups including crinoids, echinoids, scallops, bryozoans and some reef organisms. May be only regional in extent and confined to Europe.

◆ **Sm/Sp** — Smithian-Spathian boundary: Extinction of ammonites, conodonts and bivalves.

PERMIAN (after Erwin [this volume] compiled by O.H. Walliser)

◆ **P/T** — Permian-Triassic Boundary Event; End-Permian Extinction Event. Highest order extinction event. First and foremost decimation of the Palaeozoic evolutionary fauna, which suffered a familial extinction of about 80 %, relative to the 27 % family-level extinction amoung the modern fauna. There is growing evidence that tetrapods and insects suffered considerable extinction during the late Permian and there is additional evidence for a disturbance among plants.

It is suggested that the extinction began near the close of the Guadalupian and accelerated through the Djulfian, reaching a peak with a rapid pulse near the close of the Dorashamian.

Causes: The extinction appears to have begun in the early Djulfian with the loss of many marine basins and the destruction of considerable habitat area. This regression increased exposure of Pangea, exacerbating climatic instability. Toward the end of the Permian, volcanic eruptions and perhaps an increase in atmospheric carbon dioxide further increased the climatic instability and environmental degradation. Finally, immediately prior to the boundary, the regression ended and a transgression began. It seems likely that these waters may have been dysaerobic and been a contributing factor to the extinctions. Much of the terrestrial extinction may have occurred at this time as near-shore terrestrial environments were overwhelmed by the transgression.

 **Gu/Ta** — Guadalupian/Djulfian Boundary Event; End-Guadalupian Bio-Event. Decimation of diverse Tethyan bryozoan fauna at the close of Kazanian (Guadalupian) and during early Djulfian. Similar patterns for trilobites, some articulate brachiopods and tabulate and rugose corals.

◆ **L'Sm** — Late Sakmarian Bio-Event. Disappearance of a variety of Arctic-Tethyan foraminifera and the development of an Eastern Tethyan fauna.

The event interval coincides with both the onset of the Uralian orogeny and the end of the Permo-Carboniferous glaciation.

CARBONIFEROUS AND DEVONIAN (by O.H. Walliser)

 **E'NA** — Early Namurian Event; Mid-Carboniferous Event. With respect to extinct families, it is the strongest event in the history of ammonoids. Also in other marine fossil groups, such as conodonts, crinoids, brachiopods, and foraminifers, the

loss of diversity is high. Although this turnover took place in a short time interval, it is not abrupt but rather transitional in some of the affected groups.
Causes: The Middle Carboniferous Bio-Event closely coincides with a major regression and a change from the typical Early Carboniferous shale and carbonate facies to sediments which are strongly influenced by terrigenous clastic debris. These changes assumably are connected with the final phase of orogeny in the Variscan Belt.

M'Tn — Mid-Tournaisian Event; Lower Alum Shale Event. Base of *Siphonodella crenulata* Zone. Strong bio-event with regard to ammonoids. About 50% of the genera that originated after the Hangenberg-Event became extinct. Lithologically the event is characterized by worldwide spreading of black-shale facies.
Cause: Transgression.

D/C — Devonian-Carboniferous Boundary Event; Hangenberg Event. Top of *Wocklumeria sphaeroides* Zone; about late Middle *Siphonodella praesulcata* Zone. High order bio-event after a long-lasting bio-crisis in the late Famennian. Near-to-total extinction in goniatitid and clymenid ammonoids. Heavy losses in many other fossil groups, especially those inhabiting pelagic to hemipelagic environments. Lithologically there occurs a sharp boundary between Famennian cephalopod limestones and the overlying Hangenberg Shale. Corresponding to that lithofacies change a spectacular overturn in conodont biofacies occurs in hemipelagic sequences.
Causes: Anoxic event after the long-lasting Famennian regressive tendency. Probably caused by a rapid transgression-regression couplet.

L'Fa — Late Famennian Event; *annulata* Event. *Platyclymenia annulata* Zone; Late *Palmatolepis trachytera* Zone. Short-lasting anoxic event, without unusual extinctions, but causing a blooming of a few specialized ammonoid taxa and an enlargement of their habitat areas.

M'Fa 2 — Mid-Famennian Event 2; Enkeberg Event. Base and top of *Maeneceras biferum* Zone; Early *Palmatolepis marginifera* Zone. Two-stepped major extinction event in goniatites.
Causes: First step caused by a transgressive pulse, second step by a sea-level fall.

M'Fa 1 — Mid-Famennian Event 1; Condroz Event. Base and top of *Praemeroceras petterae* Zone; Late *Palmatolepis rhomboidea* Zone. Extinction of a high percentage of Tornoceratidae and Cheiloceratidae.
Cause: Termination of black *Cheiloceras* Shale in connection with the onset of the late Famennian regression.

Fr/Fa — Frasnian-Famennian Boundary Event; F/F Event; Kellwasser Event. Boundary *Palmatolepis linguiformis-Pa. triangularis* Zones. One of the 7 big Phanerozoic bio-events. Stepped extinction event following the long-lasting late Frasnian Crisis. Acmes of extinctions at base and top of the Upper Kellwasser Horizon, i.e. during an interval of about a few 100,000 years. The Kellwasser Event affected both pelagic and neritic fossil groups.
Causes: Sudden and intensive anoxic event during a long-lasting biotic crisis. Direction and magnitude of possibly connected sea-level fluctuations are still in discussion.

L'Fr — Late Frasnian Event; Lower Kellwasser Event. Black-shale event that occurred short above the base of the Upper *Pa. rhenana* Zone. Medium order bio-event

with extinctions in trilobites, goniatites and other groups. Intensification of the late Frasnian crisis.
Cause: Short-termed anoxic event.

Gi/Fr — Givetian-Frasnian Boundary Event; Frasne Event; *Manticoceras* Event. Base of *Mesotaxis falsiovalis* Zone. Important extinction event in goniatitids, of which most pharciceratids vanished. High extinction rates also in brachiopods, corals, and stromatopors.
Cause: After a phase of regression, sudden onset of black-shales due to a major sea-level rise.

L'Gi — Late Givetian Event; *Pharciceras* Event; Thaganic Event. Base of *Schmidtognathus hermanni-Polygnathodus cristatus* Zone. Important, stepped bio-event with the disappearance of Pinacitidae and most Agoniatitidae, immediately succeeded by a radiation of Pharciceratidae. Corals and stromatopors had their highest extinction rate already during the Late *Po. varcus* Zone, whereas brachiopods had their main loss between Early and Late *hermanni-cristata* Zones.
Cause: Sea-level fluctuations, beginning with a transgressive pulse at the top of the *Po. varcus* Zone.

L'Ei — Late Eifelian Events 1 and 2; Kacak Event; *otomari* Event (= L'Ei 1). Important black-shale event with extinctions at the base (L'Ei 1) and near the top (L'Ei 2). With the onset of black shales a major turnover in conodonts occurred, characterized by the extinction of numerous typical Eifelian taxa just at the boundary between *Tortodus kockelianus* and *Po. ensensis*. Together with the black shales occurred *Nowakia otomari* (s.l.). At the upper extinction level (L'Ei 2) several goniatite genera became extinct, whereas in conodonts no major change can be observed.
Causes: Sudden onset of anoxic conditions (L'Ei 1) and, after a considerable time interval, slow termination of the black-shale sedimentation (L'Ei 2).

Em/Ei — Emsian-Eifelian boundary Event; *jugleri* Event; (basal) Chotec Event. The event level is between the base of *Po. costatus partitus* Zone and the subsequent first occurrence of *Pinacites jugleri*. The Chotec Event comprises only a gentle faunal turnover but is lithologically well documented by the relatively short-termed occurrence of dark limestones and shales.
Cause: Sea-level rise.

M'Em — Mid-Emsian Event; Daleje Event; *gracilis* or *cancellata* Event. Stepped important bio-event connected with a gradual change to the Dalejan black-shale deposition. The extinction of the early goniatites which had their origination within the Zlichovian took place between the bases of *Nowakia elegans* and *N. cancellata* Zones (late *Po. gronbergi* to early *Po. laticostatus* Zones). Also trilobites and other benthic groups lost in diversity.
Cause: Sea-level rise, partly in connection with an anoxic event.

B'Zl — Basal Zlichov Event. Described as a minor global event at the Zlichov-Praha Formations boundary in Bohemia. This horizon does not coincide with the newly defined Pragian-Emsian boundary, characterized by the occurrence of *Po. kitabicus* (a new species out of the *Po. dehiscens* lineage) and by a transgressive pulse. Further investigations are needed in order to clarify the importance and extent of these events.

Lo/Pr — Lochkovian-Pragian Boundary Event. Termination of the typical, dark and platy Lochkovian ("eγ") limestones. Minor important but worldwide well presented

geo-event with well recognized change from Lochkovian to Pragian faunal associations.
Cause: A fairly rapid but not very large lowering of the sea level.

SILURIAN (by D. Kaljo)

◆ S/D Silurian/Devonian Boundary Event. Top of the *lochkovensis-transgrediens* Standard Zone or *Monograptus transgrediens* Zone s. str.. Fifth order bio-event. Species level extinctions among brachiopods, chitinozoans and graptolites, the fully marine type eurypterids and some trilobites, including the reef-associated *Illaenidae* and most of the *Encrinuridae* disappeared near the boundary
Causes: Change of environmental parameters, including sedimentary ones.

◆ M'Pr Middle Pridoli Event. *Monograptus perneri* Zone in the pelagic facies; a slightly earlier position (top of the *M. bouceki* Zone) in the carbonate facies cannot be excluded. Third order bio-event, finishing the long declining phase in the late Silurian. Final extinctions in several lineages among corals, conodonts and especially graptolites (disappeared all but *M. transgrediens* and *Linograptus posthumus*). Chitinozoans experienced the last diversity rise.
Causes: Change of the conditions of sedimentation, diminishing of suitable habitats.

◆ L'Lu Late Ludfordian Event *(spineus* or pentamerid Event). Top of the *formosus* Standard Zone. Fifth order bio-event. Extinction of almost all of the *Pentameridae* and *Subrianidae*, extinction of a number of specialized and short-lived species among graptolites.
Causes: Change of sea-level and oceanic parameters.

◆ M'Lu Middle Ludfordian Event (*podoliensis* or *hedei* Event). Top of the *bohemicus-kozlowskii* Standard Zone and *Andreolepis hedei* Zone. Second order bio-event, profound change of the vertebrate assemblage, extinction of the earlier lineages, appearance of new ones, among graptolites extinction of neocucullograptids, considerable diversity drop among corals and conodonts.
Causes: Change of environmental conditions.

◆ E'Lu Early Ludfordian Event (*leintwardinensis, siluricus, Cardiola* or Lau Event). Top of the *leintwardinensis* Standard Zone. Third order bio-event. Final extinction of plectograptids, extinction of many specialized groups of graptolite species, several ostracoderm genera, considerable rise in the disappearance rate of acritarchs and conodonts.
Causes: Sea-level lowering and accompanying facies change.

◆ M'Go Middle Gorstian Event (Linde Event). The *scanicus* Standard Zone. Fifth order bio-event. Considerable diversity rise among conodonts and chitinozoans, also graptolites.
Causes: Deepening of the sea, maximum of the early Ludlow transgression.

◆ E'Go Early Gorstian Event. Beginning of the *nilssoni* Standard Zone. Fourth order bio-event. Origination and radiation of a new Ludlow graptolite assemblage consisting of newly appeared cucullograptid, neocucullograptid and linograptid lineages, also *Monograptus uncinatus* group.
Causes: Sea-level rise and origination of suitable habitats after the late Wenlock crisis, innovation explosion and radiation.

◆ L'Ho Late Homerian Event (*ludensis* Event). Top of the *ludensis* Standard Zone. Fifth

order bio-event. Extinction of species level taxa among agnathans, conodonts and graptolites.
Causes: Sea-level low stand.

M'Ho — Middle Homerian Event (Mulde Event). The *nassa deubeli* Standard Zone. Brachiopod evidence is located here tentatively as they cannot be dated exactly. Fifth order bio-event. Adaptive radiation of pentamerinids and subrianinids, also ostracoderms and appearance of two new lineages (*praedeubeli* and *idoneus-sherrardae*) among graptolites.
Causes: Shallow sea conditions were optimal for above brachiopods and agnathans, graptolites elaborated new morphological structures.

E'Ho — Early Homerian Event (*lundgreni* or ? Valleviken Event). Top of the *lundgreni* Standard Zone. Third order bio-event. Mass-extinction of Wenlock graptolites, final extinction of cyrtograptids and considerable drop in chitinozoan diversity.
Causes: Rapid change of oceanographic parameters, shallowing of the sea.

L'Sh — Late Sheinwoodian Event (Boge Event). The *Cyrtograptus ellesae* Zone. Fifth order bio-event. Considerable diversity rise and drop among acritarchs.
Causes: Sea-level high stand and following lowering.

E'Sh — Early Sheinwoodian Event (Ireviken or *amorphognathoides* Event). Top of the *centrifugus-murchisoni* Standard Zone and of the *Pterospathodus amorphognathoides* Zone). Second order bio-event. Mass extinction among conodonts and acritarchs, considerable extinction among chitinozoans and graptolites. Involved were mostly planktic and nektoplanktic organisms.
Causes: A short-term sea-level lowering and change of oceanographic parameters.

E'Te — Early Telychian Event. The *turriculatus-crispus* Standard Zone. Fourth order bio-event. Rapid diversity increase among conodonts and acritarchs, final extinction of diplograptids.
Causes: Beginning of late Llandovery transgression.

L'Ae — Late Aeronian Event (Sandvika Event). Top of the *Monograptus sedgwickii* Zone. Fourth order bio-event. A major dispersal event among brachiopods, several new genera of acritarchs appeared, rapid radiation and extinction among chitinozoans, the first strong extinction and beginning of the declining phase in graptolite evolution.
Causes: Sea-level lowering and beginning of new transgression, change of oceanic parameters, better ventilation of bottom waters.

E'Ae — Early Aeronian Event (Sandvika Event). The *gregarius* Standard Zone. Fourth order bio-event. Rapid radiation and following extinction among conodonts, diversity maximum of graptolites and the first appearance of *Loganellia* among thelodonts.
Causes: The first sea-level high stand during the postglacial Silurian transgression. Towards the end of the zone in the deeper parts of the sea the level of anoxic waters rose.

ORDOVICIAN

(after Barnes et al. [this volume] compiled by O.H. Walliser)

O/S — Ordovician-Silurian Boundary Event; Upper Ashgill Bio-Event (U'Al). This first order bio-event is characterized by profound extinction in virtually all known fossil groups. The peak extinction for most groups lies within the Hirnantian, but it appears to vary slightly for different groups (e.g. conodonts and graptolite), and may have a stepwise extinction pattern for some group.

The U'Al Bio-Event is close to the base of the *G. ? extraordinarius* graptolite Zone (The newly defined Ordovician-Silurian boundary is within the slightly higher and more continuous graptolite sequence at the base of the *A. acuminatus* Zone).

Causes: Late Ordovician glaciation. The glacial period ranges through the Ashgill Series with perhaps three to four main phases within the Hirnantian. This final climax phase was characterized by low sea level stands, aggressive ocean circulation, and deep ocean erosion and oxygenation.

◆ B'Cc Basal Caradoc Bio-Event. In contrast to other Ordovician bio-events, the B'Cc is more drawn out. It corresponds with the *Nemograptus gracilis* Zone which spans later Llandeilo and early Caradoc.

The B'Cc Event marks an interval in Ordovician history which heralds several major changes, thus presenting a peak enhancement of restructuring the global biota. Concerning trilobites, graptolites and conodonts, the pattern of realms and provinces is dramatically modified.

Causes: The event interval is characterized by a regression, followed by a transgression which was one of the greatest floodings of stable cratons in the Phanerozoic. Thereby extensive epeiric seas with large tracts of new habitat for benthic communities were established. In addition, the event interval was one of intensive volcanic and tectonic activity, especially in the Iapetus Ocean.

◆ B'Ln Basal Llanvirn Bio-Event. The event interval is one of profound reorganization of the Ordovician biota, marked less by profound extinctions than by the origination of many taxa, especially in trilobites, conodonts, and graptolites.

Causes: The event level corresponds to a regressive-transgressive couplet, most likely with a brief stillstand. In addition, it is approximately the time of a major excursion of 87Sr/86Sr isotope curve. This may have resulted from a combination of partial flooding of the cratons combined with increased sea-floor spreading, ridge activity, and deformation along subduction zones (e.g. early Taconic Orogeny in Iapetus Ocean).

◆ B'Ag Basal Arenig Bio-Event. One of the most dramatic extinction events in conodonts, followed by a low diversity replacement fauna and a subsequent radiation. Similar patterns are observed with graptolites and trilobites. The latter then reached a maximum in provincialism due to the wide dispersal of paleocontinents.

Causes: The extinction level appears to correspond to the peak of eustatic sea level fall in the late Tremadoc, which is followed by a transgressive pulse at the base of the Arenig. In addition, the earliest Arenig might have been a time of increased oxygenation in the deep ocean.

☆ B'Tc Basal Tremadoc Bio-Event. Origination event. After evolution of euconodonts in late Cambrian, conodonts undergo a major radiation that is particularly dramatic in platform carbonate facies. Radiation of Ellesmeroceratida as the origin of nautiloid diversification. B'Tc records an abundant appearance of new platform trilobites, but continuity in off-shelf forms. The first planktonic graptolites evolved, and underwent a rapid diversification.

Causes: Global eustatic transgression, connected with almost ubiquitous "*Dictyonema*" (i.e. *Rhabdinopora*)-bearing blackshale deposition around the world.

CAMBRIAN   (by M.D. Brasier)

◆ **C/O** — Trempealeauan/Ibexian (Tp/Ib) Event (top of Ptychaspid Biomere), i.e. Trempealeauan/Ibexian, Payntonian/Datsonian stage boundaries. Second order bio-event, progressive. Mass extinction of relatively nearshore polymeroids but also of outer shelf agnostoids.
Causes: Sea level rise, oceanographic changes, cooling over carbonate platforms, reduction in provincialism, influx of eurytopic outer shelf forms.

◆ **Dr/Fc** — Dresbachian/Franconian Event (boundary between Pterocephaliid/Ptychaspid Biomeres): *Elvinia* Zone to lower *Taenicephalus* Zones, i.e. the Dresbachian/Franconian, Idamean-Post-Idamean stages. Second order bio-event, progressive. Mass extinction preferentially affected nearshore polymeroids.
Causes: Sea level changes (major sequence boundary), oceanographic changes, cooling over carbonate platforms, reduction in provincialism, influx of eurytopic outer shelf forms.

◆ **Mi/Id** — Mindyallan/Idamean Event (boundary between Marjumiid/Pterocephaliid Biomeres): top of *Glyptagnostus stolidotus* / base of *Glyptagnostus reticulatus* Zones, i.e. the Mindyallan/Idamean, Gushanian/Changshanian and 'Marjumian'/'Steptoean' stage boundaries (but within lower Dresbachian). Second order bio-event, relatively sudden. Mass extinction of relatively nearshore polymeroid trilobites but also of outer shelf agnostoids.
Causes: Sea level changes, oceanographic changes, cooling over carbonate platforms, reduction in provincialism, influx of eurytopic outer shelf forms.

◆ **Ty/Am** — Toyonian/Amgan event. Top of *Anabaraspis spendens* / base of *Schistocephalus antiquus* trilobite Zones, i.e. top Toyonian to early Amgan. First order bio-event, culmination of M'B Event, progressing to final destruction of biohermal biota. Mass extinction of irregular archaeocyathans, disappearance of late olenellid, redlichiid and protolenid trilobites.
Causes: Sea level fall, oceanographic changes, spread of eurytopic outer shelf forms.

◆ **M'Bo** — Mid Botomian Event. Top of *Carinacyathus squamosus-Botomocyathus zelenovi* archaeocyathan Zone and *Bergeroniellus micmacciformis-Erbiella* trilobite Zone / base of *Rozanovicyathus alexi* archaeocyathan 'beds' and *Bergeroniellus gurarii* trilobite Zone, i.e. mid Botomian Stage. First order bio-event, stepwise destruction of biohermal biota over c. 3 Ma, following after a period of favourable conditions. Mass extinction of regular archaeocyathans, calcareous metaphyte flora (i.e. reefal biota), small shelly fossils of 'Tommotian Fauna', hyoliths, some nearshore olenellids and ?pelagic pagetiids.
Causes: Ecological disturbance precipitated by rapid sea level rise, accelerating subsidence rates, nutrient changes, anoxia and flooding of nearshore carbonate platform habitats, influx of eurytopic outer shelf forms.

✦ **CR4** — Cambrian Radiation Event, Phase 4. Base of Atdabanian stage (*Profallotaspis jakutensis* Zone / *Retecoscinus zegebarti* Zone) to near top of Atdabanian stage (*Judomia* Zone). First order bio-event, stepwise first appearance of biomineralized arthropods (trilobites, ostracods) and stepwise increase in the diversity of archaeocyathan sponges.
Causes: as above. Evolutionary change at a time of 'ecological escalation' and increased biological interdependence, here coupled with 'late stage biomineralization' (arthropods).

☆ CR3 Cambrian Radiation Event, Phase 3. Base of Tommotian stage *(Nochoroicyathus sunnaginicus* Zone) to base of Atdabanian stage (*Profallotaspis jakutensis* Zone / *Retecoscinus zegebarti* Zone). First order bio-event, stepwise increase in invertebrate diversity (e.g. appearance of Brachiopoda, lipped and triangulate hyoliths, multimembrate tommotiid sclerites, archaeocyathan sponges; acritarch diversification).
Causes: as above. Evolutionary change at a time of 'ecological escalation' and increased biological interdependence, here coupled with migration? from unknown regions, and habitat diversification during a transgressive pulse in sea level.

☆ CR2 Cambrian Radiation Event, Phase 2. Base of *Purella antiqua* Zone (mid Nemakit-Daldynian stage) to base of the Tommotian stage (*Nochoroicyathus sunnaginicus* Zone). First order bio-event, stepwise increase in invertebrate diversity (e.g. appearance of Mollusca, halkieriids, siphogonuchitids, simple hyoliths, ?Foraminifera).
Causes: as above. Evolutionary changes at a time of 'ecological escalation' and increased biological interdependence, enhanced by taphonomic changes associated with phosphogenesis and pyritization.

☆ CR1 Cambrian Radiation Event, Phase 1. Top of *Harlaniella podolica* ichnofossil Zone / base of *Phycodes pedum* ichnofossil Zone in Newfoundland; presumed to lie close to base of *Anabarites trisulcatus* small shelly fossil Zone, in lower Nemakit-Daldynian stage of Siberia. First order bio-event, sudden appearance of a limited fauna of invertebrate skeletons (Porifera, worms, protoconodonts) and new trace fossils (larger, deeper, more complex, branching), after long-lasting crisis in Kotlinian.
Causes: plate tectonic changes?, flooding of narrow restricted shelves, nutrient influx (phosphogenesis), 'ecological escalation', rapid evolution and migration of eurytopic Cambrian biota into empty ecospace, high provincialism. Taphonomic changes associated with phosphogenesis and pyritization.

◆ PC/C Precambrian-Cambrian Boundary Event. Top of *Harlaniella podolica* ichnofossil Zone / base of *Phycodes pedum* ichnofossil Zone; presumed to lie close to base of *Anabarites trisulcatus* small shelly fossil Zone, in lower Nemakit-Daldynian Stage. First order bio-event, mass extinction of several Vendian trace fossils (*Harlaniella, Palaeopascichnus*); possible extinction of many soft bodied fossils, large acanthomorph acritarchs during preceding Kotlin crisis.

Acknowledgements

Gabriela Meyer typed the text and Cornelia Kaubisch made the illustrations. Both are greatly thanked for their help.

References

Barnes, C.R., Fortey, R.A. and Williams, S.H., 1995. The Pattern of Global Bio-Events During the Ordovician Period. pp. 139-172. In: Walliser, O.H. (ed.), Global Events and Event Stratigraphy in the Phanerozoic. Springer, Berlin Heidelberg New York.

Brasier, M.D., 1995. The Basal Cambrian Transition and Cambrian Bio-Events (From Terminal Proterozoic Extinctions to Cambrian Biomeres). pp. 113-138. In: Walliser, O.H. (ed.), Global Events and Event Stratigraphy in the Phanerozoic. Springer, Berlin Heidelberg New York.

Erwin, D.H., 1995. Permian Global Bio-Events. pp. 251-264. In: Walliser, O.H. (ed.), Global Events and Event Stratigraphy in the Phanerozoic. Springer, Berlin Heidelberg New York.

Hallam, A., 1995. Major Bio-Events in the Triassic and Jurassic. pp. 265-283. In: Walliser, O.H.

(ed.), Global Events and Event Stratigraphy in the Phanerozoic. Springer, Berlin Heidelberg New York.

Kaljo, D., Boucot, A.J., Corfield, R.M., Le Herisse, A., Koren, T.N., Männik, P., Märss, T., Nestor, V., Shaver, R.H., Siveter, D.J. and Viira, V., 1995. pp. 173-224. In: Walliser, O.H. (ed.), Global Events and Event Stratigraphy in the Phanerozoic. Springer, Berlin Heidelberg New York.

Kauffman, E.G., 1988. Concepts and methods of high-resolution event stratigraphy. Annu. Rev. Earth Planet. Sci. 16, 605-654.

Kauffman, E.G. and Hart, M.B., 1995. Cretaceous Bio-Events. pp. 285-311. In: Walliser, O.H. (ed.), Global Events and Event Stratigraphy in the Phanerozoic. Springer, Berlin Heidelberg New York.

Walliser, O.H., 1984. Global Events and Evolution. Proceedings of the 27th Intern. Geol. Congr. 2 (Palaeontology), 183-192, VNU Science Press, Moscow.

Walliser, O.H., 1985. Natural boundaries and Commission boundaries in the Devonian. Cour. Forsch.-Inst. Senckenberg 75, 401-408, Frankfurt/Main.

Walliser, O.H., 1986. The IGCP Project 216 "Global Biological Events in Earth History". Lecture Notes in Earth Sciences 8, 1-4, Springer, Berlin Heidelberg New York.

Walliser, O.H., 1995a. Patterns and Causes of Global Events. pp. 7-19. In: Walliser, O.H. (ed.), Global Events and Event Stratigraphy in the Phanerozoic. Springer, Berlin Heidelberg New York.

Walliser, O.H., 1995b. Global Events in the Devonian and Carboniferous. pp. 225-250. In: Walliser, O.H. (ed.), Global Events and Event Stratigraphy in the Phanerozoic. Springer, Berlin Heidelberg New York.

Authors' addresses:

Christopher R. Barnes, Centre for Earth and Ocean Research, University of Victoria, P.O. Box 1700, Victoria, B.C. V8W 2Y2, Canada

Anthony Hallam, School of Earth Sciences, University of Birmingham, Edgbaston, Birmingham B15 2TT, U.K.

Dmitri Kaljo, Institute of Geology, Estonian Academy of Sciences, 7 Estonia Ave., EE 200 105 Tallinn, Estonia

Erle G. Kauffman, Department of Geological Sciences, University of Colorado, Boulder, CO 80309, U.S.A.

Otto H. Walliser, Institut und Museum für Geologie und Paläontologie, Goldschmidt-Strasse 3, D-37077 Göttingen, Germany

Springer-Verlag and the Environment

We at Springer-Verlag firmly believe that an international science publisher has a special obligation to the environment, and our corporate policies consistently reflect this conviction.

We also expect our business partners – paper mills, printers, packaging manufacturers, etc. – to commit themselves to using environmentally friendly materials and production processes.

The paper in this book is made from low- or no-chlorine pulp and is acid free, in conformance with international standards for paper permanency.